THE NORTH AMERICAN
CUP-FUNGI

The North American Cup-fungi

(Inoperculates)

BY

FRED JAY SEAVER, M.S., Ph.D., Sc.D.

VISITING PROFESSOR OF BIOLOGY, FLORIDA SOUTHERN COLLEGE
FORMERLY HEAD CURATOR IN THE NEW YORK BOTANICAL GARDEN
AND
MANAGING EDITOR OF MYCOLOGIA

ILLUSTRATED

NEW YORK
PUBLISHED BY THE AUTHOR
1951

PRINTED IN THE UNITED STATES OF AMERICA
BY THE LANCASTER PRESS, INC., LANCASTER, PA.

DEDICATED
TO MY WIFE

Finetta Fry Seaver

IN APPRECIATION OF
INSPIRATION AND COLLABORATION

TABLE OF CONTENTS

vii

PREFACE

When the North American Cup-Fungi (Operculates) was issued in 1928, it was expected that this would be followed in a reasonable time by a similar volume on Inoperculates. As the work progressed, however, it became obvious that the task was so tremendous that nothing approaching completeness could be accomplished in one short life time.

Fully realizing this, it has been decided to bring together at this time, in one volume, the results of the observations of the writer extending over a period of more than forty-five years as well as those of his colleagues and coworkers in various parts of North America.

There are many controversial questions and the writer makes no pretense that this is the last word in the treatment of species, genera or families but merely a summary of the available facts as they appear at the present time. It is hoped that this volume may be as useful as the preceding one in stimulating observation, and research in this obscure, interesting, and apparently inexhaustible field.

New York, 1948

INTRODUCTION

The writer's interest in the inoperculate cup-fungi coincides with that of the operculates. In fact, when these studies were begun as a student in Iowa, the separation of the cup-fungi or Discomycetes into the operculate and inoperculate forms had not come to be generally recognized although the importance that should be attached to the dehiscence of the asci in the classification of the Discomycetes was announced by Boudier more than twenty years before (Grevillea **8**: 45. 1879.). The steps leading up to the present work will be discussed here in some detail.

Although the operculates may seem to have absorbed most of the writer's attention up to the present time, some of the first novelties encountered in student days were of the inoperculate group. One of these will be mentioned here. In the spring of 1905 the writer's attention was called to a small disco appearing under the trees of wild cherry across the river from the University of Iowa, the region at that time consisting of wooded hills and open fields constituted one of our favorite collecting grounds. Careful search revealed many of the small discs each attached by a stem to a partially buried seed of the wild cherry, *Prunus serotina*. This was at once recognized as belonging to the genus *Sclerotinia* as at that time known. Being unable to find any record of a *Sclerotinia* on this host, in his youthful enthusiasm the writer transmitted a specimen to Dr. H. Rehm of Germany, then the outstanding world authority on this group of fungi. When Dr. Rehm later reported this to be a new species and named it in honor of the collector *Sclerotinia Seaveri* (Plate 85) it was an event of no small importance in the writer's young life and furnished an added incentive to continue the researches started on this group of fungi.

A few years later G. M. Reade of Cornell University published a paper (Ann. Myc. **6**: 109–115. 1908.) entitled "Preliminary notes on some species of *Sclerotinia*" in which he listed the writer's namesake and recorded with it the conidial stage which had later been discovered and recorded as *Monilia Seaveri*. In the meantime both stages of the fungus had been collected at various places in New York State. Reade at the same time

1

recorded another of the writer's early collections made at Mt. Pleasant, Iowa, on the seeds of basswood as *Sclerotinia Tiliae*.

As a result of the work done by Reade at Cornell University, the writer some time after coming to The New York Botanical Garden invited Professor H. H. Whetzel to monograph the genus *Sclerotinia* for North American Flora, thinking at the time that this little task could be accomplished over a summer vacation. This, however, proved to be the beginning of a lifelong research problem which has been prosecuted by Whetzel up to the time of his death and by his students and collaborators, Honey, Drayton, White and others up to the present time. Unfortunately, much of Whetzel's work was still unpublished, at the time of his death.

Although the task of monographing the genus *Sclerotinia* was assigned to Whetzel, the writer never entirely lost interest in the genus and its allies, and other members of the genus or tribe as it is now regarded kept forcing them onto his attention in such a manner that they could not be disregarded. One illustration of this should be mentioned.

During the early days at The New York Botanical Garden occasional afternoons were spent in field work in the suburbs of New York City, especially the Van Cortlandt Park section when that region was more natural and unsophisticated than it is at the present time. For several seasons the writer had collected every spring on the rootstocks of wild geranium a discomycete growing in large clumps which could not be identified. No record of any such fungus growing on this host could be found. Finally it was decided to name and describe the fungus so as to have a record of it. It was regarded as a *Sclerotinia* and knowing from experiences previously recorded that some of these had conidial stages on the foliage of the host, a careful search was made for such connections. About this time a letter was addressed to Professor Whetzel as follows: "For several years past I have been collecting each spring a *Sclerotinia* on the rootstocks of wild geranium. Perhaps you recall that I referred to this species while you were here. During the present spring I have found it as usual and have been following it up in the field with a hope of finding its conidial stage. I have been unable to find anything which appeared to be its conidial stage on the leaves of the same host. However, I have in culture a number of rootstocks which were infected with apothecia and in about a week I had the most beautiful growth of a *Botrytis*. I am sending a small specimen of this to you for examination. While the occurrence of *Botrytis* might be purely coincidence, it occurs with such great abundance in connection with the apothecia and rootstocks that I am inclined to suspect that there might be some connection between the two. If you can give me any light on the *Botrytis* as to its identity, the favor will be very greatly appreciated."

To this Professor Whetzel replied as follows: "I very much doubt that the *Botrytis* can be a conidial stage of your fungus on wild geranium. Nevertheless this is quite possible, and I shall see what I can make out of it. However, I could come a great deal nearer answering the question if you could get me apothecia of the same fungus from which I could make ascospore cultures. The probability is that cultures from the *Botrytis* will give only *Botrytis*."

Since Professor Whetzel seemed doubtful of any connection between the *Botrytis* and *Sclerotinia*, it was cultured at The New York Botanical Garden and the connection established. On June 28, 1917 the following report was made to our Cornell associate: "Since writing you we have cultured out the *Sclerotinia* on wild geranium and proved beyond the question of a doubt that the *Botrytis* is the conidial stage of this fungus. Professor Horne is here for a time and being interested he offered to culture the fungus. Excellent growth of conidia can be produced from the ascospores in four or five days. I thought you might be interested in knowing our results. We will publish a short paper in the near future. I would be glad to send you some of the perfect stage of the fungus but the last time I was out I could find none. It lasts only about a month in the spring. Unless we can decide on the specific name of the *Botrytis* we will call the fungus *Sclerotinia Geranii*. According to our rules the specific name of the *Botrytis* provided it has one would be tenable for the species."

On July 2, 1917 a letter was received from Professor Whetzel containing the following statements: "I want to congratulate you upon having made the connection between a *Botrytis* and the *Sclerotinia* . . . So far as I am aware this is the first authentic case of the connecting up of the *Botrytis* with the *Sclerotinia*, and it will be a very important contribution to the general subject."

Our Cornell associate still seemed doubtful of our results, and in the spring of 1919 by previous arrangement came to New York with an assistant to check up our conclusions for his own personal satisfaction. The work was done in New York since at that time the fungus had not been collected in any other place, and it was desired to make first hand field observations. No difficulty was encountered in collecting the material, and after careful culture work our results were substantiated.

The importance of this contribution lies in the fact that De Bary (Comparative Morphology and Biology of the Fungi 238. 1887.) claimed a connection between *Botrytis cinerea* and *Sclerotinia Fuckeliana*, a claim that could never be substantiated. To use his exact words: "the conidiophores of *Sclerotinia Fuckeliana* were made a species under the name of *Botrytis cinerea*." Neither had any other connection been made between a *Sclerotinia* and a *Botrytis*, so far as we are aware, until the connection was made in The New York Botanical Garden between *Sclerotinia Geranii* and an unnamed *Botrytis*. However, when we refer to *Sclerotinia* this is used in a broad sense as it was then applied. It is now used in a much more restricted sense. Since the establishment of the one connection several others have been made, which will be reported in the proper place.

Habits and Life History

As stated in the volume on operculates, very few of those were even suspected of being parasitic in their habits. At the same time the conidial stages are either lacking or obscure and inconspicuous. On the other hand, many of the inoperculates are destructive parasites and in these the conidial stages are often much more highly developed and conspicuous. From these facts we conclude that the high development of the conidial stages in these forms is in some way associated with their parasitic habits.

Take as an illustration the species just discussed, *Sclerotinia Seaveri*. In this species the ascigerous stage appears the early part of May, at which time the asci discharge their spores vigorously just about the time young foliage of the host plant begins to appear. The spores germinate on the young leaves producing monilioid spores in great abundance and forming conspicuous blotches on the leaves and inflorescence. The conidia thus constitute the summer or repeating stage, the mycelium passing to the seeds which fall to the ground and carry the fungus over the winter. Here the high development of the conidial stage is obviously an adaptation to the parasitic habits of the fungus. Other similar illustrations might be cited.

From this high development of the conidial stages in some of the fungi of this group and their adaptation to their hosts have resulted the most complicated pleomorphic life cycles. To trace these requires the most careful laboratory studies, and it is needless to say that only a small percentage of the forms have been so studied. Many do not respond easily to this kind of treatment, or it may be possible that the right technique has not been employed. And since the classification of these forms is often based on their conidial stages, it follows that any treatment of genera adopted now is purely tentative.

While many of the inoperculates are parasitic, many more are strictly saprophytic, and in many of these the conidial stages are as yet unknown, or possibly lacking a condition similar to that found in the operculates. The types of conidia found in the inoperculates are most varied and no attempt will be made to enumerate them here. They will be treated with the species to which they are organically connected.

Sex in the Cup-fungi

Definite sex organs were observed by Tulasne and DeBary at an early date, but whether these actually functioned has remained a controversial question. Tulasne, as early as 1865, observed minute bodies in certain of the discomycetes which he called spermatia or conidiola, which he believed would be found to function like that of the pollen. They were believed to be male sperms, like the spermatia of the lichens. Recent authors held the same view.

Spermatization

In 1932, F. L. Drayton (Mycologia **24**: 345–348.) proved that microconidia, which were found in many species of *Sclerotinia* and other fungi, and which were probably morphologically identical with the spermatia or conidiola of Tulasne, stimulated the production of apothecia when placed on certain structures which develop on another thallus of the same fungus. To this process he applied the term "spermatization."

His preliminary experiments were carried on with *Sclerotinia Gladioli*. Isolates of this fungus transferred to suitable culture

media produced layers of sclerotized tissue with scattered sclerotia. From this crust-like growth so called receptive bodies are
formed which are more or less columnar and about 1 mm. in
height. These structures when spermatized with microconidia
from another thallus finally produce mature apothecia. The
receptive bodies of each isolate do not, however, react with
their own microconidia.

Heterothallism

While from the above we appear to have a condition of
heterothallism in *Sclerotinia Gladioli*, Drayton (Mycologia **26**:
69. 1934) believes that while this is in a sense heterothallism, it
is not true heterothallism as the term is usually understood
since while microconidia and receptive bodies must be produced
on different thalli, these may have arisen from a single hyphal
tip, while in true heterothallism the series are segregated on
different thalli. However, the term heterothallism is used
differently by the various authors and there is a chance for a
difference of opinion on this point.

Order **PEZIZALES** (continued)

(For diagnosis see North American Cup-fungi. Operculates)

Asci inoperculate (opening by a simple pore at
the apex).   Section 2.   INOPERCULATES
   Ascophores stipitate, clavate, spathulate,
or pileate, rarely discoid or turbinate,
on the ground or more rarely on plant
tissues.   Family 3.   GEOGLOSSACEAE.
   Ascophores typically cup-shaped or discoid, not as above, usually on living
or dead plant tissues, never or rarely
on the ground.
    Apothecia for the most part soft and
fleshy or waxy, superficial from the
first, or erumpent on herbaceous
plant tissues.   Family 4.   HELOTIACEAE.
    Apothecia soft or more often hard,
leathery, or horny, erumpent through
the substratum, often cespitose, usually on living or recently killed woody
plant tissues.   Family 5.   CENANGIACEAE.

Family 3. **GEOGLOSSACEAE**

Ascophores mostly club-shaped, or clavate, more rarely discoid or turbinate, the hymenium arising directly from the surface
of the clavula and forming a uniform stratum over it, usually
covering only a part of the ascophore but in some cases extending over the whole of it, or pileate, the hymenium occupying the
upper surface only; consistency fleshy but usually elastic or
fibrous rather than brittle, a few leathery, others distinctly
gelatinous; asci usually 8-spored; spores varying from ellipsoid
to fusiform, hyaline, or colored.

Ascophore stipitate, clavate, or spathulate, the ascigerous portion usually more or less compressed,
rarely subglobose.
   Clavate, the ascigerous portion not or only slightly
decurrent on opposite sides of the stem.
    Spores small, ellipsoid, cylindric, or fusiform,
simple.   1. MITRULA.

7

Spores long-elliptic to cylindric, 3–many-septate when mature.
Ascophores bright-colored.   2. MICROGLOSSUM.
Ascophores black or blackish.
Spores hyaline.   3. CORYNETES.
Spores fuliginous, or brown.
Hymenium without spines or setae.
Apothecia viscid.   4. GLOEOGLOSSUM.
Apothecia not viscid.   5. GEOGLOSSUM.
Hymenium beset with spines or setae.   6. TRICHOGLOSSUM.
Spathulate or fan-shaped, ascigerous portion decurrent on opposite sides of the stem.   7. SPATHULARIA.
Ascophores stipitate, pileate.
Spores ellipsoid-fusiform; ascophore gelatinous.   8. LEOTIA.
Spores filiform, or filiform-clavate.
Ascophores fleshy-gelatinous; asci very narrow; spores filiform; plants aquatic or semi-aquatic.   9. VIBRISSEA.
Ascophores fleshy-leathery; asci broadly clavate; spores filiform-clavate; plants terrestrial.   10. CUDONIA.

Note. The treatment of the Geoglossaceae is based on Durand's monograph (Ann. Myc. **6**: 388–482. 1908.) with additions and such alterations as seemed necessary to make it conform with the style of the present work. While the nomenclature does not in some cases conform to the present international rules, the writer does not feel justified in diverging from his conclusions.

1. **MITRULA** Pers. Neues Mag. Bot. **1**: 116.   1794.

Heyderia (Fries) Boud. Bull. Soc. Myc. Fr. **1**: 110.   1885.   Not C. Koch 1873.
Microglossum Boud. Bull. Soc. Myc. Fr. **1**: 110.   1885.   Not Gill.   1879.

Ascophores fleshy, erect, stipitate, clavate, the hymenium covering only the upper portion, or the ascigerous portion ellipsoid to subglobose, usually sharply delimited from the stem and often slightly free from it below, bright-colored (usually yellow or brownish); asci clavate-cylindric, opening by a pore, 8-spored; spores 1-seriate, or 2-seriate in the ascus, ellipsoid to narrowly fusiform, hyaline, continuous, or rarely 1-septate when mature; paraphyses present or absent.

Type species, *Mitrula Heyderi* Pers.

Spores broadly ellipsoid; paraphyses absent.
  Ascophores very irregular and contorted; spores 4–5
    × 6–10 μ.   1. *M. irregularis.*
  Ascophores regularly clavate; spores 3–4 × 4–6 μ.   2. *M. vitellina.*
Spores narrowly ellipsoid, clavate, or fusiform; para-
    physes present.
  Ascigerous portion vitelline-yellow; stem satiny white.   3. *M. phalloides.*
  Ascigerous portion cream-color to brownish; stem
      darker, at least not white.
    Growing on coniferous leaves; ascigerous portion
      cream-buff when fresh; stem darker.   4. *M. cucullata.*
    Growing on living moss.
      Hymenium nearly even; when dry ascigerous
        portion orange-brown; stem paler.   5. *M. gracilis.*
      Hymenium more or less convoluted.   6. *M. muscicola.*

1. **Mitrula irregularis** (Peck) Durand, Ann. Myc. **6**: 398. 1908.

Geoglossum irregulare Peck, Ann. Rep. N. Y. State Mus. **32**: 45. 1879.
Mitrula vitellina irregularis Sacc. Syll. Fung. **8**: 36. 1889.
Mitrula luteola Ellis, Am. Nat. **17**: 192. 1883.
Mitrula crispata Berk. Grevillea **3**: 149. 1875.
Spragueola americana Massee, Jour. Bot. **34**: 150. 1896.

Ascophores usually cespitose, rarely solitary, clavate, or irregular in form, usually twisted or contorted, compressed, obtuse, sometimes lobed, 1.5 cm. high; ascigerous portion vitelline-yellow, commonly occupying one-half to two-thirds the total length, up to 15 mm. broad; stem tapering downward, satiny white, pruinose, fibrillose or thinly tomentose, up to 1.5 cm. high, 2–5 mm. thick, sometimes absent; substance yellowish white; asci clavate-cylindric, the ascigerous hyphae repeatedly branched below, the apex rounded, not or slightly blue with iodine, variable in length, reaching a length of 90–150 μ and a diameter of 5–6 μ, 8-spored; spores 1-seriate, hyaline, continuous, smooth, ellipsoid, often slightly reniform, 4–5 × 6–10 μ (majority 5 × 8); paraphyses absent.

On bare soil, mossy damp soil or often among pine leaves.

Type locality: Sandlake, New York.

Distribution: Ontario and New Brunswick south to District of Columbia, and west to Colorado.

Illustrations: Bull. N. Y. State Mus. **1**²: *pl. 1, f. 5–7;* Ann. Rep. N. Y. State Mus. **48**: *pl. 5, f. 8–14;* Ann. Bot. **11**: *pl. 12, f. 23–24a;* Ann. Myc. **6**: *pl. 5, f. 7–8;* Jour. Bot. **34**: *pl. 357, f. 8, 9.*

Exsiccati: Ellis, N. Am. Fungi *978.*

2. **Mitrula vitellina** (Bres.) Sacc. Syll. Fung. **8**: 36.   1889.

Geoglossum vitellinum Bres. Rev. Myc. **4**: 212.   1882.
Microglossum vitellinum Boud. Bull. Soc. Myc. Fr. **1**: 110.   1885.

Ascophores clavate, 2–3 cm. high, "when fresh creamy-yellow," when dry yellowish cream-colored or brownish in one specimen, the ascigerous portion occupying about one-half the total length, about 3 mm. wide, compressed, obtuse, not sharply differentiated from the stem; stem terete, somewhat flexuous, equal, about 1 mm. in diameter, whitish, fibrillose; asci slender, cylindric, branched below, apex rounded, not blue with iodine, reaching a length of 75–80 μ and a diameter of 5–6 μ, 8-spored; spores 1-seriate, hyaline, continuous, smooth, broadly ellipsoid, 3–4 × 4–6 μ; paraphyses absent.

On rotten wood.

TYPE LOCALITY: Tyrol.

DISTRIBUTION: Tennessee; also in Europe.

ILLUSTRATIONS: Bres. Fungi Trid. *pl. 45, f. 1;* Rab. Krypt.-Fl. **1**³: 1143 *f. 1–3;* Ann. Bot. **11**: *pl. 12, f. 3–4a;* Ann. Myc. **6**: *pl. 5, f. 5–6.*

3. **Mitrula phalloides** (Bull.) Chev. Fl. Paris **1**: 114.   1826.
   (PLATE 76.)

?Helvella laricina Vill. Hist. Pl. Dauph. **4**: 1045.   1789.
Clavaria phalloides Bull. Hist. Champ. Fr. 214.   1791.
Mitrula paludosa Fries, Syst. Myc. **1**: 491.   1821.
Leotia uliginosa Grev. Scot. Crypt. Fl. **6**: *pl. 312.*   1828.
Leotia elegans Berk. London Jour. Bot. **5**: 6.   1846.
Mitrula elegans Fries, Nov. Symb. Myc. 103.   1851.
Mitrula laricina Massee, Ann. Bot. **11**: 271.   1897.

Ascophores solitary, or usually more or less densely gregarious, sometimes as many as fifteen to twenty closely aggregated and cohering at their bases, 2–6 cm. or more high, ascigerous portion clear vitelline-yellow, sharply differentiated from the stem, at first solid, becoming vesiculose or inflated and hollow when old, in outline ellipsoid, ellipsoid-obovoid, or pyriform, the apex rounded, obtuse, even or somewhat longitudinally furrowed especially below, often somewhat compressed, rarely more than one-fifth the total height of the plant, .5–2 cm. high, 4–10 mm. wide; stem terete, often flexuous, pure satiny white or sometimes with a pinkish tint, 1.5–2 mm. thick, smooth, when moist translucent and viscid; the whole plant soft and subtremellose; asci

1. MITRULA PHALLOIDES
2. OMBROPHILA CLAVUS

clavate, apex much narrowed, acute, very slightly blue with iodine, long-stipitate, reaching a length of 60–150 μ and a diameter of 6–8 μ, 8-spored; spores 2-seriate, hyaline, continuous, smooth, contents granular, cylindric to clavate-cylindric, 2.5–3 × 10–18 μ; paraphyses filiform, usually branched, septate, scarcely thickened above.

On decaying vegetation, often on or among *Sphagnum*, in wet places, pools and ditches.

TYPE LOCALITY: Europe.

DISTRIBUTION: Ontario to Alabama and west to British Columbia.

ILLUSTRATIONS: Bull. Herb. Fr. *pl. 463, f. 3;* Grev. Scot. Crypt. Fl. *pl. 312;* Phill. Brit. Discom. *pl. 2, f. 6;* Gill. Champ. Fr. Discom. *pl. 28, f. 1;* Cooke, Mycographia, *pl. 45, f. 175; pl. 46, f. 182;* Ann. Bot. **11**: *pl. 13, f. 69;* Rab. Krypt.-Fl. **1**[3]: 1143, *f. 1–4;* Ann. Myc. **6**: *pl. 17, f. 185.*

EXSICCATI: Rav. Fungi Car. **5**: *36;* Ellis, N. Am. Fungi *433;* Wilson & Seaver, Ascom. *38.*

4. Mitrula cucullata (Batsch) Fries, Epicr. Myc. 584. 1838.

Elvela cucullata Batsch, Elench. Fung. Cont. **1**: 189. 1786.
Mitrula Heyderi Pers. Neues Mag. Bot. **1**: 116. 1794.
Leotia Mitrula Pers. Syn. Fung. 611. 1801.

Ascophores small, solitary, or gregarious, very slender, 1–2.5 cm. high; ascigerous portion ellipsoid, ovate, or obovate, rounded above, distinct from the stem below and free from it for a slight distance, scarcely compressed, 1–3 mm. long, .5–2 mm. thick, cream-colored to yellowish-ochraceous; stem terete, equal, or tapering slightly upward, yellowish-brown above, darker below, granular-pruinose, the lower end frequently enveloped in and attached by a yellowish-brown tomentum, .5–.75 mm. thick (all parts become slightly darker in drying); asci clavate, apex narrowed, the pore blue with iodine, reaching a length of 45–70 μ and a diameter of 6 μ, 8–spored; spores obliquely 1-seriate, or 2-seriate, hyaline, continuous, smooth, narrowly fusiform, straight, or curved, 2–3 × 13–16 μ; paraphyses rather stout, gradually thickened upwards, septate, brownish, 3–4 μ thick.

On fallen leaves of conifers.

TYPE LOCALITY: Europe.

DISTRIBUTION: Northeastern United States to Idaho; also in Europe.

ILLUSTRATIONS: Batsch, Elench. Fung. Cont. 1: *f. 132;* Neues Mag. Bot. 1: *pl. 3, f. 12;* Cooke, Mycographia *pl. 45, f. 176;* Gill. Champ. Fr. Discom. *pl. 28, f. 2;* Sow. Engl. Fungi *pl. 84;* Ann. Bot. 11: *pl. 12, f. 39–41;* Ann. Myc. 6: *pl. 5, f. 1.*

5. **Mitrula gracilis** Karst. Hedwigia 22: 17.   1883.

Mitrula gracilis var. *flavipes* Peck, Ann. Rep. N. Y. State Mus. 49: 32   1896.

Ascophores solitary, slender, 1–1.5 cm. high; ascigerous portion obovate-globose, rounded above, not or very slightly free from the stem below, about 1 mm. diameter, orange-brown, even or nearly so; stem slender, flexuous, .5–.75 mm. thick, smooth, pale brownish-yellow, nearly translucent; asci clavate, apex rounded, pore blue with iodine, reaching a length of 65–80 μ and a diameter of 6–8 μ; spores 2-seriate, hyaline, smooth, oblong-fusiform to fusiform, continuous, or possibly becoming 1-septate, 2–3 × 10–14 μ; paraphyses filiform, 1.5–2 μ thick.

Attached to and evidently parasitic on *Paludella squarrosa* in bogs.

TYPE LOCALITY: Europe.

DISTRIBUTION: Labrador, Newfoundland, and Colorado (about 10,000 feet elevation); also reported from Greenland.

ILLUSTRATIONS: Ann. Myc. 6: *pl. 5, f. 3–4.*

6. **Mitrula muscicola** P. Henn. Oefv. Sv. Vet.-Akad. Förh. 42[5]: 71.   1885.

Ascophores cespitose, erect; pileus yellowish to tan-colored, obtuse, slightly *Morchella*-like, with a tendency to be ribbed or rugose, glabrous, about 4 mm. long; stipe lighter colored to whitish, glabrous, solid, about 10 mm. long; asci clavate, apex narrowed, pore blue with iodine, reaching a length of 65–75 μ and a diameter of 6–8 μ, 8-spored; spores 2-seriate, hyaline, continuous, smooth, narrowly oblong-ellipsoid, 2–3 × 10–13 μ; paraphyses filiform, very slightly thickened upward, hyaline, 2 μ thick.

On moss stems (*Webera nutans*), in moist spruce and balsam woods at about 7000 feet elevation.

TYPE LOCALITY: Europe.

DISTRIBUTION: Alberta; also in Europe.

ILLUSTRATIONS: Oefv. Sv. Vet.-Akad. Förh. 42[5]: *pl. 8, f. 6–8;* Ann. Myc. 6: *pl. 5, f. 2.*

Durand (Mycologia **13**: 185. 1921.) states that this is "doubtfully distinct" from the preceding *M. gracilis*, which has been found by the writer in quantity in Colorado (see Mycologia **3**: 57. 1911.).

Mitrula inflata Fries, Elench. Fung. **1**: 234. 1828.
    This is *Physalacria inflata* (Fries) Peck, a basidiomycete.
Mitrula roseola Morgan, Jour. Cin. Soc. Nat. Hist. **18**: 42. 1895.
    "Mr. Morgan later concluded that this was a lichen."

2. **MICROGLOSSUM** Gill. Champ. Fr. Discom. 25. 1879.

Helote Hazsl. Magyar. Akad. Ertek. Termesz. Kör. **11**[19]: 8. 1881.
Geoglossum subg. *Leptoglossum* Cooke, Mycographia 250. 1879. (in part.)

Ascophores fleshy, erect, stipitate, clavate, ascigerous only in the upper portion, bright-colored (usually yellow, brown or green); asci clavate-cylindric, 8-spored; spores 2-seriate, hyaline, smooth, ellipsoid, fusiform, or cylindric, becoming 3–many-septate; paraphyses present.

Type species, *Geoglossum viride* Pers.

Paraphyses strongly curved or uncinate and slightly
    thickened at the tips.
    Spores of one kind in the ascus.
        Ascophores bright-yellow.   1. *M. rufum.*
        Ascophores yellow-clay or tawny.   2. *M. fumosum.*
    Spores of two kinds in the ascus, the larger 60–100 μ
        long, 14–16-septate; ascophores cinnamon-
        brown.   3. *M. longisporum.*
Paraphyses straight or flexuous; ascophores usually with
    a greenish tint.
    Stem smooth, fibrous.   4. *M. olivaceum.*
    Stem squamulose, scarcely fibrous.   5. *M. viride.*

1. **Microglossum rufum** (Schw.) Underw. Minn. Bot. Studies **1**: 496. 1896.

Geoglossum rufum Schw. Trans. Am. Phil. Soc. II. **4**: 181. 1832.
Clavaria contorta Schw. Trans. Am. Phil. Soc. II. **4**: 182. 1832.
Geoglossum luteum Peck, Ann. Rep. N. Y. State Mus. **24**: 94. 1872.
Mitrula lutescens Berk.; Cooke, Hedwigia **14**: 9. 1875.
Geoglossum pistillare Berk. & Cooke; Cooke, Mycographia 206. 1878.
Mitrula rufa Sacc. Syll. Fung. **8**: 38. 1889.
Mitrula pistillaris Sacc. Syll. Fung. **8**: 38. 1889.
Leptoglossum luteum Sacc. Syll. Fung. **8**: 48. 1889.
Xanthoglossum luteum Kuntze, Rev. Gen. **2**: 875. 1891.
Leptoglossum lutescens mitruloides Rehm, Ann. Myc. **2**: 32. 1904

Ascophores solitary, gregarious or subcespitose, clavate, 2–5 cm. high, rather slender; ascigerous portion one-third to one-half the total length, ellipsoid-obovate, ellipsoid to subcylindric, obtuse, usually compressed or longitudinally furrowed, .5–2 cm. long, 6–12 mm. wide, sharply differentiated from the stem, usually clear vitelline or orange-yellow, rarely of a duller shade; stem terete, yellow, paler than the ascigerous portion, usually prominently squamulose, 1.5–3 cm. high, 2–3 mm. thick; whole ascophore becoming dingy-yellowish or reddish-brown when dry; asci clavate, apex narrowed, pore blue with iodine, reaching a length of 100–140 μ and a diameter of 10–12 μ, 8-spored; spores 2-seriate, hyaline, smooth, cylindric or slightly narrowed toward the obtuse ends, straight or curved, for a long time continuous, then multiguttulate, finally 5–10-septate, 5–6 $\times$ 18–38 μ (25–35); paraphyses filiform, 2–3 μ thick below, the apices slightly thickened and strongly curved or uncinate.

On rotten wood and humus, rarely on mossy banks, frequent.

TYPE LOCALITY: New Jersey.

DISTRIBUTION: New Hampshire to Minnesota and south to Louisiana.

ILLUSTRATIONS: Peck, Ann. Rep. N. Y. State Mus. **24**: *pl. 3, f. 20–24;* Cooke, Mycographia *pl. 3, f. 12; pl. 45, f. 178; pl. 96, f. 346 & 348;* Ann. Bot. **11**: *pl. 12, f. 28–30.*

EXSICCATI: Ellis, N. Am. Fungi *58.*

2. **Microglossum fumosum** (Peck) Durand, Ann. Myc. **6**: 408. 1908.

Leptoglossum luteum var. *fumosum* Peck, Ann. Rep. N. Y. State Mus. **43**: 40. 1889.

Leptoglossum fumosum Peck, Bull. N. Y. State Mus. **116**: 25.   1907.

Ascophores solitary, or more often densely cespitose, clavate, 2–6 cm. high, robust; ascigerous portion obovate, ellipsoid, or oblong, rounded above, more or less compressed and longitudinally furrowed, about one-third to one-half the total length of the ascophore, but little distinct from the stem, smoky-yellowish, clay-colored, or tawny, .75–3 cm. long, 3–12 mm. thick, rarely twisted or contorted, more or less puckered below where it joins the stem; stem terete, or slightly compressed, 1–3 cm. long, 3–5 mm. thick, clay-colored, slightly squamulose, or sometimes smooth and shining, with several longitudinal cavities, rather fibrous; asci clavate, the apex narrowed, pore blue with iodine, reaching a length of 100–150 μ and a diameter of 10–12 μ, 8-spored; spores 2-seriate, hyaline, smooth, cylindric, or slightly narrowed toward the obtuse ends, usually slightly curved, 5 × 20–50 μ (35–43), at first continuous, finally 7–15-septate; paraphyses filiform, septate, 2 μ thick, slightly thickened and strongly curved above.

On much decayed rotten logs and about the bases of stumps.

TYPE LOCALITY: New York.

DISTRIBUTION: New York and Massachusetts.

ILLUSTRATIONS: Ann. Myc. **6**: *pl. 5, f. 15, 16; pl. 19, f. 203, 204, 205.*

3. **Microglossum longisporum** Durand, Ann. Myc. **6**: 409. 1908.

Ascophores solitary, gregarious, clavate, often curved or contorted, rich cinnamon-brown, 3–6 cm. high; ascigerous portion occupying about one-third to one-half the total length, slightly differentiated from the stem, oblong to ellipsoid, obtuse, more or less compressed and longitudinally furrowed, slightly darker than the stem, often with an umber tint, 1–2.5 cm. long, 4–10 mm. wide, flesh yellowish-brown; stem terete, squamulose, sometimes later becoming nearly smooth and hygrophanous, clammy or slightly viscid below, 2–4 cm. long, 2–4 mm. thick, equal; asci cylindric-clavate, apex rounded, only slightly narrowed, the pore blue with iodine, reaching a length of 100–140 μ and a diameter of 12–15 μ, 8-spored; spores of two kinds in the ascus; the first 2 in number (very rarely 3 or 4), lying side by side and nearly as long as the ascus, hyaline, smooth, cylindric or a very little broader in the middle, ends rounded, straight or curved, at first continuous and multiguttulate, finally becoming about 14–16-septate, 4–5 × 60–100 μ; the second 6 (rarely less) usually placed irregularly near the apex of the ascus, similar to the first kind but smaller, 3 × 12–18 μ, rarely longer, usually continuous; paraphyses filiform, 2 μ thick, hyaline, the apices slightly thickened and strongly curved or uncinate.

On the ground among leaves in the midst of grasses and sedges, in rich woods and ravines.

TYPE LOCALITY: New York.

DISTRIBUTION: New York to Michigan and North Carolina.

ILLUSTRATIONS: Ann. Myc. **6**: *pl. 5, f. 17, 18; pl. 19, f. 206, 206a.*

4. **Microglossum olivaceum** (Pers.) Gill. Champ. Fr. Discom. 25. 1879.

Geoglossum olivaceum Pers. Obs. Myc. **1**: 40.   1796.
Geoglossum olivaceum Cooke, Mycographia 250.   1879.
Mitrula olivacea Sacc. Syll. Fung. **8**: 38.   1889.
Microglossum contortum Peck, Bull. Torrey Club **25**: 328.   1898.
Microglossum obscurum Peck, Bull. Torrey Club **26**: 71.   1899.

Ascophores solitary, or clustered, clavate, either regular, twisted, or contorted, 2–8 cm. high; ascigerous portion occupying about one-third to one-half the total length, continuous with the stem but sharply delimited by color, greenish-brown, often with a yellow or buff tint, compressed, obtuse, up to 3 cm. long, 10 mm. wide; stem terete, or compressed above, perfectly smooth, shining, hygrophanous, yellowish-buff, tawny-buff, smoky-olive-buff or brownish-cervine, fibrous, solid, 2–8 mm. thick; asci cylindric-clavate, apex rounded, rarely narrowed, pore blue with iodine, reaching a length of 75–100 μ and a diameter of 9–10 μ, 8-spored; spores 2-seriate above, 1-seriate below, hyaline, smooth, oblong-fusiform to fusiform, straight, curved, or sigmoid, for a long time simple, finally becoming 3-septate, 4–6 $\times$ 12–18 μ; paraphyses filiform, often branched, septate, straight or slightly curved, or flexuous, not thickened above, hyaline.

On the ground among leaves or in grassy places in rich woods.

TYPE LOCALITY: Europe

DISTRIBUTION: Northeastern United States and Ontario; also in Europe.

ILLUSTRATIONS: Pers. Obs. Myc. **1**: *pl. 5, f. 7;* Cooke, Mycographia *pl. 4, f. 13;* Pat. Tab. Fung. *f. 65;* Berk. Outl. Brit. Fung. *pl. 22, f. 3;* Ann. Myc. **6**: *pl. 5, f. 19, 20; pl. 20, f. 209.*

EXSICCATI: Ellis & Ev. Fungi Columb. *1745.*

5. **Microglossum viride** (Pers.) Gill. Champ. Fr. Discom. 25. 1879.

?Clavaria mitrata viridis Holmsk. Coryph. 24. *fig.*   1790.
?Clavaria viridis Schrad. Gmel. Linn. Syst. Nat. **2**: 1443.   1791.
Geoglossum viride Pers. Neues Mag. Bot. **1**: 117.   1794.
Mitrula viridis Karst. Myc. Fenn. **1**: 29.   1871.
Helote viridis Hazsl. Magyar Akad. Ertek. Termesz. Kör. **11**[19]: 9.   1881.
Leptoglossum alabamense Underw. Bull. Torrey Club **24**: 82.   1897.

Ascophores solitary, or more often gregarious, or cespitose, clavate, up to 5 cm. high; ascigerous portion about one-half the total length, sharply delimited from the stem, lanceolate to ellipsoid, obtuse, strongly compressed and deeply longitudinally furrowed in older specimens 3–10 mm. wide, olive-buff, or olive-ochraceous; stem terete, or slightly compressed, 2–5 mm. thick, conspicuously squamulose, pale pea-green; dried plants are darker, sometimes almost black; crushed flesh pea-green to olive, usually pale; asci clavate-cylindric, apex rounded or slightly narrowed, pore blue with iodine, slender below, reaching a length of 110–150 μ and a diameter of 8–10 μ, 8-spored; spores 2-seriate above, 1-seriate below, hyaline, smooth, cylindric-oblong, ellipsoid-oblong, or oblong-clavate, ends obtuse, straight, curved, or sigmoid, for a long time continuous, finally becoming clearly 3–4-septate, 5–6 × 14–22 μ; paraphyses filiform, branched hyaline, the apices often slightly pyriform-thickened and tinged green, forming a green epithecium.

On the ground, in moist woods, often along the borders of old wood roads.

TYPE LOCALITY: Europe.

DISTRIBUTION: Eastern United States; also in Europe.

ILLUSTRATIONS: Cooke, Mycographia *pl. 4, f. 14;* Gill. Champ. Fr. Discom. *pl. 26, f. 2;* Ann. Bot. **11**: *pl. 13, f. 68;* Rab. Krypt.-Fl. **1³**: 1144, *f. 1–4:* Phill. Brit. Discom. *pl. 2, f. 8;* Ann. Myc. *pl. 5, f. 23–24; pl. 20, f. 208.*

EXSICCATI: Ellis & Ev. N. Am. Fungi *2030;* Fungi Columb. *1746.*

3. **CORYNETES** Hazsl. Magyar. Akad. Ertek. Termesz. Kör. 11¹⁹: 7. 1881.

Geoglossum subg. *Leptoglossum* Cooke, Mycographia 250. 1879, (in part).
Microglossum Sacc. Bot. Cent. **18**: 214. 1884. (Not Gill. 1879.)
Leptoglossum Sacc. Bot. Cent. **18**: 214. 1884. (Not Karsten, 1879.)
Thuemenidium Kuntze, Rev. Gen. **2**: 873. 1891.
Xanthoglossum Kuntze, Rev. Gen. **2**: 875. 1891.

Ascophores fleshy, erect, stipitate, clavate, the hymenium covering only the upper portion, black, brownish-black or purplish-black; asci clavate-cylindric, opening by a pore, 8-spored; spores usually 2-seriate, hyaline, smooth, cylindric, ends rounded, becoming 3–many-septate; paraphyses present.

Type species, *Geoglossum microsporum* Cooke & Peck.

Paraphyses hyaline, or only slightly purplish tinted at
　　the tips.
　Asci and paraphyses agglutinated at the tips into a
　　　conspicuous, vinous-brown epithecium.
　　　Paraphyses clavate to pyriform, thickened at the
　　　　tips.   1. *C. purpurascens.*
　　　Paraphyses not or only very slightly thickened
　　　　at the tips.   2. *C. atropurpureus.*
　Epithecium lacking or inconspicuous; ascophores
　　　robust; paraphyses slightly clavate-thickened and
　　　more or less curved at the tips.   3. *C. robustus.*
Paraphyses conspicuously brown.   4. *C. arenarius.*

1. **Corynetes purpurascens** (Pers.) Durand, Ann. Myc. **6**: 413.
　1908.

?Clavaria mitrata Holmsk. Coryph. 21.   1790.
Geoglossum purpurascens Pers. in Holmsk. Coryph. 171.   1797.
Leotia atropurpurea Corda, Ic. Fung. **5**: 79.   1842.
Geoglossum atropurpureum Cooke, Mycographia 10.   1875.   (Not Pers.)
Microglossum atropurpureum Sacc. Syll. Fung. **8**: 40.   1889.
Thuemenidium atropurpureum Kuntze, Rev. Gen. Pl. **2**: 873.   1891.
Mitrula purpurascens Massee, Ann. Bot. **11**: 266.   1897.

　　Ascophores solitary, or cespitose, 3–6 cm. high, every part
with a distinct purplish-brown tint when fresh, blackish when
dry, the crushed flesh vinous-brown; ascigerous portion irregular,
clavate, sometimes lobed or forked at the apex, more or less
compressed, 1–2 cm. long, up to 1 cm, broad, occupying about
one-third to one-half the total length; stem cylindric, minutely
squamulose; asci clavate, apex narrowed but rounded, pore blue
with iodine, reaching a length of 105–120 μ and a diameter of
10–12 μ, 8-spored; spores 2-seriate, hyaline, smooth, cylindric,
slightly narrowed from the middle to the rounded ends, straight,
or curved, at first simple, then multiguttulate, finally 6-septate,
5–6 × 20–37 μ (20–30); paraphyses filiform, hyaline, 3 μ thick,
the apices straight and abruptly obovate-pyriform thickened,
8–10 μ thick, agglutinated with amorphous matter to form a
vinous-brown epithecium.

　　On the ground or humus.

　　TYPE LOCALITY: Europe.

　　DISTRIBUTION: Maine and New York; also in Europe.

　　ILLUSTRATIONS: Corda, Ic. Fung. **5**: *pl. 9, f. 71;* Cooke, Mycographia *pl. 4, f. 16;* Ann. Bot. **11**: *pl. 12, f. 27;* Ann. Myc. **6**:
pl. 6, f. 36–39.

asci clavate-cylindric, pore blue with iodine, 8-spored; spores fascicled, or multiseriate in the ascus, cylindric or clavate-cylindric, 3–15-septate (rarely simple in one species), brown or fuliginous; paraphyses numerous, septate, thickened, or coiled and brown above, not confined to the hymenium but continued down the stem to its base.

Type species, *Geoglossum difforme* Fries.

Spores 0–7-septate when mature.

Spores 0–7-septate, clavate-cylindric, 55–102 μ long.	1. *G. glutinosum.*
Spores 7-septate, clavate, 43–65 μ long.	2. *G. affine.*

Spores 15-septate when mature. 3. *G. difforme.*

1. **Gloeoglossum glutinosum** (Pers.) Durand, Ann. Myc. **6**: 419. 1908.

Geoglossum glutinosum Pers. Obs. Myc. **1**: 11. 1796.
?Geoglossum viscosum Pers. in Holmsk. Coryph. 39. 1797.
Geoglossum glutinosum lubricum Pers. Myc. Eur. **1**: 197. 1822.

Ascophores solitary or clustered, 5–8 cm. high, viscid-gelatinous; ascigerous portion clavate, or narrowly elliptic, more or less compressed, apex obtuse, 1.5–2.5 cm. long, 5–10 mm. thick, black, not sharply differentiated from the stem; stem 4–6 cm. long, 3–4 mm. thick, terete or slightly compressed, brown or brownish-black, very smooth and viscid, covered to the base by the paraphyses; flesh brown, composed of parallel septate hyphae, rather looser in the center; asci narrowly clavate, narrowed from the middle toward the apex, the pore blue with iodine, reaching a length of 250 μ and a diameter of 12–15 μ, 8-spored; spores multiseriate in the ascus, cylindric, or slightly narrowed toward the rounded ends, fuliginous, at first simple, then 3-septate, finally in most cases 7-septate, 5–6 × 55–102 μ (75–85); paraphyses cylindric, septate, 3 μ thick, longer than the asci, the apices pale-brown and abruptly pyriform to globose thickened, 8–10 μ thick.

On the ground and on rotten wood, in rich woods.

TYPE LOCALITY: Europe.

DISTRIBUTION: Ontario to North Carolina; also in Europe.

ILLUSTRATIONS: Cooke, Mycographia *pl. 2, f. 6; pl. 3, f. 10;* Gill. Champ. Fr. Discom. *pl. 25, f. 2; pl. 26, f. 1;* Grev. Scot. Crypt. Fl. *pl. 55;* Massee, Ann. Bot. **11**: *pl. 13, f. 66, 67;* Ann. Myc. **6**: *pl. 8, f. 70–72; pl. 14, f. 149–155.*

2. **Gloeoglossum affine** Durand, Ann. Myc. **6**: 420.   1908.

Ascophores solitary, or gregarious, black, viscid-gelatinous when fresh, clavate, 1.5–2.5 cm. high; ascigerous portion narrowly clavate-oblong, obtuse, compressed, occupying about one-half the total length up to 3 mm. thick, not sharply differentiated from the stem; stem slender, smooth; asci clavate, the apex rounded, reaching a length of 130–160 μ and a diameter of 15 μ, 8-spored; spores fasciculate, smoky-brown, clavate, 7-septate, 5–6 $\times$ 43–65 μ; paraphyses slenderly cylindric below, 2.5 μ thick, slightly and gradually thickened above, the terminal portion septate, brownish, the terminal cell abruptly ellipsoid to globose thickened, 7–8 μ in diameter, the lower ones either cylindric, or nodulose, or abruptly and conspicuously thickened below the septa, in extreme cases almost globose, either straight or more less curved, or coiled above, and continuing down the stem to its base.

On soil or humus in swamps.

TYPE LOCALITY: New York.

DISTRIBUTION: New York.

ILLUSTRATIONS: Ann. Myc. **6**: *pl. 8, f. 73–74; pl. 16, f. 165–167.*

3. **Gloeoglossum difforme** (Fries) Durand, Ann. Myc. **6**: 421. 1908.

Geoglossum difforme Fries, Obs. Myc. **1**: 159.   1815.   (Not of authors.)
Geoglossum Peckianum Cooke, Hedwigia **14**: 10.   1875.

Ascophores solitary, or gregarious, sometimes two or three together, black, smooth, viscid, especially below, evenly clavate, with no line of demarcation between ascigerous portion and stem, strongly compressed, apex obtuse, 3–6 cm. high, 8–14 mm. wide; flesh brown, composed of parallel, cylindric, septate hyphae, rather looser in the center but not hollow; asci clavate, gradually narrowed from the middle toward the apex, the latter rounded, the apical plug blue with iodine, reaching a length of 240–275 μ and a diameter of 18–25 μ, 8-spored; spores in a parallel fascicle in the ascus, brownish-fuliginous, clavate-cylindric, slightly narrowed from the middle each way, ends rounded, straight, or slightly curved, smooth, 15-septate at maturity, the cells about as long as wide, 6–7 $\times$ 95–125 μ; paraphyses longer than the asci, slender, septate, tips slightly thickened, brownish, much coiled and twisted, continued down the stem to the base.

On soil, humus, decayed wood, or among pine needles in rich woods.

TYPE LOCALITY: Europe.

DISTRIBUTION: Maine to Ontario, Florida and Michigan; also in Europe.

ILLUSTRATIONS: Cooke, Mycographia *pl. 2, f. 5;* Massee, Ann. Bot. **11**: *pl. 12, f. 42–43;* Ann. Myc. **6**: *pl. 8, f. 75–77; pl. 15, f. 156–160.*

EXSICCATI: Ellis & Ev. Fungi Columb. *1731.*

5. **GEOGLOSSUM** Pers. Neues Mag. Bot. **1**: 116.   1794.

Ascophores fleshy, erect, stipitate, clavate, hymenium covering only the upper portion, black, or brownish-black; asci clavate-cylindric, 8-spored; spores fascicled, or multiseriate in the ascus, cylindric or clavate-cylindric, 3–15-septate, fuliginous; paraphyses numerous, septate, usually brownish above, confined to the ascigerous portion.

Type species, *Geoglossum glabrum* Pers.

Paraphyses and asci free above, the apices not agglutinated to form a brown epithecium; spores early becoming brown.

 Spores 7- or 7–12-septate at maturity.

  Mature spores uniformly 7-septate; ascophore black or brownish-black; paraphyses brown above.

   Paraphyses normally clavate in the distal third, closely septate, usually constricted or moniliform. — 1. *G. glabrum.*

   Paraphyses remotely septate, scarcely moniliform, the thickening confined to the terminal cell. — 2. *G. nigritum.*

  Mature spores 7–12-septate; ascophore tawny-brown; paraphyses nearly hyaline. — 3. *G. fallax.*

  Mature spores 15-septate. — 4. *G. pygmaeum.*

Paraphyses and asci agglutinated above forming a conspicuous brown epithecium; spores usually remaining for a long time hyaline.

 Spores 0–7-septate; paraphyses brown, the tips pyriform. — 5. *G. cohaerens.*

 Spores 7–12-septate; paraphyses nearly hyaline. — 6. *G. intermedium.*

 Spores 15-septate, clavate-cylindric.

  Ascophores large, up to 5 cm. high. — 7. *G. alveolatum.*

  Ascophores small, usually 1 cm. high or less. — 8. *G. pumilum.*

1. **Geoglossum glabrum** Pers. Neues Mag. Bot. 1: 116.   1794.

?Clavaria ophioglossoides L. Sp. Pl. **2**: 1182.   1753.
Geoglossum sphagnophilum Ehrb. Sylv. Myc. Ber. 30.   1818.
Geoglossum difforme sensu Cooke, Mycographia 6, *pl. 2, f. 7*.   1875.
Geoglossum simile Peck, Bull. Buff. Soc. Nat. Sci. **1**: 70.   1873.

Ascophores solitary, or rarely clustered, gregarious, 3–7 cm. high; ascigerous portion black, about one-third the total length of the ascophore, lanceolate, compressed, 1–3 cm. long, 3–8 mm. thick, obtuse or acute, not sharply delimited from the stem; stem terete or slightly compressed, brownish-black, rather slender, densely squamulose, 2–5 cm. high, 1.5–5 mm. thick, but commonly about 2 mm., the tissue composed of a fascicle of parallel, septate, brown hyphae 10–12 μ in diameter, those in the center of the ascigerous part loose, those on the surface of the stem projecting to form flexuous, septate hairs up to 100 μ long, single or agglutinated into groups; asci rather stout, clavate-lanceolate, apex narrowed but rounded, pore blue with iodine, short-stipitate, reaching a length of 170–200 μ and a diameter of 20–22 μ, 8-spored; spores in parallel fascicle in the ascus, clavate, dark-fuliginous, 7-septate, usually slightly curved, 7–9 $\times$ 60–105 μ (the majority 75–95); paraphyses slightly longer than the asci, typically clavate in the distal third, closely septated into cells rarely more than twice as long as broad, usually constricted at the septa and the cells often swollen so as to give a moniliform appearance, brown above, straight, or curved, about 3 μ thick below, 5–9 μ thick above.

On very rotten wood, logs, stumps, or rich humus, rarely on soil.

TYPE LOCALITY: Europe.

DISTRIBUTION: Newfoundland to Florida and California; also in Europe.

ILLUSTRATIONS: Cooke, Mycographia *pl. 2, f. 7;* Gill. Champ. Fr. Discom. *pl. 24, f. 3;* Rab. Krypt.-Fl. **1**³: 1145, *f. 1–4;* Ann. Bot. **11**: *pl. 12, f. 44–46;* Boud. Ic. Myc. *pl. 423;* Ann. Myc. **6**: *pl. 7, f. 50–56; pl. 12, f. 121–129.*

2. **Geoglossum nigritum** (Pers.) Cooke, Mycographia 205. 1878.

Clavaria nigrita Pers. Syn. Fung. 604.   1801.

Ascophores usually solitary, clavate, black, or brownish-black, up to 8 cm. high; ascigerous portion lanceolate, compressed, rather acute, one-half the total length of the ascophore or less, 2–5 mm. thick; stem terete, slender, 1–2 mm. thick, furfuraceous, minutely squamulose, or almost smooth; asci clavate or clavate-lanceolate, apex narrowed but rounded, reaching a length of 150–175 μ and a diameter of 18 μ, 8-spored; spores in a parallel fascicle in the ascus, clavate, fuliginous, 7-septate, 6 $\times$ 54–85 μ; paraphyses rather longer than the asci, cylindric, septate, the cells 2–10 times as long as wide, not constricted, rarely slightly swollen below the septa, pale-brown above, either only slightly thickened, or the apex of the terminal cell pyriform, usually more or less curved.

On wet ground, banks, or among leaves on rich humus, rarely on rotten ground.

TYPE LOCALITY: Europe.

DISTRIBUTION: Maine to North Carolina and California and Bermuda; also in Europe.

ILLUSTRATIONS: Cooke, Mycographia *pl. 96, f. 345;* Peck, Ann. Rep. N. Y. State Mus. **29**: *pl. 1, f. 20–22;* Ann. Myc. **6**: *pl. 7, f. 57–59; pl. 12, f. 130–132.*

This species has been found to be very common in Bermuda, having been collected there by the writer in 1912, 1926, 1938, and 1940.

3. **Geoglossum fallax** Durand, Ann. Myc. **6**: 428.   1908.

?Geoglossum glabrum paludosum Pers. Myc. Eur. **1**: 194.   1822.

Ascophores solitary, clavate, 2–8.5 cm. high, entirely tawny-brown to umber-brown; ascigerous portion one-fifth to one-half the total length of the ascophore, lanceolate, obtuse, slightly compressed, about 8–15 mm. long, 3–5 mm. thick; stem short, or elongated and slender, squamulose especially above, slightly thickened upward, 1–2 mm. thick below, 2 mm. thick above, terete; asci clavate-cylindric, the apex narrowed, pore blue with iodine, reaching a length of 150–175 μ and a diameter of 18–20 μ, 8-spored; spores 2-seriate to multiseriate in the ascus, clavate-cylindric, straight, or curved, at first continuous and multi-guttulate, then 3– finally 7–12-septate, 5–7 $\times$ 65–105 μ (80–100), for a long time hyaline, finally becoming fuliginous; paraphyses entirely hyaline, cylindric, not closely septate, 5–6 μ thick, usually strongly curved, or circinate above, the apex abruptly ellipsoid to globose thickened.

On clay or loamy soil in woods or on slopes of ravines.

TYPE LOCALITY: Ithaca, New York.

DISTRIBUTION: New York to Michigan.

ILLUSTRATIONS: Ann. Myc. **7**: *pl. 7, f. 61–64; pl. 13, f. 133–137.*

4. **Geoglossum pygmaeum** Gerard; Durand, Ann. Myc. **6**: 429. 1908.

Ascophores very small, .5–2 cm. high, brownish-black when dry; ascigerous portion occupying about one-half the total length of the ascophore, narrow, compressed; stem terete, very slender, minutely hairy; asci clavate, apex narrowed, reaching a length of 175–200 μ and a diameter of 17–18 μ, 8-spored; spores in a parallel fascicle in the ascus, brown or fuliginous, cylindric-clavate, tapering slightly each way from above the middle, 15-septate, 6–7 $\times$ 122–140 μ; paraphyses straight, about as long as the asci, cylindric, 3 μ thick, pale, and sparingly septate below, the apex brown, septated into cells about 12–14 μ long, slightly constricted at the septa, the terminal two or three cells clavate-thickened, the apical one usually more swollen and pyriform to ellipsoid in outline, 7–8 $\times$ 10–14 μ.

On ligneous earth.

TYPE LOCALITY: Poughkeepsie, New York.

DISTRIBUTION: Known only from the type locality.

ILLUSTRATIONS: Ann. Myc. **6**: *pl. 7, f. 60; pl. 13, f. 140–141.*

5. **Geoglossum cohaerens** Durand, Ann. Myc. **6**: 430.   1908.

Ascophores clavate, 2.5–3.5 cm. high, black; ascigerous portion about one-half the total length, terete or compressed, obtuse; stem squamulose; asci clavate, reaching a length of 150 μ and a diameter of 12–15 μ, 8-spored; spores multiseriate in the ascus, cylindric-clavate, straight, or curved, for a long time hyaline, later becoming pale-brown, from simple to 7-septate, 5 $\times$ 40–55 μ; paraphyses very numerous, rather longer than the asci, cylindric, sparingly but irregularly septate, the tip rather abruptly clavate to pyriform thickened, conspicuously brown above, the apices cohering with amorphous matter to form a brown epitheceium above the asci.

On sandy soil in a dooryard.

TYPE LOCALITY: Newfield, New Jersey.

DISTRIBUTION: Known only from the type locality.

ILLUSTRATIONS: Ann. Myc. **6**: *pl. 8, f. 65; pl. 13, f. 138–139.*

6. **Geoglossum intermedium** Durand, Ann. Myc. **6**: 431. 1908.

Ascophores solitary, black, 3–4.5 cm. high; ascigerous portion one-fourth to one-third the total length, oblong to ovate-lanceolate, compressed, obtuse or rounded above, usually abruptly narrowed to the stem below, .75–1.5 cm. long, 2–6 mm. broad, the hymenium prettily alveolate when dry, the meshes less than .5 mm. across; stem slender, terete, usually more or less flexuous, brownish-black, .75–1 mm. thick, nearly smooth below, provided toward the top with slender, flexuous, sparingly septate, obtuse, brown hairs, up to 100 μ long; asci clavate-cylindric, very short-pediceled, apex narrowed but rounded, reaching a length of 150–200 μ and a diameter of 20 μ, pore deep-blue and entire ascus slightly blue with iodine, 8-spored; spores multiseriate, stout, clavate, but not much narrowed toward the lower end, ends rounded, somewhat curved, at first hyaline, becoming 1–3-septate, finally rather pale-brown, 7–11-septate, 6 × 55–75 μ; paraphyses cylindric, nearly or quite hyaline, straight or only slightly curved above, sparingly septate, very gradually and slightly thickened upward, the apex tending to become pyriform to globose thickened, the apical portion of the asci and paraphyses agglutinated by an amorphous brown mass.

In swamp and on rotten wood.

TYPE LOCALITY: Knoxboro, New York.

DISTRIBUTION: New York and Ontario.

ILLUSTRATIONS: Ann. Myc. **6**: *pl. 8, f. 66–67; pl. 13, f. 142–144.*

7. **Geoglossum alveolatum** Durand, Ann. Myc. **6**: 432. 1908.

Ascophores solitary, or gregarious, with the aspect of *G. glabrum*, slenderly clavate, 1.5–5 cm. high; ascigerous portion about one-third the total length, lanceolate, obtuse, compressed, black, 4–12 mm. long, 3–4 mm. thick, when moist with an even surface, which on drying often becomes pitted or distinctly ridged in an alveolate manner; stem slender, terete, 1–2 mm. thick, slightly thickened upwards, distinctly squamulose or hairy; flesh black, composed of a fascicle of parallel, septate hyphae 8–10 μ in diameter, the ectal ones of the stem being produced to form slender, brown, flexuous, obtuse, septate hairs up to 125 μ long; asci clavate-cylindric, apex slightly narrowed but rounded, pore blue with iodine, reaching a length of 150–170 μ and a diameter of 15 μ, 8-spored; spores fascicled, or multiseriate in the ascus, narrowly cylindric, straight, or curved, ends obtuse, one end sometimes narrower than the other, contents granular, at first continuous, then 7– finally 15-septate, 4–5 × 60–95 μ, for a long time hyaline but ultimately becoming pale-brown or fuliginous; paraphyses conspicuously brown above, cylindric, septate, 3 μ thick, the tips abruptly ellipsoid to globose thickened, 6–8 μ thick and agglutinated together into a brown epithecium.

On very rotten wood and logs, in ravines and moist woods.

TYPE LOCALITY: Canandaigua, New York.

DISTRIBUTION: New York and Idaho?

ILLUSTRATIONS: Ann. Myc. **6**: *pl. 8, f. 68–69; pl. 14, f. 145–148.*

8. **Geoglossum pumilum** Winter, Grevillea **15**: 91.   1886.

Ascophores very small, .5–2 cm. high, usually not exceeding 1 cm., slender, black; ascigerous portion distinct from the stem, clavate-ellipsoid to oblong spherical, 1.5–3 mm. long, 1–2 mm. thick when dry, rounded above; stem very slender, brownish-black, squamulose, especially above, .5 mm. thick when dry; asci clavate, stout, reaching a length of 185–200 μ and a diameter of 20 to 27 μ, 8-spored; spores fasciculate in the ascus, clavate-cylindric, tapering each way from above the middle, 15-septate, 6 × 104–125 μ (majority 110–115 μ long), deeply colored; paraphyses longer than the asci, pale-brown above, nearly hyaline below, the distal end stout, clavate, rather remotely septate, usually nearly straight but sometimes strongly curved, inclined to be constricted at the septa, 8–12 μ thick.

On soil.

TYPE LOCALITY: Brazil.

DISTRIBUTION: Virginia, Bermuda and Porto Rico.

Two minute fruiting bodies of this species were collected by the writer in Bermuda (Britton, Brown & Seaver *1364*) and two in Porto Rico (Sci. Survey Porto Rico and the Virgin Islands **8**: 74. 1926.). It will be noted that only two fruiting bodies of this species were found at one time in Bermuda and in Porto Rico.

DOUBTFUL AND EXCLUDED SPECIES

Geoglossum album A. E. Johnson, Bull. Minn. Acad. Nat. Sci. **1**: 341. 1878; *Mitrula Johnsonii* Sacc. Syll. Fung. **8**: 36. 1889; *Microglossum album* Underw. Minn. Bot. Stud. Bul. **9**: 495. 1896; *Mitrula alba* Massee, Ann. Bot. **11**: 284. 1897. This species was recorded from Minnesota but is exceedingly doubtful. The name was changed by Saccardo to avoid making a homonym.

Geoglossum farinaceum Schw. Schr. Nat. Ges. Leipzig **1**: 113. 1822. The type is missing from Schweinitz's herbarium and no specimens are known to exist elsewhere. It is probably not a discomycete.

6. **TRICHOGLOSSUM** Boud. Bull. Soc. Myc. Fr. **1**: 110. 1885.

Geoglossum Pers. in part.

Ascophores fleshy, erect, stipitate, clavate, hymenium covering only the upper portion, black; asci clavate or ligulate, 4–8-spored; spores fasciculate, or multiseriate in the ascus, clavate-cylindric, up to 15-septate, fuliginous, or brown; paraphyses numerous, septate, brown above, confined to the ascigerous portion, both the stem and hymenium beset with black, thick-walled, acicular spines or cystidia which are usually longer than the asci; paraphyses usually stout, straight or strongly curved.

Type species, *Geoglossum hirsutum* Pers.

Spores normally 100–170 μ long, narrowed each way
    from above the middle.
  Spores 4 in each ascus.
    Spores 8–11-septate.   1. *T. velutipes.*
    Spores 15-septate.   2. *T. tetrasporum.*
  Spores 8 in each ascus.
    Spores 15-septate.   3. *T. hirsutum.*
    Spores 8–9-septate.   4. *T. Wrightii.*
    Spores 7-septate.   5. *T. octopartitum.*
Spores normally 45–100 μ long.
  Spores 0–5-septate, clavate-cylindric.   6. *T. Farlowi.*
  Spores 7-septate.
    Spores 55–73 μ long, clavate.   7. *T. confusum.*
    Spores 75–100 μ long, clavate-cylindric.   8. *T. Walteri.*

1. **Trichoglossum velutipes** (Peck) Durand, Ann. Myc. **6**: 434. 1908. (PLATE 77.)

Geoglossum hirsutum var. *americanum* Cooke, Mycographia 3. 1875.
Geoglossum americanum Sacc. Syll. Fung. **8**: 46. 1889.
Geoglossum velutipes Peck, Ann. Rep. N. Y. State Mus. **28**: 65. 1876.

Ascophores solitary, or gregarious, sometimes cespitose, up to 10 cm. high, black or brownish-black; ascigerous portion lanceolate, ellipsoid, or subrotund, one-fifth to one-third the total length of the fruiting body, 3–15 mm. long, 4–10 mm. thick, more or less compressed, rounded above, usually rather sharply delimited from the stem; stem terete, somewhat flexuous, 2–3 mm. thick, equal, black, velvety; asci short-stipitate, clavate, apex narrowed, reaching a length of 175–210 μ and a diameter of

18 μ, 4-spored; spores in a parallel fascicle in the ascus, smoky-brown, narrowed each way from above the middle, 8–11-septate when mature, reaching a length of 115–166 μ and a diameter of 6–7 μ; paraphyses pale-brown, cylindric, sparingly septate, slightly thickened and curved or uncinate at the tips; spines variable in length, usually projecting one-third to one-half their length beyond the asci, sometimes shorter, acicular, black.

On soil, humus, or rotten wood.

TYPE LOCALITY: Northville, New York.

DISTRIBUTION: New York to North Carolina and Minnesota.

ILLUSTRATIONS: Cooke, Mycographia *pl. 1, f. 1;* Peck, Ann. Rep. N. Y. State Mus. **29**: *pl. 1, f. 16–19;* Ann. Myc. **6**: *pl. 9, f. 86–88; pl. 16, f. 169–173;* Mycologia **22**: *pl. 13, f. 1, 2.*

2. **Trichoglossum tetrasporum** Sinden & Fitzp. Mycologia **22**: 60. 1930.

Ascophores black, 3–8 cm. high; ascigerous portion ellipsoid to subrotund, not more than one-fifth the total length of the entire ascophore, more or less compressed, rounded above, rather sharply delimited from the stem; stem terete, rather flexuous, 1–2 mm. thick, equal, black, velvety; asci clavate, apically narrowed, reaching a length of 175–220 μ and a diameter of 20–25 μ, 4-spored; spores in a fascicle, brown, cylindric-clavate, broadest above the middle, tapering each way to subobtuse ends, normally 15-septate at maturity, 6–7 × 110–160 μ (mostly 125–150); paraphyses smoky-brown, cylindric, septate; tips somewhat curved, slightly thickened, 3 μ thick below to 7 μ thick above; setae black, projecting slightly beyond the hymenium.

TYPE LOCALITY: Labrador Lake near Apulia, New York.

DISTRIBUTION: Known only from the type locality.

ILLUSTRATIONS: Mycologia **22**: *pl. 13, f. 5–8.*

3. **Trichoglossum hirsutum** (Pers.) Boud. Hist. Class. Discom. Eu. 86. 1907. (PLATE 78.)

Geoglossum hirsutum Pers. Neues Mag. Bot. **1**: 117. 1794.
?Geoglossum capitatum Pers. Obs. Myc. **1**: 11. 1796.

Ascophores solitary, or gregarious, black, 3–8 cm. high; ascigerous portion more or less ellipsoid, or lanceolate in outline, hollow, obtuse, more or less compressed, up to 1.5 cm. long, .5–.75 cm. thick, usually not more than one-fifth as long as the stem from which it is rather sharply delimited; stem terete,

TRICHOGLOSSUM VELUTIPES

equal, up to 6 cm. long, 2–3 mm. thick, densely velvety, composed of a fascicle of parallel, brown hyphae, 5–8 μ thick, looser in the center, those at the surface projecting as short, septate hairs, giving rise also to numerous acicular, black spines up to 225 μ long; asci broadly clavate, apex narrowed, the pore blue with iodine, reaching a length of 210–225 μ and a diameter of 20–22 μ, 8-spored; spores in a parallel fascicle in the ascus, brown, cylindric-clavate, broadest above the middle, tapering each way to the obtuse ends, 15-septate at maturity, 6–7 × 100–160 μ (120–150); paraphyses brown, cylindric, septate, usually strongly curved, or coiled at the slightly thickened tips, 3 μ thick below, up to 8 μ thick above; hymenial spines numerous, usually projecting about one-third their length beyond the hymenium, straight, acute, black, opaque, 8–10 μ thick, variable in length.

On rotten wood or on humus among leaves.

TYPE LOCALITY: Europe.

DISTRIBUTION: Ontario to California and south to Louisiana and the West Indies.

ILLUSTRATIONS: Cooke, Mycographia *pl. 1, f. 3;* Gill. Champ. Fr. Discom. *pl. 24, f. 2;* Phill. Brit. Discom. *pl. 2, f. 9;* Rab. Krypt.-Fl. **1³**: 1145, *f. 5–6;* Ann. Bot. **11**: *pl. 12, f. 31–32;* Ann. Myc. **6**: *pl. 9, f. 78–80; pl. 17, f. 176–181;* Mycologia **22**: *pl. 13, f. 3.*

EXSICCATI: Ellis & Ev. Fungi Columb. *1729.*

A capitate form of this species was received from Mr. Fred L. Lewis which seemed almost distinct enough to be regarded as a distinct species (PLATE 78). However, since the ascophores are very variable in form it is recorded simply as a capitate form of the above species.

4. **Trichoglossum Wrightii** Durand, Mycologia **13**: 187. 1921. (PLATE 79.)

Trichoglossum hirsutum f. *Wrightii* Durand, Ann. Myc. **6**: 438. 1908.

Ascophores clavate, or flattened, black, velvety with numerous black cystidia, variable in size but often 2–3 cm. high; ascigerous portion occupying one-third the entire length; asci clavate-cylindric, reaching a length of 250–265 μ and a diameter of 20–25 μ, 8-spored; spores fasciculate, reaching a length of 105–145 μ and a diameter of 7 μ, brown, clavate, broadest about the middle, mostly 8–9-septate, rarely 5–6- or 7-septate, stout; paraphyses cylindric, septate, pale-brown above, only slightly thickened and strongly curved; spines black, acute, projecting only slightly above the hymenium.

On damp soil in open places.

TYPE LOCALITY: Cuba.

DISTRIBUTION: Cuba and Bermuda.

ILLUSTRATIONS: Mycologia **22**: *pl. 13, f. 4;* **32**: 390, *f. 1;* Ann. Myc. **6**: *pl. 9, f. 83; pl. 16, f. 174.*

This form was first reported by Durand on two specimens from Cuba. Later collections of the author from Bermuda in 1912 (Britton, Brown & Seaver *1404*) convinced Durand that it was a distinct species. Abundant material was collected in Bermuda in 1926 by the author and H. H. Whetzel, and again in 1940 by the author and J. M. Waterston. It is one of the commonest species in Bermuda.

5. **Trichoglossum octopartitum** Mains, Am. Jour. Bot. **27**: 325. 1940.

Ascophores clavate, 1.5–4 cm. high, slender, 2–4 mm. thick above, black, hirsute; stem slender 1–1.5 mm. thick; asci clavate, reaching a length of 175–200 μ and a diameter of 18–20 μ; spores fusoid-clavate, attenuated toward both ends from above the middle, 6 $\times$ 100–140 μ, usually 7-septate; setae brown, acuminate, reaching a length of 240 μ, projecting considerably above the hymenium; paraphyses slender slightly enlarged above and curved at their apices.

On the ground.

TYPE LOCALITY: British Honduras.

DISTRIBUTION: Smoky Mountains, National Park, Tennessee; also in British Honduras.

ILLUSTRATIONS: Am. Jour. Bot. **27**: 324, *f. 10.*

6. **Trichoglossum Farlowi** (Cooke) Durand, Ann. Myc. **6**: 438. 1908. (PLATE 80.)

Geoglossum Farlowi Cooke, Grevillea **11**: 107. 1883.
Geoglossum velutipes Peck, Ann. Rep. N. Y. State Mus. **28**: 65. 1876.

Ascophores solitary, or clustered, two or three together, 2–6 cm. high; ascigerous portion lanceolate in outline, not sharply distinguished from the stem, brownish-black, about one-third the total length of the plant, 1–3 cm. long, 3–5 mm. or more thick; stem terete, equal, often flexuous, rarely somewhat compressed, 2 mm. thick, 2–4 cm. long, black, densely velvety with

TRICHOGLOSSUM HIRSUTUM

acicular spines; asci clavate, apex somewhat narrowed but rounded, pore blue with iodine, reaching a length of 170–200 μ and a diameter of 15–18 μ, 8-spored; spores multiseriate in the ascus, clavate-cylindric, tapering very little or not at all above the middle, fuliginous, or brownish, 6 × 48–85 μ (60–75), either continuous, or 1–3–5-septate, paraphyses cylindric, septate, curved to circinate at the somewhat thickened tips, brownish above; spines projecting about one-third to one-half their length beyond the hymenium.

In open grassy woods, on humus among leaves, or in moss.

TYPE LOCALITY: Massachusetts.

DISTRIBUTION: New York and New Hampshire to Florida and Mississippi.

ILLUSTRATIONS: Ann. Myc. **6**: *pl. 10, f. 89–92; pl. 18, f. 186–189.*

7. **Trichoglossum confusum** Durand, Mycologia **13**: 185. 1921.

Ascophores solitary, when dry 1.5–2.5 cm. high; ascigerous portion one-third to one-half the total length, obovate, even or longitudinally furrowed, rather irregular; stem terete, 1–2 cm. high, 1–1.5 mm. thick, velvety with black spines; asci narrowly clavate, apex rounded, reaching a length of 175 μ and a diameter of 12 μ, 8-spored; spores multiseriate, clavate-cylindric, straight, or curved, fuliginous, at first 3-septate, finally 7-septate, 4–5 × 55–73 μ (60–68); paraphyses pale-brown, slightly thickened above, straight, or curved; spines projecting but little beyond the hymenium.

On soil.

TYPE LOCALITY: Brazil.

DISTRIBUTION: North Carolina; also in South America.

ILLUSTRATIONS: Ann. Myc. **6**: *pl. 10, f. 93; pl. 16, f. 168.*

This species was reported by Durand (Ann. Myc. **6**: 439. 1908.) as *Trichoglossum Rehmianum* (P. Henn.) Durand. Later observations showed it to be different.

8. **Trichoglossum Walteri** (Berk.) Durand, Ann. Myc. **6**: 440. 1908.

Geoglossum Walteri Berk. Hedwigia **14**: 39. 1875.
Geoglossum Rehmianum P. Henn. Hedwigia **39**: (80). 1900.
Trichoglossum Rehmianum Durand, Ann. Myc. **6**: 439. 1908.

Ascophores solitary, or aggregated, 3–7 cm. or more high, brownish-black; ascigerous portion narrowly ellipsoid to lanceolate, obtuse, not distinct from the stem, one-third to one-half the total length of the plant, 1–2 cm. long, 3–5 mm. or more wide, compressed; stem terete or compressed, about 2 mm. in diameter, densely velvety with black, acicular spines; asci clavate, apex narrowed, obtuse, pore blue with iodine, reaching a length of 175–200 μ and a diameter of 18–20 μ, 8-spored; spores multiseriate in the ascus, clavate-cylindric, not narrowed above the middle, fuliginous, or pale-brown, 6 × 82–107 μ (87–100); paraphyses cylindric, septate, curved at the tips which are slightly thickened and brown; spines acicular, projecting about one-fourth to one-third their length beyond the hymenium.

On rotten wood or humus.

TYPE LOCALITY: Australia.

DISTRIBUTION: New Hampshire to Minnesota and south Alabama; also in Europe and Australia.

ILLUSTRATIONS: Cooke, Mycographia *pl. 1, f. 4;* Ann. Myc. 6: *pl. 10, f. 94–97; pl. 18, f. 190–193.*

7. **SPATHULARIA** Pers. Neues Mag. Bot. 1: 116. 1794.

Ascophores fleshy, erect, stipitate; ascigerous portion spathulate, much compressed, fan-shaped, decurrent on opposite sides of the stem from which it is sharply delimited; bright-colored; asci clavate, 8-spored; spores fasciculate in the ascus, filiformclavate, multiseptate; paraphyses present.

Spathularia differs from other genera of the family in having bright-colored, spathulate or fan-shaped ascophores. The consistency and spore characters are those of *Cudonia.*

Type species, *Spathularia flavida* Pers.

Stem pallid or yellowish-pallid; mycelium pale-yellow.   1. *S. clavata.*
Stem bay-brown, minutely velvety; mycelium orange.   2. *S. velutipes.*

1. **Spathularia clavata** (Schaeff.) Sacc. Michelia 2: 77. 1882.
(PLATE 81, FIG. 2.)

Elvela clavata Schaeff. Fung. Bavar. 4: Ind. 59. 1774.
Spathularia flavida Pers. Neues Mag. Bot. 1: 116. 1794.
Spathularia flava Pers. in Holmsk. Coryph. 166. 1797.
Spathularia flavida var. *rugosa* Peck, Ann. Rep. N. Y. State Mus. 39: 58. 1887.
Spathularia rugosa Peck, Ann. Rep. N. Y. State Mus. 50: 118. 1897.
Mitruliopsis flavida Peck, Bull. Torrey Club 30: 100. 1903.

TRICHOGLOSSUM WRIGHTII

Ascophores solitary, or usually gregarious, rarely cespitose, with two or three together, sometimes growing in lines or circles, fleshy up to 10 cm. high, whole plant pallid when young and fresh, then becoming yellowish, or brownish; ascigerous portion darker than the stem, much compressed, fan-shaped, obtuse or rounded, decurrent on opposite sides of the stem, even, undulate, or radiately rugose, sometimes contorted, clavate, or almost capitate, occupying about one-third to one-half the total length of the ascophore, up to 2.5 cm. wide; stem hollow, smooth or farinose, terete or somewhat compressed, tapering slightly upward, often swollen or bulbous below, slightly brownish at the base, up to 1 cm. in diameter, attached by a pallid, or yellowish mycelium; flesh white; ascophores usually becoming yellowish-brown when dry; asci clavate, apex conspicuously narrowed, often submammiform, not blue with iodine, reaching a length of 100–125 μ and a diameter of 12–14 μ, 8-spored; spores in a parallel fascicle in the ascus, often twisted together above, smooth, clavate-filiform, multiseptate, hyaline, 2.5–3 $\times$ 35–65 μ (40–50); paraphyses filiform, branched, hyaline, much curled, or coiled at the apices, not thickened.

On soil or humus, but most commonly under pines among the needles.

TYPE LOCALITY: Europe.

DISTRIBUTION: New Brunswick to California; also in Europe.

ILLUSTRATIONS: Schaeff. Ic. Fung. *pl. 149;* Cooke, Mycographia *pl. 95, f. 342;* Gill. Champ. Fr. Discom. *pl. 27, f. 1;* Phill. Brit. Discom. *pl. 2, f. 7;* Grev. Scot. Crypt. Fl. *pl. 165;* Berk. Outl. Brit. Fung. *pl. 21, f. 7;* Sow. Engl. Fung. *pl. 35;* Seaver, Bull. Lab. Nat. Hist. State Univ. Iowa *pl. 1, f. 1;* Ann. Myc. **6**: *pl. 10, f. 98–100; pl. 22, f. 220.*

EXSICCATI: Ellis, N. Am. Fungi *1268;* Clements, Crypt. Form. Colo. *127.*

2. **Spathularia velutipes** Cooke & Farlow; Cooke, Grevillea **12**: 37. 1883. (PLATE 81, FIG. 1.)

Ascophores solitary, or usually gregarious, occasionally cespitose several being united by a common base, up to 5 cm. high, 1–3 cm. wide, fan-shaped; ascigerous portion much compressed, yellowish to brownish-yellow, decurrent on opposite sides of the stem, the margin rounded, even, or often wavy or incised or lobed, sometimes contorted, about 1 cm. high at the top; stem varying from nearly terete or compressed to broadly expanded and flattened above, rather rounded in outline at its junction with the ascigerous part, 2–4 cm. high, up to 1.5 cm. broad above, 3–5 mm. thick at the base, solid, bay-brown and minutely velvety, attached by an orange mycelium; the ascophore shrinks but little in drying, the color becomes brownish and the stem more or less longitudinally rugose or striate; asci clavate, apex narrowed, not blue with iodine, reaching a length of 80–105 μ and a diameter of 10 μ, 8-spored; spores in a parallel fascicle in the ascus, hyaline, smooth, clavate-filiform, straight, or curved, becoming multiseptate, 2 × 33–43 μ (35–40); paraphyses filiform, hyaline, branched, strongly curved, or coiled at the tips.

On rotten logs, humus among leaves, or especially on the ground under pines.

TYPE LOCALITY: New Hampshire.

DISTRIBUTION: New Hampshire to North Carolina and Minnesota, and Idaho?

ILLUSTRATIONS: Ann. Bot. **11**: *pl. 13, f. 85–88;* Ann. Myc. **6**: *pl. 11, f. 101–102; pl. 22, f. 221–222.*

DOUBTFUL SPECIES

Spathularia linguatus A. E. Johnson, Bull. Minn. Acad. Nat. Sci. **1**: 370. 1880. "Head tongue-shaped, flat, thin, nearly even, white, or white tinged with yellow or buff; stem white or yellowish white, thick, solid; asci very long, clavate; sporidia filiform, nearly as long as the asci, straight or curved, multinucleate. Gregarious, seldom solitary, one to two inches high; head as long or longer than the stem, one-fourth to three-fourths of an inch broad. On moss in tamarck swamps. October. Scarce." The above is all that is known about this species.

8. **LEOTIA** Pers. Neues Mag. Bot. **1**: 97.   1794.

Hygromitra Nees, Syst. Pilz. Schw. 157.   1816.
Cudoniella Sacc. Syll. Fung. **8**: 41.   1889.

Ascophores more or less gelatinous, stipitate, erect; ascigerous portion pileate, horizontal, supported in the center, bearing the hymenium spread over its upper convex surface, sterile beneath; asci clavate; spores hyaline, oblong-fusiform, at first simple, finally 3–5-septate; paraphyses present.

Type species, *Elvela lubrica* Scop.

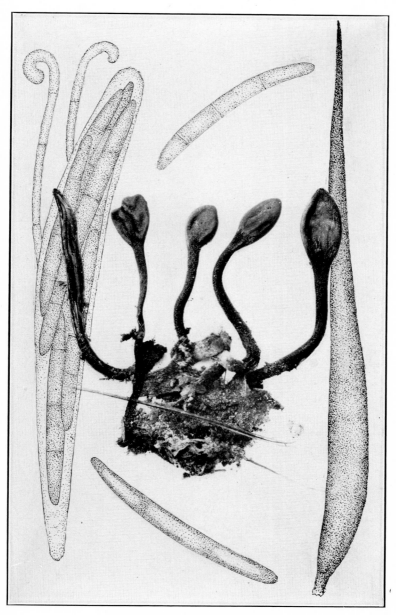

TRICHOGLOSSUM FARLOWII

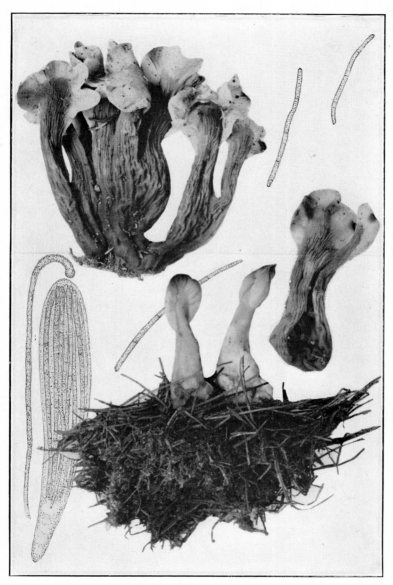

1. SPATHULARIA VELUTIPES
2. SPATHULARIA CLAVATA

Ascophores entirely yellowish-ochraceous, sometimes olivaceous.   1. *L. lubrica.*
Hymenium aeruginous-green; stem white to yellowish.   2. *L. stipitata.*
Ascophores entirely green; stem furfuraceous, substance
firm, stems slender.   3. *L. chlorocephala.*

According to Durand the type of *Cudoniella* is a *Leotia.*

The color varies much in the species listed above, but the three listed seem to be reasonably distinct.

1. **Leotia lubrica** (Scop.) Pers. Neues Mag. Bot. **1**: 97. 1794. (PLATE 82, FIG. 1.)

Elvela lubrica Scop. Fl. Carn. **2**: 477. 1772.
Helvella gelatinosa Bull. Hist. Champ. Fr. 296. 1791.

Ascophores usually densely cespitose, stipitate, more or less viscid-gelatinous, ochraceous-yellow, often with a greenish or olive tint, 3–6 cm. or more high; ascigerous portion pileate, convex above, the surface often irregularly furrowed, with a recurved margin, wrinkled or nodulose, 1–1.5 cm. or more broad; stem terete or somewhat compressed, usually tapering slightly upward, the adjacent ones often coalescing below, about 1 cm. thick below, .5 cm. above, up to 5 cm. or more high, minutely squamulose, sometimes with innate, greenish granules; asci narrowly clavate, apices rounded, slightly narrowed, not blue with iodine, reaching a length of 130–160 μ and a diameter of 10–12 μ, 8-spored; spores 2-seriate above, 1-seriate below, hyaline, smooth, cylindric-oblong to fusiform, ends obtuse, or sub-acute, straight or curved, 5–6 $\times$ 18–28 μ (20–23), at first simple, then with three to eight oil-drops, finally becoming 5–7-septate; paraphyses filiform, branched, the apices clavate to pyriform, hyaline, the tips agglutinated by amorphous matter.

On rich humus or sandy soil rarely on rotten wood, in woods.

TYPE LOCALITY: Europe.

DISTRIBUTION: Ontario to Alabama and Iowa; also in Europe.

ILLUSTRATIONS: Bull. Herb. Fr. *pl. 473, f. 2;* Sow. Engl. Fungi *pl. 70:* Fl. Dan. *pl. 719;* Berk. Outl. Brit. Fung. *pl. 22, f. 1;* Grev. Scot. Crypt. Fl. *pl. 56;* Cooke, Mycographia *pl. 44, f. 171;* Gill. Champ. Fr. Discom. *pl. 23;* Rab. Krypt.-Fl. **1**[3]: 1161, *f. 1–4;* Ann. Bot. **11**: *pl. 13, f. 61–64;* Atk. Mushrooms *f. 221;* Ann. Myc. **6**: *pl. 11, f. 106; pl. 20, f. 213;* Seaver, Bull. Lab. Nat. Hist. State Univ. Iowa *pl. 2, f. 1;* Mycologia **2**: *pl. 17, f. 1.*

EXSICCATI: Ellis, N. Am. Fungi *57;* Ellis & Ev. Fungi Columb. *1738.*

2. **Leotia stipitata** (Bosc.) Schröt. E. & P. Nat. Pfl. 1^1: 166. 1897.

Tremella stipitata Bosc. Berl. Mag. **5**: 89.   1811.
Leotia viscosa Fries, Syst. Myc. **2**: 30.   1822.

Ascophores solitary, or clustered, stipitate, viscid-gelatinous, 3–6 cm. or more high; ascigerous portion 1–2 cm. or more broad, margin incurved toward the stem, even or irregularly nodulose, hymenium clear deep aeruginous-green, whitish below; stem terete or tapering slightly upward, usually pure-white, or less often with an ochraceous or orange tint, 2–4 cm. high, .5–1 cm. thick, often beset, especially above, with minute, green squamules; asci narrowly clavate-cylindric, not or only slightly narrowed, not blue with iodine, reaching a length of 118–150 μ and a diameter of 10 μ, 8-spored; spores 2-seriate above, 1-seriate below, hyaline, smooth, at first simple, finally 5- or more-septate, ends obtuse, or subacute, straight or curved, 5–6 × 16–28 μ (20–24); paraphyses filiform, branched, the apices pyriform, intense green when fresh.

In rich humus or soil, especially among leaves in woods or on the slopes of ravines.

TYPE LOCALITY: South Carolina.

DISTRIBUTION: Maine to Florida and California; also in Europe.

ILLUSTRATIONS: Berl. Mag. **15**: *pl. 6, f. 14;* Cooke, Mycographia *pl. 44, f. 173;* Ann. Bot. **11**: *pl. 13, f. 65;* Ann. Myc. **6**: *pl. 11, f. 109; pl. 20, f. 210.*

EXSICCATI: Ellis, N. Am. Fungi *663;* Ellis & Ev. N. Am. Fungi *2032;* Rav. Fungi Car. **4**: *22;* Fungi Am. *173.*

3. **Leotia chlorocephala** Schw. Schr. Nat. Ges. Leipzig **1**: 114. 1822.

Ascophores solitary to densely clustered, subgelatinous, entirely green, 1–5 cm. high; ascigerous portion hemispherical, convex, margin incurved, obtuse, hymenium smooth or furrowed, the margin often lobed or nodulose, pea-green to aeruginous, 2–10 mm. wide; stem terete, firm, the middle layer green, surface densely squamose, or furfuraceous with green granules, 1–4.5 cm. high, 2–4 mm. thick, shrinking much in drying, the color changing but little; asci narrowly clavate, apex rounded not blue with iodine, reaching a length of 125–150 μ and a diameter of 10–12 μ, 8-spored, short-stipitate; spores subbiseriate above.

1. LEOTIA LUBRICA
2. CUDONIA CIRCINANS

1-seriate below, hyaline or with a faint greenish tint, narrowly ellipsoid to ellipsoid-fusiform, ends obtuse or subacute, at first simple, finally about 5-septate, straight or curved, 5–6 × 18–20 μ; paraphyses filiform, branched, the apices pyriform, green.

On sandy soil in rich woods, on knolls, or along wood roads or among mosses in ravines.

TYPE LOCALITY: Salem, North Carolina.

DISTRIBUTION: New Hampshire to Alabama.

ILLUSTRATIONS: Cooke, Mycographia *pl. 44, f. 174;* Ann. Myc. **6**: *pl. 11, f. 110; pl. 20, f. 211, 212;* Mycologia **2**: *pl. 17, f. 2.*

DOUBTFUL AND EXCLUDED SPECIES

Leotia exigua Schw. Schr. Nat. Ges. Leipzig **1**: 113.   1822; *Mitrula exigua* Fries, Elench. Fung. **1**: 235.   1828.   Durand states: "No specimens are known to be in existence, certainly not in the herbaria of Schweinitz or Fries. It was probably Helotiaceous."

Leotia infundibuliformis (Schaeff.) Fries, Obs. Myc. **2**: 299.   1818; *Elvela infundibuliformis* Schaeff. Fung. Bavar. **4**: Ind. 111.   1774.   Durand states: "Whatever may be thought of the European specimens the one preserved in the Schweinitzian herbarium, from New York, is an *Helvella* related to *H. elastica.*"

Leotia marcida (Müll.) Pers. Syn. 613.   1801; *Phallus marcidus* Müll. Fl. Dan. Fasc. **11**: 7, *pl. 654, f. 1.*   1777.   Durand states: "This species has been reported several times from the United States, but all the specimens which I have been able to examine, both from Europe and America, are indistinguishable internally from *L. lubrica.*"

Leotia rufa Rostrup, Med. Grønl. **3**: 536.   1888.   "Pileus repandus, margine revoluto, latit. 1–2 mm. rufus; stipes inaequaliter teres, furo-ferrugineus, altit. 5–6 mm. Asci cylindraceo-clavati, pedicellati, long. 60–70 μ crass. 2 μ.   Inter. muscos. Agdluitsok [Greenland] (Vahl)."   Durand states: "The above is Rostrup's original description.   I saw a small fragment of the original collection in the Botanical Museum Copenhagen, but did not examine it microscopically.   It is probably Helotiaceous."

Leotia chlorocephala f. *Stevensoni* (Berk. & Br.) Massee, Brit. Fungus—Fl. **4**: 472.   1895; *Leotia Stevensoni* Berk. & Br. Ann. Mag. Nat. Hist. V. **3**: 212. 1879.   This form has been reported from Missouri by J. B. Routien (Mycologia **34**: 579. 1942). The writer has no knowledge of the form.

9. **VIBRISSEA** Fries, Syst. Myc. **2**: 31.   1822.

Ascophores stipitate, pileate, soft, waxy, or subgelatinous; ascigerous portion hemispherical, bearing the hymenium on its upper, convex surface, sterile below; asci long-cylindric, narrow, 8-spored, spores in a parallel fascicle and nearly as long as the ascus, hyaline, filiform, many-septate; paraphyses slender.

Type species, *Leotia truncorum* Alb. & Schw.

Spores 175–250 μ long.   1. *V. truncorum.*
Spores 85–100 μ long.   2. *V. foliorum.*

1. **Vibrissea truncorum** (Alb. & Schw.) Fries, Syst. Myc. **2**: 31. 1822.

Leotia truncorum Alb. & Schw. Consp. Fung. 297. 1805.
Leotia Clavus Pers. Myc. Eur. **1**: 200. 1822.
Vibrissea truncorum var. *albipes* Peck, Ann. Rep. N. Y. State Mus. **44**: 37. 1891.

Ascophores solitary, gregarious, or in clusters of two or three each, 4–5 mm. high; ascigerous portion hemispherical-convex, margin obtuse, 3–5 mm. in diameter, yellow, orange, or reddish-orange, rarely pallid; stem terete, up to 1.5 cm. high, 1–2 mm. thick, white to bluish-gray, or brownish, minutely squamulose, due to minute, spreading hyphae, rather darker below; asci slenderly cylindric, apex rounded, not blue with iodine, reaching a length of 200–325 μ and a diameter of 5–6 μ, 8-spored; spores in a parallel fascicle in the ascus, hyaline, very slenderly filiform, slightly narrowed toward each end, nearly as long as the ascus, multiseptate, up to 1×250 μ; paraphyses filiform, occasionally forked, the apices somewhat clavate-thickened and colored.

On wholly or partly submerged sticks etc., in brooks, mostly in higher altitudes.

TYPE LOCALITY: Europe.

DISTRIBUTION: Laborador to Alaska and West Virginia.

ILLUSTRATIONS: Trans. Linn. Soc. II. **2**: *pl. 1, f. 1–9;* Phill. Brit. Discom. *pl. 10, f. 60;* Rab. Krypt.-Fl. **1**[3]: 1164, *f. 1–4;* Ann. Bot. **11**: *pl. 12, f. 15–17a;* Gill. Champ. Fr. Discom. *pl. 29, f. 1;* Kromb. Myk. Heft *pl. 5, f. 34–36;* Ann. Myc. **6**: *pl. 11, f. 111, 112;* Pers. Myc. Eu. **1**: *pl. 11, f. 9.*

EXSICCATI: Ellis, N. Am. Fungi *134.*

2. **Vibrissea foliorum** Thaxter; Durand, Ann. Myc. **6**: 454. 1908.

Ascophores solitary, or gregarious, stipitate, of soft consistency; ascigerous portion convex, 1–1.5 mm. in diameter, yellowish-orange; stem pallid, 2–3 times as long as the diameter of the head, slender, translucent, slightly furfuraceous with projecting hairs below; asci narrowly cylindric, apex rounded, not blue with iodine, reaching a length of 150–180 μ and a diameter of 5–6 μ, 8-spored; spores in a parallel fascicle in the ascus, hyaline, slenderly filiform, $1 \times 85–100$ μ; paraphyses filiform, hyaline, not exceeding the asci, simple, or occasionally forked near the distal ends, very slightly pyriform-thickened at the tips.

On dead oak leaves, acorn-cups etc. in a wet place.

TYPE LOCALITY: West Haven, Conn.

DISTRIBUTION: Known only from the type locality.

ILLUSTRATIONS: Ann. Myc. 6: *pl. 11, f. 113.*

10. **CUDONIA** Fries, Summa Veg. Scand. 348. 1849.

Ascophores stipitate, erect, fleshy-leathery; ascigerous portion pileate; hymenium on the upper, convex surface, sterile beneath, margin acute, incurved toward the stem; asci clavate, 8-spored; spores hyaline, clavate-filiform, multiseriate, or fasciculate in the upper part of the ascus, multiseptate; paraphyses present.

Type species, *Leotia circinans* Pers.

Spores 30–45 μ long; ascophores pale-brownish to buff, often with a rosy tint.	1. *C. circinans.*
Spores 45–78 μ long; ascophores entirely yellowish.	2. *C. lutea.*
Spores 18–25 μ long.	
Caps ochraceous, or reddish-brown.	
Caps 5 mm. in diameter.	3. *C. ochroleuca.*
Caps 10–30 mm. in diameter.	4. *C. monticola.*
Caps gray.	5. *C. grisea.*

1. **Cudonia circinans** (Pers.) Fries, Summa Veg. Scand. 348. 1849. (PLATE 82, FIG. 2.)

Leotia circinans Pers. in Holmsk. Coryph. 163. 1797.

Vibrissea circinans Hazsl. Magyar. Akad. Ertek. Termesz. Kör. 11[19]: 9. 1881.

Ascophores solitary, or gregarious, fleshy, becoming more distinctly leathery in drying, 2–6 cm. high; ascigerous portion .5–2 cm. broad, rather thin, margin acute and recurved, even or undulate, hymenium convex, even, wrinkled, or convoluted, cream-buff with a faint rosy tint, or sometimes yellowish, or pale-brownish; stem tapering slightly upward, often stout below where it is 2–10 mm. thick, 1.5–5 mm. thick above, darker than the hymenium especially below, often longitudinally striate especially above, the striae being prolonged as radiating veins on the lower side of the cap, somewhat farinaceous, sometimes becoming hollow in age; asci clavate, apex narrowed, not blue with iodine, reaching a length of 85–130 μ and a diameter of 8–10 μ, 8-spored; spores fasciculate, hyaline, smooth, clavate-filiform, broadest above the middle or at the distal end, 2 × 30–45 μ

(34–40); paraphyses hyaline, filiform, strongly curved above, often branched, tips only slightly thickened, 2 μ thick.

On rotten wood or humus among leaves, often under coniferous trees.

TYPE LOCALITY: Europe.

DISTRIBUTION: Newfoundland to Colorado, Idaho, and Alberta; also in Europe.

ILLUSTRATIONS: Pers. Ic. Descr. Fung. *pl. 5, f. 5–7;* Bres. Fungi Trident. *pl. 145;* Rab. Krypt.-Fl. **1**³: 1163, *f. 1–4;* Cooke, Mycographia *pl. 44, f. 172;* Phill. Brit. Discom. *pl. 2, f. 5;* Ann. Myc. **6**: *pl. 11, f. 103; pl. 21, f. 214, 215.*

2. **Cudonia lutea** (Peck) Sacc. Atti Inst. Venet. VI. **3**: 725.   1885.

Vibrissea lutea Peck, Bull. Buffalo Soc. Nat. Sci. **1**: 70.   1873.
Leotia lutea Cooke, Bull. Buffalo Soc. Nat. Sci. **2**: 287.   1875.

Ascophores solitary or gregarious, rarely clustered, 1–6 cm. high, but usually about 3 cm., fleshy-leathery; ascigerous portion pileate, convex, the margin acute, reflexed, hymenium when young covered by a volva-like membrane which later cracks irregularly and falls away leaving remnants attached to the margin, usually slightly depressed above, sometimes furrowed, beautiful flesh-color to orange-buff, the under surface often with radiating striae which continue down the stem, 5–15 mm. in diameter; stem terete, or slightly compressed, sometimes inflated below, even or longitudinally striate, mealy, pale-yellow, 1–5 cm. high, 2–5 mm. thick, the whole ascophore when dry becoming about the color of chamois skin or of *Otidea leporina;* asci clavate, apex narrowed, not blue with iodine, variable in size, reaching a length of 100–170 μ and a diameter of 10–12 μ, 8-spored; spores in a fascicle in the upper part of the ascus, hyaline, smooth, with a hyaline, gelatinous sheath, clavate-filiform, lower end acute, not narrowed above the middle, 2 × 45–78 μ (55–65); paraphyses filiform, branched, not thickened but strongly circinate at the tips.

On decaying leaves in thickets, rarely on rotten wood, chestnut burrs etc.

TYPE LOCALITY: North Elba, New York.

DISTRIBUTION: Ontario to Tennessee.

ILLUSTRATIONS: Ann. Rep. N. Y. State Mus. **25**: *pl. 1, f. 19–23;* Ann. Bot. **11**: *pl. 12, f. 5–7;* Ann. Myc. **6**: *pl. 11, f. 105; pl. 21–22, f. 216–219.*

EXSICCATI: Ellis & Ev. N. Am. Fungi *3533.*

3. **Cudonia ochroleuca** Cooke & Hark.; Durand, Ann. Myc. **6**: 461. 1908.

Leotia ochroleuca Cooke & Hark. Grevillea **9**: 8. 1880.
Vibrissea ochroleuca Massee, Ann. Bot. **11**: 262. 1897.

Ascophores scattered, stipitate, pileate; ascigerous portion convex, ochroleucous, 6 mm. in diameter when dry; stem slender, flexuous, white, longitudinally striate, or rugulose, 8–10 mm. high; whole ascophore when dry dark reddish-brown, substance apparently not subgelatinous; asci clavate, slenderly stipitate, apex narrowed, not blue with iodine, reaching a length of 75–100 μ and a diameter of 8–9 μ, 8-spored; spores multiseriate in the ascus, clavate-filiform, acute at each end, hyaline, 3–more-septate, 2 × 18–25 μ; paraphyses filiform, very slender.

On damp ground.

TYPE LOCALITY: San Rafael, California.

DISTRIBUTION: Known only from the type locality.

ILLUSTRATIONS: Ann. Bot. **11**: *pl. 13, f. 70–72;* Ann. Myc. **6**: *pl. 11, f. 104.*

4. **Cudonia monticola** Mains, Am. Jour. Bot. **27**: 322. 1940.

Ascophores single, gregarious, or cespitose, pileate, fleshy-leathery, 3–10 cm. high; pileus variable in form, convex, irregularly hemispherical, laterally compressed or subspathulate, 10–30 mm. broad, incarnate-cinnamon; stem 5–7 mm. thick below, somewhat attenuated upward, wood-brown; asci clavate, reaching a length of 90–100 μ and a diameter of 8–10 μ, attenuated below, 8-spored; spores acicular, 2 × 20–24 μ, hyaline; paraphyses filiform, hyaline, curved at their apices.

On spruce needles and coniferous debris.

TYPE LOCALITY: Lake Crescent, Washington.

DISTRIBUTION: Washington.

ILLUSTRATIONS: Am. Jour. Bot. **27**: 324, *f. 1.*

5. **Cudonia grisea** Mains, Am. Jour. Bot. **27**: 322. 1940.

Ascophores gregarious, pileate, stipitate, fleshy, 1.5–5 cm. high; pileus convex, thick, 5–15 mm. broad, drab or dark-gray, smooth; stem 3–8 mm. thick, attenuated above, smooth, brown; asci clavate, reaching a length of 70–90 μ and a diameter of 6–8 μ, attenuated below; spores acicular, 1.5–2 × 18–22 μ, hyaline; paraphyses filiform, hyaline, curved at their apices.

On rotten coniferous wood.

TYPE LOCALITY: Hoh River, Washington.

DISTRIBUTION: Washington.

ILLUSTRATIONS: Am. Jour. Bot. **27**: 324, *f. 2.*

EXCLUDED GENERA AND SPECIES

Cudoniella fructigena Rostrup. Med. Grønl. **3**: 605. 1891. According to Durand this belongs with the Helotiaceae. He states further that he believes the genus *Cudoniella* to be ill founded since the species thus far referred to it might better be placed elsewhere.

Helotium aciculare (Bull.) Pers. Syn. Fung. 677. 1801; *Helvella acicularis* Bull. Hist. Champ. Fr. 296. 1791; *Peziza acicularis* Fries, Syst. Myc. **2**: 156. 1822; *Cudoniella acicularis* Schröt. in E. & P. Nat. Pfl. **1**¹: 166. 1897.

Roesleria hypogaea Thüm. & Pass. Durand states: "I have seen this species growing on buried grape canes in New York and Missouri. Schröter placed it in the Geoglossaceae, while Rehm and others regard it as synonymous with *Coniocybe pallida* (Pers.) Körb., and refer it to the Calicieae. Judging from the specimens seen its affinities do not seem to be at all close to the Geoglossaceae, so that I exclude it from that family."

Family 4. **HELOTIACEAE**

Apothecia extremely variable, ranging in size from a fraction of a millimeter to 2 or 3 centimeters in diameter, sessile to long-stipitate, varying much in the same species, superficial, or more rarely erumpent on herbaceous stems, colors varying from white, bright-yellow, or red to green, olive, blue, brown, or nearly black, smooth, or hairy; asci cylindric to broad-clavate; spores simple or compound, globose to long-filiform, hyaline or more rarely colored; paraphyses hyaline or more rarely colored, filiform, or rarely with lanceolate, pyriform, or subglobose ends.

Apothecia not distinctly hairy.	
Apothecia arising from a sclerotium or sclerotium-like substratum.	Tribe 1. SCLEROTINEAE.
Apothecia not springing from a sclerotium, but occasionally on a subiculum.	
Apothecia fleshy, waxy, leathery, or sub-cartilaginous.	
Stipitate or sessile, usually bright-colored, more rarely dark-brown or blackish.	Tribe 2. HELOTIEAE.
Sessile, sordid, gray or blackish.	Tribe 3. MOLLISIEAE.
Apothecia cartilaginous or gelatinous to subtremelloid.	Tribe 4. ASCOTREMELLEAE.
Apothecia hairy.	Tribe 5. LACHNELLEAE.

Tribe 1. SCLEROTINEAE. Apothecia arising from a definite sclerotium of variable form, usually on or in the tissues of a living or recently killed host (more rarely isolated or in one case formed on the dung of animals) or from a stromatized portion of the host; conidial stage present and well developed, or unknown; apothecia cup-shaped to subdiscoid, or more rarely verpoid, usually stipitate, the length of the stem variable; asci usually 4–8-spored; spores ellipsoid, hyaline or subhyaline, or occasionally brown; spermatia often present.

The following genera have been adopted from H. H. Whetzel who spent many years in critical research on this group of fungi, the results of which were posthumously published by Dr. H. M. Fitzpatrick (Mycologia **37**: 648–714. 1945).

Conidial stage present, well developed.
  Conidial stage monilioid or botryoid.
    Conidial stage a *Monilia;* on woody plants.   1. MONILINIA.
    Conidal stage of the *Botrytis* type.
      Conidiospores rough.   2. SEAVERINIA.
      Conidiospores smooth.
        Conidiophores twisted or kinked.   3. STREPTOTINIA.
        Conidiophores not twisted or kinked.   4. BOTRYOTINIA.
  Conidial stage not monilioid or botryoid.
    Conidial stage a *Gloeosporium*.   5. SEPTOTINIA.
    Conidial stage not a *Gloeosporium*, conidia
    large-ovoid.   6. OVULINIA.
Conidial stage absent, unknown or obscure.
  Ascospores brown.
    Sclerotia indefinite or diffused.   7. LAMBERTELLA.
    Sclerotia definite, hemispherical.   8. MARTINIA.
  Ascospores hyaline or subhyaline.
    On plant tissues.
      Sclerotia discoid or subdiscoid, foliicolous.
        Apothecia cup-or saucer-shaped.   9. CIBORINIA.
        Apothecia verpoid.   10. VERPATINIA.
      Sclerotia not discoid or subdiscoid.
        Sclerotia definite, isolated or associated
        with the host.   11. SCLEROTINIA.
        Sclerotia indefinite on or within the host.
          In flowers or fruit of the host.   12. CIBORIA.
          On stems or other host tissues.   13. STROMATINIA.
    On dung of animals.   14. COPROTINIA.

Tribe 2. HELOTIEAE. Apothecia stipitate, or sessile, usually highly colored, bright-yellow, light-brown, blue, or some shade of green or rarely dark-colored, brownish-black, regular or irregular in form; stem very variable in length even in the same species; asci cylindric or clavate, usually 8-spored; spores ellipsoid to fusoid, fusiform, or filiform, simple or septate, hyaline or more rarely colored; paraphyses present and variable.

Spores hyaline.
    Apothecia usually, elongated on one side.   15. MIDOTIS.
    Apothecia not elongated.
        Color yellow-olive or greenish.
            Opening with a stellate aperture.   16. PODOPHACIDIUM.
            Not opening as above.
                Bright-green, medium large, 5 mm.–1
                    cm. broad.   17. CHLOROCIBORIA.
                Yellowish to olive, small, less than 5
                    mm. broad.
                    Spores large, 20 μ or more long.   18. KRIEGERIA.
                    Spores small, 10 μ or less long.   19. CHLOROSPLENIUM.
        Color varied but never green.
            Apothecia medium large, 4–5 mm. or more
                broad.
                On over wintering buds; conidial stage,
                    an *Acarosporium*.   20. PYCNOPEZIZA.
                On plant debris; conidial stage not as
                  above.
                  Spores simple.   21. CIBORIELLA.
                  Spores compound.   22. CALYCINA.
            Apothecia small usually less than 4 mm.;
                bright-colored, white, yellow or red.
            Margin toothed.   23. CYATHICULA.
            Margin not toothed.
                Spores simple.
                  Apothecia stipitate or substipitate.   24. HELOTIUM.
                  Apothecia entirely sessile.   25. ORBILIA.
                Spores becoming septate.
                Spores fusiform.
                  Apothecia entirely sessile.
                    Seated on a subiculum.   26. TRICHOBELONIUM.
                    Without definite subiculum.
                        Spores 1-septate.   27. CALLORIA.
                      Spores 3- or more-septate.   28. BELONIUM.
                  Apothecia stipitate.   29. BELONIOSCYPHA.
                Spores filiform.
                Asci not strongly protruding.
                  Apothecia sessile.   30. GORGONICEPS.
                  Apothecia stipitate.   31. POCILLUM.
                Asci strongly protruding.   32. APOSTEMIDIUM.
Spores brown.   33. PHAEOHELOTIUM.

Tribe 3. MOLLISIEAE. Apothecia entirely sessile, mostly minute, sordid-gray to black, superficial, or erumpent, occurring on living plants or more often on dead plant debris; asci cylindric to clavate, usually 8-spored; spores simple, or septate, hyaline, or faintly colored; paraphyses filiform to clavate.

Occurring on living plants.
  Apothecia erumpent.
    Spores simple.   34. PSEUDOPEZIZA.
    Spores septate.   35. FABRAEA.
  Apothecia superficial.   36. PESTALOPEZIA.
Occurring on dead plant debris.
  Spores simple.
    Apothecia subdiscoid, superficial.
      Seated on a subiculum.   37. TAPESIA.
      Not seated on a subiculum.
        Spores hyaline; apothecia minute.
          Spores ellipsoid, on woody tissues.   38. MOLLISIA.
          Spores globose, on ascomycetous
            fungi.   39. MOLLISIELLA.
        Spores green; apothecia medium large,
          up to 1 cm.   40. CATINELLA.
    Apothecia, cupulate, often erumpent.   41. PYRENOPEZIZA.
  Spores with 3 or more septa.   28. BELONIUM.

Tribe 4. ASCOTREMELLEAE. Apothecia more or less gelatinous and *Tremella*-like, ranging in size from 1 mm. to several cm., stipitate, or sessile; spores from ellipsoid to filiform, hyaline, or colored, simple, or septate; paraphyses filiform to clavate.

Apothecia not on a stromatic base.
  Apothecia small and delicate, usually less than 1
    cm. broad.
    Erumpent, sessile or subsessile.   42. STAMNARIA.
    Superficial, stipitate.   43. OMBROPHILA.
  Apothecia large, usually exceeding 1 cm. broad.
    Spores simple.
      Spores brown.   44. PHAEOBULGARIA.
      Spores hyaline.   45. ASCOTREMELLA.
    Spores septate.
      Spores fusoid to fusiform.   46. CORYNE.
      Spores filiform or vermiform.   47. HOLWAYA.
Apothecia on a stromatic base, large 4 cm. in diameter. 48. ACERVUS.

Tribe 5. LACHNELLEAE. Apothecia ranging in size from minute bodies less than 1 mm. in diameter to nearly 1 cm., externally clothed with well developed hairs which are sometimes closely adpressed but often presenting a wooly appearance; hairs rigid or more often flexuous, thin-walled and often delicately roughened, more rarely smooth, usually septate, varying in color from white (hyaline) to yellow, green, purple or pale- to dark-brown; asci cylindric to clavate, usually 8-spored; spores hyaline or subhyaline, globose to ellipsoid, fusiform, or filiform; paraphyses filiform, clavate, or lanceolate, in one genus with conidium-like apices.

Spores simple or very rarely sparingly septate.
    Apothecia not seated on a definite mycelial subiculum.
        Spores ellipsoid to fusoid.

Paraphyses filiform-clavate, or lanceolate, without conidium-like apices.	49. LACHNELLA.
Paraphyses with easily detached conidium-like apices.	50. DIPLOCARPA.
Spores globose.	51. LACHNELLULA.
Apothecia seated on a definite mycelial subiculum.	52. ERIOPEZIZA.

Spores definitely septate, fusoid to filiform.

Apothecia seated on a definite subiculum.	53. ARACHNOPEZIZA.
Apothecia not on a definite subiculum.	
Spores 1-septate.	54. HELOTIELLA.
Spores 3–many septate.	
Hairs hyaline.	55. ERINELLINA.
Hairs dark-brown or black.	56. ECHINELLA.

1. **MONILINIA** Honey, Mycologia **20**: 153.   1928.

Apothecia of variable size, occurring singly, or several developing from pseudosclerotia which are commonly formed within fruits of the higher plants; conidial stage consisting of a *Monilia;* asci clavate 8-spored; spores simple, hyaline or subhyaline; paraphyses slender, often slightly enlarged above.

Type species, *Ciboria fructicola* Wint.

I am indebted to Dr. Edwin E. Honey for the following notes on the genus: "All known North American species are vernal. During the life-history of members of *Monilinia* two types of stroma are developed (1) the ectostroma which is developed first, and which functions in the rupture of the epidermis of the host, and upon which develop the typical monilioid conidia and (2) as a result of the initial levy on the food supply of the newly invaded host or some other factor, a change takes place and in affected fruits, under favorable environmental conditions, there results the development of an entostroma, which after overwintering may give rise to apothecia. The entostroma is a composite stroma, that is, it is an admixture of living fungous hyphae and dead host cells in contrast to the typical sclerotium characteristic of the genus *Sclerotinia* as represented by *Sclerotinia sclerotiorum* in which the sclerotium is composed entirely of fungous hyphae. The term pseudosclerotium has been used to designate this composite stroma which is so characteristic of the monilioid group.

"Emphasis has been placed on the life-history of the various species in showing relationships. It is felt that the study of any one stage, as for example, the stromatic, the apothecial or the conidial in the life-history of these species is not sufficient to adequately show true phylogenetic relationships. In order to obtain either the specific or the generic concept, all stages developed during the life-history must be given consideration. On the basis of life-histories which appear to be correlated with certain morphological characteristics, the genus *Monilinia* may be divided into two general groups:—

"(1) Those species in which in the spring ascospores infect the young leaves and stems resulting in the development of an ectostroma. Upon these host organs the ectostroma gives rise to conidia which in turn function as inoculum for blossom infection. Young developing fruits become infected through the blossom, and as a result pseudosclerotia are formed which commonly drop to the ground and overwinter, giving rise in the spring to apothecia which produce ascospores (the primary inoculum which again infects the young leaves and stems). Commonly but a single cycle occurs. The majority of the North American representatives fall in this sub-group, the members of which are also characterized by the presence of disjunctors within the conidial chains (a specialization for the dissemination of conidia). They are furthermore characterized by a marked sweetish odor (mandelic acid; benzaldehyde and hydrocyanic acid) emitted from infected parts of the host at about the time of conidial production and they appear to be more specialized and limited in their host range. In the *Vaccinium*-inhabiting members there is a tendency to suppress the development of the ectostroma in infected fruits, and to further segregate the conidial production from the apothecial production which, according to certain European workers (Fischer, Ed., etc.) reaches its climax in such forms as *Monilinia Ledi* (= *Sclerotinia heteroica* Woronin and Nawaschin) and *Monilinia Rhododendri* (*Sclero-*

tinia Rhododendri Fischer) in which heterocism is said to occur and in which the conidial stage occurs upon one host and the pseudosclerotial and apothecial stages upon another host.

"(2) Those members in which ascospores normally infect blossoms, upon which conidia then develop and function as inoculum for secondary cycles on fruits (chiefly). Pseudosclerotia develop within infected fruits, overwinter upon the ground and give rise to apothecia, thus furnishing ascosporic inoculum for blossom infection in the spring. Commonly many successive secondary cycles may occur. In this subgroup the sweetish odor is not evident, disjunctors between the conidia are lacking, and the members appear to be less specialized as to host. North American representatives of this group are *Monilinia fructicola* and *Monilinia laxa.*

"All known members of this genus occur as parasites on members of the three following families of plants: the Rosaceae, the Cornaceae, and the Ericaceae."

On plants of the family Rosaceae.
  On plants of the genus *Prunus.*
    Conidial stage forming brown spots on fruit.
      Conidial stage *Monilia cinerea americana.*   1. *M. fructicola.*
      Conidial stage *Monilia cinerea.*   2. *M. laxa.*
    Conidial stage not forming brown-rots.
      Conidia large, 10–20 μ long.   3. *M. Padi.*
      Conidia medium 7–15 μ long.
        On *Prunus serotina.*   4. *M. Seaveri.*
        On *Prunus demissa.*   5. *M. demissa.*
  Not on *Prunus.*
    On *Amelanchier.*   6. *M. Amelanchieris.*
    On *Crataegus.*
      Ascospores 6–8 × 12–15 μ.   7. *M. Johnsoni.*
      Ascospores 2–3.5 × 8–12 μ.   8. *M. gregaria.*
On plants of the family Ericaceae.
  On *Polycodium.*   9. *M. Polycodii.*
  On *Vaccinium.*
    On blueberry.   10. *M. Vacciniicorymbosi.*
    On cranberry.   11. *M. Oxycocci.*
  On *Azalea.*   12. *M. Azaleae.*

1. **Monilinia fructicola** (Wint.) Honey, Mycologia **20**: 153. 1928. (PLATE 83.)

Ciboria fructicola Winter, Hedwigia **22**: 131. 1883.
Sclerotinia fructicola Rehm; Sacc. Syll. Fung. **18**: 41. 1906.
Monilia cinerea forma *americana* Wormald, Ann. Bot. **34**: 168. 1920.
Sclerotinia americana Norton & Ezek. Phytopathology **14**: 31. 1924.

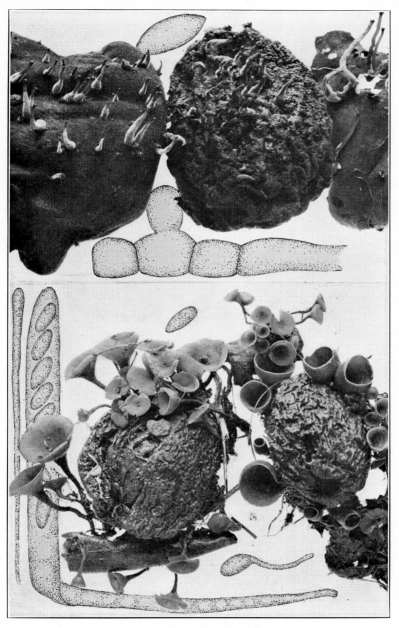

MONILINIA FRUCTICOLA

Conidiophores commonly arising from a well-developed ecto-stroma which ruptures the epidermis of the host, ash-gray or dark-olivaceous, producing simple, or branched conidial chains; conidia ellipsoid, elongate-ellipsoid, or rarely spherical, developed in chains, hyaline, or light-colored; microconidia present small, spherical, hyaline on short-clavate conidiophores.

Apothecia occurring in considerable numbers from a mummi-fied fruit, stipitate, shallow cup-shaped, or nearly plane at maturity, reaching a diameter of 1–1.5 cm., pale-brown; hy-menium usually concave, darker than the outside of the apo-thecium; stem of variable length but often reaching 2–3 cm.; asci cylindric-clavate, 8-spored; spores ellipsoid, containing one or more oil-drops, 3–3.5 × 6–7 μ, hyaline or slightly colored; paraphyses simple, or branched near their bases.

On overwintering orchard fruits.

TYPE LOCALITY: Europe.

DISTRIBUTION: Eastern North America; also in Europe.

ILLUSTRATIONS: Mycologia **20**: 149, *f. 1–2;* 154, *f. 3–4; pl. 17–19;* **37**: 671, *f. 8–10;* Phytopathology **32**: 635, *f. 1.*

The cause of brown-rot of fruits throughout Eastern North America. Not to be confused with *Sclerotinia fructigena* (Pers.) Schröt. of Europe, with which it was at first thought to be identical.

2. **Monilinia laxa** (Ehrenb.) Honey, Am. Jour. Bot. **23**: 105. 1936.

Oidium laxum Ehrenberg, Sylvae Myc. Ber. 22. 1818.
Acarosporium laxum Pers. Myc. Eu. **1**: 25. 1822.
Oospora laxa Wallr. in Bluff & Fing. Fl.-Crypt. Ger. **4**: 183. 1883.
Monilia cinerea Bonord. Handbk. Myk. 76. 1851.
Monilia laxa Sacc. & Vogl.; Sacc. Syll. Fung. **4**: 35. 1886.
Sclerotinia cinerea Schröt. in Cohn, Krypt.-Fl. Schles **3**²: 67. 1893.
Sclerotinia laxa Aderh. & Ruhl, Gesundheits Arbeit. Land.-Forstw. Berlin **4**: 437. 1905. (Citation from Whetzel.)
Sclerotinia cinerea Schröt.; Wormald, Ann. Bot. **35**: 134. 1920. (in part.)
Stromatinia laxa Chifflot. Ann. Ephip. **7**: 317. 1921.
Monilia oregonensis Barss & Posey; Barss, Oregon Exp. Sta. Cir. **53**: 5. 1923.

Conidial stage consists of a *Monilia* of the cinerea type; conidia 9–15 × 12–23 μ.

Apothecia springing from a mummified fruit, usually several from the same stroma, stipitate, at first clavate, expanding and becoming cup-shaped, finally nearly plane with an umbilicate center, reaching a diameter of 4–9 mm., gray, or almost white;

hymenium concave, or nearly plane; asci cylindric, narrowed below, reaching a length of about 175–180 μ and a diameter of 8–9 μ, 8-spored; spores ellipsoid, 1-seriate, or becoming partially 2-seriate, about 6 × 12 μ in diameter; paraphyses slender, slightly enlarged above, 2–2.5 μ in diameter.

On mummified plums, *Prunus*.

TYPE LOCALITY: Europe.

DISTRIBUTION: Wisconsin and the Pacific Coast; also in Europe, and reported from Japan.

ILLUSTRATIONS: Ann. Bot. **35**: *pl. 6, 7.*

3. **Monilinia Padi** (Wor.) Honey, Am. Jour. Bot. **23**: 105.   1936.

Sclerotinia Padi Woronin, Mem. Acad. Sci. St. Petersburg VlII. **2**: 3–14.   1895.
Sclerotinia angustior Reade, Ann. Myc. **6**: 113.   1908.
Monilia Peckiana angustior Sacc. Syll. Fung. **10**: 517.   1902.
Monilia angustior Reade, Ann. Myc. **6**: 113.   1908.

Conidial stage (*Monilia*) effused, ash-gray, occurring on the stems, petioles, sometimes on the principal veins on the backs of the leaves, later on the immatured fruit in minute, scattered cespitulae, the conidia citron-shaped, hyaline, simple, 10–20 μ long, in long di- or trichotomously branched chains with slender, fusiform disjunctors 2–3 μ long; microconidia in clusters, globose, 2.5–3 μ; sclerotia in mummified fruits.

Apothecia one or two from a single mummy, 5–20 mm. high, at first closed, expanding to saucer-shaped, 2–5 mm. in diameter, Isabel-colored; stem smooth, slender, cylindric, slightly tapering toward the base, reaching a length of 3–15 mm. and a diameter of 1 mm.; asci cylindric-clavate, reaching a length of 150–160 μ and a diameter of 8–10 μ, 8-spored; spores obliquely 1-seriate, ellipsoid, simple, 5–6 × 10–11 μ; paraphyses scattered, slender, 2 μ thick below, enlarged above to 4 μ.

Parasitic on twigs, leaves, and fruits of *Prunus virginiana*, apothecia appearing during the latter part of April and the conidial stage on leaves in May and on fruit in June.

TYPE LOCALITY: Ithaca, New York.

DISTRIBUTION: Known only from the type locality.

This species is said to be closely related to *Monilinia Seaveri* but differs in having larger conidia borne upon the twigs and petioles rather than upon the upper surface of the leaves and in the occurrence on a different host.

4. **Monilinia Seaveri** (Rehm) Honey, Am. Jour. Bot. **23**: 105. 1936. (PLATE 84.)

Sclerotinia Seaveri Rehm, Ann. Myc. **4**: 66. 1906.
Monilia Seaveri Reade, Ann. Myc. **6**: 112. 1908.

Conidial stage (*Monilia*) effused, ash-gray, epiphyllous or sometimes on the twigs also, still later in minute cespitulae on immature fruits, the conidia citron-shaped, simple, hyaline, 7–15 μ long, in long di- or trichotomously branched chains with slender, fusiform disjunctors 3 or 4 μ long; sclerotia formed in mummified fruits.

Apothecia one or two from a single mummy, about 1 cm. high, long-stipitate, at first closed then expanding and becoming saucer-shaped to convex and umbilicate, reaching a diameter of 11 mm.; stem slender, reaching a length of 5–20 mm.; asci cylindric-clavate, reaching a length of 155–180 μ and a diameter of 8–11 μ, 8-spored; spores obliquely 1-seriate, ellipsoid, hyaline, simple, 5–8 $\times$ 11–17 μ; paraphyses sparse, filiform, slightly enlarged above.

Parasitic on leaves, twigs, and fruits of *Prunus serotina*.

TYPE LOCALITY: Iowa City, Iowa.

DISTRIBUTION: New York to Iowa, and south to Georgia.

ILLUSTRATIONS: Phytopathology **30**: 89, *f. 1*.

EXSICCATI: Rehm, Ascom. *1633*.

This species has been reported as destructive to young cherry seedlings by John C. Dunegan of Fayetteville, Arkansas (Phyto-pathology **30**: 89. 1940.), as follows: "*A Blight of Wild Cherry Seedlings.*—A blighting of wild cherry (*Prunus serotina.*) seed-lings has been under observation since 1924. The disease, caused by *Sclerotinia Seaveri* Rehm, appears each spring about the time the second pair of true leaves unfolds. The first symptom is the development of a brown water-soaked region near the apex of the stem. This condition is accompanied by a loss of turgor and the infected seedlings are readily detected by the character-istic drooping of the affected portion of the stem. The infection spreads from the stem into the leaves through the petiole and midrib. The basal portion of the leaf turns brown and finally the whole leaf is affected, assuming a bleached grey color. Conidial masses frequently develop on the leaves. The fungus continues to spread down the stem and, when it reaches the ground line, the young plant dies.

"The disease was observed in Fort Valley, Georgia, from 1924 to 1928 and subsequently has been observed in the vicinity of Fayetteville, Arkansas.

"In 1928, 3 quadrats, each 1 sq. m. in area, were laid out at random under a large tree near Fort Valley, Georgia. Although the total number of seedlings blighted during the period of seedling germination was not ascertained, the counts made on April 4 showed from 34 to 62.5 per cent of the seedlings affected on that date. It is evident from these figures that the disease must be considered as a factor limiting the reproduction of *Prunus serotina* in the South."

5. **Monilinia demissa** (Dana) Honey, Am. Jour. Bot. **23**: 105. 1936.

Sclerotinia demissa Dana, Phytopathology **11**: 106.   1921.

Conidia (*Monilia*) produced in unbranched chains, ovoid to globose, hyaline, cream-colored in mass, simple, 3–9 × 7–14 μ; sclerotia in mummied fruits.

Apothecia mostly solitary, sometimes two from a mummy, long-stipitate, brown, glabrous, at first cup-shaped, later flat; stem 1–3.5 cm. long; asci hyaline, 8-spored, reaching a length of 150–160 μ and a diameter of 7 μ; spores ellipsoid, 1-seriate, hyaline, one end narrower than the other, 5–6 × 9–15 μ; paraphyses slender.

Conidial stage on living leaves, twigs, and fruits of *Prunus demissa;* the perfect stage on overwintered mummies of the same host.

TYPE LOCALITY: Pullman, Washington.

DISTRIBUTION: Known only from the type locality.

ILLUSTRATIONS: Phytopathology **11**: *pl. 8, f. 4–6.*

6. **Monilinia Amelanchieris** (Reade) Honey, Mycologia **34**: 575. 1942.

Sclerotinia Amelanchieris Reade, Ann. Myc. **6**: 114.   1908.

Entostroma formed within the fallen, overwintering, mummied fruits; microconidia small, 2.5–3.5 μ in diameter, globose, hyaline, associated with mummied fruits containing entostroma, produced at one or both poles of the ascospores remaining in old apothecia, on conidia, on mycelium directly, or on single or clustered, flask-shaped spermatiophores.

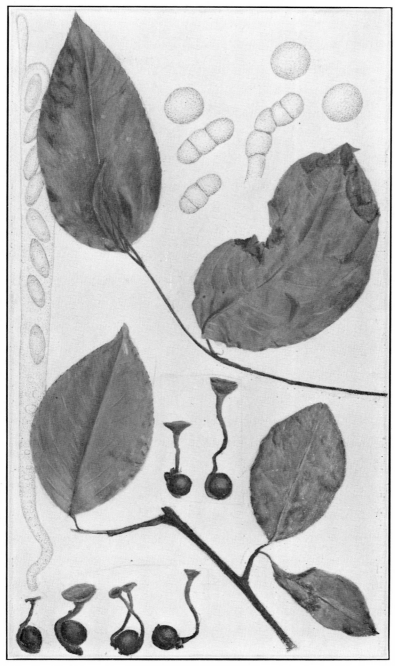

MONILINIA SEAVERI

Apothecia one to three, generally one arising as small, pointed fundaments from the entostroma, commonly from the blossom end but may appear at any point, reaching a height of 3–4 cm., cyathoid, 1.5–9 mm. in diameter, stipitate; stem smooth, slender, cylindric, tapering below, 1 mm. thick and reaching a length of 3.5–4 cm.; asci cylindric-clavate, reaching a length of 117–188 μ and a diameter of 6–13 μ, 8-spored; spores commonly obliquely 1-seriate, ellipsoid, 5.5–9 $\times$ 10–15 μ; paraphyses moderately abundant, filiform, slightly swollen above, 3.3–5.5 μ in diameter.

Apothecial stage on overwintering mummied fruits of *Amelanchier intermedia* on the ground during April and May; conidial stage parasitic on *Amelanchier intermedia, A. canadensis. A. oblongifolia* and unnamed species of *Amelanchier* about the middle of June.

TYPE LOCALITY: New York.

DISTRIBUTION: New York.

7. **Monilinia Johnsoni** (Ellis & Ev.) Honey, Am. Jour. Bot. **23**: 105. 1936.

Ciboria Johnsoni Ellis & Ev. Proc. Acad. Phila. **1894**: 348. 1894.
Monilia Crataegi Diedicke, Ann. Myc. **2**: 529. 1904.
Sclerotinia Crataegi Magnus, Ber. Deutsch. Bot. Ges. **23**: 197. 1905.
Sclerotinia Johnsoni Rehm, Ann. Myc. **4**: 338. 1906.

Conidial stage (*Monilia*) on leaves of the host, epiphyllous, effused, ash-gray to olive-buff, the conidia citron-shaped to globose, hyaline, 11–20 μ in diameter, in long di- and trichotomously branched chains with fusiform disjunctors 5–6 μ long; sclerotia formed in mummified fruits.

Apothecia one to several from a single mummy, 1–4.5 cm. high, long-stipitate, fawn-colored, darker with age; at first closed, then expanding, becoming cup-shaped or saucer-shaped, umbilicate, reaching a diameter of 3–10 mm.; stem smooth, slender, cylindric, tapering slightly below, reaching a length of 1–4 cm. and a diameter of 1 mm.; asci cylindric-clavate, reaching a length of 140–150 μ and a diameter of 8–10 μ, 8-spored; spores obliquely 1-seriate, ellipsoid, hyaline, simple, 6–8 $\times$ 12–15 μ; paraphyses filiform, clavate, hyaline.

Parasitic on leaves of *Crataegus punctata*.

TYPE LOCALITY: Ann Arbor, Michigan.

DISTRIBUTION: New York to Michigan.

ILLUSTRATIONS: Ber. Deutsch. Bot. Ges. **23**: *pl. 5*.

EXSICCATI: Ellis & Ev. N. Am. Fungi *3131*.

8. **Monilinia gregaria** (Dana) Honey, Am. Jour. Bot. **23**: 105. 1939.

Sclerotinia gregaria Dana, Phytopathology **11**: 106.   1921.

Conidia light-gray in mass, globose to lemon-shaped, 4–13 × 5–13 μ; microconidia globose produced singly on short sporophores 2–3.5 μ in diameter.

Apothecia gregarious pale-gray in color at first, later becoming darker, at first cup-shaped later flat, circular, 1–5 mm. in diameter, stipitate; stem 1–3 mm. in length; asci clavate, reaching a length of 48–57 μ and a diameter of 5–6 μ, 8-spored; spores 1-seriate, hyaline, containing two oil-drops, pointed at one end, 2–3.5 × 8–12 μ; paraphyses simple, or branched reaching a diameter of 2 μ.

Conidial stage on living leaves and fruits of *Amelanchier Cusickii;* the ascigerous stage on mummied fruits of the same host.

Type locality: Pullman, Washington.

Distribution: Known only from the type locality.

Illustrations: Phytopathology **11**: *pl. 8, f. 1–3.*

While this species has been recorded, Honey expresses some doubt, suspecting that the wrong fungus may have been reported as the perfect stage of the *Monilia*.

9. **Monilinia Polycodii** (Reade) Honey, Am. Jour. Bot. **23**: 106. 1936.

Sclerotinia Polycodii Reade, Ann. Myc. **6**: 110.   1908.
Monilia Polycodii Reade, Ann. Myc. **6**: 110.   1908.

Conidial stage (*Monilia*) on blighted shoots of the host, effused, powdery, ash-gray to olive-buff, on stems, petioles, and midribs at the base of the leaf-blades, or in minute cespitulae on mummified fruits, the conidia citron-shaped, simple, hyaline 11–13 × 15–18 μ in long di- or trichotomously branched chains with fusiform disjunctors 3–5 μ long; sclerotia formed in the mummified fruits; microconidia on mummified fruits in the open and on spores and mycelium in cultures, sphaerical, 2.5–3 μ in diameter, hyaline with a central refractive spot.

Apothecia single, or three or four from a mummy, 1–3.5 cm. high, at first closed then expanding, cup-shaped or sometimes convex and umbilicate .5–1 mm. in diameter, stipitate; hymenium wood-brown, darker outside; stem smooth, cylindric, 1–3 mm. thick and 1–3 cm. long; asci cylindric-clavate, reaching a length of 200–240 μ and a diameter of 10–14 μ, 8-spored; spores obliquely 1-seriate, ellipsoid, hyaline, simple, with two oil-drops, 10–12 $\times$ 15–20 μ; paraphyses slender, simple, 2 μ thick, slightly enlarged above.

Parasitic on twigs and fruits of *Polycodium stamineum*.

TYPE LOCALITY: Ithaca, New York.

DISTRIBUTION: New York.

10. **Monilinia Vaccinii-corymbosi** (Reade) Honey, Am. Jour. Bot. **23**: 105. 1936.

Sclerotinia Vaccinii-corymbosi Reade, Ann. Myc. **6**: 109. 1908.
Monilia Vaccinii-corymbosi Reade, Ann. Myc. **6**: 109. 1908.

Conidial stage (*Monilia*) on blighted shoots of the host, effused, powdery, ash-gray to olive-buff, on stems, petioles and midribs at the base of the leaf-blades or on peduncles of blighted flowers; citron-shaped, continuous, hyaline, the conidia 19–25 $\times$ 23–32 μ, in long di- or trichotomously branched chains, with fusiform disjunctors 2 $\times$ 3–5 μ; sclerotia formed on mummified fruits, microconidia on spores and mycelium in cultures 2.5–3 μ, globose, hyaline with a central refractive spot.

Apothecia one to several from a single mummy, 1–3.5 cm. high and 5–10 mm. in diameter, cup-shaped, somewhat convex, or cyathiform, stipitate; hymenium fawn-colored, outer surface darker and satiny; stem smooth, slender, cylindric, reaching a length of 1–3 cm. and a diameter of 1–2 mm., slightly tapering and shading into clove-brown below; asci cylindric-clavate, reaching a length of 200–260 μ and a diameter of 10–12 μ, 8-spored; spores obliquely 1-seriate, ellipsoid, 9–10 $\times$ 14–18 μ, simple, hyaline, containing a few granules; paraphyses slender, simple, 2–3 μ in diameter, slightly swollen at the tips,

Parasitic on twigs and fruits of *Vaccinium corymbosum*.

TYPE LOCALITY: Malloryville, New York.

DISTRIBUTION: New York and Massachusetts.

ILLUSTRATIONS: Mycologia **20**: *pl. 13, f. d, e, i.*

11. **Monilinia Oxycocci** (Woron.) Honey, Am. Jour. Bot. **23**: 105. 1936.

Sclerotinia Oxycocci Woronin, Mem. Acad. Sci. St.-Petersburg. **36**[6]: 28. 1888.

Conidial stage as in *Monilinia Vaccinii-corymbosi*.

Apothecia arising from sclerotia in the mummified fruits, usually one from each sclerotium, stipitate, the stem often 4–5 cm. long and slender, about 1 mm. in diameter, gradually expanding above into the cup which is at first rather deep, becoming shallow cup-shaped, or often reflexed, reddish-brown, about 5 mm. in diameter and often as deep; hymenium concave, plane, or convex, similar in color to the outside of the apothecium; asci clavate, reaching a length of 150 μ and a diameter of 5–6 μ, 8-spored; spores ellipsoid, 6 × 12–14 μ; paraphyses gradually enlarged above.

On fallen fruits of cranberry, *Vaccinium Oxycoccos.*

TYPE LOCALITY: Europe.

DISTRIBUTION: Where the cranberry is cultivated.

ILLUSTRATIONS: Mem. Acad. Sci. St. Petersburg **36**[6]: *pl. 7.*

12. Monilinia Azaleae Honey, Phytopathology **30**: 537.   1940.

Ectostroma developed beneath the epidermis, particularly on the leaves, the young succulent shoots and fruits forming as ash-gray coating of the conidial fructification, commonly on the upper surface of the midrib of the leaf and the surface of the fruits; conidia limoniform, simple, hyaline, 8.5–19 × 5.5–14.5 borne on long di- and trichotomously branched chains, disjunctors commonly present between the conidia; microconidia not observed; pseudosclerotia developing in the infested capsules, at maturity filling the loculi of the immature fruit with a solid mass of thick-walled, hyaline hyphae, falling to the ground and overwintering.

Apothecia one or two arising as small fundaments from the outer surface of the pseudosclerotia, reaching a height of .8–3.5 cm., cyathoid to patelliform, .2–1.4 cm. in diameter, stipitate; stem smooth, slender, cylindric, tapering downward, reaching a length of .4–3 cm. and a diameter of .5–2 mm.; asci cylindric-clavate, reaching a length of 178–258 μ and a diameter of 11–16.5 μ, 8-spored; spores obliquely 1-seriate, or occasionally irregularly arranged, ellipsoid, 5–14 × 9–20 μ; paraphyses filiform slightly swollen above.

Apothecia in May on overwintering fruits of *Rhododendron roseum;* conidial stage on leaves, young shoots and on the young fruits of *Rhododendron* spp. in June and July.

TYPE LOCALITY: Ithaca, New York.

DISTRIBUTION: New York and Georgia.

DOUBTFUL SPECIES

Monilinia Corni (Reade) Honey, Am. Jour. Bot. **23**: 105.   1936; *Sclerotinia Corni* Reade, Ann. Myc. **6**: 113.   1908; *Monilia Corni* Reade, Ann. Myc. **6**: 113.   1908.   According to Reade the apothecial stage of this species is un-

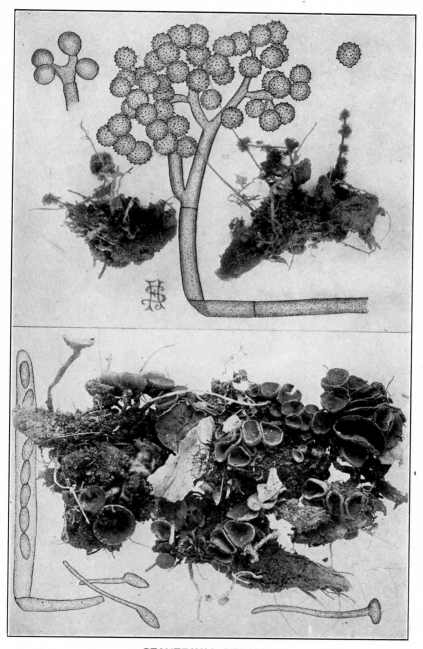

SEAVERINIA GERANII

known.   Honey believes that this is a good species even though the connection with a perfect has not actually been proven.

In addition to the species listed above, two others were listed by Edwin E. Honey (Am. Jour. Bot. **23**: 105.   1936) as follows: *Monilinia emarginata* on *Prunus emarginata* and *Monilinia Aroniae* on *Aronia* spp.   So far as we are aware these are still unpublished and must remain in doubt.

Monilinia Urnula (Weinm.) Whetzel, Mycologia **37**: 673.   1945; *Peziza Urnula* Weinm. Flora **15**: 455.   1832; *Peziza* (Phialea) *Urnula* Weinm. Hymeno-Gastero-Mycetes 459.   1836; *Sclerotinia Vaccinii* Woron. Mem. Acad. Sci. St. Petersb. **36**: 3.   1888.   This species has been reported from Washington by Dr. B. Kanouse on *Vaccinium* sp.   There seemed to have been some doubt in Whetzel's mind as to the identity of *Phialea Urnula* Weinm. and *Sclerotinia Vaccinii* Woron.

2. **SEAVERINIA** Whetzel, Mycologia **37**: 703.   1945.

Stroma substratal, poorly developed, perhaps vestigial, not a definite sclerotium, formed in the rhizomes of the host and visible on its surface usually as a narrow, black line; spermatia not observed; conidial stage a *Botrytis;* conidiophores botryose 1 mm. or more in length, pale-brown, sparingly septate, formed in tufts on the rhizome and roots of the host and, under moist conditions, profusely developed, bearing conidia in rather dense clusters, conidia pale-brown, minutely, but definitely tuberculate, subglobose, tapering somewhat to the point of attachment.

Apothecia arising from the partially decayed rhizome, stipitate, the length of the stem varying considerably depending on the depth to which the rhizome is buried, shallow cup-shaped, reaching a diameter of 15 mm.; asci cylindric or subcylindric, 8-spored; spores ellipsoid, hyaline.

Type species, *Sclerotinia Geranii* Seaver & Horne.

1. **Seaverinia Geranii** (Seaver & Horne) Whetzel, Mycologia **37**: 705.   1945.   (PLATE 85.)

Sclerotinia Geranii Seaver & Horne, Mem. Torrey Club **17**: 205.   1918.
Stromatinia Geranii Seaver & Horne, Mem. Torrey Club **17**: 206.   1918.

Conidial stage (*Botrytis*) occurring on the roots and rootlets of the host, being especially abundant when left in moist chamber for a few days and even developing on the outside of the apothecia, usually appearing in tufts and often springing from minute, sclerotium-like bodies, although the latter are not always present, dark-brown in mass at maturity; conidiophores reaching a length of 1 mm. or more and a diameter of 10–15 μ, pale-brown, sparingly septate and branched, the conidia borne in rather large masses like bunches of grapes; conidia subglobose or pyriform, the small end representing the point of attachment, reaching a diameter of 10 μ or rarely as large as 12 μ, slightly longer than broad, at first smooth, becoming quite strongly roughened, pale-brown with transmitted light.

Apothecia springing from the partially decayed rootstocks in clusters of variable numbers, stipitate, shallow-cup-shaped, or subdiscoid, reaching a diameter of 1 cm. or rarely larger, pale-brown externally; hymenium concave or nearly plane, a little darker than the outside of the apothecium; stem reaching a diameter of 2 mm. and often reaching a length of several cm., though often short and occasionally almost wanting, the length varying with the depth to which the rootstocks of the host are buried; asci cylindric or subcylindric, 8-spored, reaching a length of 120–140 μ and a diameter of 8–10 μ; spores hyaline, ellipsoid, or almond-shaped, 4–5 μ × 12 μ, usually containing two very small oil-drops.

On the rootstocks of wild geranium (*Geranium maculatum*).

TYPE LOCALITY: Van Cortlandt Park, New York, N. Y.

DISTRIBUTIONS: New York and Wisconsin.

ILLUSTRATIONS: Mem. Torrey Club **17**: *pl. 3;* Mycologia **37**: 704, *f. 32, 33;* 706, *f. 34–36;* **39**: 117, *f. 1.*

This fungus occurs regularly season after season in abundance in the same region in the suburbs of New York City.

3. **STREPTOTINIA** Whetzel, Mycologia **37**: 684. 1945.
(PLATE 86.)

Stroma a small, black sclerotium, flattened, or hemispherical, firmly attached to the substratum and flat or flattish on the attached surface; conidiophores as in *Botrytis* except that the branches are strikingly and characteristically twisted tightly as in *Streptothrix;* conidia smooth, hyaline or nearly so; apothecia minute, short-stipitate; asci 8-spored; spores hyaline, ellipsoid.

Type species, *Streptotinia Arisaemae* Whetzel.

Streptotinia Arisaemae Whetzel, Mycologia **37**: 686. 1945.

Stroma and conidia as above.

Apothecia stipitate, 1 mm. or less in diameter; asci reaching a length of 100–150 μ and a diameter of 8–10 μ; ascospores, 4–6 × 8–14 μ.

On disintegrating leaves of *Arisaema triphyllum*.

TYPE LOCALITY: Ithaca, New York.

DISTRIBUTION: Known only from the type locality.

ILLUSTRATIONS: Mycologia **37**: 685, *f. 22–24.*

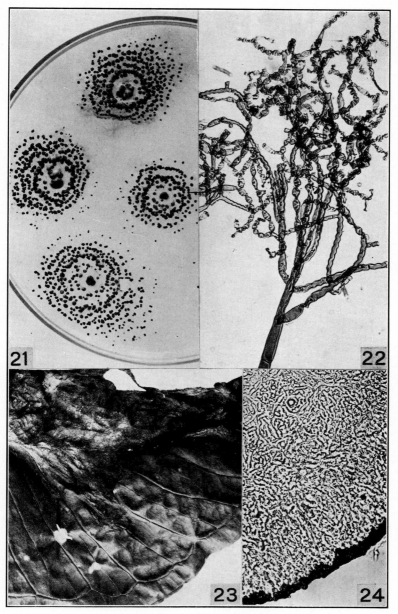

STREPTOTINIA ARISAEMAE

4. **BOTRYOTINIA** Whetzel, Mycologia **37**: 678.   1945.

Stroma consisting of a black, flattened, or irregularly hemi-spherical sclerotium, usually formed just beneath the cuticle of the epidermis, finally erumpent, remaining firmly attached to the host; spermatia globose, borne on branching spermatio-phores, enveloped in a mucilaginous matrix; conidial stage a *Botrytis;* conidia on short sterigmata in dense clusters, smooth, subglobose to pyriform.

Apothecia cupulate, stalked, some shade of brown, infundib-uliform to discoid, or margin reflexed; asci cylindric, or subcy-lindric; spores hyaline, simple, ellipsoid.

Type species, *Sclerotinia convoluta* Drayton.

On garden *Iris*.	1. *B. convoluta.*
On *Ricinus*.	2. *B. Ricini.*
In stems of *Gladiolus*.	3. *B. Draytoni.*
On *Narcissus*.	4. *B. narcissicola.*

1. **Botryotinia convoluta** (Drayton) Whetzel, Mycologia **37**: 679. 1945.

Botrytis convoluta Whetzel & Drayton, Mycologia **24**: 475.   1932.
Sclerotinia convoluta Drayton, Mycologia **29**: 314.   1937.

Sclerotia shining-black, convolute, agglomerated, reaching a size of 16 × 18 mm., frequently hollow in the center; conidio-phores brown, erect, fasciculate, branched at the apex, about 1 mm. tall and 9–12 μ thick at the base, tapering toward the apex; conidia light-brown, simple, smooth, ovoid to slightly pyriform, borne in dense clusters on sterigmata produced from the ultimate branchlets of the conidiophores, averaging 9 × 11 μ; micro-conidia present, 4–4.5 μ in diameter.

Apothecia densely gregarious, arising from the sclerotial agglomerations, infundibuliform to cyathiform, becoming discoid, reaching a diameter of 2.5–4 mm., stipitate; stems reaching a length of 3–6 mm.; asci cylindric, reaching a length of 150–195 μ and a diameter of 9–13 μ, 8-spored; spores 1-seriate, ellipsoid, hyaline, simple, at maturity with several oil-drops, averaging 7 × 15 μ; paraphyses abundant, filiform, hyaline, 2.5–3 μ, slightly enlarged above.

On garden *Iris*.

TYPE LOCALITY: Central Experimental Farm, Ottawa, Canada.

DISTRIBUTION: New York and Canada.

ILLUSTRATIONS: Mycologia **24**: 473, *f. 1; pl. 15, f. 1, 2; pl. 16, f. 3–5;* Mycologia **29**: 307, *f. 1, 2;* 309, *f. 3–5;* 312, *f. 6;* 313, *f. 7–9;* **37**: 681, *f. 19, 20.*

2. **Botryotinia Ricini** (Godfrey) Whetzel, Mycologia **37**: 680. 1945.

Sclerotinia Ricini Godfrey, Phytopathology **9**: 566. 1919.

Conidial stage (*Botrytis*) forming wide-spread, cobwebby or somewhat woolly masses, pale drab-gray, dried specimens dark olive-gray; fertile hyphae long slender, olivaceous when mature, dichotomously branched, terminal branches compact, collapsing when conidia fall; conidia borne on sterigmata, globose, smooth, hyaline, 6–12 μ in diameter; microconidia globose, hyaline, 2–3.5 μ in diameter, on short, obclavate conidiophores from the sides of the hyphae or tips of special branches; sclerotia black, rough, elongate, irregular, 1–25 mm. in diameter.

Apothecia one to several from a single sclerotium, 5–30 mm. high, usually 6–15 mm., infundibuliform to cyathiform and discoid, stipitate, cinnamon-brown to chestnut-brown, becoming saucer-shaped, with the margin somewhat recurved, 1–7 mm. in diameter; stem cylindric, slender, smooth, flexuous; asci cylindric-clavate, reaching a length of 50–110 μ and a diameter of 6–10 mm., 8-spored; spores ellipsoid to subfusoid, hyaline, simple, containing two oil-drops, 4–5 × 9–12 μ; paraphyses abundant, filiform, septate, hyaline, 1.5–2 mm. in diameter.

Parasitic on *Ricinus communis*, on inflorescence, or on leaves.

TYPE LOCALITY: Orlando, Florida.

DISTRIBUTION: Florida to Texas and Cuba.

ILLUSTRATIONS: Phytopathology **9**: *pl. 40, 41.*

3. **Botryotinia Draytoni** (Buddin & Wakef.) Seaver, comb. nov.

Sclerotinia Draytoni Buddin & Wakefield, Trans. Brit. Myc. Soc. **29**: 150. 1946.

Sclerotia black, smooth, applanate, then convex-lenticular, about 3–7 × 8–12 mm., in irregular masses about 2.5 cm. long, agglutinate, white within; microconidia globose, 2–2.5 μ in diameter; macroconidiophores, springing from the sclerotium, erect, brown below, 12–15 μ in diameter, above hyaline, two to three branched, the branches capitate, with numerous minute sterigmata; conidia hyaline, or pale-brown, ellipsoid, or broadly ovoid, 5–7.5 × 8–16 μ.

Apothecia long-stipitate, occurring singly, or several from each sclerotium, at first infundibuliform with the margin incurved, then umbilicate-discoid, finally slightly convex with the margin reflexed, 2.5–5 mm. in diameter; hymenium yellowish-olivaceous, the margin at first brown, then pallid; stem 10–13 mm. long, flexuous, enlarged above, same color as the hymenium, brown below; asci cylindric, 8-spored, reaching a length of 140–190 μ and a diameter of 9–10 μ; spores 1-seriate, broad-ellipsoid, or subfusiform, hyaline, often with two oil drops, 6–8 × 12–17 μ; paraphyses filiform, enlarged above.

On the stems of cultivated *Gladiolus*.

TYPE LOCALITY: Europe.

DISTRIBUTION: Known only from the type locality.

This species is included since it was named in honor of an American mycologist, Dr. F. L. Drayton, although it has not been reported from America. It is not unlikely that it will be found here also.

ILLUSTRATIONS: Trans. Brit. Myc. Soc. **29**: 151, *f. 11*.

4. Botryotinia narcissicola (Gregory) Seaver, comb. nov.

Sclerotinia narcissicola Gregory, Trans. Brit. Myc. Soc. **25**: 37. 1941.

Conidial stage a *Botrytis* (*Botrytis narcissicola* Kleb.); sclerotia black; smooth, more or less globose, 1–1.5 mm.

Apothecia occurring singly on the sclerotium, cup-shaped, becoming funnel-shaped, or expanded and plane, reaching a diameter of 2.5 mm., sepia to raw-umber when moist; stem 1.5–5 mm. or more long, tapering below; asci reaching a length of 126–140 μ and a diameter of 8 μ; spores 1- or 2-seriate, hyaline, navicular, often with two oil-drops, 5–9 × 10–20 μ; paraphyses filiform.

On old leaves and bud scales of species of *Narcissus*.

TYPE LOCALITY: Europe.

DISTRIBUTION: Conidial stage widely distributed in the United States. The apothecia stage has not been seen but doubtless could be found.

ILLUSTRATIONS: Trans. Brit. Myc. Soc. **25**: *pl. 2, f. 2–8*.

5. SEPTOTINIA Whetzel, Mycologia **29**: 134. 1937.

Conidia on massed, branched conidiophores, hyaline, elongate, septate; sclerotia angular, elongate, or circular, thin, black formed in the tissues of the host usually after they have fallen to the ground; spermatia ovoid, very minute, produced on short, indian-club-shaped spermatiophores, clustered to form minute spermadochia on decaying tissues, accompanying formation of sclerotia.

Apothecia shallow cup-shaped, stipitate, arising in the spring from sclerotia in the soil or leaf-mold; asci cylindric; spores hyaline, ovoid; paraphyses simple or branched with swollen tips.

Type species, *Gloeosporium podophyllina* Ellis & Ev.

1. **Septotinia podophyllina** (Ellis & Ev.) Whetzel, Mycologia **29**: 135. 1937.

Gloeosporium podophyllinum Ellis & Ev. Jour. Myc. **4**: 103. 1888.
Septogloeum podophyllinum Sacc. Syll. Fung. **10**: 497. 1892.

Conidia hyaline, often becoming 0–4-septate, straight or curved, often breaking apart at the septa when mature, truncate at the base, attenuated toward the apex, very variable in length; about 6 μ wide, borne on branched, hyaline, septate conidiophores which are densely clustered to form minute, white, sporodochia which become horny and often amber-colored when dry; sporodochia gregarious or scattered over both sides of the leaf; usually most numerous on the upper surface; sclerotia formed in the tissues of the dead leaves and stalks on the ground, mostly along the veins, angular elongated, or circular, thin, black, 3–5 mm. long, 1 mm. broad; spermatia accompanying the formation of sclerotia, minute, ovoid, 1–2 $\times$ 2–3.6 μ, hyaline.

Apothecia shallow cup-shaped, stipitate, arising from sclerotia on or in the soil or leaf-mold, 1 μ mm. in diameter, pale fawn-colored; stem tapering downward; asci cylindric, 8-spored, reaching a length of 148 μ and a diameter of 10 μ; spores ovoid, hyaline, 5–6 $\times$ 10–16 μ; paraphyses slender, swollen above.

On leaves of *Podophyllum peltatum*.

TYPE LOCALITY: Concordia, Missouri.

DISTRIBUTION: New York to Delaware and Missouri.

ILLUSTRATIONS: Mycologia **29**: 137, *f. 9–10;* 141, *f. 11–18*.

EXSICCATI: Ellis & Ev. N. Am. Fungi *2442*.

6. **OVULINIA** Weiss, Phytopathology **30**: 242.   1940.

Sclerotia irregularly discoid to shallow cup-shaped, thin, black, formed within but distinct from the host tissue; micro-conidia minute, globose, produced in chains on short, fusoid hyphae forming tufts on the surface of the host accompanying the sclerotia; conidia, large, obovoid, hyaline, produced singly on short, simple, or branched conidiophores, forming a thin mat on the surface of the host.

Apothecia arising singly or in groups from the sclerotium; asci slender, cylindric or subcylindric, 8-spored; spores ellipsoid; paraphyses mostly simple.

Type species, *Ovulinia Azaleae* Weiss.

Conidia large, 21–36 × 40–60 μ; on Ericaceae.	1. *O. Azaleae.*
Conidia small, 6–10 × 8–15 μ; on vegetables.	2. *O. perplexa.*

1. **Ovulinia Azaleae** Weiss, Phytopathology **30**: 243.   1940.

Sclerotia, microconidia and conidia as above.

Apothecia two or three or occasionally as many as eight in a group, stipitate, urceolate, or cyathiform, expanding at maturity, becoming subdiscoid, ochraceous to brown, 2–5 mm. in diameter; stem typically 2–3 mm. long and 1–1.5 mm. thick, but occasionally 15–18 mm. long; asci cylindric, reaching a length of 140–260 μ and a diameter of 9–14 μ, 8-spored; spores 1-seriate, 8–10 × 10–18 μ, hyaline; paraphyses simple enlarged above.

On flowers of cultivated species of *Azalea* and *Rhododendron* causing blight, the apothecia appearing in late winter or spring from sclerotia lying on or in the soil.   It is said also to infect *Kalmia* and *Vaccinium*.

TYPE LOCALITY: North Carolina.

DISTRIBUTION: Southeastern United States from North Carolina to Texas.

ILLUSTRATIONS: Phytopathology **30**: 238, *f. 1*, 239, *f. 2;* 240, *f. 3.*

2. **Ovulinia perplexa** (Lawrence) Seaver, comb. nov.

Sclerotinia perplexa Lawrence, Western Wash. Exp. Sta. Bull. **107**: 10.   1912.

Sclerotia gray when young, becoming dull-black, snow-white within, becoming pink or brown with age, depressed-subglobose, reaching a diameter of 1–3 mm., forming thin, black crusts of considerable extent; conidial stage on or accompanying the sclerotia, consisting of straight, or branched conidiophores often several mm. long, producing ovoid conidia 6–10.5 × 8–15 μ.

Apothecia stipitate one to many from a single sclerotium, flesh-colored, reaching a diameter of 2–8 mm.; stem reaching a length of 1–2 mm.; asci cylindric, or subcylindric, reaching a length of 115–145 μ and a diameter of 5–7 μ, 8-spored; spores subellipsoid, 3.5–5 × 8–10 μ; paraphyses, filiform.

On Jerusalem artichoke, onions, cucumber, cabbage and a large number of cultivated vegetables.
TYPE LOCALITY: Western Washington.
DISTRIBUTION: Known only from the type locality.
ILLUSTRATIONS: Western Wash. Exp. Sta. Bull. **107**: *f. 1–9*.

7. **LAMBERTELLA** Höhn. Sitz.-ber. Akad. Wien I. **127**: 375. 1918.

Apothecia springing from a more or less definitely outlined, dark-colored, stromatic base consisting of a single layer of cells and loosely interwoven hyphae, stipitate, gregarious, or scattered, fleshy, becoming coriaceous, or corneous on drying, some shade of brown, or yellowish-brown when fresh; hymenium slightly darker; stem relatively stout, variable in length, or occasionally wanting, hirsute, or furfuraceous; asci cylindric to clavate, attenuated below, rounded or truncate at the tip, 8-spored; spores usually 1-seriate, or occasionally becoming partially 2-seriate, simple, broadly ellipsoid, ovoid, or lunate, usually unequal sided, or lunate, smooth, or rough, golden-brown, or olivaceous when mature; paraphyses often branched, hyaline, slightly enlarged above. Spermatia usually present. No conidial stage observed.

Type species, *Lambertella Corni-maris* Höhn.

Spores small, not over 10–11 μ long.
  On stromatized hulls of *Carya*.   1. *L. Hicoriae*.
  On mummied berries and leaves.
    On *Jasminum* and *Citharexylum* in Bermuda.   2. *L. Jasmini*.
    On *Coccoloba?* leaves in British Honduras.   3. *L. tropicalis*.
Spores large, 12–18 μ long.
  On mummied fruits of *Prunus*.   4. *L. Pruni*.
  On fallen or hanging leaves.
    On leaves of *Viburnum*.   5. *L. Viburni*.
    On leaves of *Cephalanthus*.   6. *L. Cephalanthi*.

1. **Lambertella Hicoriae** Whetzel, Lloydia **6**: 33. 1943.

Apothecia arising from stromatized hickory-nut hulls partially buried in soil or leaf-mould, about 2 mm. in diameter, fleshy, stipitate, brown or tawny-olive to dark vinaceous-brown; hymenium olive-brown or fuscous; stem stout, light above, dark-brown toward the base; asci cylindric-clavate, attenuated below, rounded or truncate above, reaching a length of 101–127 μ and a diameter of 6–8 μ; spores obliquely 1-seriate, later becoming 2-

seriate, broadly ellipsoid, flattened on one side, with two oil-drops, and hyaline when young, becoming olivaceous-brown when mature, 4–6 × 7–10 μ; paraphyses slender, branched, slightly enlarged above. Spermatia globose, 2–3 μ in diameter.

On stromatized hulls of *Carya ovata*.

TYPE LOCALITY: Woods east of Cayuta Lake, New York.

DISTRIBUTION: Known only from the type locality.

ILLUSTRATIONS: Lloydia 6: *pl. 3, f. 1–5;* p. 35, *f. 2.*

2. **Lambertella Jasmini** Seaver & Whetzel, Lloydia 6: 37. 1943.

Apothecia one to several arising from the stroma in fruits, pedicels, or leaves of the host, 1–5 mm. in diameter, fleshy, stipitate, dark olive-buff, outside of the apothecium fibrillose, the short tips of the hyphae forming a fringe about the margin; hymenium darker brown; stem stout to slender, usually rela-tively short, cinnamon-brown below, lighter above, more or less hairy; asci cylindric, attenuated below, rounded above, reaching a length of 95–135 μ and a diameter of 7–8 μ; spores 1-seriate, broadly ellipsoid, flattened on one side, slightly roughened, with two prominent oil-drops, becoming dark golden-brown at ma-turity, about 4–5 × 6–9 μ; paraphyses hyaline, slender, branched, slightly enlarged above.

On mummied berries and leaves of *Jasminum gracile* and on leaves of *Citharexylum spinosum*.

TYPE LOCALITY: Walsingham, Bermuda.

DISTRIBUTION: Bermuda.

ILLUSTRATIONS: Lloydia 6: 36, *pl. 4, f. 1–3;* p. 38, *f. 3.*

3. **Lambertella tropicalis** (Kanouse) Whetzel, Lloydia 6: 49. 1943.

Ciboria tropicalis Kanouse, Mycologia 33: 463. 1941.

Apothecia scattered over a stromatic area on the upper sur-face of the leaf, the stroma forming an irregular area along the midrib and side veins of the leaf, the individual apothecia small, about 2 mm. in diameter, very short-stipitate or nearly sessile, waxy, white when fresh, alutaceous when dry, minutely furfuraceous beneath; hymenium nearly plane; stem very short and thick; asci cylindric-clavate, broadest at the apex, the tip rounded and thickened, reaching a length of 90–100 μ and a diameter of 8–11 μ; spores 1-seriate or irregularly 2-seriate, almost completely filling the ascus, inequilateral, slightly allantoid, granular, becoming golden-brown, 4–5 × 11–14 μ; paraphyses filiform, sparingly branched.

On leaves of *Coccoloba*?

TYPE LOCALITY: British Honduras.

DISTRIBUTION: Known only from the type locality.

ILLUSTRATIONS: Lloydia **6**: 50, *f. 7.*

4. Lambertella Pruni Whetzel & Zeller, Lloydia **6**: 40.   1943.

Apothecia gregarious, as many as a hundred springing from a single stromatized fruit, the stroma consisting of a thin, wrinkled, black crust surrounding the host, the individual apothecia 1.5 mm. in diameter when mature, short-stipitate to sessile, externally furfuraceous, the margin fringed with short hyphal tips, brown; hymenium plane or convex, pale-brown, darker with age; stem very short, sharply constricted at point of attachment and black at the base; asci cylindric, slightly narrowed above, attenuated at the base, reaching a length of 115–150 μ and a diameter of 10–14 μ; spores 1-seriate, lunate, with broad, blunt ends, smooth, containing two oil-drops, golden-brown, becoming dark olivaceous-brown at maturity, 7–8 × 12–15 μ; paraphyses branched, septate, slightly enlarged above.

On mummied fruits and seedlings of sweet cherry, *Prunus avium.*

TYPE LOCALITY: Reynold's Estate near Salem, Oregon.

DISTRIBUTION: Known only from the type locality.

ILLUSTRATIONS: Lloydia **6**: 36, *pl. 4, f. 4;* p. 41, *f. 4.*

5. Lambertella Viburni Whetzel, Lloydia **6**: 43.   1943.

Apothecia scattered, stipitate, arising from a subcuticular stromatic base, restricted largely to the veins of the leaves, individual apothecia small, .5–2 mm. in diameter, externally fibrillose, light-buff, the margins dark-brown, incurved, rimmed with hyphal tips; hymenium shallow-concave to plane, light-brown; stem stout, cinnamon-brown, lighter above, hirsute; asci slender-cylindric, tapering slightly below, rounded above, reaching a length of 60–114 μ and a diameter of 8–9 μ; spores ellipsoid, bluntly apiculate, flattened on one side, light-brown, becoming darker after discharge, 4–6 × 12–18 μ, with two oil-drops; paraphyses apparently simple, actually fasciculately 3-branched near the base, septate, slightly enlarged at the tips.

On hanging dead leaves of *Viburnum cassinoides.*

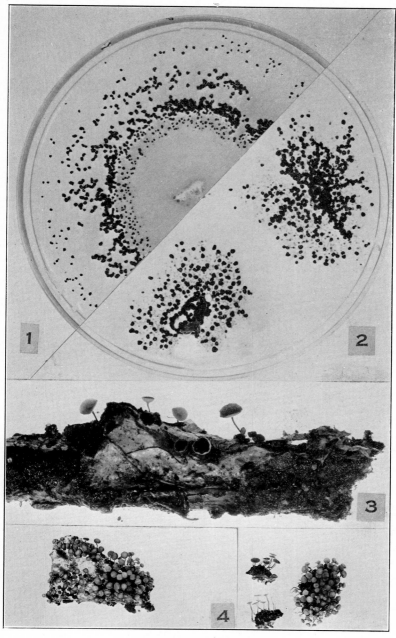

MARTINIA PANAMAENSIS

TYPE LOCALITY: The "Bottomless Pit" near Hanover, New Hampshire.

DISTRIBUTION: Known only from the type locality.

ILLUSTRATIONS: Lloydia **6**: 44, *pl. 5;* p. 45, *f. 5.*

6. **Lambertella Cephalanthi** Whetzel, Lloydia **6**: 47. 1943.

Apothecia scattered, arising from the surface of the leaf without any distinct stromatization, minute, not exceeding 1 mm. in diameter, stipitate, pale-brown, the margin fringed with short hyphal tips; hymenium dark-brown, deeply concave, paler after spore discharge; stem dark-brown below, pruinose and colored like the outside of the apothecium; asci short, stout, broadly cylindric, tapering slightly toward the base, slightly narrowed above, the apex rounded or more or less truncate, reaching a length of 90–117 μ and a diameter of 10–12 μ; spores 1-seriate, or becoming 2-seriate, nearly filling the entire ascus, broadly ellipsoid, olivaceous-brown with a thickened, darker band on one side, containing two oil-drops, 5–8 × 12–17 μ; paraphyses 3-branched, septate, swollen above.

On fallen leaves of *Cephalanthus occidentalis.*

TYPE LOCALITY: West Roxbury, Massachusetts.

DISTRIBUTION: Massachusetts and Cayuta Lake, New York.

ILLUSTRATIONS: Lloydia **6**: 48, *f. 6.*

8. **MARTINIA** Whetzel, Mycologia **34**: 585. 1942.

Apothecia arising singly or several from minute, hemispherical sclerotia on the surface of the substratum, thin, membranous, shallow cup-shaped, reaching a diameter of 2–3 mm., stipitate; hymenium olivaceous to olive-brown when spores are mature; stem long, slender, hair-like; asci 8-spored; spores simple, ellipsoid, olive-brown.

Type species, *Martinia panamaensis* Whetzel.

This name is apparently untenable having previously been used. See Vaniot, Acad. Geogr. Bot. **12**: 31. 1903.

1. **Martinia panamaensis** Whetzel, Mycologia **34**: 586. 1946.

Stromata and apothecia as above; asci cylindric above, attenuated below, reaching a length of 35–40 μ and a diameter of 5 μ; spores ellipsoid, slightly flattened on one side, 2.5 × 4.5 μ; paraphyses branched near the base, scarcely enlarged above.

On the surface of bark and wood of some undetermined tree.

TYPE LOCALITY: Near Balboa, Panama.
DISTRIBUTION: Known only from the type locality.
ILLUSTRATIONS: Mycologia **34**: 587, *f. 1–4*.

9. **CIBORINIA** Whetzel, Mycologia **37**: 667.   1945.

Stroma a definite sclerotium of the discoid type, black, circular, subcircular, or occasionally elongated, thin, flat, or on drying somewhat concavo-convex, foliicolous, persistent, or deciduous; spermatia present, globose or ovoid, hyaline, in mass pale-yellow.

Apothecia, one to several arising from each sclerotium, cup-shaped to shallow saucer-shaped or discoid, small to medium, 1–5 mm. in diameter, reddish to brown, or nearly white; asci usually 8-spored; spores simple, hyaline, ellipsoid; paraphyses slender, usually swollen at the tips.

Type species, *Sclerotinia Whetzelii* Seaver.

On leaves of *Populus tremuloides*.
    Apothecia usually one to each sclerotium, large, 2–10
      mm. in diameter.   1. *C. Whetzelii.*
    Apothecia several to each sclerotium small, .5–1.5 mm.
      in diameter.   2. *C. bifrons.*
Not on *Populus tremuloides*.
    On leaves of *Salix*.   3. *C. foliicola.*
    On *Erythronium*.   4. *C. Erythronii.*
    On petioles of *Magnolia*.   5. *C. gracilipes.*

1. **Ciborinia Whetzelii** (Seaver) comb. nov.   (PLATE 89.)

Sclerotium bifrons Ellis & Ev. Fungi Columb. *155;* Sacc. Syll. Fung. **14**: 1169.
    1899.   (in part).
Sclerotinia bifrons Whetzel, Mycologia **32**: 126.   1940.   Not *Sclerotinia
    bifrons* Seaver & Shope.   1930.
Sclerotinia Whetzelii Seaver, Mycologia **32**: 127.   1940.
Ciborinia bifrons Whetzel, Mycologia **37**: 668.   1945.

Sclerotia formed in the living leaves, circular, or irregularly subcircular in form, persistent or dehiscent.

Apothecia occurring on sclerotia lying on the ground, stipitate, at first cup-shaped, expanding and becoming subdiscoid, brownish, 2–10 mm. in diameter; stems 5–25 mm. long; asci clavate or subclavate, reaching a length of 150–200 μ and a diameter of 9–12 μ; spores 1-seriate, ellipsoid, 4–7 $\times$ 11–16 μ.

On sclerotia dehisced from the leaves of *Populus tremuloides* and lying on the ground.

TYPE LOCALITY: New York.

DISTRIBUTION: New York and Quebec, Canada.

ILLUSTRATIONS: Mycologia **37**: 669, *f. 6, 7;* Canadian Jour. Res. **18**: *pl. 1, f. 1–10.*

2. **Ciborinia bifrons** (Ellis & Ev.) Seaver, comb. nov. (PLATE 75 (frontispiece), 88.)

Sclerotium bifrons Ellis & Ev. N. Am. Fungi *155;* Sacc. Syll. Fung. **14**: 1169. 1899. (in part).
Sclerotinia bifrons Seaver & Shope, Mycologia **22**: 3. 1930.
Sclerotinia confundens Whetzel, Mycologia **32**: 126. 1940.
Ciborinia confundens Whetzel, Mycologia **37**: 668. 1945.

Sclerotia formed in the living leaves of the host, irregularly circular in form, persistent, or dehiscent, black, thin; apothecia stipitate, several to each sclerotium, minute, whitish, or slightly yellowish, .5–1.5 mm. in diameter; stems variable in length, usually 2–3 mm.; asci clavate, or subcylindric, 8-spored, reaching a length of 50–75 μ and a diameter of 4–7 μ; spores fusoid, or ellipsoid, hyaline, 2–3 $\times$ 7–10 μ; paraphyses slender.

On sclerotia dehisced from the leaves of *Populus tremuloides* and lying on the ground in moist places, usually embedded in humus.

TYPE LOCALITY: University of Colorado summer camp near Boulder, Colorado.

DISTRIBUTION: Known only from the type locality.

ILLUSTRATIONS: Mycologia **22**: *pl. 4; 37*: No. 6, frontispiece, 645, *f. 2.*

EXSICCATI: Ellis & Ev. Fungi Columb. *155;* Barth. Fungi Columb. *2554.*

3. **Ciborinia foliicola** (Cash & Davidson) Whetzel, Mycologia **37**: 668. 1945.

Sclerotinia foliicola Cash & Davidson, Mycologia **25**: 269. 1933.

Sclerotia surrounding the midribs of overwintering leaves of the host, .5–1 mm. wide and 1–4 cm. long, thick and appearing on both sides of the leaf-blade.

Apothecia stipitate at first cup-shaped, expanding and becoming almost discoid at maturity, up to 1 cm. in diameter, grayish-brown; stems .5–1 mm. thick, reaching a length of 1–3 cm., rather stout when young, becoming slender at maturity, asci cylindric, or subcylindric, attenuated below, reaching a length of 120–140 μ and a diameter of 8–10 μ; spores 1-seriate, ellipsoid, hyaline, 5 × 9–13 μ; paraphyses filiform, septate, gradually enlarged above to 3–4 μ thick.

On leaves of *Salix* sp.

TYPE LOCALITY: Mesa Lakes, Colorado.

DISTRIBUTION: Known only from the type locality.

ILLUSTRATIONS: Mycologia **25**: *pl. 37, f. 8, 9.*

4. Ciborinia Erythronii Whetzel, Mycologia **37**: 668.   1945.

?Sclerotinia gracilis Clements; Bessey, Contr. Bot. Dep. Univ. Nebr. **3**: 47. 1892.

Sclerotinia Erythronii Whetzel, Mycologia **18**: 232.   1926.

?Ciborinia gracilis Whetzel, Mycologia **37**: 668.   1945.

Sclerotia usually more or less pointed, black, deeply buried, or flat and ovate lying on the surface of the soil, 1–1.5 × 5–10 mm.

Apothecia slightly elevated above the surface of the soil, cup-shaped to shallow saucer-shaped, finally expanded and sub-discoid, or convex, brown, 3–5 mm. in diameter; stems often very long, reaching a length of 5–10 cm. the length varying according to the depth at which the sclerotia are buried; asci cylindric, or subclavate, reaching a length 200–275 μ and a diameter of 12 μ, 8-spored; spores long-ellipsoid, 7–10 × 20–30 μ; paraphyses filiform, only slightly swollen above.

Parasites on *Erythronium americanum*, occurring on shaded knolls or wooded hillsides.

TYPE LOCALITY: Labrador Lake, near Tulley, New York.

DISTRIBUTION: New York and Nebraska?

ILLUSTRATIONS: Mycologia **18**: 232. *f. 1, pl. 27.*

5. Ciborinia gracilipes (Cooke) Seaver, comb. nov.

Peziza gracilipes Cooke, Bull. Buffalo Soc. Nat. Sci. **2**: 294.   1875.

Sclerotinia gracilipes Sacc. Syll. Fung. **8**: 200.   1889.

Sclerotia thin, rounded, or ellipsoid, formed on the petioles of the host.

Apothecia stipitate, at first cup-shaped, finally becoming discoid, submembranaceous, yellowish-brown, reaching a diameter of 2–3 cm.; asci cylindric, or subcylindric, 8-spored; spores ellipsoid, 4 × 8 μ.

From sclerotia on the petioles of *Magnolia glauca*.

TYPE LOCALITY: Newfield, New Jersey.

CIBORINIA BIFRONS

DISTRIBUTION: Known only from the type locality.
EXSICCATI: Ellis, N. Am. Fungi. *390.*
Specimens in the herbarium of the New York Botanical
Garden show excellent sclerotia and apothecia.

10. **VERPATINIA** Whetzel & Drayton, Mycologia **37**: 690. 1945.

Stroma an elongated, black sclerotium of the discoid type
identical with that of *Ciborinia*, foliicolous, formed beneath the
cuticle of the host; conidial stage believed to be wanting.

Apothecia arising from the sclerotia singly or in pairs, stipitate, campanulate to cylindric or subturbinate; the hymenial
surface pitted or often longitudinally furrowed; asci 8-spored;
spores ellipsoid to fusiform.

Type species, *Verpatinia calthicola* Whetzel.

On old leaves of *Caltha palustris*.	1. *V. calthicola.*
On old leaves of *Betula lutea*.	2. *V. duchesnayensis.*

1. **Verpatinia calthicola** Whetzel, Mycologia **37**: 692. 1945.

Stroma and sclerotia as above.

Apothecia campanulate, or turbinate, 2–3 mm. long and 1–2
mm. thick, clay-colored to pale-brown, long-stipitate; stem
slender, slightly swollen at the base, 5–15 mm. long and .5 mm.
in diameter; asci cylindric, gradually tapering below, reaching a
length of 30–38 μ and a diameter of 3–6 μ; spores subellipsoid
with one side flattened, 2–3 $\times$ 6–10 μ.

On overwintering petioles of *Caltha palustris*.
TYPE LOCALITY: Ithaca, New York.
DISTRIBUTION: Known only from the type locality.
ILLUSTRATIONS: Mycologia **37**: 687. *f. 25–28.*

2. **Verpatinia duchesnayensis** Whetzel, Mycologia **37**: 694. 1945.

Sclerotia elongated, at maturity as long as 25 mm. though
often much shorter, varying from short-fusiform in early stages
to long, slender-cylindric, rather abruptly tapering toward the
ends, dull to shining, formed in the midrib or one of the primary
ribs of the host.

Apothecia, stipitate, arising singly, or in pairs from the embedded sclerotium, cylindric to barrel-shaped, or turbinate, ap-

próximately 2 mm. long, 1 mm. thick, pale ashy-gray, surface irregularly furrowed or wrinkled; stem of uniform diameter throughout, brownish, smooth, paler above; asci cylindric, or sub-cylindric; spores, fusoid, one side often flattened, 3–4 × 9–12 μ; paraphyses thick, enlarged at their apices.

On disintegrating leaves of *Betula lutea*.

TYPE LOCALITY: Duchesnay, Quebec, Canada.

DISTRIBUTION: Known only from the type locality.

ILLUSTRATIONS: Mycologia **37**: 691, *f. 30, 31*.

11. **SCLEROTINIA** Fuckel, Symb. Myc. 330. 1871.

Apothecia arising from a definite tuberoid, elongated, or irregular sclerotium, usually formed in the living or recently killed tissues of the host, more rarely apparently disconnected with any host, at first rounded, becoming cup-shaped, urn-shaped or discoid; asci clavate-cylindric, usually 8-spored; spores ellipsoid to fusoid, simple, hyaline; paraphyses filiform, or slightly enlarged above.

Type species, *Peziza sclerotiorum* Lib.

Sclerotia occurring unattached to any host.	1. *S. tuberosa.*
Sclerotia closely associated with a host plant.	
On monocotyledons, grasses or sedges.	
On seeds of *Secale.*	2. *S. temulenta.*
On stems or leaves.	
On *Juncus.*	3. *S. juncigena.*
On *Carex.*	
Apothecia large, 2–4 cm. in diameter.	4. *S. Caricis-ampullaceae.*
Apothecia small, less than 1 cm.	
Sclerotia small, .3–1 mm. in diameter on leaves of host.	5. *S. paludosa.*
Sclerotia large, 7 mm.–2 cm. long.	
Sclerotia 1–7 mm. long.	6. *S. longisclerotialis.*
Sclerotia 2–2.5 cm. long.	7. *S. Duriaeana.*
On dicotyledons, various species.	
On flowers or seeds.	
On seeds.	
On *Tilia americana.*	8. *S. Tiliae.*
On *Nyssa sylvatica.*	9. *S. nyssaegena.*
On flowers of *Camellia.*	10. *S. Camelliae.*
On leaves or stems.	
On leaves.	
On *Potentilla.*	11. *S. fallax.*
On lettuce, *Lactuca* and other garden plants.	12. *S. minor.*

CIBORINIA WHETZELII

On stems or seed pods.
　　On *Veratrum californicum.* 13. *S. Veratri.*
　　On stems or seed-pods of *Pedicularis.* 14. *S. coloradensis.*
On bulbs, tubers or roots.
　　On roots of *Tragopogon* and *Daucus.* 15. *S. intermedia.*
　　On bulbs or tubers.
　　　　On tubers of *Panax.* 16. *S. Panacis.*
　　　　On bulbs of *Tulip* and *Medicago.* 17. *S. sativa.*

1. Sclerotinia tuberosa (Hedw.) Fuckel, Symb. Myc. 331. 1869. (PLATE 90.)

Octospora tuberosa Hedw. Descr. **2**: 39. 1788.
Peziza tuberosa Bull. Hist. Champ. Fr. 266. 1791.
Macroscyphus tuberosus S. F. Gray, Nat. Arr. Brit. Pl. **1**: 672. 1821.
Rutstroemia tuberosa Karst. Myc. Fenn. **1**: 105. 1871.
Phialea tuberosa Gill. Champ. Fr. Discom. 97. 1882.
Hymenoscypha tuberosa Phill. Brit. Discom. 113. 1887.

Sclerotia buried in the ground, forming irregular nodules 5–7 mm. in diameter, occasionally larger, externally black, internally white, strongly convoluted when dry.

Apothecia infundibuliform to cup-shaped, more or less expanded at maturity, margin often inturned, pale umber-brown, stipitate, 1 cm. or more in diameter; hymenium slightly darker than the outside of the apothecium, even, or occasionally slightly convoluted; stem slender, about 2 mm. in diameter, gradually expanding into the cup above, and with the base more or less swollen, reaching a length of several cm., usually one fruiting body from each sclerotium but occasionally several; asci cylindric-clavate, reaching a length of 150 μ and a diameter of 8–10 μ, 8-spored; spores ellipsoid, the ends slightly narrowed, 6 × 12–14 μ; paraphyses filiform, the ends slightly enlarged above, 3–4 μ in diameter.

On sclerotia buried in the ground, the stem of the apothecia being long enough to reach the surface of the ground.

TYPE LOCALITY: Europe.

DISTRIBUTION: New York to North Dakota and south to Tennessee, probably distributed in North America; also in Europe.

ILLUSTRATIONS: Hedwig, Descr. **2**: *pl. 10, f. B;* Bull. Herb. Fr. *pl. 485, f. 2–3;* Rab. Krypt.-Fl. **1**[3]: 802, *f. 1–3;* E. & P. Nat. Pfl. **1**[1]: 193, *f. 156 K.*

European authors report this species as occurring on the rhizomes of *Anemone nemorosa.* The writer has collected hundreds of sclerotia and apothecia of this species in the suburbs of

New York City, over a period of years, and in no case has the
fungus been associated with the rhizomes of this or any other
host.   While there might be a mycelial connection, none was
apparent.   This leads us to suspect that our American form may
be distinct from the European.   However, the two agree so well
in other characters that for the time being our species are regarded as identical with the European.

2. **Sclerotinia temulenta** (Prill. & Delacr.) Seaver, comb. nov.

Phialea temulenta Prill. & Delacr. Bull. Soc. Myc. Fr. **8**: 23.   1892.
?Endoconidium temulentum Prill. & Delacr. Bull. Soc. Myc. Fr. **7**: 116.   1891.

Conidial stage consisting of globose, or subglobose conidia
produced in chains within the branches and near the tips of the
conidiophores.

Apothecia solitary, or gregarious on a sclerotium formed in
the seed of the host, stipitate, at first partly closed, pallid, from
ochraceous to honey-colored, .5–7 μ in diameter; hymenium
plane to slightly convex; stem slender, 7–10 mm. long and .5–1
mm. thick; asci cylindric, reaching a length of 130 μ and a
diameter of 5 μ, 8-spored; spores, 1-seriate, fusoid, 4–5 $\times$ 10 μ;
paraphyses filiform, slightly enlarged above 1.5–2 μ thick.

On fruits of *Secale cereale* to which it gives poisonous proper-
ties.   Said to be a very destructive parasite.

TYPE LOCALITY: Europe.

DISTRIBUTION: Williamette Valley, Oregon; also in Europe.

ILLUSTRATIONS: Bull. Soc. Myc. Fr. **7**: 116, *f. 1* (conidia).

If Prillieus and Delacroix are correct in assuming that this is
the perfect stage of *Endoconidium temulentum* and we follow the
Whetzel scheme of separating genera on the basis of their
conidial stages this should be made a separate genus since it is
the only member of the tribe which has an *Endoconidium* as the
conidial stage.

3. **Sclerotinia juncigena** (Ellis & Ev.) Whetzel, Farlowia **2**:
432.   1946.

Ciboria juncigena Ellis & Ev. Proc. Acad. Sci. Phila. **1894**: 348.   1894.

Sclerotia one, possibly more within a diseased culm, slender,
cylindric, with truncate, slightly rounded ends, reaching a length
of 15 mm. and a diameter of 2 mm., externally black, sulcate,
internally white when mature, enclosed in a cavity in the pith
region of the culm.

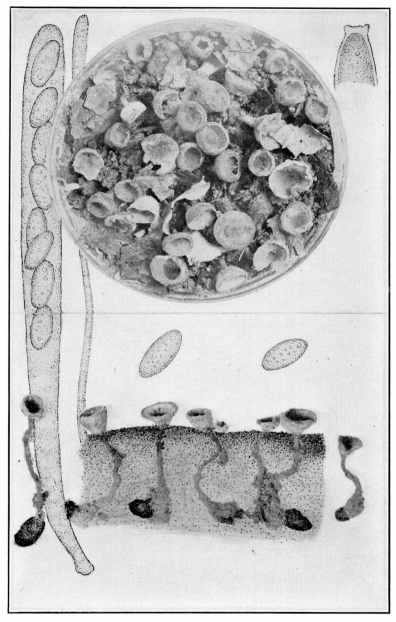

SCLEROTINIA TUBEROSA

Apothecia usually but one from each sclerotium, long-stipitate, thin membranous, shallow, cup-shaped, subumbilicate, 4–5 μ in diameter, reddish-brown; stem relatively long, reaching a length of 1 cm. and a diameter of 1 mm. attached below by a spreading, dark hyphal mat; asci clavate-cylindric, reaching a length of 60 μ and a diameter of 4–5 μ, 8-spored; spores allantoid, slender, distinctly curved, hyaline, 1–1.5 × 8–9 μ; paraphyses not apparent.

In culms of *Juncus* (probably *J. effusus* var. *californica*).
TYPE LOCALITY: Falcon Valley, Washington State.
DISTRIBUTION: Known only from the type locality.

4. **Sclerotinia Caricis-ampullaceae** Nyberg, Mem. Soc. Fauna Fl. Fenn. **10**: 22. 1933. (PLATE 90.)

Sclerotia externally black (inside white), longitudinally sulcate, bowed or S-shaped, usually broader below and tapering above, the bow often variable, reaching a length of 1–10 cm. and a diameter of 1 mm. to 1 cm., each producing one to twenty apothecia.

Apothecia stipitate, often funnel-shaped at first, later cup-shaped, finally expanding and becoming nearly flat and umbilicate, reaching a diameter of 2–4 cm., externally finely tomentose, some shade of brown; hymenium concave to plane, brown; stems 1–2 mm. thick and reaching a length of 5–10 cm., gradually expanding above into the cup, darker below, almost black; asci cylindric, rounded above, reaching a length of 200–300 μ and a diameter of 11 μ; spores 1-seriate, ellipsoid, smooth, without oil-drops, 6–9 × 11–15 μ; paraphyses thread-like, branching below, 1–2 μ thick, scarcely enlarged above.

In culms of *Carex aquatilis*, *C. ampullacea* (*C. rostrata*) and *C. inflata*.

TYPE LOCALITY: Europe.
DISTRIBUTION: Medicine Bow Mountains, Wyoming; also in Europe.
ILLUSTRATIONS: Mycologia **35**: 387, *f. 1;* 393, *f. 2–3;* 395, *f. 4–6;* Farlowia **2**: 406, *pl. 6,f. 23–25.*

5. **Sclerotinia paludosa** Cash & Davidson, Mycologia **25**: 271. 1933.

Sclerotia on leaves of host, often on edges, inconspicuous, 3–1 mm. in diameter, sometimes elongated up to 1.7 mm., black.

Apothecia brown, small, cupulate, then becoming almost plane, reaching a diameter of 1.2–2 mm.; stem reaching a length of 2–4 mm. and a thickness of .3 mm.; asci cylindric, attenuated at the base, with a short, stem-like base, 8-spored; spores 1–2-seriate, simple, containing two oil-drops, 5×12–$14\ \mu$; paraphyses numerous, filiform, septate, agglutinated and brown at the tips, 1–4 μ thick.

On overwintering leaves of *Carex exsiccata*.

TYPE LOCALITY: Grand Mesa, Colorado.

DISTRIBUTION: Grand Mesa National Park, Colorado.

ILLUSTRATIONS: Mycologia 25: *pl. 37, f. 6–7*.

6. Sclerotinia longisclerotialis Whetzel, Mycologia 21: 28.   1929.

Sclerotia usually one or two or sometimes three in each culm, remaining enclosed even at maturity, not discharging by the breaking of the culm as in *Sclerotinia Duriaeana*, somewhat exposed by a narrow slit in the epidermis, at maturity black, oblong, truncate, uniform in thickness, more or less 3-angled, striate due to the pressure of the vascular strands of the culm during development, originating as a loose, cottony weft of white mycelial hyphae in the interior of the culm, gradually thickening, forming a firm body at first of a pinkish color, gradually changing to a smoky-gray on the exterior, finally black, the ends capped by a long, pointed weft of white mycelium continuous with the mycelium in the tissues of the culm, at maturity varying in size with the slenderness of the host culm, reaching a length of 1–7 cm. and a diamater of 1–2 mm.

Apothecia 2–5 mm. broad, goblet-shaped with mouth constricted, stipitate, dark fawn-colored, usually one to each sclerotium; stem slender, reaching a length of 15–30 mm.; asci cylindric, attenuated below, reaching a length of 170–230 μ and a diameter of 9–15 μ, 8-spored; spores strongly unequilateral, flat, or incurved on one side, averaging $8 \times 18\ \mu$; paraphyses simple, very slender, slightly clavate.

On species of *Carex*, *C. prairea*, *C. interior*, *C. crinita*, *C. vesicaria* and *C. retrorsa*.

TYPE LOCALITY: New York.

DISTRIBUTION: New York and Maine.

ILLUSTRATIONS: Mycologia 21: *pl. 4, pl. 5, f. 20–22;* Farlowia 2³: *pl. 5, f. 19–22.*

SCLEROTINIA CARICIS-AMPULLACEAE

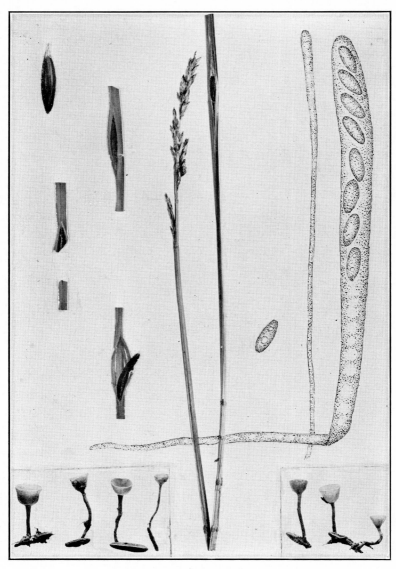

SCLEROTINIA DURIAEANA

7. **Sclerotinia Duriaeana** (Tul.) Rehm, Hedwigia **21**: 66. 1882. (PLATE 92.)

Sclerotium sulcatum Roberge; Desm. Ann. Sci. Nat. III. **16**: 329. 1851.
Epidochium ambiens Desm. Ann. Sci. Nat. III. **20**: 231. 1853.
Epidochium affine Desm. Ann. Sci. Nat. III. **20**: 232. 1853. (in part).
Peziza Duriaeana Tul. Fung. Carp. **1**: 103. 1861.
Sphacelia ambiens Sacc. Syll. Fung. **4**: 666. 1886.
Hymenoscypha Duriaeana Phill. Brit. Discom. 115. 1887.

Sclerotia one to three in each culm, when mature exposed by the rupture of the epidermis along one or two faces of the culm which thus weakened soon breaks over widening the slit and allowing the sclerotia to fall, black, fusiform, much resembling ergot, the sclerotia of *Claviceps*, often inequilateral, or slightly curved in larger specimens, of variable size according to the host, reaching a length of 20–25 mm. and a diameter of 3–4 mm. but often smaller, 3-angled.

Apothecia stipitate, reaching a diameter of 2–10 mm., deep cup-shaped when young, expanded when mature, fawn-colored, varying in size with the sclerotia from which they arise; stem short, 5–20 mm. long; asci slightly attenuated below, reaching a length of 150–180 μ and a diameter of 7–11 μ, 8-spored; spores 1-seriate, long-ovoid, often inequilateral, 5–9 $\times$ 8–18 μ; paraphyses slender, slightly swollen at their apices.

On *Carex stricta, Carex prairea, Carex hystericina, Carex riparia* var. *lacustris. Carex flava, Carex rostrata, Carex crinita,* and *Carex nebraskensis.*

TYPE LOCALITY: Ithaca, New York.

DISTRIBUTION: New York and Oregon; also known in Europe.

ILLUSTRATIONS: Mycologia **21**: 21, *f. 1; pl. 2, 3, 5, f. 14, 15, 16;* Farlowia **2**³: *pl. 1, 2, 3, 4* (as *S. sulcata*).

Whetzel (Farlowia **2**: 397) treats *S. sulcata* as distinct from *S. Duriaeana*.

8. **Sclerotinia Tiliae** Reade, Ann. Myc. **6**: 114. 1908.

Apothecia mostly solitary, cyathoid, long-stipitate, .5–1 cm. high, Isabela color, 1–3 mm. in diameter; stem smooth, slender, cylindric, .5 mm. or less thick; asci clavate-cylindric, reaching a length of 140–170 μ and a diameter of 8–10 μ, 8-spored; spores obliquely 1-seriate, ellipsoid, hyaline, simple, 4–5 $\times$ 9–11 μ; paraphyses sparse, filiform, hyaline.

From sclerotia in seeds of *Tilia americana* lying on the ground.

TYPE LOCALITY: Mt. Pleasant, Iowa.

DISTRIBUTION: Known only from the type locality.

This species was collected in the spring of 1906. So far as we know it has not been reported from any other place.

9. **Sclerotinia nyssaegena** (Ellis) Rehm, Ann. Myc. **4**: 338. 1906.

Peziza nyssaegena Ellis, Bull. Torrey Club **8**: 73.   1881.
Ciboria nyssaegena Sacc. Syll. Fung. **8**: 207.   1889.
Helotium nyssaegenum Sacc. Syll. Fung. **8**: 207.   1889.   (as synonym)

Apothecia arising singly, gregarious, or cespitose, concave, becoming nearly plane, pale-yellow to yellowish-brown, reaching a diameter of 3–5 mm.; hymenium concave, or becoming nearly plane, similar in color on the outside of the apothecium; stem very variable in length but often reaching 1–2 cm., about 1 mm. in thickness, often branched, usually imbedded in soil; asci clavate, reaching a length 100–110 μ and a diameter of 6–8 μ at their apices; spores partially 2-seriate, or irregularly crowded, fusoid, narrowed toward the base, 3–4 × 12–14 μ; paraphyses slender, reaching a diameter of 1–2 μ.

On seeds of *Nyssa sylvatica*, buried under dead leaves.

TYPE LOCALITY: Newfield, New Jersey.

DISTRIBUTION: New Jersey, Pennsylvania and New York.

EXSICCATI: Ellis, N. Am. Fungi *389*.

A small collection of this species was obtained by the author in The New York Botanical Garden, September, 1930, growing with *Helotium nyssicola* from which it differs in size, color and in its spore measurements.

10. **Sclerotinia Camelliae** Hansen & Thomas, Phytopathology **30**: 170.   1940.

Sclerotia dark-brown to black, usually compound, impregnating and surrounding the petal tissue, variable in form, reaching a size of 12 × 30 mm.; conidia none; microconidia globose to pyriform, 2.5–3.5 μ in diameter, produced in long chains on conidiophores which are produced in clusters.

Apothecia occurring singly or in groups, buff-olive, becoming darker with age, scantily pubescent, cyathiform to discoid, reaching a diameter of 5–20 mm.; stem 2–3 mm. thick and reaching an extreme length of 40 mm., tapering below; asci cylindric, reaching a length of 100–125 µ and a diameter of 4–6 µ, 8-spored; spores 1-seriate, ellipsoid, simple hyaline, 2.5–3.5 × 5–7 µ; paraphyses filiform, 1.2–2.5 µ thick, slightly swollen above.

On diseased flowers of *Camellia japonica*.

TYPE LOCALITY: Hayward, California.

DISTRIBUTION: California and Ontario, Canada.

ILLUSTRATIONS: Phytopathology **30**: 167, *f. 1;* 169, *f. A–D*.

11. Sclerotinia fallax (Sacc.?) Cash & Davidson, Mycologia **25**: 270. 1933.

?Sclerotium fallax Sacc. Nuovo Giorn. Bot. Ital. **23**: 197. 1916.

Sclerotia on leaf, reaching a length of .3–2 mm., a width of 1 mm. and a thickness of .5 mm., falling out and leaving holes in the leaf.

Apothecia small, patellate, 1–1.5 mm. in diameter, pale-brown, long-stipitate; stem reaching a length of 1.4 cm.; asci cylindric, attenuated at the base, reaching a length of 55 µ and a diameter of 5–6 µ, 4-spored; spores 1-seriate in the upper half of the ascus, narrowed at the lower end, 3–4 × 9–12 µ; paraphyses filiform, septate, simple, 2 µ thick.

On *Potentilla* sp.

TYPE LOCALITY: Mesa Lakes, Colorado.

DISTRIBUTION: Known only from the type locality.

12. Sclerotinia minor Jagger, Jour. Agr. Res. **20**: 333. 1920.

Sclerotia formed on the recently killed leaves of the host, small, .5–2 mm. in diameter; microconidia present, 3–4 µ in diameter.

Apothecia occurring singly, or rarely more than one to each sclerotium, stipitate, .5–2 mm. in diameter; stem cylindric, slender, flexuous, attenuated downward, .5–12 mm. long; asci cylindric, or subcylindric, reaching a length of 125–175 µ and a diameter of 8–11 µ, 8-spored; spores ellipsoid to ovoid, 6–8 × 12–16 µ; paraphyses filiform, enlarged above, 3–4 µ in diameter.

Parasitic on lettuce, *Lactuca*, celery, *Apium* and other plants.

TYPE LOCALITY: New York .

DISTRIBUTION: New York, Massachusetts, Pennsylvania, and Florida.

ILLUSTRATIONS: Jour. Agr. Res. **20**: 332, *f. 1, pl. 59, f. C*.

13. **Sclerotinia Veratri** Cash & Davidson, Mycologia **25**: 267. 1933.

Sclerotia flat, elliptic, or irregularly elongate, embedded in stems and when infection is severe, diffused over considerable area, dark reddish-brown to black, white within, reaching a length of 3–7 mm. and a width of 1–3 mm. and a thickness of 1 mm.

Apothecia reddish-brown, cupulate at first with inrolled margins, becoming almost flat, usually reaching a diameter of 3–7 mm., or occasionally as large as 1 cm., wrinkled when dry; hymenium brown, becoming lighter, grayish-brown at maturity; stem black, swollen toward the base, .5–1 cm. long and .5–1 mm. thick; asci cylindric, attenuated near the base, reaching a length of 140–150 μ and a diameter of 11–13 μ, 8-spored; spores irregularly 1-seriate, oblong-elliptic, simple, hyaline, usually with two oil-drops, 5–6.5 × 15–17.6 μ; paraphyses filiform, simple or branched near the base, pale-brown, 2–2.5 μ thick at the apex.

On *Veratrum californicum*.

TYPE LOCALITY: Mesa Lakes Reservoir, Colorado.

DISTRIBUTION: Known only from the type locality.

ILLUSTRATIONS: Mycologia **25**: *pl. 36, f. 4.*

14. **Sclerotinia coloradensis** Cash & Davidson, Mycologia **25**: 268.   1933.

Sclerotia on stems and seed-pods, thin, flat, elongated, sometimes confluent, black, white within, .2–3 cm. long and 2–5 mm. broad, .5 mm. thick, inconspicuous on weathered material.

Apothecia one to several from each sclerotium, cup-shaped, becoming flat, margin inrolled when dry, pale-brown, reaching a diameter of 2–3.5 mm.; hymenium pale-brown; stem brown, reaching a length of 4–7 mm. and a diameter of .5 mm.; asci cylindric, short-stipitate, reaching a length of 135–155 and a diameter of 7.5–9.5 μ, 8-spored; spores ellipsoid, hyaline, simple, 4–5 × 10–12 μ; paraphyses filiform, septate, simple, 2.5 μ in diameter at their apices.

On *Pedicularis groenlandica* and *Pedicularis bracteosa*.

TYPE LOCALITY: Grand Mesa, Colorado.

DISTRIBUTION: Known only from the type locality.

ILLUSTRATIONS: Mycologia **25**: *pl. 38, f. 12.*

15. **Sclerotinia intermedia** Ramsey, Phytopathology **14**: 323. 1924.

Sclerotia black, irregular, 1–3 mm. in diameter, often joined together in long chains; microconidia globose, hyaline, 3.8 μ in diameter formed on short, flask-shaped sterigmata.

Apothecia funnel-shaped to discoid, one to several from a single sclerotium, stipitate, reaching a diameter of 6 mm.; stem 7–12 mm. long, 1–1.25 mm. in diameter, expanding upward into the apothecium; hymenium tawny-olive at first, changing to snuff-brown at maturity, buff after the ejection of the spores; asci cylindric to cylindric-clavate, reaching a length of 130 μ and a diameter of 7–8 μ; spores ellipsoid to ovoid, with one oil-drop, hyaline, 4–6 $\times$ 10–15 μ; paraphyses filiform, simple.

On roots of salsify (*Tragopogon porrifolius*) and carrot (*Daucus Carota*) from the markets.

TYPE LOCALITY: Chicago, Illinois.

DISTRIBUTION: Known only from the type locality.

ILLUSTRATIONS: Phytopathology **14**: 324, *f. 1; 325, f. 4–6.*

16. **Sclerotinia Panacis** Rankin, Phytopathology **2**: 30. 1912.

Sclerotia large, .3–1 cm. in diameter, irregularly depressed-globose, solitary, or aggregated, black; conidia in potato-agar cultures, globose, small, 3–5.5 μ in diameter, borne on verticillate-branching conidiophores.

Apothecia scattered, or gregarious and sometimes cespitose. fleshy to subcoriaceous, closed at first, then expanding, finally flat with a distinct depression in the center from which radiate folds in the hymenium, more or less irregular, reaching a diameter of 1.5–2.5 cm., brown, stipitate; stem smooth, variable in length, tortuous, 2–3 mm. thick; asci narrowly cylindric, reaching a length of 125–137 μ and a diameter of 6.5 μ; spores obliquely 1-seriate, hyaline, with two oil-drops, simple, smooth, ellipsoid, 5–7 $\times$ 11–16 μ; paraphyses scarce, slightly thickened above.

On sclerotia of "black-rotted" tubers of cultivated ginseng, *Panax quinquefolium.*

TYPE LOCALITY: Labrador Lake, New York.

DISTRIBUTION: Known only from the type locality.

ILLUSTRATIONS: Phytopathology **2**: 29, *f. a–c, pl. 3.*

17. **Sclerotinia sativa** Drayton & Groves, Mycologia **35**: 526. 1943.

Sclerotia black, irregular in shape, usually more or less circular to elongated, 1–5 mm. in length or rarely longer and 1–2 mm. thick, frequently becoming laterally fused and forming irregular crusts not adhering closely to the substratum.

Apothecia appearing singly, or in clusters, stipitate, reaching a diameter of 1–5 mm., shallow cup-shaped to almost plane, or finally convex, slightly hairy to almost glabrous; hymenium concave to plane or convex, brown; stem reaching a length of 2–6 mm., tapering downward less than 1 mm. thick, dark-brown to almost black at the base; asci cylindric with a slender tapering base, reaching a length of 100–130 μ and a diameter of 7.5–10 μ; spores ellipsoid to ovoid, hyaline, 1-seriate, 4–7 $\times$ 9–12 μ; paraphyses filiform, septate, simple or branched, 2.5–3.5 μ thick, the tips enlarged to 4–5 μ; spermatia present.

On bulbs of *Tulipa* and roots of *Medicago sativa*, *Melilotus alba*, and *Melilotus officinalis*.

TYPE LOCALITY: New York.

DISTRIBUTION: New York to Quebec and Alberta, Canada.

ILLUSTRATIONS: Mycologia **35**: 520, *f. 1, 7;* 525, *f. 8.*

DOUBTFUL SPECIES

Sclerotinia incondita (Ellis) Sacc. Syll. Fung. **8**: 200.   1889; *Peziza incondita* Ellis, Bull. Torrey Club **8**: 73.   1881.   Apothecia occurring on sclerotia the origin of which was undetermined; asci 4-spored; spores 7–8 $\times$ 10–12 μ. Distributed by Ellis (N. Am. Fungi *391*).   Nothing more is known of this species.

Sclerotinia wisconsinensis Rehm, Ann. Myc. **6**: 317.   1908.   Occurring on unattached sclerotia in the soil in Wisconsin.   This appears to be close to *Sclerotinia tuberosa*.

12. **CIBORIA** Fuckel, Symb. Myc. 311.   1870.

Stroma a sclerotium, mummiform, occurring in the flowers or fruits of the host which retain much their original form scarcely suggesting a sclerotium; spermidium often present, manteling the developing sclerotium; spermatia globose, or ovoid, hyaline, in mass faintly yellowish; conidia unknown.

Apothecia cupulate to shallow saucer-shaped, often becoming flat, expanded or even reflexed, usually some shade of brown, rarely red, yellow, or whitish, medium sized; asci cylindric or

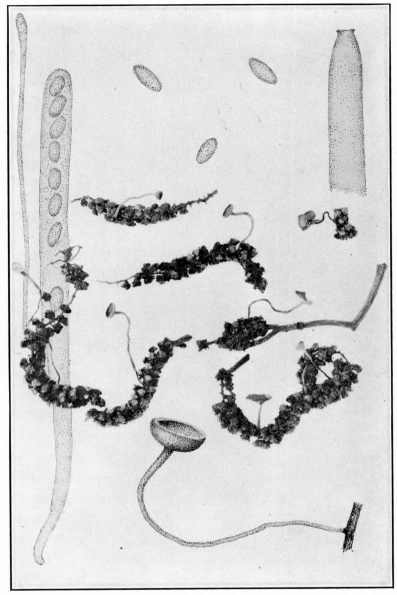

CIBORIA AMENTACEA

subcylindric, usually 8-spored, rarely 4-spored; spores ellipsoid, or subellipsoid, simple, hyaline, smooth, or minutely roughened; paraphyses, hyaline, or colored, filiform, slightly thickened above.

Type species, *Peziza Caucus* Rebentich.

On mummified flowers.
  On overwintering flowers of *Acer*.   1. *C. acerina.*
  On catkins of *Salix*, and *Populus*.
    Apothecia minute, 1 mm. in diameter, whitish.   2. *C. amenti.*
    Apothecia medium, 2–8 mm. in diameter, brown.
      Spores 4–5 × 9–13 μ.   3. *C. amentacea.*
      Spores 5–6 × 9–10 μ.   4. *C. Caucus.*
On fruits.
  On mulberry fruits, *Morus*.   5. *C. carunculoides.*
  On old acorns, *Quercus*.   6. *C. pseudotuberosa.*
  On mummified apples and other fruits.   7. *C. aestivalis.*
  On involucres of *Castanea*.   8. *C. americana.*

1. **Ciboria acerina** Whetzel & Buchw. Mycologia **28**: 516. 1936.

Apothecia stipitate, from one to several from the black stromatized male (rarely the female) inflorescence of the host, small, 1–4 mm. in diameter, funnel-shaped to shallow cup-shaped, finally expanded and subdiscoid; stem variable in length, 1–10 mm. long, somewhat darker than the apothecium, slightly pubescent when young; asci clavate, reaching a length of 75–100 μ and a diameter of 7–9 μ, 4-spored; spores ellipsoid, slightly flattened on one side, 1-seriate, hyaline, smooth; paraphyses unbranched, slender, gradually enlarged above.

On overwintering male and female flowers of *Acer rubrum* and on male flowers of *Acer saccharinum*, *Myrica Gale*, *Salix discolor* and *Ostrya virginiana*.

TYPE LOCALITY: Cornell University Campus, Ithaca, New York.

DISTRIBUTION: New York and New Jersey.

ILLUSTRATIONS: Mycologia **28**: 517, *f. 7–12;* 521, *f. 13–19.*

2. **Ciboria amenti** (Batsch) Whetzel, Mycologia **37**: 675. 1945.

Peziza amenti Batsch, Elench. Fung. Cont. **1**: 211. 1786.
Helotium amenti Fuckel, Symb. Myc. 313. 1870.
Phialea amenti Quél. Bull. Soc. Bot. Fr. **26**: 234. 1879.
Hymenoscypha amenti Phill. Brit. Discom. 143. 1887.

Apothecia scattered, stipitate, whitish, or grayish, reaching a diameter of nearly 1 mm.; hymenium pallid, concave, or nearly plane; stem slender, reaching a length of 5 mm., similar in color to the apothecium; asci cylindric-clavate, reaching a length of 60–70 μ and a diameter of 6–7 μ, 8-spored; spores ellipsoid, or ovoid, 3–4 × 7–10 μ; paraphyses filiform, reaching a diameter of 3 μ.

On fallen catkins of *Salix* and *Populus*.

TYPE LOCALITY: Europe.

DISTRIBUTION: New York and Oregon; also in Europe.

ILLUSTRATIONS: Batsch, Elench. Fung. Cont. **1**: *pl. 27, f. 148;* Boud. Ic. Myc. *pl. 496;* Gill. Champ. Fr. Discom. *pl. 90, f. 1.*

3. **Ciboria amentacea** (Balbis) Fuckel, Symb. Myc. 311.   1870. (PLATE 93.)

Peziza amentacea Balbis, Mem. Acad. Sci. Turin II. **2**: 79.   1805.
Rutstroemia amentacea Karst. Myc. Fenn. **1**: 106,   1871.
Hymenoscypha amentacea Phill. Brit. Discom. 120.   1887.

Apothecia springing from the sclerotium formed in the mummified catkins of the host, gregarious, stipitate, yellowish-brown, shallow cup-shaped, then expanded, reaching a diameter 1 cm. or rarely larger; stem variable in length, often reaching a length of 2–4 cm.; asci cylindric, or subcylindric, reaching a length of 110–120 μ and a diameter of 7–10 μ, 8-spored; spores 1-seriate, ellipsoid, or subellipsoid, 4–5 × 9–13 μ; paraphyses filiform, 1–2 mm. in diameter, slightly enlarged above.

On catkins of *Alnus incana* and *Salix discolor*.

TYPE LOCALITY: Europe.

DISTRIBUTION: New York to Oregon.

ILLUSTRATIONS: Mycologia **37**: 677, *f. 11–13;* Rab. Krypt.-Fl. **1**³: 750, *f. 1–3;* Friesia **3**: 247, *f. 6;* 249, *f. 7;* 250, *f. 8.*

4. **Ciboria Caucus** (Reb.) Fuckel, Symb. Myc. 311.   1870.

Peziza Caucus Rebentisch, Prodr. Fl. Neomarch 386.   1804.
Phialea Caucus Gill. Champ. Fr. Discom. 110.   1882.
Hymenoscypha Caucus Phill. Brit. Discom. 120.   1887.

Apothecia stipitate, springing from a mummified sclerotium within the tissues of the host, cup-shaped with margin incurved, finally expanding and becoming saucer-shaped, or subdiscoid, 2–8 mm. in diameter, brown; stem 2–8 mm. long, 5 mm. thick; asci cylindric-clavate, reaching a length of 130 μ and a diameter of 9 μ, 8-spored; spores ellipsoid, or subellipsoid, 5–6 × 9–10 μ, hyaline; paraphyses filiform, slightly enlarged above.

On fallen catkins of *Populus tremuloides* and *Salix*.

TYPE LOCALITY: Europe.

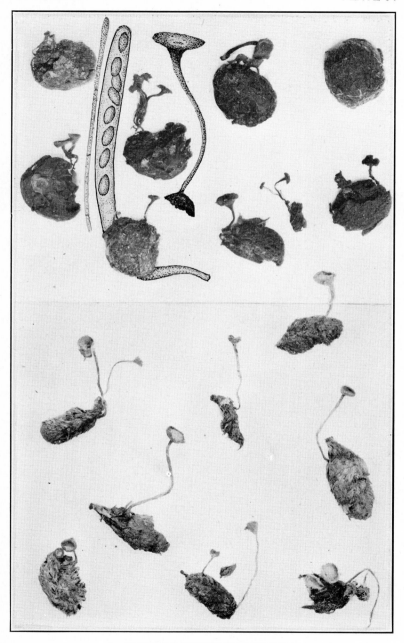

1. CIBORIA PSEUDOTUBEROSA
2. CIBORIA CAUCUS

DISTRIBUTION: New York to Oregon and Winnipeg, Canada; also Europe.

ILLUSTRATIONS: Mycologia **37**: 677, *f. 14, 15*.

This and the preceding species are very close both in habitat and in morphological characters. They should be studied more carefully from living material.

5. **Ciboria carunculoides** (Siegler & Jenkins) Whetzel & Wolf, Mycologia **37**: 676.   1945.   (PLATE 95.)

Sclerotinia carunculoides Siegler & Jenkins, Jour. Agr. Res. **23**: 835.   1923.

Sclerotia formed in the ovaries and calyx lobes of mulberry, preventing the formation of normal fruits, black, fairly regular, subspherical; microconidia produced in great numbers, subglobose 2–4 × 2–3.2 μ.

Apothecia one to several from a single sclerotium, cupulate, or subcupulate, 4–12 mm. in diameter, inside snuff-brown, outside Prout's brown, stipitate; stem cylindric, flexuous, smooth, attenuated upwards, 15–42 mm. long, reaching a diameter of 1.5 mm.; asci cylindric, or cylindric-clavate, reaching a length of 104–123 μ and a diameter of 6–8 μ. 8-spored, spores 1-seriate, reniform, hyaline, 2.4–4 × 6–9.6 μ, with two bodies on the concave surface, one more or less rhombic as seen from above about 2 × 4 μ and adjoining it a more or less hemispherical body 3 μ in its longest diameter; paraphyses filiform to cylindric-clavate.

On mulberry fruits (*Morus alba*).

TYPE LOCALITY: Scranton, South Carolina.

DISTRIBUTION: South Carolina; also reported from other southern states.

ILLUSTRATIONS: Jour. Agr. Res. **23**: 834, *f. 1, pl. 1, 2;* Mycologia **37**: 479, *f. 1;* 481, *f. 2;* 482, *f. 3;* 485, *f. 4.*

6. **Ciboria pseudotuberosa** Rehm, Ascom. *106*: 1872; Ber. Nat. Ver. Augsburg **26**: 28.   1881.

Hymenoscypha pseudotuberosa Phill. Brit. Discom. 119.   1887.
Sclerotinia pseudotuberosa Rehm in Rab. Krypt.-Fl. **1**[3]: 809.   1893.
Stromatinia pseudotuberosa Boud. Ic. Myc. 278.   1911.

Apothecia springing from a sclerotium-like growth within the seeds of the host, stipitate, at first closed, opening and becoming cup-shaped, later expanding and becoming nearly plane, or umbilicate, reaching a diameter of 1–2 cm. or rarely larger, brown or brownish; stem variable in length, often reaching 1–2 cm.;

asci cylindric or subcylindric, reaching a length of 120–150 μ and a diameter of 6–10 μ, 8-spored; spores ellipsoid, 5–6 $\times$ 8–10 μ; paraphyses slender, 3 μ in diameter, slightly enlarged above.

On old acorns, *Quercus*.

TYPE LOCALITY: Europe.

DISTRIBUTION: New Jersey to Iowa; also in Europe.

ILLUSTRATIONS: Rab. Krypt. Fl. 1³: 800, *f. 1–5;* Phill. Brit. Discom. *pl. 5; f. 25;* Boud. Ic. Myc. *pl. 480.*

EXSICCATI: Ellis, N. Am. Fungi *983* (as *Peziza pseudotuberosa* Rehm).

7. **Ciboria aestivalis** (Pollock) Whetzel; Harrison, Mycologia **27**: 317. 1935.

Sclerotinia aestivalis Pollock; Harrison, Mycologia **27**: 303. 1935.

Apothecia from one to fifty on a single mummified apple, stipitate, light reddish-brown, at first cup-shaped, discoid at maturity, 1–7 mm. in diameter, with radiating ridges and furrows outside; stem rarely reaching 1 cm. long, .5 mm. thick; asci clavate or nearly cylindric, reaching a length of 50–85 μ and a diameter of 6–8 μ, 8-spored; spores narrow ellipsoid, 2–3.5 $\times$ 6–12 μ.

On mummified fruits, especially apples, *Malus* sp. in the early spring.

TYPE LOCALITY: Ann Arbor, Michigan.

DISTRIBUTION: Michigan and Maryland?; also in Australia on various fruits.

ILLUSTRATIONS: Mycologia **27**: 310, *f. A;* 312, *f. B.*

8. **Ciboria americana** Durand, Bull. Torrey Club **29**: 461. 1902.

Rutstroemia americana White, Lloydia **4**: 188. 1941.

Apothecia solitary, stipitate, cup-shaped, usually becoming plain or with the margin reflexed and umbilicate, thin, waxy-membranous, pale-cinnamon to brown, 3–10 mm. in diameter; stem 2–10 mm. long, slender; asci clavate, usually curved, reaching a length of 75 μ and a diameter of 9 μ, 8-spored; spores 2-seriate, hyaline, smooth, ellipsoid to ovoid, slightly unsymmetrical, 4–5 $\times$ 9–12 μ; paraphyses filiform.

On stromatized involucres of *Castanea vesca* lying on the ground among leaves.

TYPE LOCALITY: Ithaca, New York.

DISTRIBUTION: New York and Pennsylvania to Ontario.

CIBORIA CARUNCULOIDES

Ciboria Liquidambaris Ellis & Ev. Proc. Acad. Sci. Phila. **1895**: 428. 1895. Apothecia stipitate, 2 mm. in diameter; hymenium convex, brownish; stem whitish, 4 mm. long; asci clavate, 7–8 × 75–80 μ; spores ellipsoid, 3 × 8–11 μ. On decaying capsules of *Liquidambar*, Naamans Creek, Delaware. Type material very scant.

Ciboria rufofusca (Weberb.) Sacc. Syll. **8**: 203. 1889; *Peziza rufofusca* Weberb. Pilze Norddeuts. 7. 1875. On scales of cones of *Abies*, Washington. Whether this species is a *Ciboria* in the sense here used is uncertain.

13. **STROMATINIA** Boud. Hist. Class. Discom. Eu. 108. 1907.

Stroma of two types, a thin, black, subcuticular sclerotium covering or manteling the affected portion of the host from which the apothecia are produced and small, black sphaerules borne free on the mycelium and not producing apothecia; spermatia globose.

Apothecia as in *Sclerotinia;* spores hyaline, simple.

Type species, *Peziza Rapulum* Bull.

On dead rhizomes of *Smilacina*.   1. *S. Smilacinae.*

On *Gladiolus* and other ornamentals.   2. *S. Gladioli.*

1. **Stromatinia Smilacinae** (Durand) Whetzel, Mycologia **37**: 674. 1945. (PLATE 96.)

Sclerotinia Smilacinae Durand, Bull. Torrey Club **29**: 462. 1902.

Sclerotia small, not exceeding 1–2 mm. in diameter, irregularly globose, aggregated and sometimes coalesced into a thin, crust-like mass 1–2 cm. in diameter.

Apothecia scattered, or gregarious, long-stipitate, fleshy-leathery, closed and subglobose at first, expanding to cup-shaped, finally becoming campanulate, usually with a depression in the center, sometimes contorted or irregular, reaching a diameter of 1–3 cm., bright cinnamon-brown, externally smooth; stem reaching a length of 2–6 cm. and a thickness of 2–3 mm., tapering downward, sometimes tomentose below; asci narrowly cylindric, apex rounded, reaching a length of 120–140 μ and a diameter of 6–8 μ, 8-spored; spores obliquely 1-seriate, hyaline, simple, smooth, narrowly ellipsoid, often with two oil-drops, 4–5 × 12–15 μ; paraphyses filiform, scarcely thickened above.

On dead rhizomes of *Smilacina racemosa* buried in rich humus.

TYPE LOCALITY: Fall Creek, Ithaca, New York.

DISTRIBUTION: New York.

ILLUSTRATIONS: Jour. Agr. Res. 5: *pl. 29.*

Dr. Edwin E. Honey (Mycologia **20**: 139.   1923.) states that it is believed by him and others that Durand misidentified the host of his fungus, and that his species is identical with *Stromatinia Rapulum* (Bull.) Boud. which occurs on *Polygonatum*.

2. **Stromatinia Gladioli** (Massey) Whetzel, Mycologia **37**: 674. 1945.

Sclerotium Gladioli Massey; Drayton, Phytopathology **18**: 521.   1928.
Sclerotinia Gladioli Drayton, Phytopathology **24**: 400.   1934.

Sclerotia black, 90–300 μ in diameter; microconidia globose, 1.2–1.8 μ in diameter, apparently functioning as spermatia.

Apothecia densely cespitose, stipitate, 3–7 mm. broad, 6–10 mm. high; hymenium umbilicate, convex-discoid, sometimes convolute, cinnamon-brown; stem chestnut-brown; asci cylindric to cylindric-clavate, reaching a length of 190–235 μ and a diameter of 8–10 μ, 8-spored; spores 1-seriate, ellipsoid, hyaline, usually with one oil-drop, 5.6–9.5 $\times$ 10–16 μ; paraphyses abundant, filiform, slightly clavate, 2.8–3.2 μ in diameter at their apices.

Parasitic on species of *Gladiolus*, *Freesia* and *Crocus*.

TYPE LOCALITY: Ithaca, New York.

DISTRIBUTION: United States and Canada; also in Europe and New Zealand.

ILLUSTRATIONS: Phytopathology **24**: 401, *f. 1;* 402, *f. 2;* 403, *f. 3.*

14. **COPROTINIA** Whetzel, Farlowia **1**: 484.   1944.

Stroma not observed in nature, of indefinite form, 1–2 mm. thick and several cm. in diameter, consisting of one to several layers of black hyphae; spermatia not seen; conidial stage wanting.

Apothecia gregarious, numerous, long, slender-stipitate, some shade of brown, relatively small; stem hair-like, minutely roughened by the ends of the hyphal tips; asci very small, clavate; spores 2-seriate, crowded near the end of the ascus, hyaline.

Type species, *Coprotinia minutula* Whetzel.

1. **Coprotinia minutula** Whetzel, Farlowia **1**: 484.   1944.

Stroma as above.

Apothecia arising in large numbers from all over the surface of the stroma, very long-stipitate, 1–2 mm. in diameter, chestnut

SCLEROTINIA SMILACINA

brown; stems very slender, hair-like, reaching a length of 20 mm. and one-fourth of a mm. thick; asci short, stout, clavate, reaching a length of 31–47 μ and a diameter of 3–5 μ, 8-spored; spores ellipsoid, 1.5–2 $\times$ 4–6 μ; paraphyses cylindric, 3–3.5 μ thick.

On a small dung ball of some unknown animal. Found only once.

TYPE LOCALITY: Malloryville, New York.

DISTRIBUTION: Known only from the type locality.

ILLUSTRATIONS: Farlowia **1**: 485, *f. 1–3;* 486, *f. 4–6.*

15. **MIDOTIS** Fries, Syst. Orbis Veg. 363.　1825; Elench. Fung. **2**: 29.　1828.

Rutstroemia Karst. Myc. Fenn. **1**: 105.　1871.
Ionomidotis Durand, Proc. Am. Acad. Sci. **59**: 8.　1923.

Apothecia superficial, solitary, or several springing from the same substratum, at first subglobose and closed, expanding and usually becoming vertically elongated on one side when mature, yellow, brown, dark-blue or olivaceous; the excipulum subcoriaceous, composed of interwoven hyphae which often project from the surface as short, stout, septate, hair-like hyphae, often violet with transmitted light, or when treated with K. O. H.; asci cylindric, or clavate, 8-spored; spores small, hyaline, simple; paraphyses filiform, or in one species lanceolate above.

Type species, *Midotis Lingua* Fries.

This genus corresponds with *Scodellina* or *Otidea* of the operculates on the one hand, and *Wynnea* and *Phillipsia* on the other, but differs from both in the inoperculate asci and the relatively minute size of its asci and spores. Saccardo combined *Wynnea* and *Midotis* although they are widely separated, the former being an operculate and the latter an inoperculate.

Rutstroemia Karst. was founded on *Peziza bulgarioides* Rab. which in the opinion of the writer is a *Midotis*. Karsten's genus contained several other species not now regarded as congeneric which has resulted in much confusion. We cannot accept White's version of the genus in which the last species mentioned by Karsten is adopted as the type.

Apothecia dark-blue, violet, or purple, nearly black in
　mass.
　Paraphyses with lance-like tips; apothecia large,
　　several cm. in diameter.　　　　　　　　1. *M. irregularis.*

Paraphyses filiform, without lance-like tips.
    Apothecia large 1–2 cm.; spores 5–7 × 15–18 μ.   2. *M. nicaraguensis.*
    Apothecia small 1–2 mm.; spores 2–3 × 5–8 μ.   3. *M. plicata.*
Apothecia not blue, violet, or purple.
  Apothecia olivaceous.
    Apotheica large, 1–3 cm.   4. *M. versiformis.*
    Apothecia small, not exceeding 5 mm. in diameter.
      Spores medium, 3 × 7–8 μ.   5. *M. olivascens.*
      Spores minute, 1 × 3–4 μ.   6. *M. floridana.*
  Apothecia yellow, reddish, or brownish.
    Apothecia externally bright-yellow; hymenium
      darker.   7. *M. heteromera.*
    Apothecia pale- to dark-brown.
      Occurring singly usually.   8. *M. occidentalis.*
      Occurring in dense cespitose clusters.   9. *M. Westii.*

1. **Midotis irregularis** Cooke; Sacc. Syll. Fung. **11**: 422.   1895.

Peziza irregularis Schw. Trans. Am. Phil. Soc. II. **4**: 171.   1832.
Cordierites irregularis Cooke, Bull. Buffalo Acad. Sci. **3**: 26.   1875.
Peziza doratophora Ellis & Ev. Jour. Myc. **1**: 90.   1885.
Otidea doratophora Sacc. Syll. Fung. **8**: 96.   1889.
Ionomidotis irregularis Durand, Proc. Am. Acad. Sci. **59**: 9.   1923.

Apothecia occurring singly, or more often in cespitose clusters from a common stem-like base up to 2 cm. in length which penetrates the substratum, at first closed then opening and becoming much elongated on one side and irregularly lobed or lacerated, reaching a length and width of 3 cm., the entire cluster reaching a diameter of 5–7 cm., externally scurfy, dark chestnut-brown; hymenium darker than the outside of the apothecium, almost black; asci cylindric-clavate, the apex truncate-rounded, reaching a length of 50–70 μ and a diameter of 4–5 μ, 8-spored; spores 1-seriate, or partially 2-seriate, hyaline, or subhyaline, smooth, containing two or three oil-drops, 3–4 × 8–10 μ; paraphyses with a violet tint, each terminated by a lanceolate, 1–3-septate head 3–4 × 18–30 μ which project beyond the hymenium and is easily detached.

On rotten wood and branches lying on the ground.

TYPE LOCALITY: Bethlehem, Pennsylvania.

DISTRIBUTION: New Hampshire to Oregon, Pennsylvania and Ohio.

ILLUSTRATIONS: Proc. Am. Acad. Sci. **59**: *pl. 1.*

A specimen in the herbarium from Ohio is labeled *Diplocarpa tinctoria* Massee & Morgan (type).   The writer could find no

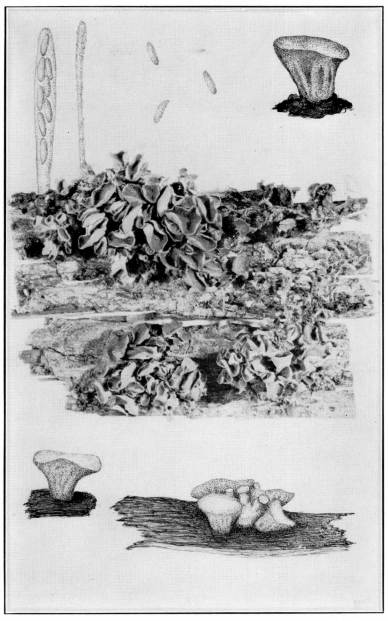

MIDOTIS WESTII

record of its publication but we record it here in order to call attention to the name.

2. **Midotia nicaraguensis** (Durand) Seaver, comb. nov.

Ionomidotis nicaraguensis Durand, Proc. Am. Acad. Sci. **59**: 11. 1923.
Ciboria caespitosa Seaver, Mycologia **17**: 48. 1925.
?Ionomidotis portoricensis Seaver, Mycologia **17**: 50. 1925.

Apothecia closely clustered, or coalesced at the base, at first closed, then opening and expanding so as to become applanate (when dry), sessile, attached at one side of the base, reaching a diameter of 1–2 cm., externally dark ferruginous-brown, verrucose; hymenium blackish-brown, becoming dark reddish-purple when wet; asci cylindric-clavate, the apex rounded, reaching a length of 135–150 and a diameter of 8 μ, 8-spored; spores 1-seriate, hyaline, smooth, 5–7 $\times$ 15–18 μ; paraphyses slender, hyaline, below, slightly thickened and brownish at their apices.

On rotten logs.

TYPE LOCALITY: Nicaragua.

DISTRIBUTION: Nicaragua and Porto Rico.

ILLUSTRATIONS: Proc. Am. Acad. Sci. **59**: *pl. 1*.

Our species from Porto Rico agrees so closely with Durand's description that we feel safe in combining them. *Ionomidotis portoricensis* is close to *Ciboria caespitosa* and may be identical. The material is scant.

3. **Midotis plicata** Phill. & Hark. Bull. Calif. Acad. Sci. **1**: 24. 1884.

Ionomidotis plicata Durand, Proc. Am. Acad. Sci. **59**: 12. 1923.

Apothecia small, clustered, or several arising from a common base, at first closed, then opening by a pore, becoming unsymmetrically goblet-shaped, resembling a minute *Urnula Craterium*, reaching a height and width of 1–2 mm., externally black when moist, brownish-black when dry, externally granular-furfuraceous, the substance thin, dark violet-brown and opaque under the microscope; asci cylindric-clavate, the apex rounded, reaching a length of 65–70 μ, 8-spored; spores irregularly 2-seriate, smooth, hyaline, narrowly ellipsoid, straight or strongly curved, 2–3 $\times$ 5–8 μ; paraphyses abundant and often forked, the contents pale violet-brown, the tips slightly swollen and darker.

On dead *Umbellularia californica*.

TYPE LOCALITY: Sausalito, California.
DISTRIBUTION: Known only from the type locality.
ILLUSTRATIONS: Proc. Am. Acad. Sci. **59**: *pl. 1.*

4. **Midotis versiformis** (Pers.) Seaver, comb. nov. (PLATE 99, FIG. 1.)

Peziza versiformis Pers. Ic. Descr. 25. 1798.
Chlorosplenium versiforme DeNot. Comm. Critt. Ital. **1**: 376. 1864.
Helotium versiforme Berk. Outl. Brit. Fung. 372. 1860.
Helotium rugipes Peck, Ann. Rep. N. Y. State Mus. **26**: 82. 1874.
Lanzia rugipes Sacc. Syll. Fung. **8**: 480. 1889.
Coryne versiformis Rehm in Rab. Krypt.-Fl. **1³**: 492. 1891.
Chlorociboria versiformis Seaver, Mycologia **28**: 393. 1936.

Apothecia short-stipitate, becoming expanded and sub-discoid, or more often elongated on one side, entirely light-green, or olivaceous, occasionally brownish or orange, reaching a diameter of 1–3 cm.; stem short, usually not exceeding 4 or 5 mm., rather stout; asci clavate, reaching a length of 100–110 μ and a diameter of 5–7 μ, 8-spored; spores irregularly long-ellipsoid, straight, or curved, 3–4 × 9–14 μ occasionally becoming 1-septate; paraphyses strongly enlarged above where they reach a diameter of 2–3 μ, containing a greenish coloring matter.

On decaying wood.

TYPE LOCALITY: Europe.

DISTRIBUTION: New York and Massachusetts to Iowa and south to Mexico.

ILLUSTRATIONS: Boud. Ic. Myc. *pl. 486;* Bull. Lab. Nat. Hist. State Univ. Iowa **6**: *pl. 24, f. 2;* Pers. Ic. Descr. *pl. 7, f. 7;* Berk. Outl. Brit. Fung. *pl. 2, f. 6;* Mycologia **28**: 392 (upper figure).

EXSICCATI: Ellis, N. Am. Fungi *988.*

5. **Midotis olivascens** (Durand) Seaver, comb. nov.

Ionomidotis olivascens Durand, Proc. Am. Acad. Sci. **59**: 13. 1923.

Apothecia stipitate, solitary, gregarious, or rarely cespitose, at first closed, then opening by a pore, finally becoming discoid, nearly plane with a slightly upturned margin, reaching a diameter of 1.5–3 mm., becoming unsymmetrical and *Otidea*-like, entirely blackish-olive, externally granular, or nearly smooth; stem slender, reaching a length of 1–2 mm. and a diameter of .5 mm. attached at one side; hymenium nearly plane with a yellowish tint; asci clavate, the apex rounded, reaching a length of 45–50 μ, 8-spored; spores 2-seriate, hyaline, smooth, straight, or slightly curved, narrow-ellipsoid, 3 × 7–8 μ; paraphyses filiform, scarcely thickened above.

On rotten wood.

TYPE LOCALITY: Coconut Grove, Florida.

DISTRIBUTION: Known only from the type locality.

ILLUSTRATIONS: Proc. Am. Acad. Sci. **59**: *pl. 1.*

6. **Midotis floridana** Seaver, sp. nov.

Apothecia occurring singly, or often several in cespitose clusters from a common base, stipitate, at first regular in form, soon becoming strongly one-sided, reaching a diameter of 3–4 mm., externally olive-green and pruinose; stem slender, gradually expanding into the apothecium, about 2 mm. long, more or less longitudinally striated; hymenium slightly darker than the outside of the apothecium and stem; asci clavate, reaching a length of 30–55 μ and a diameter of 4 μ, 8-spored; spores minute, baccilloid, often slightly curved, 1 × 4–6 μ; paraphyses filiform, about 1 μ in diameter.

Apotheciis sparsis aut caespitosis, stipitatis, difformis, 3–4 mm. diam., extus viridis; ascis clavatis, 8-sporis, 4 × 30–55 μ; sporis minutis, 1 × 4–6 μ; paraphysibus filiformibus, 1 μ diam.

On rotten wood.

TYPE LOCALITY: Gainesville, Florida.

DISTRIBUTION: Florida.

7. **Midotis heteromera** (Mont.) Fires, Summa Veg. Scand. 362. 1849.

Peziza heteromera Mont. Ann. Sci. Nat. II. **13**: 206. 1840.
Midotis verruculosa Berk. & Curt. Jour. Linn. Soc. **10**: 370. 1869.

Apothecia occurring singly, or several arising from a common, stem-like base, irregularly elongated on side, tough, corky to leathery when dry, reaching a length and width of 2–3 cm., externally verrucose, rhubarb-colored to citrine; hymenium reddish-brown; asci clavate-cylindric, the apex rounded, reaching a length of 80–90 μ and a diameter of 5 μ, 8-spored; spores 1-seriate, hyaline, smooth, ellipsoid to cymbiform, 3.5–4 × 8–10 μ; paraphyses slender projecting above the asci.

On rotten wood.

TYPE LOCALITY: South America.

DISTRIBUTION: West Indies and Costa Rica; also in South America.

ILLUSTRATION: Ann. Sci. Nat. II. **13**: *pl. 6, f. 3.*

8. **Midotis occidentalis** Durand, Proc. Am. Acad. Sci. **59**: 6. 1923.

Apothecia small, scattered, or closely gregarious, stipitate, coriaceous, elongated on one side and split to the stem on the other, reaching a length and width of 3–5 mm., externally blackish-brown, vertically striated; hymenium brownish; stem reaching a length of 3–7 mm. and a thickness of 1 mm., gradually expanding above into the apothecium; asci clavate-cylindric, apex rounded, reaching a length of 80 μ, 8-spored; spores 1-seriate, hyaline, smooth, ovoid-ellipsoid, 3 × 5–6 μ; paraphyses slender, hyaline, not thickened above, longer than the asci.

On a decaying log.

TYPE LOCALITY: Mabess River, Jamaica.

DISTRIBUTION: Known only from the type locality.

9. **Midotis Westii** Seaver, sp. nov.   (PLATE 97.)

Apothecia in dense, semiglobose clusters springing from a common base, the clusters reaching a diameter of 1–2 cm. and consisting of twelve to fifteen apothecia closely crowded together, individual apothecia subturbinate and vertically striated outside, the striations very prominent in dried specimens, the flesh thick and coriaceous when fresh, corky when dry, becoming more or less one sided, externally pale-brown, lighter below, reaching a diameter of 7 mm.; hymenium slightly concave, darker than the outside of the apothecium; asci clavate, reaching a length of 60 μ and a diameter of 5–6 μ, 8-spored; spores minute, about 2 × 4–5 μ; paraphyses slender, slightly enlarged above, not over 2 μ in diameter at their apices.

Apotheciis dense caespitosis, subturbinatis, extus verticalis striatis, 7 mm. diam., hymenio concavo, dilute bruneo; ascis clavatis, 8-sporis, 5–6 × 60 μ; sporis 2 × 4–5 μ; paraphysibus clavatis, 2 μ diam.

On wood of *Ostrya.*

TYPE LOCALITY: Sanchez Hammock, near Gainesville, Florida.

DISTRIBUTION: Known only from the type locality.

PODOPHACIDIUM XANTHOMELUM

Ionomidotis fulvotingens (Berk. & Curt.) Cash, Jour. Wash. Acad. Sci. **29**: 50. 1939; *Cenangium fulvotingens* Berk. & Curt. Grevillea **4**: 4. 1875. According to Miss Cash, this species belongs with the present genus. One specimen collected by L. O. Overholts (*16856*) in Pennsylvania has been referred to this species by him. No authentic material has been seen.

16. **PODOPHACIDIUM** Niessl in Rab. Fungi Eu. *1153*. 1868.

Melachroia Boud. Hist. Class. Discom. Eu. 96. 1907.

Apothecia contracted at the base, substipitate, obconic to turbinate, opening with a laciniate aperture; hymenium freely exposed at maturity and bright-colored, yellow-olivaceous; asci clavate, 8-spored; spores simple, hyaline.

Type species, *Podophacidium terrestre* Niessl.

Only one species of the genus known to the author.

1. **Podophacidium xanthomelum** (Pers.) Kavina, Crypt. Cech. Exsicc. *217*. 1936. (PLATE 98.)

Peziza zanthomela Pers. Syn. Fung. 665. 1801.
Peziza xanthomela Pers. Myc. Eu. **1**: 296. 1822.
Podophacidium terrestre Niessl in Rab. Fungi Eu. *1153*. 1868.
Aleuria xanthomela Gill. Fr. Champ. Discom. 207. 1886.
Humaria xanthomela Sacc. Syll. Fung. **8**: 128. 1889.
Urnula terrestris Sacc. Syll. Fung. **8**: 550. 1889.
Melachroia xanthomela Boud. Hist. Class. Discom. Eu. 97. 1907.
Melachroia terrestris Boud. Hist. Class. Discom. Eu. 97. 1907.

Apothecia thickly gregarious, occasionally a few closely crowded, the hymenium bright-yellow with a slightly olive tint, surrounded with a dark-brownish or almost black laciniate border, the outside of the apothecium dark-brownish or nearly black, reaching a diameter of 3–4 mm.; asci clavate, reaching a length of 90–125 μ and a diameter of 7–9 μ; spores ellipsoid, the ends slightly attenuated, usually with two oil drops, 5–6 $\times$ 10–17 μ; paraphyses very slender, branched.

On soil in coniferous woods.

TYPE LOCALITY: Europe.

DISTRIBUTION: Washington to northern New York, Toronto and Quebec; also in Europe.

ILLUSTRATIONS: Rab. Fungi Eu. *1153;* Niessl Beit. *pl. 7, f. 50;* Boud. Ic. Myc. *pl. 449* (as *Melachroia xanthomela* (Pers.) Boud.); Cooke, Mycographia, *pl. 11, f. 41* (as *Peziza xanthomela* Pers.), Papers Mich. Acad. Sci. **22**: *pl. 15, f. 2;* Mycologia **31**: 351, *f. 1.*

17. **CHLOROCIBORIA** Seaver, Mycologia **28**: 390. 1936.

Chlorosplenium DeNot. Comm. Critt. Ital. **1**: 376. 1864. Not Fries 1849.

Apothecia stipitate, or substipitate, often reaching a diameter of 1 cm., or rarely as large as 2 cm., the stem usually about half as long as the diameter of the apothecium, resembling *Ciboria*, color green, or olivaceous, staining the substratum green; asci usually 8-spored; spores irregularly ellipsoid to vermiform, simple, hyaline; paraphyses slender, clavate.

Type species, *Elvela aeruginosa* Oed.

The reasons for the establishment of this genus are given in detail in Mycologia (**28**: 390. 1936). This genus has some characters in common with *Midotis*, one species, *C. versiformis*, having been transferred to that genus.

Occurring on rotten wood, staining the wood verdigris
green.   1. *C. aeruginosa.*
Occurring on spruce cones, not staining the substratum.   2. *C. strobilina.*

1. **Chlorociboria aeruginosa** (Oed.) Seaver, Mycologia **28**: 391. 1936.

Elvela aeruginosa Oed. Fl. Dan. **9**: 7. 1770.
Peziza aeruginosa Pers. Obs. Myc. **1**: 27. 1796.
?Cantharellus viridis Schw. Trans. Am. Phil. Soc. II. **4**: 153. 1832.
?Peziza chlorascens Schw. Trans. Am. Phil. Soc. II. **4**: 175. 1832.
Helotium aeruginosum Fries, Summa Veg. Scand. 355. 1849.
Chlorosplenium aeruginosum DeNot. Comm. Critt. Ital. **1**: 376. 1864.
Peziza aeruginescens Nyl. Not. Fauna Fl. Fenn. **10**: 42. 1869.
Chlorosplenium aeruginescens Karst. Myc. Fenn. **1**: 103. 1871.
Chlorosplenium viride Morgan, Jour. Myc. **8**: 185. 1902.
Chlorociboria aeruginescens Kanouse, Mycologia **39**: 641. 1947.

Apothecia gregarious, stipitate, or subsessile, at first cup-shaped, becoming expanded and subdiscoid with the margin slightly elevated, verdigris-green and producing a similar color in the wood on which it grows, reaching a diameter of 5 mm.; hymenium plane or slightly concave, similar in color to the outside of the apothecium, or lighter and sometimes yellowish; stem darker, scarcely exceeding in length one-half the diameter of the apothecium and about 1 mm. thick; asci clavate, reaching a length of 45–50 μ and a diameter of 3–4 μ; spores 2-seriate, or irregularly crowded, narrow ellipsoid, 2–2.5 $\times$ 5–7 μ (rarely 10–12); paraphyses very slender, about 1.5 μ in diameter, scarcely enlarged above.

On dead wood.

1. MIDOTIS VERSIFORMIS
2. CHLOROCIBORIA STROBILINA

TYPE LOCALITY: Europe.

DISTRIBUTION: New York to Colorado, south to Mexico and the West Indies; also in South America, Europe, Asia and Australia.

ILLUSTRATIONS: Boud. Ic. Myc. *pl. 485;* Fl. Dan. *pl. 534, f. 2;* Cooke, Austr. Fungi *pl. 20, f. 158;* Gill. Champ. Fr. Discom. *pl. 88, f. 1;* Grev. Scot. Crypt. Fl. *pl. 241;* Bull. Lab. Nat. Hist. State Univ. Iowa **6**: *pl. 24, f. 1;* Phill. Brit. Discom. *pl. 5, f. 28;* Rab. Krypt.-Fl. **1**³: 749, *f. 1–5;* E. & P. Nat. Pfl. **1**¹: *f. 155, H–L;* Massee, Brit. Fungus-Fl. **4**: 156, *f. 41, 42;* Sow. Engl. Fungi *pl. 347.*

EXSICCATI: Ellis, N. Am. Fungi *987;* Ellis & Ev. N. Am. Fungi *2047;* Rav. Fungi Car. **5**: *40.*

Some European authors recognize *Chlorosplenium aeruginescens* as distinct from *Chlorosplenium aeruginosum.* The writer has been unable to detect any difference of specific importance in the material which he has examined.

2. **Chlorociboria strobilina** (Alb. & Schw.) Seaver, comb. nov. (PLATE 99, FIG. 2.)

Peziza tuberosa strobilina Alb. & Schw. Consp. Fung. 313. 1805.
Peziza versiformis livida Alb. & Schw. Consp. Fung. 314. 1805.
Peziza Abietis strobilina Alb. & Schw. Consp. Fung. 342. 1805.
Chlorosplenium versiforme nigrescente-olivacea Weinm. Hymeno-Gastro-Mycetes 467. 1836.
Pezia ciborioides strobilaria Nyl. Not. Fauna Fl. Fenn. **10**: 36. 1869.
Cenangium strobilinum Sacc. Fung. Ital. *pl. 1306.* 1883.
Chlorosplenium lividum Karst. Acta Soc. Fauna Fl. Fenn. II. **6**: 124. 1885.
Peziza bulgarioides Rehm in Rab. Fungi Eu. *1311;* Hedwigia **9**: 136. 1870.
Rustroemia bulgarioides Karst. Myc. Fenn. **1**: 105. 1871.
Ombrophila strobilina Rehm in Rab. Krypt.-Fl. **1**³: 482. 1891.
Ciboria strobilina Bresadolae Boud. Ic. Myc. **4**: 279. 1907.

Apothecia stipitate, or subsessile, at first concave, becoming expanded and subdiscoid or shallow cup-shaped, occasionally repand, reaching a diameter of 1 cm. or rarely larger, regular or slightly irregular in form, brownish-black or with a slightly olive tint; hymenium similar in color to the outside of the apothecium; stem short, scarcely exceeding one-half the diameter of the apothecium, slightly lacunose; asci clavate, reaching a length of 85 μ to 100 and a diameter of 5–7 μ; spores irregularly ellipsoid, 3–4 × 7 μ; paraphyses filiform.

On spruce cones, *Picea* sp.

TYPE LOCALITY: Europe.

DISTRIBUTION: Nova Scotia to Michigan; also in Europe.

ILLUSTRATIONS: Sacc. Fungi Ital. *pl. 1306;* Boud. Ic. Myc. *pl. 480* bis; Mycologia **28**: 392 (lower figure).

EXCLUDED SPECIES

Chlorosplenium striisporum Ellis & Dearn. See *Ascobolus striisporus* (Ellis & Dearn.) Seaver, N. Am. Cup-fungi Operculates 90. 1928.

Chlorosplenium epimyces Cooke. See *Ascobolus epimyces* (Cooke) Seaver, N. Am. Cup-fungi Operculates 91. 1928.

Chlorosplenium canadense Ellis & Ev. See *Holwaya gigantea.*

18. **KRIEGERIA** Rab. Hedwigia **17**: 32. 1878.

Chloroscypha Seaver, Mycologia **23**: 248. 1931.

Apothecia gregarious, or scattered, sessile, or stipitate, minute, or of medium size, yellowish-green to blackish when dry, the substance yellowish-green by transmitted light and resembling that of *Ascobolus*, occurring on the foliage of conifers, *Thuja, Sequoia, Libocedrus,* and *Juniperus* and apparently parasitic; asci when young greenish, normally 8-spored; spores comparatively large, at first granular and appearing greenish, but hyaline when mature, typically fusiform, or more rarely broad-ellipsoid, usually simple, rarely 1-septate; paraphyses slender, simple, or branched, surrounded by a greenish matrix.

Type species, *Ombrophila Kriegeriana* Rab.

Occurring on foliage of *Thuja.*
  Apothecia subsessile; spores broad-fusoid.   1. *K. Seaveri.*
  Apothecia stipitate; spores narrow-fusoid.   2. *K. enterochroma.*
Not on *Thuja.*
  On *Sequoia;* spores fusoid, 5 × 10 μ.   3. *K. chloromela.*
  On *Juniperus;* spores broad-ellipsoid or fusoid, 10 ×
      20 μ.
    On leaves; spores simple.   4. *K. juniperina.*
    On branches; spores often 1-septate.   5. *K. cedrina.*
  On *Libocedrus;* spores 4–5 × 18–19 μ.   6. *K. alutipes.*

1. **Kriegeria Seaveri** (Rehm) Seaver, Mycologia **35**: 493. 1943.
   (PLATE 100.)

?Ombrophila thujina Peck, Bull. N. Y. State Mus. **150**: 60. 1911.
Helotium Seaveri Rehm (in litt.). 1912.
Chloroscypha Seaveri Seaver, Mycologia **23**: 249. 1931.

Apothecia minute, scarcely exceeding .5 mm. in diameter, short-stipitate, occurring singly, or in small, cespitose clusters

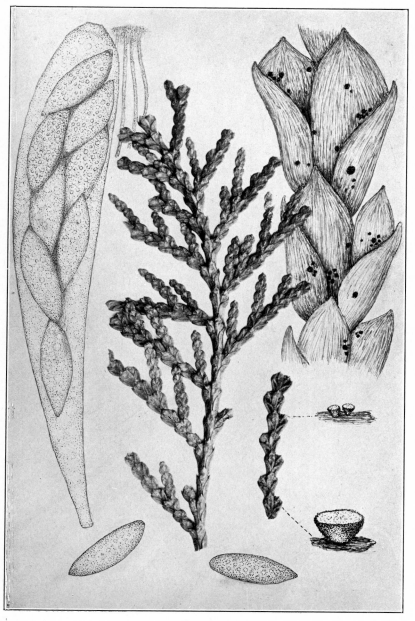

KRIEGERIA SEAVERI

from the leaves of the host, turbinate, greenish, becoming almost black in dried material; hymenium plane or nearly so, lighter than the outside of the apothecium, the substance of the apothecium pale olivaceous-green when teased out and viewed by transmitted light; asci clavate, reaching a length of 100–135 μ and a diameter of 25–30 μ, the contents greenish; spores irregularly 2–3-seriate, fusoid to fusiform, the lower end more pointed than the upper, densely filled with granules and slightly yellowish-green, simple, about 8–9 $\times$ 25–28 μ, smooth, or very minutely roughened; paraphyses filiform, scarcely enlarged above.

On foliage of white cedar, *Thuja plicata*.

TYPE LOCALITY: Libby, Montana.

DISTRIBUTION: Montana and (New York?).

ILLUSTRATIONS: Mycologia 23: *pl. 23*.

According to the late J. R. Weir this fungus is the cause of a very destructive blight. *Ombrophila thujina* Peck apparently differs only in the smaller size of the spores. The type has been examined.

2. **Kriegeria enterochroma** (Peck) Seaver, comb. nov. (PLATE 101.)

Peziza enterochroma Peck, Ann. Rep. N. Y. State Mus. **32**: 47. 1879.
Ombrophila enterochroma Sacc. Syll. Fung. **8**: 619. 1889.
?Helotium limonicolor Bres. Fungi Trid. **2**: 81. (1898?).
Chloroscypha Jacksoni Seaver, Mycologia **23**: 249. 1931.
Kriegeria Jacksoni Seaver, Mycologia **38**: 493. 1943.

Apothecia scattered, stipitate, at first closed, gradually opening and becoming shallow cup-shaped, then discoid, externally yellowish, becoming darker with age, reaching a diameter of 2 mm.; hymenium concave, plane, or slightly convex, yellowish with a greenish tint, often becoming nearly black with age; stem slender, reaching a length of 2 mm., similar in color to the outside of the apothecium; asci clavate, reaching a length of 100–130 μ and a diameter of 12–14 μ, 8-spored; spores irregularly 2-seriate, fusoid, or fusiform, often with two distinct oil-drops, or granular, apparently greenish when young, usually hyaline when mature, often minutely roughened or smooth, simple, 7–8 $\times$ 20–38 μ; paraphyses slender, rather abruptly enlarged above and surrounded with a greenish-yellow substance.

On *Thuja occidentalis*.

TYPE LOCALITY: Adirondack Mountains, New York.

DISTRIBUTION: New York to Ontario.

ILLUSTRATIONS: Mycologia **23**: *pl. 24;?* Bres. Fungi Trid. *pl. 195, f. 3;* Ic. Myc. *pl. 1235, f. 2.*

This species differs from the preceding which also occurs on *Thuja* in the much narrower spores and stipitate apothecia; whether this species is also parasitic has not been determined.

3. **Kriegeria chloromela** (Phill. & Hark.) Seaver, Mycologia **35**: 493. 1943.

?Peziza sphaerophoroides Phill. & Hark. Bull. Calif. Acad. Sci. **1**: 21.   1884.
Peziza chloromela Phill. & Hark. Grevillea **13**: 22.   1884.
?Helotium sphaerophoroides Sacc. Syll. Fung. **8**: 236.   1889.
Chlorosplenium chloromelum Sacc. Syll. Fung. **8**: 319.   1889.
Chloroscypha chloromela Seaver, Mycologia **35**: 493.   1943.

Apothecia scattered, or gregarious, short-stipitate, externally smooth, greenish-black, reaching a diameter of .6 mm.; hymenium becoming nearly plane, yellowish-green; stem reaching a length of 1 mm., a little paler than the outside of the apothecium; asci clavate-cylindric; spores clavate, or fusiform, usually curved, at first hyaline, becoming greenish, simple, 4–5 × 20–25 μ; paraphyses filiform, indistinct, adhering together.

On leaves of *Sequoia sempervirens.*

TYPE LOCALITY: California.

DISTRIBUTION: Known only from the type locality.

A note from the Royal Botanic Garden states that the material of *Peziza chloromela* at Kew is very scanty. Through the kindness of the Director of that institution a microscopic slide has been examined. The spores, as indicated in the description, are smaller than in the species on white cedar.

In 1932 specimens of what appears to be this, collected by H. E. Parks at Trinidad, California, were sent to the writer by Dr. Lee Bonar. This is the first material of this species seen. In 1935 it was collected by H. E. Parks in Humboldt Co., California.

4. **Kriegeria juniperina** (Ellis) Seaver, Mycologia **35**: 493.   1943.

Dermatea juniperina Ellis, Am. Nat. **17**: 192.   1883.
Chloroscypha juniperina Seaver, Mycologia **23**: 250.   1931.

Apothecia gregarious, at first rounded, expanding and becoming turbinate, tapering into a stem-like base, black to the unaided eye, greenish with transmitted light, reaching a diameter of .25 mm.; asci clavate, reaching a length of 130 μ and a diameter of 20 μ, tapering rather abruptly below; spores ellipsoid, or

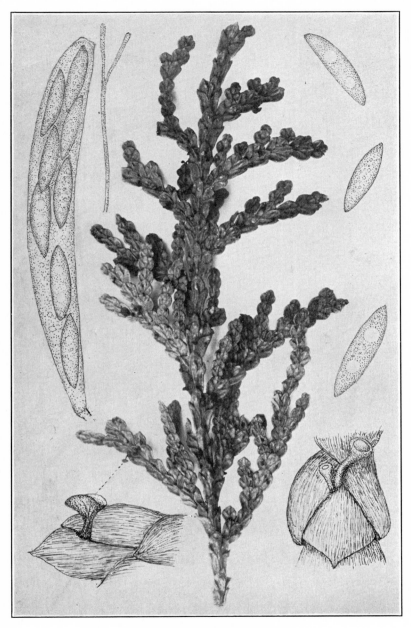

KRIEGERIA ENTEROCHROMA

fusoid, about 9–10 × 18–20 μ, granular within, often appearing greenish from the greenish material which surrounds the asci and paraphyses, simple; paraphyses slender, enlarged above, reaching a diameter of 4 μ, adhering together at their tips, yellowish-green.

On leaves of *Juniperus communis.*

TYPE LOCALITY: Decorah, Iowa.

DISTRIBUTION: Iowa and New Jersey.

This species is quite similar in general appearance to *Kriegeria Seaveri* but differs in the form and size of the spores as well as in the host.

5. **Kriegeria cedrina** (Cooke) Seaver, Mycologia **35**: 493. 1941. (PLATE 102.)

Peziza cedrina Cooke, Bull. Buffalo Soc. Nat. Sci. **2**: 294. 1875.
Lachnella cedrina Sacc. Syll. Fung. **8**: 395. 1889.
Chloroscypha cedrina Seaver, Mycologia **30**: 594. 1938.

Apothecia gregarious, sessile, reaching a diameter of 1 mm. and nearly as deep, black and vertically striated with dark mycelium; hymenium concave, dark; asci cylindric, or sub-clavate, reaching a length of 140–160 μ and a diameter of 12–14 μ, 8-spored but some of the spores often immature; spores usually 1-seriate, ellipsoid with the ends lightly attenuated, densely filled with granules and oil-drops, usually simple but occasionally becoming tardily 1-septate, 10 × 20 μ; paraphyses slender, about 3 μ in diameter below, strongly enlarged above where they reach a diameter of 5 μ, the ends strongly curved, becoming greenish-brown.

On branches of *Juniperus virginiana.*

TYPE LOCALITY: New York.

DISTRIBUTION: New York and North Carolina.

ILLUSTRATIONS: Mycologia **30**: 595, *f. 1.*

6. **Kriegeria alutipes** (Phill.) Seaver, comb. nov.

Peziza alutipes Phill. Grevillea **7**: 23. 1878.
Phialea alutipes Sacc. Syll. Fung. **8**: 266. 1889.

Apothecia gregarious, stipitate or subsessile, yellowish to pale-brown, darker, almost black in dried specimens, yellowish-green with transmitted light, 1–2 mm. in diameter; hymenium similar in color to the outside of the apothecium; stem very short, or 1–2 mm. long, gradually expanding into the apothecium; asci clavate, reaching a length of 90 μ and a diameter of 10–12 μ;

spores fusoid, 5–8 × 18–20 μ, multinucleate; paraphyses filiform, rather abruptly enlarged above, the ends agglutinated together and surrounded with a yellowish substance.

On dead foliage of *Libocedrus decurrens*.

TYPE LOCALITY: Blue Canon, California.

DISTRIBUTION: California and Oregon.

Authentic material collected by Dr. Harkness in California has been studied.

DOUBTFUL SPECIES

Rutstroemia elatina (Alb. & Schw.) Rehm in Rab. Krypt.-Fl. **1**³: 767. 1893; *Peziza elatina* Alb. & Schw. Consp. Fung. 330. *pl. 2, f. 3.* 1805; *Chlorosplenium elatinum* Sacc. Syll. Fung. **8**: 318. 1889. This species is regarded by White as synonymous with *Ombrophila Kriegeriana* on which this genus was founded. It has been doubtfully recorded from North America.

19. **CHLOROSPLENIUM** Fries, Summa Veg. Scand. 356. 1849.

Apothecia sessile, or short-stipitate, not exceeding 1–3 mm. in diameter with the margin upturned, yellowish; hymenium concave, becoming olivaceous green; spores simple, hyaline; paraphyses filiform, slightly clavate.

Type species, *Peziza chlora* Schw.

Apothecia yellowish.	1. *C. chlora.*
Apothecia sage-green or pea-green.	
Spores 1–1.25 × 3–4 μ.	2. *C. salviicolor.*
Spores 3–4 × 9–10 μ.	3. *C. olivaceum.*

1. **Chlorosplenium chlora** (Schw.) Massee, Jour. Linn. Soc. **35**: 116. 1901.

Peziza chlora Schw. Schr. Nat. Ges. Leipzig **1**: 122. 1822.
Chlorosplenium Schweinitzii Fries, Summa Veg. Scand. 356. 1849.
Peziza crocitincta Berk. & Curt.; Cooke & Peck, Grevillea **1**: 6. 1872.
?Peziza pomicolor Berk. & Rav.; Berk. Grevillea **3**: 157. 1875.
Pezizella crocitincta Sacc. Syll. Fung. **8**: 286. 1889.
?Pseudohelotium pomicolor Sacc. Syll. Fung. **8**: 300. 1889.

Apothecia thickly gregarious, sessile, or subsessile, at first closed and subglobose, becoming shallow cup-shaped with the margin slightly incurved, externally slightly roughened but not hairy, bright-yellow, reaching a diameter of 1–2 mm.; hymenium concave, or nearly plane, at first yellow, gradually becoming green; asci clavate, reaching a length of 40–50 μ and a diameter of 6–8 μ; spores 2-seriate above, narrow-ellipsoid, or fusoid, 2 × 8 μ; paraphyses slender, reaching a diameter of 2 μ.

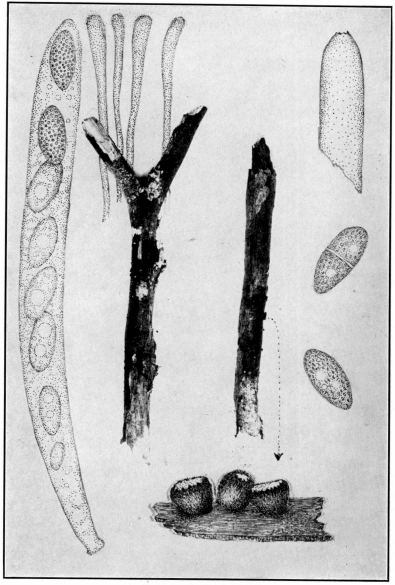

KRIEGERIA CEDRINA

On rotten wood.

TYPE LOCALITY: North Carolina.

DISTRIBUTION: New York and Connecticut to Alabama and North Dakota.

ILLUSTRATIONS: Grevillea 1: *pl. 1, f. 5;* Bull. Lab. Nat. Hist. State Univ. Iowa 6: *pl. 23, f. 3.*

EXSICCATI: Ellis, N. Am. Fungi *664;* Ellis & Ev. Fungi Columb. *249;* Rav. Fungi Car. 5: *39;* Wilson & Seaver, Ascom. *1;* ? Brenckle, Fungi Dak. *667* (as *Phialea sordida*); Reliq. Farlow. *107.*

2. **Chlorosplenium salviicolor** Ellis & Dearn.; Ellis & Ev. Proc. Acad. Sci. Phila. **1893**: 146. 1893.

Apothecia at first subhemispheric with the margin incurved, expanding and becoming subdiscoid, externally dark sage-green, reaching a diameter of 1.5–2 mm., contracted below into a very short, stem-like base; asci clavate-cylindric, reaching a length of 30 μ and a diameter of 2.5–3 μ; spores 2-seriate, minute, elongated, hyaline, 1–1.25 × 3–4 μ; paraphyses slender, branched above.

On dead stems of *Vitis vulpina.*

TYPE LOCALITY: St. Martinsville, Louisiana.

DISTRIBUTION: Known only from the type locality.

3. **Chlorosplenium olivaceum** Seaver, sp. nov.

Apothecia thickly gregarious, entirely sessile, shallow cup-shaped to scutellate, the margin often irregularly wavy, externally slightly furfuraceous, dark olive-green, reaching a diameter of 2–3 mm.; hymenium concave, light pea-green; asci clavate, reaching a length of 55–60 and a diameter of 6–7 μ, 8-spored; spores irregularly 2-seriate, narrow-ellipsoid, 3–4 × 9–10 μ; paraphyses slender, slightly enlarged above, reaching a diameter of 2–3 μ.

Apotheciis gregariis, sessile, scutellatis, margine plicato-crenato, extus pallide viridis, 2–3 mm. diam.; hymenio palidiore; ascis clavatis 8-sporis, 6–7 × 50–60 μ; sporis ellipsoideis, 3–4 × 9–10 μ; paraphysibus filiformibus, sursum clavatis, 2–3 μ diam.

On rotten wood of *Quercus* sp.

TYPE LOCALITY: Princeton, Georgia.

DISTRIBUTION: Georgia and North Carolina.

DOUBTFUL SPECIES

Chlorosplenium ? *atrovirens* (Pers.) DeNot. Comm. Critt. Ital. **1**: 377. 1864; *Peziza atrovirens* Pers. Syn. Fung. 635.   1801; *Calloria atrovirens* Fries, Summa Veg. Scad. 359.   1849; *Coryne virescens* Tul. Fung. Carp. **3**: 193. 1865; *Coryne atrovirens* Sacc. Syll. Fung. **8**: 641.   1889.   This species has been reported from North America but no material has been seen which could be ascribed to it with any certainty. This species was treated by DeNotaris under *Chlorosplenium* but he did not actually make the combination attributed to him.

Chlorosplenium tortum (Schw.) Sacc. Syll. Fung. **8**: 320.   1889; *Peziza torta* Schw. Trans. Am. Phil. Soc. II. **4**: 175.   1832.   Described as 8 mm. broad and high, externally aeroginous-green.   Identity uncertain.

20. **PYCNOPEZIZA** White & Whetzel, Mycologia **30**: 187.   1938.

Pycnidial stage an *Acarosporium.*

Apothecia small, not over 5 mm. in diameter, solitary, or gregarious, very short-stipitate to practically sessile, brown or brownish, at first closed, opening irregularly or by a pore, finally expanded, the margin stellate or circular; hymenium light-brown or buff; asci clavate or cylindric-clavate; spores small, ellipsoid, hyaline, simple; paraphyses filiform, simple.

Type species, *Pycnopeziza sympodialis* White & Whetzel.

Apothecia opening irregularly, becoming stellate.   1. *P. sympodialis.*
Apothecia opening by a circular pore, becoming scutellate.   2. *P. quisquiliaris.*

1. **Pycnopeziza sympodialis** White & Whetzel, Mycologia **30**: 190.   1938.

Acarosporium sympodiale Bubak & Vlengel, Ber. Deuts. Bot. Gesell. **29**: 385. 1911.

Apothecia solitary, scattered, short-stipitate, at first erect, pyriform and closed, opening by splitting *Geaster*-like into 4–6 stellate rays, 3–5 mm. in diameter; stem short but usually distinct, .5–1.5 mm. long and 1 mm. thick, black; hymenium nearly plane, buff to ochraceous, slightly umbilicate; asci cylindric-clavate, reaching a length of 75–85 μ and a diameter of 6.5–7.5 μ, 8-spored; spores narrow-ellipsoid, 3–3.5 × 7–9 μ irregularly 1-seriate; paraphyses filiform, scarcely enlarged above, 2.5 μ in diameter.

For description of the pycnidial stage see Mycologia.

On buds and staminate flowers of *Acer rubrum*, catkins of *Populus canadensis*, and buds and catkins of *Populus tremuloides*.

TYPE LOCALITY: Malloryville Bog, near Ithaca.

DISTRIBUTION: Known only from the type locality.

ILLUSTRATIONS: Mycologia **30**: 189, *f. 1–6;* 194, *f. 11–13.*

2. **Pycnopeziza quisquiliaris** (Ellis & Ev.) White & Whetzel, Mycologia **30**: 192. 1938.

Cyathicula quisquiliaris Ellis & Ev. Proc. Acad. Sci. Phila. **1893**: 451. 1893.

Apothecia solitary, or gregarious, attached by a narrow stipe-base, opening by a circular pore, finally saucer-shaped, 1–4 mm. in diameter, with a few dark squamules about the base, furfuraceous above, yellowish or cream-colored; hymenium concave, same color as the outside of the apothecium; asci clavate-cylindric, reaching a length of 75–85 μ and a diameter of 6–7 μ, 8-spored; spores ellipsoid, irregularly 1-seriate, 3–4 × 7–9 μ; paraphyses filiform, 2.5 μ in diameter, scarcely enlarged above.

For description of pycnidial stage (*Acarosporium quisquiliaris*) see Mycologia.

On various kinds of over-wintering buds; also on leaf debris, apparently of some herbaceous plants.

TYPE LOCALITY: Nuttalburg, West Virginia.

DISTRIBUTION: West Virginia and New York.

ILLUSTRATIONS: Mycologia **30**: 189, *f. 7–9;* 194, *f. 10;* 197, *f. 14.*

21. **CIBORIELLA** Seaver, gen. nov.

Apothecia stipitate, or substipitate, as in *Ciboria*, not stromatic; asci clavate, 8-spored; spores simple, ellipsoid to fusoid, hyaline; paraphyses filiform.

Apotheciis stipitatis vel substipatitis; *Ciboria* sine sclerotio; ascis clavatis, 8-sporis; sporis ellipsoideis vel fusoideis, hyalinis; paraphysibus filiformibus.

Type species, *Ciboria rufescens* Kanouse.

The genus *Ciboria* Fuckel has been restricted by H. H. Whetzel to those forms occurring on flowers and fruits and included in the Sclerotineae. The present genus is to include similar forms which would be excluded from the Sclerotineae.

1. **Ciboriella rufescens** (Kanouse) Seaver, comb. nov.

Ciboria rufescens Kanouse, Mycologia **33**: 463. 1941.

Apothecia short-stipitate, firm-fleshy, gregarious, pale-orange when fresh, blood-red when dry, exuding a red juice when bruised in water, 2–5 mm. in diameter; stem slender, 1–2 mm. long; asci clavate, reaching length of 35–60 μ and a diameter of 7–8 μ, 8-spored; spores fusoid, hyaline, simple, 3.5 × 7–8 μ; paraphyses filiform.

On decaying leaves of *Alnus*.

TYPE LOCALITY: Hoh River, Oregon.

DISTRIBUTION: Known only from the type locality.

DOUBTFUL SPECIES

Discinella washingtonensis Kanouse, Mycologia **39**: 650. 1947. Apothecia described as 6–12 mm. in diameter with a stem 3–5 mm. long. Spores fusoid, 4 × 12–14 μ; paraphyses filiform. It would seem to the writer that the species might be included here.

22. **CALYCINA** S. F. Gray, Nat. Arrang. Brit. Pl. **1**: 669.   1821.

Ciboria Fuckel, Symb. Myc. 311. 1869. (in part only).
Rutstroemia White, Lloydia **4**: 169. 1941 (in part). Not *Rutstroemia* Karst. 1871.

Apothecia medium large, up to 1 cm. in diameter stipitate, or subsessile, the length of the stem varying with the conditions, light-colored or dull, externally smooth, or with very poorly developed, hair-like structures; asci cylindric or clavate, usually 8-spored; spores ellipsoid to fusoid, hyaline, for a long time simple, later often becoming septate with one to several septa.

Type species, *Peziza firma* Pers.

This is a *Ciboria*-like fungus with a septate spore.

Spores comparatively small, not over 20 μ long.
   On wood or twigs.
      Substance of the apothecium not golden-yellow.   1. *C. firma*.
      Substance of the apothecia golden-yellow.   2. *C. bolaris*.
   On overwintering leaves.   3. *C. petiolorum*.
Spores large, 30–35 μ long.   4. *C. macrosporum*.

1. **Calycina firma** (Pers.) S. F. Gray, Nat. Arrang. Brit. Pl. **1**: 670.   1821.

?Peziza ochroleuca Bolton, Fungi Halifax **3**: 105.   1789.
Peziza firma Pers. Syn. Fung. 658.   1801.
Ciboria firma Fuckel, Symb. Myc. 312.   1869.
Helotium firmum Karst. Not. Soc. Fauna Fl. Fenn. **11**: 233.   1871.
Rutstroemia firma Karst. Not. Soc. Fauna Fl. Fenn. **13**: 233.   1873.
Phialea firma Gill. Champ. Fr. Discom. 101.   1882.
Hymenoscypha firma Phill. Brit. Discom. 123.   1887.
Ciboria ochroleuca Massee, Brit. Fungus-Fl. **4**: 274.   1895.

Apothecia gregarious, infundibuliform, becoming expanded and often nearly discoid, stipitate, brownish, reaching a diameter of 1 cm. though often smaller; hymenium brown, darker than the outside of the apothecium; stem variable in length, but up to 12 mm. and .5 mm. in diameter, but gradually expanding above;

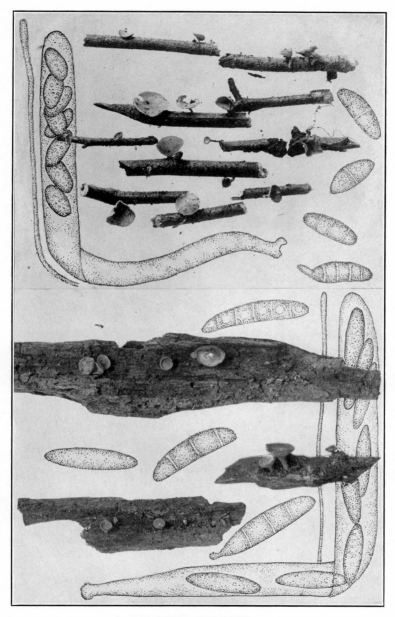

1. CALYCINA BOLARIS
2. CALYCINA MACROSPORA

asci cylindric, or subcylindric, reaching a length of 130 μ and a diameter of 9–12 μ, 8-spored; spores ellipsoid, or fusoid, obliquely 1-seriate with the ends overlapping, 4–6 × 15–20 μ, at first simple, occasionally becoming 1–3-septate; paraphyses filiform, enlarged, above, reaching a diameter of 2 μ.

On woods of various kinds.

TYPE LOCALITY: Europe.

DISTRIBUTION: Doubtfully reported from North America, also in Europe.

ILLUSTRATIONS: ?Bolton Hist. Fung. **3**: *pl. 105;* Boud. Ic. Myc. *pl. 483;* Gill. Champ. Fr. Discom. *pl. 74, f. 2;* E. & P. Nat. Pfl. 195, *f. 155, O, P;* Sow. Engl. Fungi *pl. 155;* Lloydia **4**: 174, *f. 1–9.*

For full discussion of this species see White (Lloydia **4**: 173–181.). According to W. L. White all American reports of this species are based on misdeterminations. It is not unlikely that the species will eventually be found in this country.

2. **Calycina bolaris** (Batsch) Seaver, Mycologia **26**: 346.   1934. (PLATE 103, FIG. 1.)

Peziza bolaris Batsch, Elench. Fung. Cont. **1**: 221.   1786?
Ciboria bolaris Fuckel, Symb. Myc. 311.   1869.
Phialea bolaris Boud. Bull. Soc. Myc. Fr. **1**: 116.   1885.
Hymenoscypha bolaris Phill. Brit. Discom. 124.   1887.

Apothecia gregarious, stipitate, or subsessile, expanding and becoming discoid, or slightly convex, externally yellowish, reaching a diameter of 5–6 mm., with a few club-shaped hair-like structures about the margin; hymenium slightly concave, plane, or a little convex, yellowish-brown, a little darker than the outside of the apothecium, (substance of the apothecia when crushed, golden-yellow); stem short, usually a little less than the diameter of the apothecium, about .5 mm. in diameter, expanding rather abruptly into the apothecium; asci cylindric, or clavate, 8-spored, reaching a length of 200 μ and a diameter of 12–14 μ; spores ellipsoid, usually slightly curved, 7–9 × 18–20 μ, for some time simple, finally 1-septate and often with apiculi at one or both ends, later 3-septate; paraphyses filiform, enlarged above, the contents yellow.

On twigs of various kinds.

TYPE LOCALITY: Europe.

DISTRIBUTION: New York; also in Europe.

ILLUSTRATIONS: Boud. Ic. Myc. *pl. 482;* Mycologia **26**: *pl. 40* (upper figure); Lloydia **4**: 182, *f. 10–14;* 185, *f. 15–26*.

This species is close to *Calycina firma* but seems to differ in that it is less robust and lighter colored, yellowish instead of brownish.   The only American specimen seen is one collected by H. H. Whetzel, No. *10784*.

3. Calycina petiolorum (Roberge) Seaver, comb. nov.

Peziza petiolorum Rob. Pl. Crypt. Fr. *1158;* Ann. Sci. Nat. II. **17**: 96.   1842.
Helotium petiolorum DeNot. Comm. Critt. Ital. **1**: 378.   1864.
Phialea petiolorum Gill. Champ. Fr. Discom. 102.   1882.
Calycella petiolorum Quél. Elench. Fung. 305.   1886.
Hymenoscypha petiolorum Phill. Brit. Discom. 132.   1887.
Cyathicula petiolorum Sacc. Syll. Fung. **8**: 305.   1889.
Ciboria petiolorum Schröt in Cohn, Krypt.-Fl. Schles **3**²: 61.   1893.

Apothecia solitary, scattered, becoming patellate, stipitate, reaching a diameter of 4 mm., externally showing faint longitudinal striations, pale-brown, the margin dentate; stem slender, usually tapering gradually below, becoming reddish-brown, nearly black at the base; hymenium slightly concave, ochraceous to cinnamon-brown; asci clavate, reaching a length of 95–120 μ and a diameter of 9–12 μ, 8-spored; spores 1-seriate, or 2-seriate above, oblong-reniform, 4.5–5.5 × 14–17 μ, at first 1-septate, later 2–3-septate, producing spermatia from the tips of the ascospores; paraphyses simple, enlarged above, 2.5–3.8 μ at their apices.

On rudimentary stromatic base on the petioles and midribs of overwintering leaves usually of *Fagus* and *Quercus*.

TYPE LOCALITY: Europe.

DISTRIBUTION: Iowa to Maine and North Carolina; also in Europe.

ILLUSTRATIONS: Grevillea **4**: *pl. 65, f. 301;* Ann. Myc. **4**: 228, *f. 46, 299, f. 47;* Lloydia **4**: 198, *f. 35–29*.

As will be seen, from the synonymy, this species has been placed in several genera at different times and by different authors.   Saccardo placed it in *Cyathicula* because of the slightly dentate margin.   It seems to the writer to belong to the present genus, largely on the basis of its spore characters.

4. Calycina macrospora (Peck) Seaver, Mycologia **26**: 346. 1934.
(PLATE 103, FIG. 2.)

Helotium macrosporum Peck, Ann. Rep. N. Y. State Mus. **26**: 82.   1874.
Bulgaria decolorans Berk. & Curt. Grevillea **4**: 6.   1875.

Peziza nebulosa Cooke, Mycographia 163. 1877.
Geopyxis nebulosa Sacc. Syll. Fung. **8**: 70. 1889.
Ombrophila decolorans Sacc. Syll. Fung. **8**: 615. 1889.
Ciboria Dallasiana Ellis & Ev. Jour. Myc. **9**: 165. 1903.
Tarzetta cinerascens Rehm, Ascom. *1853;* Ann. Myc. **2**: 352. 1904.
Ciboria fuscocinerea Rehm, Ann. Myc. **2**: 525. 1909.

Apothecia gregarious, stipitate, or occasionally subsessile, at first closed and subglobose, gradually expanding and becoming nearly discoid, attenuated at the base, reaching a diameter of 1 cm., cinereous to yellowish-brown; hymenium concave, or nearly plane, similar in color to the outside of the apothecium; stem reaching a length of 1 cm. and a diameter of about 1 mm.; asci cylindric, or clavate, 8-spored, reaching a length of 150 μ and a diameter of 10–12 μ; spores narrow-ellipsoid, or fusoid, straight, or slightly curved, hyaline, granular, for a long time simple, finally becoming 1–5-septate, and often with an apiculus at one or both ends and an oil-drop between each two septa, reaching a length of 30–35 μ and a diameter of 6–7 μ; paraphyses filiform, slightly enlarged above.

On rotten wood.

TYPE LOCALITY: South Carolina.

DISTRIBUTION: New York to Alabama, North Dakota and Colorado; also reported from Europe as *Ombrophila decolorans.*

ILLUSTRATIONS: Cooke Mycographia, *pl. 73, f. 281;* Seaver, Bull. Lab. Nat. Hist. State Univ. Iowa **6**: *pl. 20, f. 2;* Mycologia **26**: *pl. 40* (lower).

EXSICCATI: Ellis, N. Am. Fungi *477;* Rehm, Ascom. *1853;* Seaver, North Dakota Fungi *6.*

23. **CYATHICULA** DeNot. Comm. Critt. Ital. **1**: 381. 1863. (in part).

Hyalinia Boud. Bull. Soc. Myc. Fr. **1**: 114. 1885.
Calycella Quél Ench. Fung. 305. 1886.
Pezoloma Clements, Gen. Fungi 175. 1909; Minn. Bot. Stud. **4**: 186. 1911.
Pezizellaster Höhn. Ann. Myc. **15**: 349. 1917.

Apothecia scattered, stipitate, at first closed and subglobose, expanding, the margin beset with sharp teeth, the outside of the apothecium smooth; asci cylindric-clavate, 8-spored; spores fusoid, elongated, finally 1-septate; paraphyses filiform.

Type species, *Peziza coronata* Bull.

When the genus was established no species was designated as the type. Later authors have regarded *Peziza coronata* as the type, a species originally included by DeNotaris but not the first one mentioned. This species is therefore regarded as the type.

The genus *Pezoloma* was proposed by F. E. Clements for sessile species of *Cyathicula*. The character seems to the writer to be of no generic significance.

Occurring on herbaceous stems.   1. *C. coronata.*
Occurring on living *Marchantia.*   2. *C. Marchantiae.*

1. **Cyathicula coronata** (Bull.) Rehm in Rab. Krypt-Fl. **1**³: 740. 1893.

Peziza coronata Bull. Hist. Champ. Fr. 251.   1809.
Peziza radiata Pers. Myc. Eu. 1: 234.   1822.
Helotium coronatum Karst. Myc. Fenn. 1: 136.   1871.
Hymenoscypha coronata Phill. Brit. Discom. 127.   1887.
Peziza denticulata Fl. Dan. Fasc. 17: 10, pl. *1016, f. 3.*   1790.

Apothecia scattered, or gregarious, stipitate, at first rounded and closed, opening and becoming pitcher-shaped, finally expanded, the margin beset with long teeth, pale rose-colored or yellowish, .5–3 mm. in diameter; stem 1–6 mm. long, .2–.5 mm. thick, smooth, white, or yellowish-white; asci cylindric-clavate, reaching a length of 80–110 μ and a diameter of 8–9 μ, 8-spored; spores elongated, fusoid, simple, or 1-septate, 3–4.5 × 15–18 μ; paraphyses filiform, 2 μ thick.

On herbaceous stems of various kinds.

TYPE LOCALITY: Europe.

DISTRIBUTION: New York to Michigan; also in Europe.

ILLUSTRATIONS: Bull. Herb. Fr. *pl. 416;* Phill. Brit. Discom. *pl. 5, f. 26;* Nees, Syst. Pilze *f. 293;* E. & P. Nat. Pfl. **1**¹: 205, *f. 160,* N–O; Papers Mich. Acad. Sci. **20**: *pl. 15, f. 1.*

2. **Cyathicula Marchantiae** (Sommerf.) Sacc. Syll. Fung. **8**: 307. 1889.

Peziza Marchantiae Sommerf. Fl. Lap. Suppl. 295.   1826.

Apothecia solitary, turbinate, sessile, or with a short, thick stem, substance fleshy to waxy, thin, translucent, pallid-white, usually with a pale-lilac tint, reaching a diameter of .5–2 mm.; hymenium plane, or slightly concave, the margin ornamented with ciliate teeth composed of bundles of narrow cells; asci clavate-cylindric, not conspicuously narrowed below, apex rounded, reaching a length of 60–75 μ and a diameter of 6–8 μ;

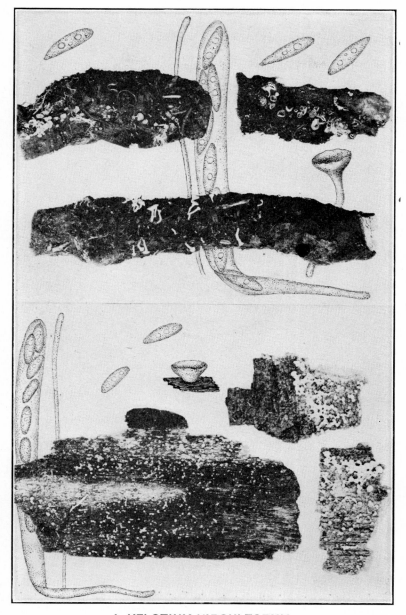

1. HELOTIUM VIRGULTORUM
2. HELOTIUM CITRINUM

spores 1- or 2-seriate, smooth, ovoid, or ellipsoid, 4 × 8–10 μ, usually with two small oil-drops; paraphyses filiform.

On living *Marchantia polymorpha*.

TYPE LOCALITY: Europe.

DISTRIBUTION: New York and Michigan; also in Europe.

Our only knowledge of this species is Durand's report in Bull. Torrey Club **29**: 463. 1902, and a report of its occurrence in Michigan by Dr. B. Kanouse.

DOUBTFUL AND EXCLUDED SPECIES

Cyathicula alpina Ellis & Ev. Proc. Acad. Sci. Phila. **1894**: 349. 1894. Apothecia minute, about .5 mm. in diameter and less than 1 mm. high; pallid, the margin fibrous, the extremities of the fibers prolonged and subfasciculate so as to make the margin obscurely dentate; asci clavate-cylindric, 8-spored; spores subcylindric, 1.5–2 × 8–10 μ; paraphyses filiform. On dead stems of *Pedicularis* and culms of *Elymus condensatus*, Colorado, alt. 9500 ft. It is doubtful if this species belongs here.

Cyathicula aquilina (Rehm) Sacc. Syll. Fung. **8**: 307. 1889; *Coronellaria Aquilinae* Rehm, Hedwigia **24**: 231. 1885. This species has been reported from Washington by Dr. B. Kanouse on fern debris and on *Gaultheria Shallon*. Rehm makes this a synonym of *Mollisia pteridina* Karst.

Cyathicula dentata (Pers.) Sacc. Syll. Fung. **8**: 307. 1889; *Peziza dentata* Pers. Ic. Descr. Fung. 5. *pl. 1, f. 6, 7.* 1798; *Niptera dentata* Fuckel, Symb. Myc. Nacht. **1**: 47. 1871; *Cistella dentata* Quél. Ench. Fung. 319. 1886; *Mollisia dentata* Gill. Champ. Fr. Discom. 124. 1882; *Calloria dentata* Phill. Brit. Discom. 332. 1887. This species has been reported from Michigan under the name *Cistella dentata*. It is also reported from Washington under the name *Pezizellaster radiostriatus* (Feltig) Höhn. Von Höhnel regarded these as identical.

Pezoloma griseum Clements, Gen. Fungi 175. 1909; Minn. Bot. Stud. **4**: 186, *pl. 25, f. 2.* 1911. The illustration shows the apothecium beset with a border of very long, tooth-like structures consisting of a fascicle of mycelial threads. The spores are described as 3–4 × 15–18 μ. The species should be further investigated. It is known only from the type collection in Colorado.

Lachnaster miniatus Kanouse, Mycologia **39**: 662. 1947. Described as having long, whiplash hairs arranged in teeth, pale-yellow; spores 1–1.5 × 6–8 μ; paraphyses lanceolate. The toothed margin would place this in the genus *Cyathicula*. On *Pteridium aquilinum*, Washington.

24. **HELOTIUM** Fries, Summa Veg. Scand. 354. 1849.

Pseudohelotium Fuckel, Symb. Myc. 298. 1869.
Phialea (Fries) Boud. Bull. Soc. Myc. Fr. **1**: 112. 1885.
Discinella Boud. Hist. Class. Discom. Eu. 96. 1907.
?*Pseudociboria* Kanouse, Mycologia **36**: 460. 1941.

Apothecia long-stipitate to sessile or subsessile, externally smooth (not hairy), usually small (not exceeding 2–3 mm. in diameter), for the most part bright-colored, whitish, yellowish, or occasionally dingy-brown, reddish, or with a greenish tint, usually with a whitish vegetative mycelium; asci usually 8-spored; spores simple (1-celled), or occasionally spuriously septate, ellipsoid, fusoid, or allantoid, hyaline or subhyaline; paraphyses very slender.

Type species, *Helvella acicularis* Bull.

The genus *Pseudociboria* is characterized by having two kinds of paraphyses, hyaline and colored, which in the opinion of the writer would scarcely warrant the establishment of a separate genus.

This is a large and cumbersome genus. Attempts to break it up have not been very successful. The separation of *Phialea* on the presence of a stem has been disregarded by recent workers. Many species have been described from scant material and never re-collected so that their value must remain uncertain at best. If it were possible to call a moratorium on the description of new species until the old ones could be better known it would be a great advantage to the monographer but that is probably too much to expect. The species are keyed out to the best of our ability.

Spores 20 μ or more long.
  Spores more than 30 μ long.
    On fallen leaves of *Alnus*.   1. *H. fastidiosum.*
    On fallen leaves of *Steironema*.   2. *H. Dearnessii.*
  Spores not usually exceeding 30 μ.
    Spores clavate.   3. *H. serotinum.*
    Spores fusoid.
      On herbaceous stems.   4. *H. Scutula.*
      On fallen leaves of *Salix*.   5. *H. salicellum.*
Spores usually 15–20 μ long.
  On fruits.
    On old acorns and hickory-nuts.   6. *H. fructigenum.*
    On seeds of *Nyssa sylvatica*.   7. *H. nyssicola.*
  Not on fruits.
    On cryptogams.
      On living mosses.   8. *H. turbinatum.*
      On old fungi, *Fomes*.   9. *H. mycetophilum.*
    On phanerogams.
      On monocot. leaves, *Iris*.   10. *H. nigromaculatum.*
      On dicots. of various kinds.
        On fallen leaves or petioles.
          On petioles of *Acer*.   11. *H. fraternum.*
          Not on petioles.

On fallen leaves of *Betula* etc.   12. *H. Linderi.*
On decaying leaves.
    Spores clavate.   13. *H. naviculasporum.*
    Spores fusoid.   14. *H. epiphyllum.*
On rotten wood.
    Apothecia cudonioid.   15. *H. cudonioides.*
    Apothecia not cudonioid.   16. *H. virgultorum.*
Spores usually 10–15 μ long.
On coniferous plants.
    On fallen leaves of *Picea.*
      Apothecia pallid, with *Fumosa.*   17. *H. fumosellum.*
      Apothecia sulphur-yellow.   18. *H. sulphuratum.*
    On fallen leaves of *Larix.*   19. *H. acicularum.*
Not on coniferous plants.
    On seeds of *Alnus oregona.*   20. *H. seminicola.*
    Not on seeds.
      On old fungus, *Dichaena strumosa.*   21. *H. strumosum.*
      Not on fungi.
        On herbaceous plants.
          Apothecia dark-brown or black.
            On stems of *Erigeron.*   22. *H. nigrescens.*
            On leaves of *Antenuaria.*   23. *H. phaeoconium.*
          Apothecia yellowish or cinereous.
            On stems of *Eschscholtzia.*   24. *H. Eschscholtziae.*
            On various plants especially
              Impatiens.   25. *H. herbarum.*
        Not on herbaceous plants.
          Spores reniform.   26. *H. renisporum.*
            Apothecia yellowish-green, on
              (usually) petioles of *Fraxinus.*   27. *H. luteovirescens.*
          Apothecia not yellowish-green.
            On rotten wood.
              Apothecia lemon-yellow.   28. *H. citrinum.*
              Apothecia pallid or yel-
                lowish.
                Spores clavate, on
                  Symphoricarpus.   29. *H. dakotense.*
                Spores not clavate.
                  Apothecia not
                    over 1 mm. in
                    diam.   30. *H. pallescens.*
                  Apothecia 3 mm.
                    in diam. on
                    Salix.   31. *H. amplum.*
          On dead leaves.
            Apothecia yellow to pale-
              orange.
              On leaves of *Fagus.*   32. *H. albopunctum.*

Not confined to
Fagus.
On leaves of
Quercus.   33. *H. puberulum.*
On leaves of va-
rious kinds.   34. *H. immutabile.*
Apothecia pallid or yel-
lowish-white.
Not over 1 mm. in
diameter.
Spores 12–15 μ
long, on *Acer*
and *Quercus.*   35. *H. translucens.*
Spores 10–13 μ
long, on *Popu-*
lus, more rare-
ly on other
kinds.   36. *H. populinum.*
Reaching 2–3 mm. in
diameter.
Apothecia creamy-
yellow.   37. *H. erraticum.*
Apothecia wa-
tery-white.   38. *H. simulatum.*
Apothecia not yellow or
orange.
Apothecia lilac, on
Conocarpus.   39. *H. Conocarpi.*
Apothecia brown or
reddish-brown.   40. *H. castaneum.*
Spores usually less than 10 μ long.
Occurring on cryptogams.
On living mosses and liverworts.   41. *H. destructor.*
On fungi, *Hypoxylon, Xylaria*, etc.   42. *H. episphaericum.*
Occurring on phanerogams.
On coniferous trees.
On needles of *Larix.*   43. *H. laricinum.*
On bark of *Picea.*   44. *H. aurantium.*
Not on coniferous trees.
On monocotyledons.
On palm, *Archontophoenix.*   45. *H. atrosubiculatum.*
On grasses or sedges.
Apothecia lemon-yellow, on *Carex.* 46. *H. citrinulum.*
Apothecia pale-yellow, on *Andro-*
pogon.   47. *H. planodiscum.*
On dicotyledons.
On herbaceous plants.
On stems of various plants.
Apothecia distinctly stipitate.   48. *H. cyathoideum.*
Apothecia subsessile.   49. *H. consanguineum.*

1. **Helotium fastidiosum** Peck, Ann. Rep. N. Y. State Mus. **27**: 107. 1875.

Calycina fastidiosa Kuntze, Rev. Gen. Pl. **3**³: 448. 1898.

Apothecia small, stipitate, pale-yellow, reaching a diameter of about 1 mm., rarely larger; hymenium plane, or slightly concave, yellow; stem slender, about equal in length to the diameter of the apothecium, brownish, or yellow with a brownish base; asci narrow-clavate, reaching a length of 80–100 μ and a diameter of 9–12 μ, 8-spored; spores crowded, or 2-seriate, elongate, subclavate, multiguttulate, about 4–4.5 × 32 μ; paraphyses slender.

On petioles and midribs of fallen leaves of *Alnus* in wet places; also reported on catkins of (*Alnus?*).

TYPE LOCALITY: Forestburgh, New York.

DISTRIBUTION: New York to Quebec and Oregon.   Probably common.

ILLUSTRATIONS: Mycologia **34**: 158, *f. 2;* 166, *f. 9;* Farlowia **1**: 153. *f. 7.*

This species seems to be distinct in its very long, slender spores.   A part of the original collection in the herbarium of The New York Botanical Garden has been examined.

2. **Helotium Dearnessii** (Ellis & Ev.) White, Mycologia **34**: 167. 1942.

Phialea Dearnessii Ellis & Ev. Proc. Acad. Sci. Phila.   **1893**: 146.   1893.
Hymenoscyphus Dearnessii Kuntze, Rev. Gen. Pl. **3³**: 485.   1898.

Apothecia scattered, stipitate, nearly sulphur-yellow, .7 mm. in diameter, at first subolivaceous, becoming yellow, the margin subfimbriate; stem short, stout, less than the diameter of the apothecium; asci cylindric-clavate, reaching a length of 100–108 μ and a diameter of 10–13 μ, 8-spored; spores 2-seriate, fusoid to clavate, slightly curved, with two large oil-drops, 4–4.5 × 30–35 μ, gradually attenuated below; paraphyses filiform, scarcely enlarged above.

On dead stems of *Steironema ciliatum.*

TYPE LOCALITY: London, Ontario, Canada.

DISTRIBUTION: Known only from the type locality.

ILLUSTRATIONS: Farlowia **1**: 615, *f. 31–34.*

EXSICCATI: Ellis & Ev. N. Am. Fungi *2624* (as *Peziza Dearnessii*).

3. **Helotium serotinum** (Pers.) Rehm in Rab. Krypt.-Fl. **1³**: 781. 1893.

Peziza serotina Pers. Syn. Fung. 661.   1801.
Hymenoscypha serotina Phill. Brit. Discom. 125.   1887.

Apothecia gregarious, stipitate, at first club-shaped and closed, finally opening and becoming cup-shaped, then plane or nearly so, .25–.5 mm. in diameter, bright-yellow; stem 2–10 mm. long, .2–.5 mm. thick; asci clavate, reaching a length of 100–150 μ and a diameter of 10–12 μ, 8-spored; spores clavate, straight, or slightly curved, sharp-pointed, 4 × 22–24 μ (Rehm states 30–36 μ long and occasionally 1-septate); paraphyses filiform, 2 μ in diameter.

On fallen leaves and branches of different kinds.

TYPE LOCALITY: Europe.
DISTRIBUTION: Ohio; also in Europe.
ILLUSTRATIONS: E. & P. Nat. Pfl. 1¹: 207. *f. 162, E;* Rab.
Krypt.-Fl. 1³: 770, *f. 1–4.*

Although this species has been reported from North America
by Saccardo the only specimen seen is one from Ohio determined
by Dr. B. Kanouse.

4. **Helotium Scutula** (Pers.) Karst. Not. Soc. Fauna Fl. Fenn.
  11: 233. 1871.

Peziza Scutula Pers. Myc. Eu. **1**: 284. 1822.
Ciboria ciliatospora Fuckel, Symb. Myc. 311. 1869.
Helotium Scutula caudatum Karst. Myc. Fenn. **1**: 112. 1871.
Helotium gracile Cooke & Peck; Peck, Ann. Rep. N. Y. State Mus. **26**: 83.
  1874.
Helotium vitellinum Rehm, Ber. Naturh. Ver. Augsburg **26**: 124. 1881.
Phialea Scutula Gill. Champ. Fr. Discom. 108. 1882.
Helotium virgultorum Scutula Rehm, Ascom. Lojk. 7. 1882.
Helotium Scutula Rubi Rehm, Hedwigia **24**: 229. 1885.
Calycella Scutula Quél. Ench. Fung. 305. 1886.
Hymenoscypha Scutula Phill. Brit. Discom. 136. 1887.
Hymenoscypha Scutula Lysimachiae Phill. Brit. Discom. 137. 1887.
Hymenoscypha Scutula Rudbeckiae Phill. Brit. Discom. 138. 1887.
Phialea vitellina Sacc. Syll. Fung. **8**: 262. 1889.
Phialea gracilis Sacc. Syll. Fung. **8**: 265. 1889.
Phialea Scutula Rudbeckiae Sacc. Syll. Fung. **8**: 266. 1889.
Helotium Verbenae Cavara, Rev. Myc. **11**: 178. 1889.
Phialea appendiculata Oud. Versl. Med. Akad. Wet. 7: 313. 1890.
Belonioscypha ciliatospora Rehm in Rab. Krypt.-Fl. 1³: 744. 1893.
Helotium Scutula vitellina Rehm in Rab. Krypt.-Fl. 1³: 794. 1893.
Helotium Scutula Lysimachiae Massee, Brit. Fungus-Fl. **4**: 254. 1895.
Helotium Scutula Rudbeckiae Massee, Brit. Fungus-Fl. **4**: 254. 1895.
Hymenoscyphus gracilis Kuntze, Rev. Gen. Pl. 3³: 485. 1898.
Helotium vitellinum pallidostriatum Fairman, Jour. Myc. **10**: 231. 1904.
Phialea vitellina pallidostriata Sacc. & D. Sacc. in Sacc. Syll. Fung. **18**: 56.
  1906.
Helotium appendiculatum Boud. Hist. Class. Discom. Eu. 114. 1907.
Helotium ciliatosporum Boud. Hist. Class. Discom. Eu. 114. 1907.
Belospora ciliatospora Clements, Gen. Fungi 175. 1909.
Hymenoscypha Scutula Grossulariae Kauffm. Papers Mich. Acad. Sci. **1**: 107.
  1921.

Apothecia thickly gregarious, stipitate, or occasionally almost
sessile, often springing from a dark-colored subiculum at first
shallow cup-shaped, becoming nearly plane, reaching a diameter
of 1–2 mm., pale-yellow; hymenium slightly concave, similar in
color to the outside of the apothecium; stem relatively slender and

variable in length, often reaching 5 mm., usually about .3 mm. thick; asci clavate, reaching a length of 90–100 μ and a diameter of 8–10 μ, 8-spored; spores irregularly 2-seriate, fusoid, or clavate, slightly curved, often spuriously 1–3-septate, 3–4 × 20–28 μ, occasionally with a cilium at one or both ends; paraphyses filiform, reaching a diameter of 7 μ.

On herbaceous stems of various kinds.

TYPE LOCALITY: Europe.

DISTRIBUTION: New York to Oregon, and Missouri, probably throughout North America; also in Europe.

ILLUSTRATIONS: Rab. Krypt.-Fl. **1**³: 771, *f. 1–5;* Trans. Brit. Myc. Soc. **18**: 76–83, *f. 1–5;* Mycologia **34**: 158, *f. 1.*

EXSICCATI: Ellis & Ev. Fungi Columb. *2339;* Clements, Crypt. Form. Colo. *80.*

5. **Helotium salicellum** Fries, Summa Veg. Scand. 356.   1849.

Peziza salicella Fries, Syst. Myc. **2**: 133.   1822.

Apothecia scattered, springing through the bark of the host, short-stipitate, at first globose, then expanding and becoming shallow cup-shaped, brownish-yellow, reaching a diameter of 2 mm.; hymenium slightly concave, or nearly plane, yellowish; stem scarcely exceeding 1 mm. in length and .3 mm. thick; asci clavate, reaching a length of 75–100 μ and a diameter of 12–25 μ, 8-spored; spores fusoid, with the ends acute, straight, or curved, usually containing one or two oil-drops, becoming spuriously 1–3-septate, 5–7 × 25–30 μ; paraphyses filiform, reaching a diameter of 2 μ.

On fallen branches of *Salix.*

TYPE LOCALITY: Europe.

DISTRIBUTION: Oregon; also in Europe.

This species has been reported from North America by Saccardo. No American material was found in the collection at The New York Botanical Garden, except the Oregon specimens. Fuckel regards this as a variety of *Helotium virgultorum.*

6. **Helotium fructigenum** (Bull.) Karst. Myc. Fenn. **1**: 113.   1871.

Peziza fructigena Bull. Hist. Champ. Fr. 236.   1791.
Phialea fructigena Gill. Champ. Fr. Discom. 99.   1882.
Hymenoscypha fructigena Phill. Brit. Discom. 135.   1887.

Apothecia gregarious, stipitate, at first closed, then opening and expanding, becoming plane or nearly so, reaching a diameter

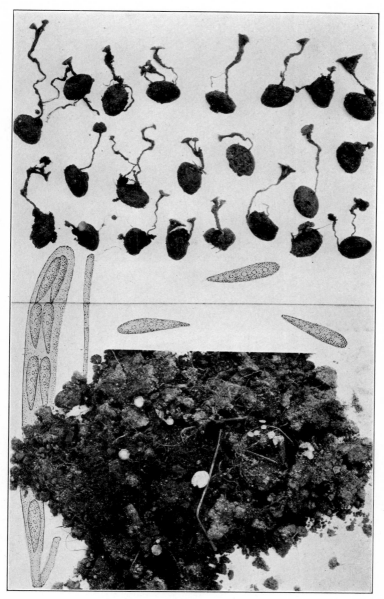

HELOTIUM NYSSICOLA

of 1–3 mm., pale-yellow; hymenium plane, or slightly concave, similar in color to the outside of the apothecium; stem slender, very variable in length, sometimes reaching nearly 1 cm., at other times so short that the apothecia appear subsessile; asci clavate, reaching a length of 80–100 μ and a diameter of 7–9 μ, 8-spored; spores fusoid, or subclavate, often with two oil-drops and numerous smaller ones, 3–5 × 12–18 μ; paraphyses filiform, slightly enlarged above, reaching a diameter of 3 μ.

On acorns and hickory-nut husks, *Quercus* and *Hicoria*.

TYPE LOCALITY: Europe.

DISTRIBUTION: New York to Iowa and Pennsylvania; also in Europe.

ILLUSTRATIONS: Bull. Herb. Fr. *pl. 228;* Gill. Champ. Fr. Discom. *pl. 73;* Bull. Lab. Nat. Hist. State Univ. Iowa **6:** *pl. 23, f. 1.*

EXSICCATI: Ellis & Ev. N. Am. Fungi *2048.*

There seems to be considerable variation in the size of the spores in this species. The spores are often smaller than indicated. As the plants are otherwise typical this may be due to immaturity. Occasionally the spores seem to be septate. This has been reported from Minnesota by Miss Daisy S. Hone as *Helotium virgultorum fructigenum,* following Rehm.

7. Helotium nyssicola Seaver, Mycologia **30:** 79. 1938. (PLATE 105.)

Apothecia gregarious, or occasionally occurring singly, stipitate, or subsessile, reaching a diameter of 2–4 mm., pale-yellow, orbicular, or with the margin irregularly split; hymenium concave, or nearly plane, pale-yellow; stem very slender, short, or reaching a length of 2 or more cm., the length depending upon the depth of the substratum; asci clavate, reaching a length of 125 μ and a diameter of 8 μ, gradually tapering below into a slender, stem-like base, 8-spored; spores fusoid, or clavate, about 5–5.5 × 15–20 μ, containing a number of small granules; paraphyses rather stout, granular, reaching a diameter of 3–4 μ.

On buried or partially buried seeds of *Nyssa sylvatica.*

TYPE LOCALITY: The New York Botanical Garden.

DISTRIBUTION: Known only from the type locality.

ILLUSTRATIONS: Mycologia **30:** 80, *f. 1.*

This species has been collected season after season under one tree in The New York Botanical Garden. It has not been found in any other locality.

8. **Helotium turbinatum** (Fuckel) Boud. Hist. Class. Discom. Eu. 113. 1907.

Leucoloma turbinata Fuckel, Symb. Myc. 318. 1869.
Helotium bryogenum Peck, Ann. Rep. N. Y. State Mus. **30**: 61. 1878.
Humaria turbinata Sacc. Syll. Fung. **8**: 127. 1889.
Plicaria turbinata Rehm in Rab. Krypt.-Fl. **1³**: 1009. 1894.
Belonium bryogenum Rehm, Ascom. *1279;* Hedwigia **38**: Beibl. (244). 1899.
Calycina bryogena Kuntze, Rev. Gen. Pl. **3³**: 448. 1898.
Calycella turbinata Höhn. Sitz.-ber. Akad. Wien **127¹**: 594. 1918.

Apothecia solitary, sparse, short-stipitate, fleshy, brown when dry, .8–1 mm. in diameter, waxy to waxy-cartilaginous; hymenium plane, or slightly concave, dark-brown in dried specimens; stem short, stout, about .3 mm. long; asci clavate, reaching a length of 70–90 μ and a diameter of 8–11 μ, 8-spored; spores elongated, straight, or very slightly curved, usually simple, or rarely 1-septate, 3–4 $\times$ 12–15 μ, somewhat attenuated at the ends; paraphyses, simple, or once branched near the base.

On living mosses, especially *Bryum*.

TYPE LOCALITY: Europe.

DISTRIBUTION: New York; also in Europe.

ILLUSTRATIONS: Mycologia **34**: 166, *f. 10*.

According to W. L. White, *Helotium bryogenum* Peck is synonymous with the above. This was described as pallid or yellowish-white when moist, livid-red or subolivaceous when dry.

9. **Helotium mycetophilum** Peck, Ann. Rep. N. Y. State Mus. **43**: 33. 1890.

Calycina mycetophila Kuntze, Rev. Gen. Pl. **3³**: 448. 1889.

Apothecia scattered, or gregarious, appearing singly, or more rarely two in close contact, substipitate, .1 to .3 mm. in diameter, more or less turbinate in form, dark-red; stem very short, tapering upward; hymenium plane, or slightly concave; asci clavate, reaching a length of 100–120 μ and a diameter of 13–15 μ; spores 1-seriate below, irregularly 2-seriate above, narrow-ellipsoid, slightly curved, or at least flattened on one side, occasionally becoming 1-septate but usually 1-celled, 5.5–7 $\times$ 17–20 μ; paraphyses simple, slightly flexuous, scarcely more than 1 μ in diameter.

On pileus of old sporophore of *Fomes fomentarius*.

TYPE LOCALITY: Rainbow Lake, New York.

DISTRIBUTION: Known only from the type locality.

ILLUSTRATIONS: Mycologia **34**: 166, *f. 16.*

10. **Helotium nigromaculatum** (Earle) Seaver, comb. nov.

Hymenoscypha nigromaculata Earle, Bull. N. Y. Bot. Gard. **2**: 338. 1902.
Phialea nigromaculata Sacc. Syll. Fung. **18**: 57. 1906.

Apothecia occurring on black spots which are .5–1 cm. in diameter, gregarious, brown, sessile, or substipitate, the margin inrolled when dry, becoming turbinate when moist, about .3 mm. in diameter and of equal height, externally fibrillose, the ends of the fibers extending slightly above the margin of the apothecium; asci broadly clavate, reaching a length of 70 μ and a diameter of 14 μ, 8-spored; spores 2-seriate, or irregularly disposed, irregularly ellipsoid, 6 × 18–20 μ; paraphyses sparse and rather stout.

On dying leaves of *Iris* sp.

TYPE LOCALITY: Palmetto, Florida.

DISTRIBUTION: Known only from the type locality.

No material has been seen.

11. **Helotium fraternum** Peck, Ann. Rep. N. Y. State Mus. **32**: 47. 1879.

Apothecia solitary, scattered, stipitate, reaching a diameter of 2 mm., at first cylindric, or columnar, finally expanding above, becoming globose, then cup-shaped and finally fully expanded, yellowish to whitish; hymenium at first concave, becoming convex when moist, wax-yellow to ochraceous-orange, when dry becoming concave and reddish-brown; asci clavate-cylindric, reaching a length of 70–100 μ and a diameter of 10–12 μ, 8-spored; spores 2-seriate or partially so, elongate-fusiform, often slightly curved, usually with an oil-drop in either end, 4–5 × 16–20 μ; paraphyses filiform, scarcely enlarged above.

On decaying petioles of various species of *Acer*.

TYPE LOCALITY: Adirondack Mountains, New York.

DISTRIBUTION: Maine to Indiana and West Virginia.

ILLUSTRATIONS: Mycologia **34**: 166. *f. 14.*

The type specimen in the herbarium of The New York Botanical Garden has been examined.

12. **Helotium Linderi** White, Farlowia **1**: 154. 1943.

Apothecia scattered, stipitate, reaching a diameter of 1 mm., pure-white when fresh, becoming cinereous to nearly black on drying; stem slender, white when fresh, drying to hyaline-ochraceous, smooth, slightly thickened above; hymenium becoming plane, or slightly concave, white, drying darker than the outside of the apothecium, often nearly black; asci clavate, reaching a length of 65–75 μ and a diameter of 9–10 μ, 8-spored; spores 2-seriate, elongated, narrowed toward the lower end, rounded above and somewhat curved, 3.5–4 × 16–18 μ, containing several oil-drops; paraphyses simple, or once forked below the middle, scarcely enlarged above, about 3 μ in diameter at their apices.

On fallen leaves of *Betula* sp. and others.

TYPE LOCALITY: Chimney Trail, Great Smoky Mts. National Park, Tennessee.

DISTRIBUTION: Tennessee and Wisconsin.

ILLUSTRATIONS: Farlowia 1: 153, *f. 9.*

13. Helotium naviculasporum Ellis, Bull. Torrey Club **5**: 46. 1874.

?Peziza caudata Karst. Fungi Fenn. *547.* 1866.
Helotium saprophyllum Cooke & Peck; Peck, Ann. Rep. N. Y. State Mus. **29**: 55. 1878.
Calycina naviculaspora Kuntze, Rev. Gen. Pl. **3**³: 448. 1898.
Calycina saprophylla Kuntze, Rev. Gen. Pl. **3**³: 449. 1898.
Helotium sparsum Boud. Hist. Class. Discom. Eu. 111. 1907.

Apothecia scattered, sparse, minute, delicate, slender-stipitate, hyaline-white when fresh, becoming opaque and more or less yellowish, or ochraceous on drying, reaching a diameter of .4 to 1.3 mm.; stem slender, delicate, smooth, similar in color to the apothecium, scarcely 1 mm. long; hymenium plane, or slightly convex, white to pale-yellow, or ochraceous; asci clavate, reaching a length of 80–90 μ, or rarely 110 μ, and a diameter of 8–11 μ; spores irregularly 2-seriate, slightly curved and attenuated at the lower end, 4–5 × 16–20 μ, granular within; paraphyses simple, scarcely enlarged above, reaching a diameter of 2–3 μ.

On fallen leaves usually from the midribs or veins, more rarely on petioles: *Acer, Alnus, Carya, Fagus,* and *Tilia.*

TYPE LOCALITY: Newfield, New Jersey.

DISTRIBUTION: New York to New Jersey and Michigan; also in Europe.

ILLUSTRATIONS: Boud. Ic. Myc. *pl. 495;* Mycologia **34**: 158, *f. 3;* 166, *f. 3;* Farlowia 1: 153, *f. 8* (as *Helotium caudatum* (Karst.) Vel.)

EXSICCATI: Ellis, Fungi Nova-Caesar. *53;* N. Am. Fungi *62.*

HELOTIUM RHIZICOLA

14. **Helotium epiphyllum** (Pers.) Fries, Summa Veg. Scand. 356. 1849.

Peziza epiphylla Pers. Tent. Disp. Fung. 72. 1797.
Calycina aurantiaca Kuntze, Rev. Gen. Pl. 3³: 448. 1898.
Hymenoscypha epiphylla Rehm; Kauffm. Papers Mich. Acad. Sci. 9: 177. 1929.

Apothecia scattered, or in groups of two to four, sessile to short-stipitate, bright-yellow, or dull-orange when fresh, the color not changing much on drying, fleshy to waxy, reaching a diameter of 4 mm.; stem when present exceedingly short, pale-yellow; hymenium slightly concave to plane, or sometimes repand, becoming yellow to orange, or occasionally dull-red; asci clavate, reaching a length of 90–130 μ and a diameter of 9–12 μ, 8-spored; spores fusoid, more or less unsymmetrical or slightly curved, simple, 1-seriate below, 2-seriate above, 4–5 $\times$ 16–20 μ, granular, often with two or three definite oil-drops; paraphyses straight, scarcely thickened above, reaching a diameter of 2–3 μ.

On leaves of various deciduous or more rarely coniferous trees; also on forest debris.

TYPE LOCALITY: Europe.

DISTRIBUTION: Probably throughout the United States and Canada; also in Europe.

ILLUSTRATIONS: Farlowia 1: 143, *f. 1*.

EXSICCATI: Ellis & Ev. N. Am. Fungi *2328*.

On Sept. 13, 1935, the writer examined a specimen of this species in the Persoon Herbarium at Leiden, Holland, and it showed three yellowish apothecia about 3 mm. in diameter on a leaf labeled *Corylus*. No microscopic examination was possible.

15. **Helotium cudonioides** Seaver, Mycologia **37**: 267. 1945.

Apothecia stipitate, single, or cespitose, at first concave, soon becoming strongly convex, reaching a diameter of 2 mm., pale- to dark-brown; stem white or nearly white, reaching a length of 2–3 mm. and less than 1 mm. in diameter; asci clavate, 8-spored, reaching a length of 100 μ and a diameter of 10–12 μ; spores 1-seriate below and irregularly 2-seriate above, fusoid, hyaline, about 5 $\times$ 16–20 μ; paraphyses filiform, 1 μ or less in diameter.

On rotten wood.

TYPE LOCALITY: Cleveland, Ohio.

DISTRIBUTION: New York and Ohio.

ILLUSTRATIONS: Mycologia 37: 268, *f. 1*.

This species has been sent in several times for determination. As the name signifies it looks like a miniature *Cudonia*.

16. **Helotium virgultorum** (Vahl.) Fries, Summa Veg. Scand. 355. 1849. (PLATE 104, FIG. 1.)

Peziza virgultorum Vahl. Fl. Dan. **17**: 10. 1790.
Hymenoscypha virgultorum Phill. Brit. Discom. 134. 1887.
Phialea virgultorum Sacc. Syll. Fung. **8**: 266. 1889.

Apothecia gregarious, stipitate, at first closed, expanding and becoming shallow cup-shaped to patellate, or occasionally even convex, reaching a diameter of 1–3 mm., whitish to pale-yellowish; hymenium usually concave, bright-yellow; stem cylindric, two to three times as long as the diameter of the apothecium, whitish, or yellowish; asci clavate, reaching a length of 100–120 μ and a diameter of 9–10 μ, 8-spored; spores partially 2-seriate, fusoid, or subclavate, usually containing two oil-drops with several smaller ones, 4–5 $\times$ 15–20 μ, occasionally spuriously septate; paraphyses filiform, slightly enlarged above.

On partially buried sticks and roots.

TYPE LOCALITY: Europe.

DISTRIBUTION: Toronto to Colorado, Washington, and California; also in Europe.

ILLUSTRATIONS: Fl. Dan. *pl. 1016, f. 2.*

Excellent material collected in Colorado on roots of *Alnus* has been studied and is the basis of the illustrations shown in this work. The distribution is difficult to determine since this species has been confused with two others which are here regarded as distinct, *Helotium fructigenum* and *Helotium Scutula*.

17. **Helotium fumosellum** (Cooke & Ellis) Seaver, comb. nov.

Peziza fumosella Cooke & Ellis, Grevillea **6**: 91. 1878.
Phialea fumosella Sacc. Syll. Fung. **8**: 269. 1889.
Allophylaria fumosella Nannf. Nova Acta Soc. Sci. Upsal. IV. **8**: 291. 1932.

Apothecia minute, stipitate, fumose, at first clavate, then expanding, becoming cyathiform, .25–.7 mm. in diameter and .5 mm. high; hymenium pallid; asci cylindric-clavate, 8-spored; spores elongated, straight, or curved, 10 μ long.

On fir leaves, *Abies*, accompanying *Hymenula fumosa* which may be its conidial stage.

TYPE LOCALITY: Newfield, New Jersey.
DISTRIBUTION: Known only from the type locality.

18. **Helotium sulphuratum** (Fries) Phill. Brit. Discom. 161. 1887.

Peziza sulphurea Schum. Pl. Saell. **2**: 428.   1803.   Not Pers.   1797.
Peziza sulphurata Fries, Syst, Myc. **2**: 72.   1822.
?Helotium Piceae (Kauffm.) Kanouse, Mycologia **33**: 465.   1941.

Apothecia scattered, or occasionally two or three together, stipitate, or subsessile, at first rounded and closed, expanding and becoming shallow cup-shaped, or nearly plane, 1–2 mm. in diameter, externally pale-yellow; hymenium sulphur-yellow; asci cylindric-clavate, 8-spored; spores fusiform, about 3–4 × 10–12 μ; paraphyses filiform.
On leaves of conifers matted on the ground, *Picea.*
TYPE LOCALITY: Europe.
DISTRIBUTION: Colorado and Manitoba, Canada.
ILLUSTRATIONS: Fl. Dan. *pl. 1915, f. 2.*
Helotium Piceae was based on specimens collected by C. H. Kauffman and regarded by him as a variety of *H. sulphuratum.* The difference, in our opinion, is too slight to segregate it as a distinct species.

19. **Helotium acicularum** (Roll.) Sacc. Syll. Fung. **10**: 8.   1892.

Calycella acicularum Rolland, Bull. Soc. Myc. Fr. **5**: 170.   1889.
Phialea acicularum Rehm in Rab. Krypt.-Fl. **1³**: 719.   1892.

Apothecia short-stipitate, at first closed and rounded, finally opening and becoming cup-shaped, then plane, externally yellowish-white, .2–.3 mm. in diameter, fleshy-waxy; hymenium concave to plane, egg-yellow; asci clavate, reaching a length of 160 μ and a diameter of 10–12 μ, 8-spored; spores ellipsoid, slightly bent, 5–8 × 10–12 μ; paraphyses filiform.
On fallen needles of larch, *Larix* sp.
TYPE LOCALITY: Europe.
DISTRIBUTION: Michigan; also in Europe.
ILLUSTRATIONS: Bull. Soc. Myc. Fr. **5**: *pl. 15, f. 3.*

20. **Helotium seminicola** Keinh. & Cash, Mycologia **29**: 82. 1937.

Apothecia stipitate, occurring singly, or two or three arising from a seed, at first cup-shaped, expanding and often plane, reaching a diameter of 6 mm.; hymenium fawn-colored, plane, or concave; stem reaching a length of 10 mm. and usually .5–.8 mm. thick, sometimes hairy toward the base when young; asci clavate, reaching a length of 75–130 μ and a diameter of 7–12 μ, 8-spored; spores fusoid-ellipsoid, with ends attenuated, with several small oil-drops, 3.5–5 × 12–20 μ; paraphyses filiform simple, or branched below, slightly enlarged above, reaching a diameter of 4 μ, hyaline.

On seeds of *Alnus oregona*.

TYPE LOCALITY: Hood River, Oregon.

DISTRIBUTION: Known only from the type locality.

ILLUSTRATIONS: Mycologia **29**: 83, *f. 1–4*.

21. **Helotium strumosum** Ellis & Ev. Jour. Myc. **4**: 56.   1888.

Pseudohelotium strumosum Sacc. Syll. Fung. **8**: 300.   1889.

Apothecia gregarious, sessile or subsessile, bright lemon-yellow, closed and subspherical at first, then opening and becoming cup-shaped and finally expanding to nearly plane, about .3 mm. in diameter, tomentose-pubescent outside and attached to the matrix by fine, white hairs at the base; asci clavate, reaching a length of 70–80 μ and a diameter of 10–12 μ, 8-spored; spores 2-seriate, ellipsoid, with two or three oil-drops, and indications of a medial septum, 3–4.5 × 10–12 μ; paraphyses stout, slightly yellowish and thickened above.

On old *Dichaena strumosa* on *Quercus coccinea*.

TYPE LOCALITY: Newfield, New Jersey.

DISTRIBUTION: New Jersey and Delaware.

22. **Helotium nigrescens** (Cooke) Rehm, Ascom. *307;* Ber. Naturh. Ver. Augsburg **26**: 77.   1881.

Peziza nigrescens Cooke, Hedwigia **14**: 83.   1875.

Apothecia stipitate, brownish-black, becoming black, at first clavate, then expanded, .3 mm. in diameter; hymenium pallid or subcinereous; stem thick, expanding into the apothecium, .5 mm. long; asci cylindric, or subcylindric, reaching a length of 60 μ and a diameter of 9 μ, 8-spored; spores elongate-fusiform, straight, or slightly curved, 3 × 12 μ; paraphyses not observed.

On dead stems of *Erigeron*.

TYPE LOCALITY: Newfield, New Jersey.

DISTRIBUTION: Known only from the type locality.

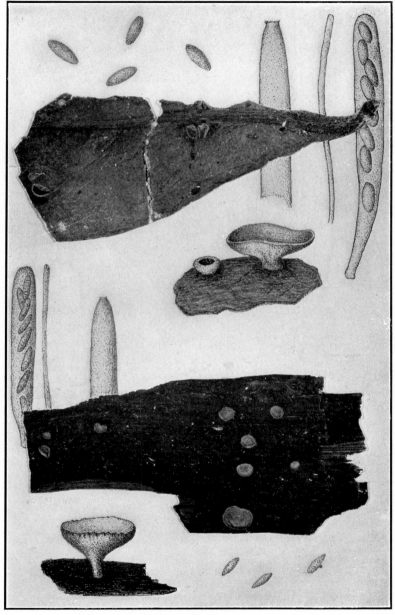

HELOTIUM CONOCARPI
HELOTIUM ATROSUBICULATUM

EXSICCATI: Ellis, N. Am. Fungi *140;* Thüm. Myc. Univ. *214;* Rehm, Ascom. *307* (both from Newfield, New Jersey).

23. **Helotium phaeoconium** (Fairman) Seaver, comb. nov.

Phialea phaeoconia Fairman, Ann. Myc. **9**: 151. 1911.

Apothecia gregarious, stipitate, occurring on the under side or projecting from the margins of the leaves, dark-brown, or black, with a light-brown, or grayish margin, globose and closed, finally opening and becoming deep cup-shaped, .25 mm. in diameter; hymenium concave, light-brown; stem about as long as the diameter of the apothecium, swollen just below the point of attachment; asci clavate-cylindric, reaching a length of 50 μ and a diameter of 10 μ, 8-spored; spores long-ellipsoid, or fusoid, straight, or curved, irregularly 1-seriate, or partially 2-seriate, 3–4 $\times$ 10–13 μ; paraphyses filiform.

On dead leaves of *Antennaria plantaginifolia.*
TYPE LOCALITY: Lyndonville, New York.
DISTRIBUTION: Known only from the type locality.
ILLUSTRATIONS: Ann. Myc. **9**: 150, *f. 6, 7.*

24. **Helotium Eschscholtziae** (Phill. & Hark.) Seaver, comb. nov.

Peziza Eschscholtziae Phill. & Hark. Bull. Calif. Acad. Sci. **1**: 22. 1884.
Phialea Eschscholtziae Sacc. Syll. Fung. **8**: 271. 1889.

Apothecia scattered, short-stipitate, cyathiform, cinereous, pruinose, .5–.7 mm. in diameter; hymenium same color as outside of apothecium; asci cylindric, 8-spored; spores fusoid, usually with an oil-drop in each end, 2 $\times$ 10 μ; paraphyses not observed.

On dead stems of *Eschscholtzia californica.*
TYPE LOCALITY: California.
DISTRIBUTION: Known only from California.
EXSICCATI: Ellis & Ev. N. Am. Fungi *2042;* Rab.-Winter, Fungi Eu. *3272* (from California).

This species seems to be very close to *Helotium cyathoideum.*

25. **Helotium herbarum** (Pers.) Fries, Summa Veg. Scand. 356. 1849.

Peziza herbarum Pers. Tent. Disp. Fung. 72. 1797.

Apothecia thickly gregarious, very short-stipitate, or sessile, at first rounded, expanding and becoming shallow cup-shaped, or discoid, yellowish, or whitish, reaching a diameter of 1–2 mm.; hymenium nearly plane, light-yellow; stem vrey short, usually less than 1 mm. long, or absent; asci clavate, gradually tapering below, reaching a length of 80 μ and a diameter of 6–7 μ, 8-spored; spores irregularly 2-seriate, fusoid, slightly curved, containing several small oil-drops, often spuriously septate, 3–4 × 10–14 μ; paraphyses filiform, reaching a diameter of 1–2 μ.

On herbaceous stems of various kinds, *Impatiens* and *Eupatorium*.

TYPE LOCALITY: Europe.

DISTRIBUTION: New York to Maine, Michigan and Kansas; also in Europe.

ILLUSTRATIONS: Boud. Ic. Myc. *pl. 497;* Gill. Champ. Fr. Discom. *pl. 89, f. 2;* E. & P. Nat. Pfl. 1¹: 207, *f. 162 C–D;* Rab. Krypt.-Fl. 1³: 770, *f. 1–4.*

EXSICCATI: Ellis, N. Am. Fungi *670;* Ellis & Ev. Fungi Columb. *740, 1309;* Reliq. Farlow. *124* (as *Helotium herbarum Rubi*).

26. **Helotium renisporum** Ellis; Cooke, Bull. Buffalo Soc. Nat. Sci. **2**: 299.   1875.

?Ciboria Sydowiana Rehm, Hedwigia **24**: 226.   1885.
Hymenoscypha renispora Phill. Brit. Discom. 143.   1887.
Ciboria renispora Sacc. Syll. Fung. **8**: 207.   1889.
Rutstroemia renispora White, Lloydia **4**: 215.   1941.

Apothecia stipitate, slightly concave, or plane, from black stromatic lines in the leaf, cinnamon, or brownish-yellow, reaching a diameter of 2 mm., externally clothed with adpressed, brown, club-shaped hair-like structures; hymenium similar in color to the outside of the apothecium, usually plane, or convex at maturity; stem blackish, reaching a length of 3–6 mm.; asci cylindric, or subcylindric, reaching a length of 65–100 μ and a diameter of 7–10 μ; spores reniform, with two or three oil-drops, 5–6 × 10–14 μ; paraphyses slender, slightly clavate.

On decaying leaves.

TYPE LOCALITY: Newfield, New Jersey.

DISTRIBUTION: New Jersey and Pennsylvania to South Carolina.

ILLUSTRATIONS: Lloydia **4**: 214, *f. 62–65.*

EXSICCATI: Ellis & Ev. N. Am. Fungi *2049.*

Phillips, in his British Discomycetes cited above, regards Rehm's European species as identical with Ellis' American form, characterized by its reniform spores.   W. L. White claims that

Phillips was in error in making *Ciboria Sydowiana* synonymous with *Helotium renisporum*.

27. **Helotium luteovirescens** (Rob.) Seaver, comb. nov.

Peziza luteovirescens Rob. in Desm. Pl. Crypt. Fr. *1541.* 1846; Ann. Sci. Nat. III. **8**: 188. 1847.
?*Peziza longipes* Peck, Bull. Buffalo Soc. Nat. Sci. **2**: 295. 1875.
Peziza pallidovirescens Phill.; Phill. & Plow. Grevillea **6**: 24. 1877.
Helotium sulfurellum Ellis & Ev. Bull. Torrey Club **10**: 98. 1883.
Ciboria tabacina Ellis & Holw.; Arthur, Bull. Geol. Nat. Hist. Surv. Minn. **3**: 35. 1887.
Ciboria luteovirescens Sacc. Syll. Fung. **8**: 206. 1889.
Ciboria sulfurella Rehm; Durand, Bull. Torrey Club **29**: 461. 1902.
?*Phialea anomala* Peck, Bull. N. Y. State Mus. **167**: 29. 1913.

Apothecia gregarious, or scattered, stipitate, reaching a diameter of 2–5 mm., sulphur-yellow when fresh with a tinge of green; hymenium concave, or plane; stem very variable in length, sometimes as long as 2 cm. or often very short; hymenium concave, plane, or slightly convex, darker than the outside of the apothecium, reddish, or reddish-brown, when dry entire apothecium almost black; asci clavate, 8-spored, reaching a length of 75–90 μ and a diameter of 7–10 μ; spores 1-seriate with the ends overlapping, ellipsoid, 3–5 $\times$ 10–12 μ; paraphyses filiform.

On sticks and leaf petioles, *Fraxinus* especially; also reported on *Acer*, *Tilia* and *Platinus*.

TYPE LOCALITY: Europe.

DISTRIBUTION: New England to North Dakota, Iowa and Pennsylvania; also in Europe.

EXSICCATI: Ellis, N. Am. Fungi *1275* (as *Helotium sulfurellum*); Rehm, Ascom. *1880.* (as *Ciboria tabacina*).

Durand has reported *Ciboria luteovirescens* from this country, but regards *Helotium sulfurellum* as distinct mainly on color. He states that the latter is never green. The writer has often collected this in Iowa on ash petioles and described it as sulphur-yellow with a tinge of green. It seems doubtful if the two forms are distinct, although for lack of material it is difficult to be certain.

28. **Helotium citrinum** (Hedw.) Fries, Summa Veg. Scand. 355. 1849. (PLATE 104. FIG. 2.)

?*Peziza lenticularis* Bull. Hist. Champ. Fr. 248. 1791.
Octospora citrina Hedw. Descr. **2**: 33. 1789.
Peziza citrina Batsch, Elench. Fung. Cont. **2**: 95. 1789.
Peziza confluens Schw. Trans. Am. Phil. Soc. II. **4**: 176. 1832.

?Helotium citrinum sublenticulare Fries, Summa Veg. Scand. 355.   1849.
Phialea citrina Gill. Champ. Fr. Discom. 109.   1882.
Calycella citrina Boud. Bull. Soc. Myc. Fr. **1**: 112.   1885.
Helotium confluens Sacc. Syll. Fung. **8**: 222.   1889.
?Helotium lenticulare Sacc. Syll. Fung. **8**: 225.   1889.
?Helotium sublenticulare Rehm in Rab. Krypt.-Fl. **1**³: 784.   1893.
Helotium alaskae Sacc.; Sacc. Peck, & Trelease in Harriman Alaska Exped. **5**:
    25.   1904.

Apothecia thickly gregarious and often running together in confluent masses, the individuals with a short and relatively thick stem, reaching a diameter of .5–3 mm.; hymenium plane or nearly so, bright lemon-yellow; stem seldom exceeding 1 mm. in length and often nearly wanting, lighter colored than the hymenium; asci clavate, 8-spored, reaching a length of 75–100 μ and a diameter of 10 μ; spores ellipsoid, or fusoid, usually with an oil-drop in either end and sometimes pseudoseptate, 3–4 × 9–14 μ; paraphyses filiform, reaching a diameter of 1.5 μ.

On decaying wood, of all kinds; also on humus.

TYPE LOCALITY: Europe.

DISTRIBUTION: Throughout North America; probably worldwide.

ILLUSTRATIONS: Hedw. Descr. **2**: *pl. 8, f. B;* ? Fl. Dan. *pl. 1971, f. 3;* Batsch Elench. Fung. Cont. **2**: *pl. 39, f. 218;* Sow. Engl. Fungi *pl. 150* (as *Peziza aurea* Bolt.), *151;* Bull. Herb. Fr. *pl. 300.*

EXSICCATI: Ellis, N. Am. Fungi *1315, 1316;* Clements, Crypt. Form. Colo. *79;* Shear, N. Y. Fungi *326;* Reliq. Farlow. *123.*

One of the most common and widely distributed of the cup-fungi.   It is quite easily recognized by the bright lemon-yellow color and short, thick stem-like base.

29. **Helotium dakotense** (Rehm) Seaver, comb. nov.

Pezizella dakotensis Rehm, Ann. Myc. **11**: 396.   1913.

Apothecia scattered, subsessile, at first globose and closed, finally patelliform, 1–1.5 μ in diameter; hymenium dilutely yellowish; asci clavate, reaching a length of 50–60 μ and a diameter of 8 μ, 8-spored; spores clavate, straight, 1-seriate, 2 × 12 μ; paraphyses filiform, 2 μ in diameter, hyaline.

On branches of *Symphoricarpos occidentalis.*

TYPE LOCALITY: Kulm, North Dakota.

DISTRIBUTION: Known only from the type locality.

30. **Helotium pallescens** (Pers.) Fries, Summa Veg. Scand. 355. 1849.

Peziza pallescens Pers. Obs. Myc. **2**: 85. 1799.
Calycella pallescens Quél. Ench. Fung. 306. 1886.

Apothecia short-stipitate, waxy, white, pallid, or pale-yellow, .5–1 mm. in diameter; hymenium concave, or plane; stem short, .2–.6 mm. in diameter, whitish; asci clavate-cylindric, reaching a length of 85–120 μ and a diameter of 6–7 μ, 8-spored; spores long-fusiform, straight, or slightly curved, with two to four oil-drops, or occasionally pseudoseptate, 3–4 $\times$ 10–15 μ; paraphyses filiform, enlarged slightly above.

On dead wood and bark of various trees.

TYPE LOCALITY: Europe.

DISTRIBUTION: New Hampshire to California and Pennsylvania.

Specimens referable to this species have been frequently collected in North America.

31. **Helotium amplum** (Ellis & Ev.) Seaver, comb. nov.

Phialea ampla Ellis & Ev. Bull. Torrey Club **24**: 135. 1897.

Apothecia gregarious, stipitate, or subsessile, obconic, or clavate at first, at length expanding, 3–4 mm. in diameter and shallow cup-shaped, externally pale-yellow, farinose-pubescent, or minutely striate, the margin involute; hymenium lemon-yellow; asci clavate-cylindric, reaching a length of 110 μ and a diameter of 6 μ; spores 1-seriate, overlapping, narrow-ovoid, 3.5–4 $\times$ 11–12 μ; paraphyses filiform.

On decaying wood of *Salix*.

TYPE LOCALITY: Mt. Paddo [Mt. Adams], Washington.

DISTRIBUTION: Known only from the type locality.

Ellis states: "Differs from *Helotium citrinum*, in its involute margin and larger size." Type in The New York Botanical has been examined. While these are described as stipitate, mature plants appear to be sessile.

32. **Helotium albopunctum** Peck, Ann. Rep. N. Y. State Mus. **31**: 47. 1879. Not *Helotium albopunctum* (Desm.) Bucknall, 1882.

Pezizella albopuncta Sacc. Syll. Fung. **8**: 276. 1889.
Hymenoscyphus albopunctus Kuntze, Rev. Gen. Pl. **3**³: 485. 1898.

Apothecia scattered, solitary, punctiform, subsessile, when dry visible to the unaided eye only as dots, waxy-cartilaginous, the color to the unaided eye similar to that of the dried leaves on which they occur, when magnified pale-yellow to pale-orange; hymenium plane to slightly concave; asci clavate, reaching a length of 60–70 μ and a diameter of 8–10 μ; spores straight, or slightly curved, clavate, the narrow end downward, 3.5–4 $\times$ 14–16 μ; paraphyses simple, slightly enlarged above, reaching a diameter of 2.5–3.5 μ.

On fallen leaves of *Fagus grandifolia.*

TYPE LOCALITY: Adirondack Mountains, New York.

DISTRIBUTION: New York; also reported from Michigan on leaves of *Cornus.* A portion of the type in The New York Botanical Garden has been examined.

ILLUSTRATIONS: Mycologia **34**: 166, *f. 13;* Farlowia **1**: 168, *f. 10.*

33. **Helotium puberulum** (Lasch) Fuckel, Fungi Rhen. *1150.* 1865.

Peziza puberula Lasch in Klotzs. Herb. Viv. Myc. *1529;* Flora **34**: 566.   1851.
    Not Berk. & Curt.   1875.
Pseudohelotium puberulum Fuckel, Symb. Myc. 298.   1870.
Lachnella puberula Phill. Grevillea **18**: 85.   1890.
Pezizella puberula Rehm in Rab. Krypt.-Fl. **1**³: 665.   1892.
Dasyscypha puberula Massee, Brit. Fungus-Fl. **4**: 355.   1895.
Belonium sulphureotinctum Rehm, Hedwigia **35**: Beibl. (146).   1896.
Urceolella puberula Boud. Hist. Class. Discom. Eu. 129.   1907.
Niptera sulphureotincta Boud. Hist. Class. Discom. Eu. 141.   1907.
Phialina puberula Höhn. Mitt. Bot. Inst. Techn. Hochsch. Wein **3**³: 106.   1926.

Apothecia rather numerous on localized areas of the leaf surface, sometimes in groups of five to fifteen with occasional scattered ones, very small, appearing sessile but with a short, obscure stem, less than 1 mm. in diameter, when fresh bright lemon-yellow, not contracting or changing color much on drying; stem very short and scarcely noticeable; hymenium convex when fresh and bright-yellow, on drying becoming plane and remaining yellow to pale-orange; asci short, clavate, reaching a length of 38 to 55 μ and a diameter of 6–8 μ, 8-spored; spores 2-seriate, or irregularly crowded, slightly irregular in outline, simple, 2–3 $\times$ 12–18 μ, often containing several small oil-drops; paraphyses rather stout, scarcely enlarged above, occasionally branched, 3–3.5 μ in diameter at their apices.

HELOTIUM CUDONIOIDES

On decaying leaves of *Quercus* and occasionally on leaves of other kinds.

TYPE LOCALITY: Europe.

DISTRIBUTION: New Hampshire, Massachusetts and New York; also in Europe.

ILLUSTRATIONS: Farlowia 1: 168, *f. 17*.

34. **Helotium immutabile** Fuckel, Symb. Myc. Nacht. **1**: 50. 1871.

Pachydisca immutabilis Boud. Hist. Class. Discom. Eu. 94. 1907.

Apothecia scattered, short-stipitate, white when fresh, becoming yellow to orange when dry, .8–1.5 mm. in diameter; stem .2–.5 mm. long, gradually expanding into the apothecium above, the color as above; hymenium when fresh white, becoming pale-yellow or in old, dried specimens pale-orange to reddish-brown, plane to slightly convex; asci clavate, reaching a length of 80–105 μ and a diameter of 8–10 μ, 8-spored; spores 1-seriate to 2-seriate, simple, angular-obovoid, 3.8–4.6 × 10–13 μ, the contents granular, or with a few minute, scattered oil-drops; paraphyses once forked below the middle, scarcely enlarged above, reaching a diameter of 3–3.5 μ.

On decaying leaves, especially on the veins and petioles of various deciduous leaves.

TYPE LOCALITY: Europe.

DISTRIBUTION: Massachusetts to New Jersey and Minnesota; also in Europe.

ILLUSTRATIONS: Farlowia 1: 143, *f. 3*.

35. **Helotium translucens** White, Farlowia 1: 149. 1943.

Apothecia scattered, or subgregarious, stipitate, subfleshy, practically hyaline when fresh, drying to pale-ochraceous, reaching a diameter of .4 mm.; stem .3 mm. long, expanding abruptly into the apothecium; hymenium becoming plane and remaining so on drying; asci cylindric-clavate, abruptly narrowed below, reaching a length of 75–95 μ and a diameter of 10–11.5, 8-spored; spores 1-seriate, or becoming partially 2-seriate, obovoid, the contents scarcely granular, 4–4.5 × 12–15 μ; paraphyses clavate, 3–3.5 μ in diameter above.

On the veins and parenchyma of leaves of *Acer* sp. and *Quercus* sp.

TYPE LOCALITY: Blue Hills near Milton, Massachusetts.

DISTRIBUTION: Massachusetts.
ILLUSTRATIONS: Farlowia 1: 153, *f. 6.*

36. **Helotium populinum** Fuckel, Symb. Myc. 316.   1869.

?Peziza punctiformis Grev. Scot. Crypt. Fl. **2**: 63.   1824.
?Helotium punctatum Fries, Summa Veg. Scand. 356.   1849.
Helotium Ilicis Phill. Brit. Discom. 164.   1887.
?Helotium punctiforme Phill. Brit. Discom. 168.   1887.
?Pseudohelotium punctiforme Sacc. Syll. Fung. **8**: 295.   1889.
?Pezizella punctiformis Rehm in Rab. Krypt.-Fl. **1**³: 664.   1892.
Pezizella populina Rehm in Rab. Krypt.-Fl. **1**³: 668.   1892.
Cálycina Ilicis Kuntze, Rev. Gen. Pl. **3**³: 448.   1898.
Calycina populina Kuntze, Rev. Gen. Pl. **3**³: 448.   1898.
Calycella Ilicis Boud. Hist. Class. Discom. Eu. 95.   1907.
Hyaloscypha punctiformis Boud. Hist. Class. Discom. Eu. 126.   1907.
Micropodia populina Boud. Hist. Class. Discom. Eu. 128.   1907.
Hymenoscypha punctiformis Schröt. Krypt.-Fl. Schles. **3**²: 71.   1893.
Calycellina populina Migula in Thome's Krypt.-Fl. **10**²: 1152.   1913.
Calycellina punctiformis Höhn. Sitz. Akad. Wissen. Wien **127**¹: 601.   1918.

Apothecia usually numerous, solitary, or more rarely in groups of two or three, unevenly distributed over large portions of the substratum, at first minute, globose, dirty yellowish-white when fresh, narrowed toward the base, subsessile, opening with a pore, becoming cup-shaped, more shallow as the cup expands, reaching a diameter of 1 mm., contracting on drying; hymenium creamy-white when fresh, plane, slightly concave, or convex, drying to dull-yellow; asci clavate, reaching a length of 50–70 μ and a diameter of 6–9 μ, 8-spored; spores 2-seriate or partially so, irregularly fusoid, slightly curved, or unequal-sided, 2.7–3.4 × 10–13 μ; paraphyses filiform, enlarged above.

On dead leaves of various kinds.

TYPE LOCALITY: Europe.

DISTRIBUTION: Massachusetts, New York and Pennsylvania.

ILLUSTRATIONS: Farlowia 1: 168, *f. 14.*

37. **Helotium erraticum** White, Farlowia 1: 606.   1944.

Apothecia scattered, or subgregarious, stipitate, when fresh white to creamy-yellow, reaching a diameter of 2.5 mm. and a height of 1 mm., when dry becoming yellow or reddish-yellow, cartilaginous; hymenium when fresh white or nearly so, plane, on drying becoming bright-yellow to reddish; stem stout, whitish to translucent-yellow; asci clavate, reaching a length of 90–120 μ

and a diameter of 9–11.5 μ, 8-spored; spores 2-seriate above, 1-seriate below, ellipsoid to ovoid, slightly curved, or flattened on one side, attenuated toward the ends, sometimes more or less slipper-shaped, 3.5–4 $\times$ 11–17 μ; paraphyses long, straight, simple, or rarely once branched.

On much decayed leaves of various trees and on old pods of *Robinia Pseudo-acacia.*

TYPE LOCALITY: Coy Glen, New York.

DISTRIBUTION: New York to Quebec.

ILLUSTRATIONS: Farlowia 1: 607, *f. 14–18.*

38. **Helotium simulatum** (Ellis) Seaver, comb. nov.

Peziza simulata Ellis, Bull. Torrey Club **8**: 73. 1881.
Phialea simulata Sacc. Syll. Fung. **8**: 254. 1889.

Apothecia gregarious, stipitate, dull watery-white, reaching a diameter of 3 mm.; hymenium plane, or nearly so, slightly umbilicate, becoming concave in drying; stem about 3 mm. long, contracted and darker below; asci cylindric, 8-spored; spores ellipsoid, hyaline, 12 μ long.

On decaying maple, *Acer*, leaves lying on the ground.

TYPE LOCALITY: Newfield, New Jersey.

DISTRIBUTION: Known only from the type locality.

Ellis states: "The stipe arises directly from the leaf, mostly from the veinlets, and is not attached to any sclerotium. When young the whole plant is milk white."

39. **Helotium Conocarpi** Seaver & Waterston, Mycologia **34**: 517. 1942. (PLATE 107, FIG. 1.)

Apothecia gregarious, at first subglobose and subsessile, becoming expanded and short-stipitate, reaching an extreme diameter of 3 mm., the hymenium usually remaining concave, the margin even or nearly so; hymenium with a delicate lilac tint, externally nearly the same, darker toward the base; stem scarcely exceeding 1 mm. in length; asci clavate, reaching a length of 50–60 μ and a diameter of 6 μ, 8-spored; spores ellipsoid, hyaline, 4 $\times$ 10 μ paraphyses filiform, scarcely enlarged above, about 2 μ in diamater.

On dead leaves of *Conocarpus erectus.*

TYPE LOCALITY: Near Hamilton, Bermuda.

DISTRIBUTION: Known only from the type locality.

ILLUSTRATIONS: Mycologia **34**: 518, *f. 1* (upper figure).

This fungus was found to be rather abundant on the fallen leaves of its host. However, the cups are often so small and, even when expanded, so nearly like the substratum in color that very close search is necessary to find them.

40. **Helotium castaneum** Sacc. & Ellis; Sacc. Michelia **2**: 572. 1882.

Apothecia thickly gregarious, on dead areas on the leaves of the host, occurring on the under side of the leaf, minute, brown, or reddish-brown with a slightly darker exterior, subsessile, scutellate, reaching a diameter of .25 mm.; asci short, stout, reaching a length of 40–50 μ and a diameter of 12–14 μ; spores 2-seriate, or irregularly disposed, ellipsoid, 3–4 × 11–14 μ, densely granular within; paraphyses slender.

On leaves of *Quercus laurifolia* and *Rhododendron maximum*.
TYPE LOCALITY: Gree Cove Springs, Florida.
DISTRIBUTION: Florida and West Virginia.
EXSICCATI: Ellis, N. Am. Fungi *994*.

The spores are described as 1-septate. The writer does not find them so. Specimen on *Rhododendron* appears to be identical.

41. **Helotium destructor** Peck; White, Mycologia **34**: 163.   1942.

Peziza subcarnea Cooke & Peck; Cooke, Bull. Buffalo Soc. Nat. Sci. **2**: 295. 1875.   Not *Peziza subcarnea* Schum.   1803.
Phialea subcarnea Sacc. Syll. Fung. **8**: 265.   1889.
Hymenoscyphus subcarneus Kuntze, Rev. Gen. Pl. **3**³: 486.   1889.

Apothecia minute, stipitate, scarcely visible to the unaided eye, solitary, or gregarious and occasionally rather numerous, reaching an extreme diameter of .5 mm. but often much smaller, gradually expanding and becoming subglobose; stem relatively long, reaching a length of .3 to .5 mm., slender and slightly broadened at the junction with the cup; hymenium similar in color to the outside of the apothecium, at maturity concave with upturned margin; asci clavate, reaching a length of 40–50 μ and a diameter of 5–6 μ, 8-spored; spores clavate, 2-seriate, the narrow end directed downward, 2–2.5 × 4.5–6 μ; paraphyses simple, or occasionally branched, scarcely enlarged above, 3–3.5 μ in diameter.

Growing on and apparently killing various species of liverworts and mosses: *Jungermania* sp. (type) and *Dicranum flagellare*.

TYPE LOCALITY: Indian Lake, New York.

DISTRIBUTION: New York and New Hampshire to Michigan and Alberta, Canada.

ILLUSTRATIONS: Mycologia **34**: 166, *f. 8.*

EXSICCATI: Ellis & Ev. N. Am. Fungi *2143.*

42. **Helotium episphaericum** Peck, Ann. Rep. N. Y. State Mus. **40**: 66. 1887.

Calycina episphaerica Kuntze, Rev. Gen. Pl. **3³**: 448. 1898.
Helotium parasiticum Ellis & Ev. Jour. Myc. **9**: 165. 1903.
Dermatea mycophaga Massee, Kew Bull. Misc. Inf. **1908**: 218. 1908.

Apothecia minute, sessile, or subsessile, reaching a diameter of .2 to .5 mm., solitary, or gregarious, occasionally several crowded together, pale-yellow to bright-orange, or slightly rufous; stem-like base V-shaped; hymenium concave, the color as above; asci clavate-cylindric, reaching a length of 48–65 μ and a diameter of 5 μ; spores clavate, 1-seriate, or partially 2-seriate, 1.8–2.5 × 4.5–7 μ; paraphyses scarcely enlarged above, reaching a diameter of 2 μ.

On stromatic pyrenomycetous fungi: *Hypoxylon, Diatrypella, Xylaria* and *Valsa*?

TYPE LOCALITY: Elizabethtown, New York.

DISTRIBUTION: New York and Ontario; also in Europe and Asia.

ILLUSTRATIONS: Mycologia **34**: 166, *f. 15.*

43. **Helotium laricinum** (Ellis & Ev.) Seaver, comb. nov.

Pseudohelotium laricinum Ellis & Ev. Proc. Acad. Sci. Phila. **1894**: 349. 1894.

Apothecia sessile or subsessile, thin, almost membranaceous, fleshy, orange-colored, flat-discoid, 3–5 mm. in diameter, the margin slightly incurved when dry, externally minutely pubescent, darker outside; asci cylindric, reaching a length of 75 μ and a diameter of 8 μ, 8-spored; spores 1-seriate, long-ellipsoid, 4–5 × 8–10 μ; paraphyses filiform, not thickened above.

On decaying needles of *Larix* in a tamarack swamp.

TYPE LOCALITY: Northfield, Michigan.

DISTRIBUTION: Known only from the type locality.

The type in The New York Botanical Garden has been examined. As stated by Ellis the color in dried specimens is pale-orange.

44. **Helotium aurantium** Seaver, nom. nov.

Pezizella aurantiaca Cash, Mycologia **28**: 301.   1936.   Not *Helotium auranti-acum* Cooke; Phill. Grevillea **19**: 106.   1891.

Apothecia sessile, or subsessile, superficial, single, or cespitose, at first subglobose, then cup-shaped to applanate, contorted by mutual pressure, waxy-fleshy, salmon-orange to orange-brown, 1–2 mm. in diameter, externally powdery, margin delicately fimbriate; hymenium concave, or plane, colored like the exterior; asci cylindric-clavate, gradually narrowed below, 8-spored, reaching a length of 38–45 μ and a diameter of 4–5 μ; spores long-ellipsoid, straight, or curved, 1–2-seriate, hyaline, or pale-greenish, usually with two oil-drops, 1.5–2 $\times$ 7–9 μ; paraphyses filiform, rarely branched, enlarged above to 1.5 μ.

On bark of *Picea*.

TYPE LOCALITY: Grand Mesa, Colorado.

DISTRIBUTION: Known only from the type locality.

ILLUSTRATIONS: Mycologia **28**: 302, *f. 5.*

The author of this species states that in general appearance it resembles the preceding, *Helotium laricinum*, from which it differs microscopically.

45. **Helotium atrosubiculatum** Seaver & Waterston, Mycologia **32**: 397.   1940.   (PLATE 107, FIG. 2.)

Apothecia thickly gregarious, occasionally forming congested masses, stipitate, gradually expanding above, becoming shallow cup-shaped, occasionally convoluted, 2–4 mm. in diameter and about 2 mm. high, externally grayish-brown and pruinose; hymenium concave, whitish, even, or in larger specimens convoluted; asci clavate, reaching a length of 60 μ and a diameter of 6 μ; spores ellipsoid, each containing two oil-drops, 2–2.5 $\times$ 6–7 μ; paraphyses filiform, about 1 μ in diameter.

On the blackened surface of leaves of *Archontophoenix Alexandrae* rotting on the ground.

Type collected at Hungry Bay, Bermuda, December 2, 1938. The black subiculum seems to be a constant character in this species.   The base of the stem of the fungus is also black and easily detached near the base leaving disc-like scars which themselves look like minute discomycetes.

DISTRIBUTION: Bermuda and British Guiana.

ILLUSTRATIONS: Mycologia **34**: 15, *f. 1,* (lower figure).

This fungus was collected in Bermuda by the writer and Mr. Waterston in 1938 and again in 1940. It was also collected by D. H. Linder in British Guiana and Trinidad on palm leaves.

46. **Helotium citrinulum** Karst. Not. Soc. Fauna Fl. Fenn. **11**: 238. 1871.

Pezizella citrinula Sacc. Syll. Fung. **8**: 288. 1889.
Helotium flexuosum Massee, Brit. Fungus-Fl. **4**: 263. 1895.
Helotium citrinulum Seaveri Rehm, Ascom. *1634;* Ann. Myc. **4**: 67. 1906.
Mollisiella citrinula Boud. Hist. Class. Discom. Eu. 142. 1907.

Apothecia scattered, sessile or subsessile, smooth, becoming discoid, lemon-yellow, reaching a diameter of 1 mm.; hymenium plane, or slightly concave; asci reaching a length of 45–60 μ and a diameter of 6 μ, 8-spored; spores fusoid, straight, or curved, 1.5–2 $\times$ 6–9 μ; paraphyses very slender, slightly enlarged above.

On stems of grasses and sedges near the base. American specimen on *Carex*.

TYPE LOCALITY: Europe.

DISTRIBUTION: Iowa and New York; also in Europe.

EXSICCATI: Rehm, Ascom. *1634.*

Two collections of this fungus have been seen from North America. Rehm made a new variety on the Iowa material but it seems to us doubtful that it should be so regarded. The type of *Helotium flexuosum* in The New York Botanical Garden seems to be identical.

47. **Helotium planodiscum** (Peck & Cooke) White, Mycologia **34**: 171. 1942.

Peziza planodisca Peck & Cooke; Peck, Ann. Rep. N. Y. State Mus. **31**: 46. 1879.
Pezizella planodisca Sacc. Syll. Fung. **8**: 281. 1889.
Hymenoscypha planodisca Lindau in E. & P. Nat. Pfl. **1¹**: 204. 1897.

Apothecia scattered, or subgregarious, pale-yellow, sessile or subsessile, attached by a relatively narrow basal portion with the margins free, reaching a diameter of .4 mm.; hymenium entirely plane from the beginning, pale-yellow; asci clavate, slightly narrowed to a thick stem-like base, reaching a length of 40–55 μ and a diameter of 5–7 μ; spores obliquely 1-seriate with ends overlapping, straight, or slightly curved, fusoid, narrower below, not granular, 2–2.5 $\times$ 7–9 μ; paraphyses simple, or sparingly branched, enlarged above, 2.5–3.5 μ in diameter.

On dead leaves of *Andropogon scoparius*.

TYPE LOCALITY: Buffalo, New York.

DISTRIBUTION: Known only from the type locality.

ILLUSTRATIONS: Mycologia 34: 166, *f. 12.*

This according to W. L. White is very close to *Helotium citrinulum* Karst. (*Helotium flexuosum* Massee).

48. **Helotium cyathoideum** (Bull.) Karst. Fungi Fenn. *836*; Not. Soc. Fauna Fl. Fenn. 11: 237. 1871.

?Peziza tenella Batsch, Elench. Fung. Cont. 1: 215. 1786.
Peziza cyathoidea Bull. Hist. Champ. Fr. 250. 1791.
Peziza Solani Pers. Obs. Myc. 2: 80. 1799.
Peziza Urticae Pers. Myc. Eu. 1: 286. 1822.
Cyathicula vulgaris DeNot. Comm. Critt. Ital. 1: 381. 1864.
Helotium Urticae Karst. Not. Soc. Fauna Fl. Fenn. 13: 234. 1873.
Helotium Limonium Cooke & Peck; Peck, Ann. Rep. N. Y. State Mus. 26: 83. 1874.
Phialea cyathoidea Gill. Champ. Fr. Discom. 106. 1882.
Calycella cyathoidea Quél. Ench. Fung. 307. 1886.
?Calycella tenella Quél. Ench. Fung. 307. 1886.
Phialea Solani Sacc. Syll. Fung. 8: 252. 1889.
Hymenoscypha cyathoidea Phill. Brit. Discom. 140. 1887.
Hymenoscypha Urticae Phill. Brit. Discom. 141. 1887.
Hymenoscypha cyathoidea Solani Phill. Brit. Discom. 141. 1887.
Calycina Limonium Kuntze, Rev. Gen. Pl. 3³: 448. 1889.

Apothecia gregarious, stipitate, at first closed, becoming expanded and shallow cup-shaped, reaching a diameter of .5 to 2 mm.; whitish, or slightly yellowish; hymenium concave, or nearly plane, a little darker than the outside of the apothecium; stem slender, nearly as long as the diameter of the cup, but occasionally much longer, similar in color to the outside of the apothecium; asci cylindric-clavate, reaching a length of 45–50 μ and a diameter of 4.5–5 μ, 8-spored; spores 2-seriate, elongated, fusoid, or subclavate, straight, or curved, usually with a small oil-drop in each end, 1.5–2 × 5–10 μ; paraphyses slender, clavate, reaching a diameter of 3 μ.

On herbaceous stems or more rarely on small twigs.

TYPE LOCALITY: Europe.

DISTRIBUTION: New York to Pennsylvania, Alabama and California?; also in Europe.

ILLUSTRATIONS: Bull. Herb. Fr. *pl. 416, f. 3;* Batsch, Elench. Fung. Cont. 1: *f. 151;* Boud. Ic. Myc. *pl. 494;* Rab. Krypt.-Fl. 1³: 703, *f. 1–5.*

EXSICCATI: Ellis, N. Am. Fungi *986;* Ellis & Ev. N. Am.

Fungi *2632* (as *Helotium fumosum*); Clements, Crypt. Form. Colo. *77;* Brenckle, Fungi Dak. *537;* Shear, N. Y. Fungi *325*.

A form *"graminicola"* is distributed by Roumeguère, Fungi Selecti Exsiccati *7158*, and ascribed to Lambotte (Mem. Soc. Sci. Liege **14**: 310. 1888.) This would be *Helotium cyathoideum* f. *graminicola*.

49. Helotium consanguineum Ellis & Ev. Fungi Columb. *1220*. 1897.

Apothecia minute, about .5 mm. in diameter, thickly gregarious, often several confluent, subsessile, yellow; hymenium plane or nearly so, bright lemon-yellow, asci clavate, reaching a length of 45 μ and a diameter of 6 μ; spores fusoid, indistinct, about 2 $\times$ 6–7 μ; paraphyses filiform.

On old stems of *Tephrosia virginiana*.

TYPE LOCALITY: Newfield, New Jersey.

DISTRIBUTION: Known only from the type locality.

EXSICCATI: Ellis & Ev. Fungi Columb. *1220*.

Two collections of this species are in the herbarium of The New York Botanical Garden, one collected in February and one in September, 1879. While distributed in exsiccati the species was not previously published so far as we can find.

50. Helotium rhizicola Seaver, sp. nov. (PLATE 106.)

Apothecia thickly gregarious, short-stipitate, or nearly sessile, yellowish, or with a tinge of salmon when moist, becoming shallow cup-shaped, or nearly discoid, pale yellowish-white, 1–2 mm. in diameter; hymenium nearly plane, yellowish, or with a tinge of salmon; stems whitish, often very short and again several mm. in length depending upon the position of the substratum; asci clavate, reaching a length of 90–100 μ and a diameter of 6–7 μ, 8-spored; spores ellipsoid, or fusoid, 3–4 $\times$ 7–10 μ, granular within and often with a small oil-drop or group of oil-drops at either end; paraphyses filiform, about 2–3 μ in diameter at their apices.

Apotheciis sessilis vel subsessilis, 1–1.5 mm. diam., extus pilis brunneis vestitis; ascis clavatis, 5–6 $\times$ 150 μ, 8-sporis; sporis fusoideis, 4 $\times$ 13–15 μ, paraphysibus, clavatis, 4 μ diam.

On exposed or partially submerged roots and basal stems of *Polygonum virginianum* and *Collinsonia canadensis*.

TYPE LOCALITY: The New York Botanical Garden.

DISTRIBUTION: Known only from the type locality.

This species was common and abundant during the month of September, 1930.

51. Helotium gemmarum Boud. Bull. Soc. Myc. Fr. **4**: 81.   1881.

Phialea gemmarum Sacc. Syll. Fung. **8**: 271.   1889.

Apothecia scattered, or subgregarious, stipitate, white when fresh, reaching a diameter of 1.4 mm., at first infundibuliform, then expanding; hymenium concave, or plane, same color as the outside of the apothecium; stem slender, slightly enlarged upward, white, reaching a length 1.5 mm. (5 mm. according to Boudier); asci clavate, reaching a length of 45–50 μ and a diameter of 5–7 μ, 8-spored; spores pyriform, small, 2.5–3 $\times$ 6–9 μ; paraphyses sparse, simple, or once branched below the middle, not enlarged above.

On the scales of buds of various species of *Populus*, in the early spring.

TYPE LOCALITY: Europe.

DISTRIBUTION: New York to Pennsylvania and Georgia; also in Europe.

ILLUSTRATIONS: Bull. Soc. Myc. Fr. **4**: *pl. 17, f. 2;* Boud. Ic. Myc. *pl. 493;* Farlowia **1**: 615, *f. 35–40.*

The only specimen of this species seen was one sent, on request, the fungus collected on the campus of the Agricultural College, Athens, Georgia, by Dr. J. H. Miller.

52. Helotium Craginianum (Ellis & Ev.) Seaver, comb. nov.

Phialea Craginiana Sacc. Syll. Fung. **8**: 258.   1889.
Peziza Craginiana Ellis & Ev. Jour. Myc. **1**: 47.   1885.

Apothecia stipitate, discoid, pale waxy-white when fresh, darker when dry; stem slender, 2–3 mm. long; asci cylindric, reaching a length of 75 μ and a diameter of 6 μ, 8-spored; spores ellipsoid, hyaline, 1-seriate, or partially 2-seriate, 2.5–3 $\times$ 5–6 μ; paraphyses rather stout, often branched above, but scarcely thickened.

On very rotten wood.

TYPE LOCALITY: Topeka, Kansas.

DISTRIBUTION: Known only from the type locality.

The type in The New York Botanical Garden is so fragmentary that nothing can be added to the above which has been compiled from the original description.

53. **Helotium lanceolato-paraphysatum** (Rehm) Seaver, comb. nov.

Pezizella lanceolato-paraphysata Rehm, Ann. Myc. **6**: 316. 1908.

Apothecia gregarious, sessile, or subsessile, at first globose-closed, finally expanding and becoming plane, .3–.8 mm. in diameter, irregularly contracted when dry; hymenium yellowish with a whitish margin; asci clavate, reaching a length of 30 μ and a diameter of 5–7 μ, 8-spored; spores clavate, 2-seriate, 2 $\times$ 6–8 μ; paraphyses lanceolate, acute, prominent.

On dried stems of *Spiraea filipendula*.

TYPE LOCALITY: Lyndonville, New York.

DISTRIBUTION: Known only from the type locality.

As indicated by the specific name this species is characterized by its lanceolate paraphyses, unusual in the plants of this genus.

54. **Helotium albuminum** (Cooke & Peck) Sacc. Syll. Fung. **8**: 214. 1889.

Peziza albumina Cooke & Peck; Peck, Ann. Rep. N. Y. State Mus. **26**: 81. 1872.

Calycina albumina Kuntze, Rev. Gen. Pl. **3**³: 448. 1898.

Apothecia stipitate, or substipitate, solitary, more often gregarious, or sometimes crowded and showing a tendency to become confluent, whitish, pale-yellow when dry, expanding and becoming nearly plane in dried specimens, rather deeply concave, even, or slightly irregular, less than 1 mm. in diameter; stem slender, often nearly lacking; hymenium similar in color to the outside of the apothecium, or yellowish; asci cylindric above and slightly narrowed below, reaching a length of 42–55 μ and a diameter of 5 μ, 8-spored; spores allantoid, 1.2–1.8 $\times$ 7–9 μ; paraphyses simple, scarcely enlarged at their apices, about 2 μ in diameter.

On wood and bark, reported on quince and maple, *Acer*.

TYPE LOCALITY: New York.

DISTRIBUTION: New York to Ontario and Oregon.

ILLUSTRATIONS: Mycologia **34**: 166, *f. 5*.

55. **Helotium ammoides** (Sacc.) Seaver, comb. nov.

Pseudohelotium ammoides Sacc. Ann. Myc. **6**: 564. 1908.

Apothecia densely or loosely gregarious, superficial, subsessile, minute, granuliform, whitish-honey-colored, waxy, at first cup-shaped, finally expanded, about .3 mm. in diameter, whitish-pruinose; asci clavate, reaching a length of 60–80 μ and a diameter of 8–9 μ, 8-spored; spores 2-seriate or partially so, ellipsoid, 2.5–3 × 8.5–9 μ; paraphyses filiform, scarcely enlarged above.

On old bark of *Carpinus*.

TYPE LOCALITY: Lyndonville, New York.

DISTRIBUTION: Known only from the type locality.

56. **Helotium propinquum** Sacc. & Ellis; Sacc. Michelia **2**: 572. 1882.

Apothecia sparse, sessile or subsessile, patelliform, yellowish, .5–.7 mm. in diameter; hymenium concave to nearly plane; asci clavate, reaching a length of 50–60 μ and a diameter of 6–7 μ, 8-spored; spores fusoid, slightly curved, 1.5–2 × 7–10 μ; paraphyses filiform.

On branches of *Cornus*.

TYPE LOCALITY: Pennsylvania.

DISTRIBUTION: Known only from the type locality.

EXSICCATI: Ellis, N. Am. Fungi *995*.

Type material in the herbarium is too scant to permit of satisfactory study and nothing can be added to the above.

57. **Helotium Cassandrae** (Kanouse) Seaver, comb. nov.

Phialea Cassandrae Kanouse, Papers Mich. Acad. Sci. **20**: 71. 1935.

Apothecia solitary, scattered, stipitate, reaching a diameter of .5 mm., translucent-white; stem extremely long and thin, flexuous, the lower half or more black, the upper half like the apothecium, sub-bulbous at the base; hymenium similar in color to the outside of the apothecium; asci clavate, reaching a length of 30–42 μ and a diameter of 5–7 μ, 8-spored; spores 2-seriate, elongated, attenuated below, 1.4–1.8 × 6–8 μ; paraphyses simple, or sparingly branched, reaching a diameter of 3–4.5 μ above.

On fallen leaves of *Chamaedaphne calyculata*.

TYPE LOCALITY: Mud Lake Bog, Michigan.

DISTRIBUTION: Michigan.

ILLUSTRATIONS: Farlowia **1**: 168, *f. 13*.

The author of the species states: "This beautiful little fungus is distinguished by the small spores and by the contrasting black and white of the stipe."

58. **Helotium Friesii** (Weinm.) Sacc. Syll. Fung. **8**: 228.   1889.

Peziza Friesii Weinm. Hymeno-Gasteromycetes 469.   1836.

Apothecia short-stipitate, or nearly sessile, reaching a diameter of 1–2 mm.; hymenium plane, or convex, pale-yellow, when dry rather deep-yellow; asci clavate, reaching a length of 65–70 μ and a diameter of 5–6 μ, 8-spored; spores slightly clavate, 3–4 $\times$ 8–9 μ; paraphyses filiform, slender.

On decaying leaves of *Populus*.

TYPE LOCALITY: Europe.

DISTRIBUTION: Iowa; also in Europe.

ILLUSTRATIONS: Bull. Lab. Nat. Hist. State Univ. Iowa **6**: *pl. 23, f. 2.*

This was originally reported on dejected leaves of *Populus Tremula* and *Betula alba* but no spore measurements given. The above description is drawn from Iowa material. Rehm treats this as a doubtful synonym of *Helotium foliicolum* Schröt. (Schles. Crypt.-Fl. **3**²: 82.   1893.) but the spores of that species are 4–5 $\times$ 20–24 μ.

59. **Helotium midlandense** White, Farlowia **1**: 605.   1944.

Apothecia (described from dried specimens), scattered to subgregarious, usually solitary, more rarely in clusters of three to eight, stipitate, yellow, or yellow-orange, .3–.7 mm. in diameter and about the same height; hymenium plane to patelliform, more rarely slightly convex, yellow, or yellow-orange; stem about equal to the diameter of the apothecium; asci clavate, reaching a length of 45–65 μ and a diameter of 6–7 μ; spores 2-seriate or partially so, irregularly ovoid, 2.6–3.2 $\times$ 7–10 μ; paraphyses filiform, septate.

On petioles and larger leaf veins of leaves of *Quercus;* also on old pods of *Gleditsia triacanthos*.

TYPE LOCALITY: Iowa.

DISTRIBUTION: Ohio to Iowa and Kansas.

ILLUSTRATIONS: Farlowia **1**: 607, *f. 10–13.*

60. **Helotium umbrinum** (Kanouse) Seaver, comb. nov.

Pseudociboria umbrina Kanouse, Mycologia **36**: 460.   1941.
?Helotium contortum White, Farlowia **1**: 147.   1943.

Apothecia foliicolous, stipitate, arising from leaf-blades or veins, externally longitudinally striate, dark-brown to brownish-black, reaching a diameter of 1 mm.; hymenium plane, similar in color to the outside of the apothecium; stem 1 mm. long, expanding rather abruptly into the apothecium; asci cylindric-clavate, reaching a length of 55–70 μ and a diameter of 6–7 μ, 8-spored; spores obliquely 1-seriate, ellipsoid to subovoid, 3.5–4 × 5–7 μ; paraphyses of two kinds, hyaline paraphyses 1 μ in diameter and colored paraphyses, stout, 4–4.5 μ in diameter and entirely filled with dark-brown coloring matter, the apices rounded.

On decaying leaves of *Alnus*.

TYPE LOCALITY: Lake Crescent, Olympic National Forest, Washington.

DISTRIBUTION: New York to Washington.

ILLUSTRATIONS: Farlowia 1: 153, *f. 4* ?

61. **Helotium arenicola** (Ellis & Ev.) Seaver, comb. nov.

Phialea arenicola Ellis & Ev. Am. Nat. 31: 426. 1897.

Apothecia stipitate, at first concave, becoming plane, or even slightly convex, externally light-yellow, subpruinose, 2–3 mm. in diameter; hymenium dull-orange; stem stout, 2–4 mm. long, substriate, gradually enlarged above, same color as the hymenium; asci cylindric, or subcylindric, reaching a length of 60–65 μ and a diameter of 4 μ, 8-spored; spores 2-seriate, or partially so, long-ellipsoid, 1.5 × 6–8 μ.

On sandy ground, probably attached to submerged roots or stems.

TYPE LOCALITY: Blackbird Landing Bridge, Delaware.

DISTRIBUTION: Known only from the type locality.

Type in The New York Botanical Garden has been examined. The material is scant.

DOUBTFUL AND EXCLUDED SPECIES

Helotium albovirens Cooke, Bull. Buffalo Soc. Nat. Sci. 2: 299. 1875. Apothecia scattered, or subgregarious, pale greenish-white, sessile, attached beneath by white, arachnoid threads, 1 mm. in diameter; hymenium plane, or convex, darker when dry; asci cylindric, 8-spored; spores long-ellipsoid, straight, or curved, 3 × 18 μ; paraphyses not observed. On wood of *Acer*, New Jersey.

Helotium alniellum (Nyl.) Karst. Not. Soc. Fauna Fl. Fenn. 11: 239. 1871; *Peziza alniella* Nyl. Not. Soc. Fauna Fl. Fenn. 10: 45. 1869; *Phialea alniella* Sacc. Syll. Fung. 8: 257. 1889. This species has been reported from Washington by Dr. B. Kanouse on maple seed. It has previously been reported on catkins of *Alnus* and would seem to belong to the genus *Ciboria* as defined by Whetzel.

Helotium aurantiacum Cooke; Phill. Grevillea **19**: 106. 1891. This orange-yellow species is described as having spores 4–5 × 14–18 μ, becoming pseudoseptate. On the under side of decaying leaves, United States. The species is in doubt.

Helotium brassicaecolum (Schw.) Sacc. Syll. Fung. **8**: 226. 1889; *Sarea brassicaecola* Schw. Trans. Am. Phil. Soc. II. **4**: 178. 1832. Saccardo places this doubrfully in the genus *Helotium*. Nothing more is known of it.

Helotium Buccina (Pers.) Fries, Summa Veg. Scand. 355. 1849; *Peziza Buccina* Pers. Syn. Fung. 659. 1801. This has been reported from North America by Saccardo on pine branches. It is usually regarded as a basidiomycete, *Guepiniopsis*.

Helotium caraborum Vel. Monog. Discom. Bohem. **1**: 208. 1934. Apothecia less than 1 mm. in diameter, milky-gray; spores 4–5 × 8–10 μ. Reported from Lake Crescent, Washington on moss by Dr. B. Kanouse. No material seen.

Phialea carneola Sacc.; Sacc. Peck, & Trelease in Harriman Alaska Exped. **5**: **25**. 1904. This is said to be close to *P. cyathoidea* but differs slightly in color and size of spores, 3–3.5 × 11–14 μ.

Helotium Cecropiae P. Henn. Hedwigia **41**: 25. 1902; *Phialea Cecropiae* Seaver, Mycologia **17**: 50. 1925. This species has been recorded from Porto Rico by the writer on leaves of *Cecropia*.

Helotium conformatum Karst. Not. Soc. Fauna Fl. Fen. **11**: 236. 1871. *Peziza conformata* Karst. Not. Soc. Fauna Fl. Fenn. **10**: 149. 1869. This species has been reported from Michigan by Dr. B. Kanouse. No material has been seen.

Phialea crocea (Schw.) Sacc. Syll. Fung. **8**: 261. 1889; *Peziza crocea* Schw. Trans. Am. Phil. Soc. II. **4**: 176. 1832. Saccardo suggests that this may be a *Guepinia*.

Helotium crocinum Berk. & Curt.; Berk. Jour. Linn. Soc. **10**: 369. 1868. Described as yellow, obconic, with a short, thick, stem, crateriform, the margin inflexed. On dead twigs, Cuba. Drawings from the type by Massee in the herbarium of The New York Botanical Garden show the spores 1-septate which would make it a *Calloria*. Specimens in the herbarium of The New York Botanical Garden from Ohio are claimed by Massee, through whom the specimens came, to agree with the type of this species.

Helotium discretum Karst. Not. Soc. Fauna Fl. Fenn. **1**: 235. 1871; *Phialea discreta* Rehm in Rab. Krypt.-Fl. **1**³: 729. 1893. This species has been reported on gooseberry (*Ribes*) cane from Washington by Dr. B. Kanouse. No material seen.

Helotium ferrugineum (Schum.) Fries, Summa Veg. Scand. 356. 1849; *Peziza ferruginea* Schum. Pl. Saell. **2**: 429. 1803. This species on dead leaves of *Quercus*, has been reported from North America. No material has been seen.

Helotium fumigatum Sacc. & Sacc. & Speg. Michelia **2**: 78. 1880; Sacc. Syll. Fung. **8**: 235. 1889. Reported from Alaska by Saccardo in Harriman Alaska Exped. **5**: 26. 1904. On decaying herbaceous stems, Yakutat Bay, Alaska.

Helotium fumosum Ellis & Ev. N. Am. Fungi *2632;* Sacc. Syll. Fung. **11**: 403. 1895. No description has been seen and apparently it was never published. On stems of *Leonurus Cardiaca*, Lyndonville, New York.

Helotium furfuraceum Phill. & Hark. Bull. Calif. Acad. Sci. **1**: 24.   1884.
Scattered, minute, short-stipitate, or sessile, cyathiform, alutaceous, furfuraceous; asci clavate, 8-spored; spores ellipsoid, slightly curved, 5 × 15 μ.   On
under side of dead leaves of *Quercus agrifolia* in California.   Nothing more
known of the species.

Helotium imberbe (Bull.) Fries, Summa Veg. Scand. 356.   1849; *Peziza
imberbis* Bull. Hist. Champ. Fr. 245, *pl. 467, f. 2.*   1791.   This white species
has been reported from Wisconsin and Michigan.   The determinations are
not at all certain.

Pseudohelotium isabellinum Clements, Bot. Surv. Nebr. **4**: 15.   1896.
Reported on wet twigs from Rock Creek, Keyapaha County, Nebraska.   No
material seen.

Pezizella leguminum (Schw.) Sacc. Syll. Fung. **8**: 290.   1889; *Peziza
leguminum* Schw. Trans. Am. Phil. Soc. II.**4**: 176.   1832.   A minute, scattered,
emergent, brown species occurring on pods of *Bignonia* and *Catalpa* in Bethlehem, Pennsylvania.   The fruit was not described and the species is in doubt.

Phialea leucopsis (Berk. & Curt.) Sacc. Syll. Fung. **8**: 264.   1889; *Peziza
leucopsis* Berk. & Curt.; Berk. Jour. Linn. Soc. **10**: 368.   1868.   Flesh-pallid,
irregular, the margin erect, then reflexed, 3–6 mm. in diameter; spores allantoid, 5 μ long.   On dead wood, Cuba.

Discinella lividopurpurea Boud. Bull. Soc. Myc. Fr. **4**: 79.   1888; *Humaria
lividopurpurea* Sacc. Syll. Fung. **8**: 124.   1889.   This species has been reported
from Michigan by Dr. B. Kanouse.   Since this was described by Boudier as an
inoperculate it should not have been placed in the genus *Humaria* as pointed
out by Dr. Kanouse.   Its position is uncertain.

Helotium microspis Karst. Myc. Fenn. **1**: 152.   1871; *Peziza microspis*
Karst. Not. Soc. Fauna Fl. Fenn. **10**: 178.   1869; *Pezizella microspis* Sacc. Syll.
Fung. **8**: 281.   1889.   Reported from Washington by Dr. B. Kanouse on
sedges.   The description would indicate that it is very close to *Helotium
citrinulum.*

Phialea microspora Seaver, Mycologia **17**: 50.   1925.   Apothecia dull-
yellow, stipitate, 1 mm. in diameter; spores 1.5–2 × 6 μ.   On leaves of
unidentified host in Porto Rico.

Rutstroemia microspora Kanouse, Mycologia **39**: 684.   1947.   This species
is reported by its author as arising from a black stroma in old branches in
Washington.   The general description would indicate that it belongs with our
genus *Helotium.*

Helotium miserum Berk. & Curt.; Berk. Jour. Linn. Soc. **10**: 369.   1868.
Minute, white, clavate, short-stipitate, subglobose; asci clavate; spores sub-
clavate, contents finally four-parted, 12 μ long.   On bark among mosses in
Cuba.

Helotium montaniense [*montanense*] Ellis & Anders. Bot. Gaz. **16**: 45.
1891.   The type specimen in the herbarium of The New York Botanical
Garden is too fragmentary to permit of study.

Helotium monticola Berk. Grevillea **4**: 1.   1875.   On dead wood, North
Carolina.   Described as "crowded, pale tawny, obovate; disc plane; sporidia
biseriate, subfusiform."   A minute specimen, apparently from the type, in the
herbarium of The New York Botanical Garden is too meager for study.

Helotium nigripes (Pers.) Sacc. Syll. Fung. **8**: 215.   1889; *Peziza nigripes*

Pers. Syn. Fung. 661. 1801. This species has been recorded from North America and Cuba. No material has been seen.

Phialea olympiana Kanouse, Mycologia **39**: 681. 1947. Apothecia described as stipitate, 1 mm. in diameter, yellowish-citrine, drying black; spores 1 × 6–8 μ; paraphyses filiform. Described from material collected in Washington. On herbaceous stems.

Phialea pallida Kanouse, Mycologia **39**: 681. 1947. Apothecia described as pale olive-buff, .5 mm. in diameter, stipitate; spores 1.5 × 6–9 μ; paraphyses filiform. Reported from Washington on old stems of *Delphinium*.

Pezizella Pastinacae (Schw.) Sacc. Syll. Fung. **8**: 290. 1889; *Peziza Pastinacae* Schw. Trans. Am. Phil. Soc. II. **4**: 176. 1832. The brief description suggests a *Helotium*. On *Pastinaca*. No material has been seen.

Helotium Phiala (Vahl.) Fries, Summa Veg. Scand. 355. 1849; *Peziza Phiala* Vahl. Fl. Dan. **18**: 8, *pl. 1078, f. 2.* 1792. This has been reported from Washington on leaves of *Sambucus callicarpa* by Dr. B. Kanouse. Apothecia described as buff-yellow, stout with a stem 7 mm. long; spores 7–8 × 14–16 μ, finally becoming 1-septate. No material has been seen.

Helotium pullatum Gerard; Cooke, Bull. Buffalo Soc. Nat. Sci. **2**: 298. 1875. On stems of *Vitis*. No fruit was described and the identity of the species is uncertain. Looks like an *Orbilia*.

Helotium rhytidodes Berk. & Curt.; Berk. Jour. Linn. Soc. **10**: 369. 1868. Described as sessile, cup-shaped, below rugose-plicate; spores ellipsoid, smooth. Cuba. Habitat not given.

Helotium rubicolum (Fries) Fuckel, Symb. Myc. 314. 1869; *Peziza fructigena rubicola* Fries, Syst. Myc. **2**: 119. 1822; *Helotium virgultorum rubicolum* Fries, Summa Veg. Scand. 355. 1849; *Phialea rubicola* Sacc. Syll. Fung. **8**: 253. 1889. This has been reported from Michigan under the latter name but no American material seen.

Helotium scrupulosum Karst. Myc. Fenn. **1**: 152. 1871; *Peziza scrupulosa* Karst. Not. Soc. Fauna Fl. Fenn. **10**: 178. 1869. This species is reported on old wood from Washington by Dr. B. Kanouse under the name of *Unguicularia scrupulosa* (Karst.) Höhn. Nothing more is known of it.

Helotium sordidatum Karst. & Starb.; Karst. Hedwigia **26**: 124. 1887; *Phialea sordidata* Sacc. Syll. Fung. **8**: 271. 1889. This species has been recorded from Michigan on leaves and petioles of *Quercus*. The septate spores would place it in the genus *Calloria*.

Helotium sparsum Boud. Hist. Class. Discom. Eu. 111. 1907. Reported by Dr. B. Kanouse from Michigan on rotting leaves. Apothecia described as minute but the spores relatively large 7–8 × 18 μ. See p. 124.

Phialea subgalbula Rehm in Rab. Krypt.-Fl. **1³**: 711. 1892. Reported from Michigan by Dr. B. Kanouse and said to agree with Rehm, Ascom. *1981*. No American material seen.

Helotium tumidulum (Rob.) Massee; Massee & Crossl. Fungus Fl. Yorshire 285. 1905; *Peziza tumidula* Rob.; Desm. Ann. Sci. Nat. III. **16**: 325. 1851; *Pezizella tumidula* Sacc. Syll. Fung. **8**: 276. 1889. This species has been reported from Michigan by Dr. B. Kanouse but according to White (Farlowia **1**: 167. 1943) is based on incorrect determination.

Helotium turgidellum Karst. Myc. Fenn. **1**: 150. 1871; *Peziza turgidella* Karst. Not. Soc. Fauna Fl. Fenn. **10**: 179. 1869; *Pezizella turgidella* Sacc.

Syll. Fung. **8**: 281.   1889.   Reported from Michigan on *Carex* and *Typha*.
No material seen.

   Helotium vitigenum DeNot. Comm. Critt. Ital. **1**: 377.   1864.   This
species originally recorded from Italy has been reported from Ann Arbor,
Michigan.   Spores described as 4–6 × 17–20 μ.   It would seem to be close
to *Helotium virgultorum*.

25. **ORBILIA** Fries, Summa Veg. Scand. 357.   1849.

Pezizella Fuckel, Symb. Myc. 299.   1869.
Myridium Clements, Gen. Fungi 174.   1909.

Apothecia sessile, membranous, subgelatinous, bright-colored, white, yellow, or red, typically smooth, subcorneous when dry; asci cylindric to clavate, usually 8-spored; spores bacilliform, or ellipsoid to subglobose; paraphyses filiform the apices often clavate to subglobose.

Type species, *Peziza leucostigma* Fries.

The plants of this genus are usually minute but because of their light or bright color they are more easily seen and consequently often collected.   The apothecia are typically thin and membranous, the substance subcartilaginous and often translucent.   The spores are minute and often difficult to diagnose. A number of the species have paraphyses with subglobose or pyriform apices.

Apothecia white to golden-yellow.
   Color golden-yellow.   1. *O. chrysocoma*.
   Color white or whitish when fresh.
      Occurring on rotten wood.
         Spores ellipsoid.
            Spores large, 8 × 17–20 μ.   2. *O. diaphanula*.
            Spores small, 3 × 6–7 μ.   3. *O. Fairmani*.
         Spores bacilliform.
            Spores 3–4 μ long.   4. *O. leucostigma*.
            Spores 5–8 μ long.   5. *O. inflatula*.
            Spores 10–15 μ long.   6. *O. curvatispora*.
      Not on rotten wood.
         On rinds of squash, *Cucurbita*.   7. *O. Cucurbitae*.
         On old fungi, *Polyporus*.   8. *O. epispora*.
Apothecia some shade of red, orange to scarlet.
   Asci 8-spored.
      Spores 12 μ or more long.
         Spores bacilliform, 1.5–2 × 12–17 μ.   9. *O. vinosa*.
         Spores ellipsoid to fusoid.
            On reeds in California.   10. *O. phymatodes*.
            On bark of *Gleditsia*.   11. *O. rubrococcinea*.

Spores mostly less than 12 μ long.
  Spores subglobose, or broad-ellipsoid.
    Paraphyses with globose apices.   12. *O. coccinella.*
    Paraphyses filiform.
      Spores subglobose, 2–3 μ in diameter.   13. *O. rubinella.*
      Spores ellipsoid, 3 × 5–7 μ.   14. *O. Eucalypti.*
  Spores bacilliform.
    Spores 4–5 μ long, on bark.   15. *O. cruenta.*
    Spores 5–6 μ long, on *Caulophyllum.*   16. *O. Caulophylli.*
    Spores 6–12 μ long.
      Apothecia yellowish-red.   17. *O. luteorubella.*
      Apothecia flesh-red to dark-red.   18. *O. rubella.*
    Spores minute, spermatoid, on *Phytolacca.*   19. *O. pulviscula.*
Asci many-spored.   20. *O. myriospora.*

1. **Orbilia chrysocoma** (Bull.) Sacc. Syll. Fung. **8**: 624. 1889.

Peziza chrysocoma Bull. Hist. Champ. Fr. 254. 1791.
Calloria chrysocoma Fries, Summa Veg. Scand. 359. 1849.
Dacryomyces chrysocomus Tul. Ann. Sci. Nat. III. **19**: 211. 1853.
Guepiniopsis chrysocomus Brasfield, Am. Mid. Nat. **20**: 226. 1938.

Apothecia gregarious, superficial, at first globose, soon expanded, golden-yellow, when dry subcorneous, flexuous, scarcely .5 mm. in diameter; asci cylindric, subsessile, reaching a length of 40–45 μ and a diameter of 5 μ, 8-spored; spores bacilliform, curved, 1 × 14–15 μ; paraphyses filiform, slightly enlarged above.

On old wood and pasteboard.

TYPE LOCALITY: Europe.

DISTRIBUTION: Vermont to California, Porto Rico and Bermuda; probably throughout North America; also in Europe.

ILLUSTRATIONS: Bull. Herb. Fr. *pl. 376, f. 2;* Cooke, Austr. Fungi *pl. 20, f. 161.*

2. **Orbilia diaphanula** (Cooke) Seaver, comb. nov.

Peziza diaphanula Cooke, Hedwigia **14**: 84. 1875.

Apothecia gregarious, minute, transparent-white, at first hemispherical, then expanded, .1 mm. in diameter; asci clavate; spores elongate-ellipsoid, 8 × 17–20 μ; paraphyses filiform, scarcely enlarged above.

On old bark and wood of decaying *Magnolia.*

TYPE LOCALITY: Newfield, New Jersey.

DISTRIBUTION: Known only from the type locality.

3. **Orbilia Fairmani** (Rehm) Seaver, comb. nov.

Pezizella Fairmani Rehm, Ann. Myc. **5**: 519.   1907.

Apothecia scattered, sessile, first globose, then expanded, externally smooth, .25 mm. in diameter, hyaline to slightly yellowish; asci clavate, reaching a length of 30–35 μ and a diameter of 6–8 μ, 8-spored; spores ovoid, 2-seriate, 3 × 6–7 μ; paraphyses filiform, 3 μ in diameter.

On wood of *Tsuga canadensis.*

TYPE LOCALITY: Lyndonville, New York.

DISTRIBUTION: Known only from the type locality.

4. **Orbilia leucostigma** Fries, Summa. Veg. Scand. 357.   1849.

Peziza leucostigma Fries, Obs. Myc. **1**: 165.   1815.
Peziza xanthostigma Fries, Syst. Myc. **2**: 146.   1818.
Mollisia leucostigma Gill. Champ. Fr. Discom. 126.   1882.
Mollisia xanthostigma Gill. Champ. Fr. Discom. 125.   1882.
Calloria xanthostigma Phill. Brit. Discom. 329.   1887.
Calloria leucostigma Phill. Brit. Discom. 330.   1887.

Apothecia gregarious, or scattered, sessile, becoming plane, submembranaceous, whitish, becoming yellowish, translucent, reaching a diameter of .5 to 1 mm.; asci cylindric to cylindric-clavate, reaching a length of 30–35 μ and a diameter of 3–3.5 μ, 8-spored; spores ellipsoid, 1–1.5 × 3–4 μ; paraphyses slender with subglobose apices.

On dead wood.

TYPE LOCALITY: Europe.

DISTRIBUTION: New Jersey to California and West Virginia; probably widely distributed in North America; also in Europe.

The writer can see no reason for separating *Peziza leucostigma* and *Peziza xanthostigma.* Rehm recognizes a varietal difference only.

5. **Orbilia inflatula** Karst. Not. Soc. Fauna Fl. Fenn. **11**: 248. 1871.

Peziza inflatula Karst. Not. Soc. Fauna Fl. Fenn. **10**: 175.   1869.
Calloria inflatula Phill. Brit. Discom. 335.   1887.

Apothecia patellate, subtremelloid, 1–1.5 mm. in diameter, whitish-translucent, when dry becoming subglobose, or angular, sordid-yellowish, subcartilaginous; asci cylindric, or subcylindric, reaching a length of 18–24 μ and a diameter of 2–3 μ, 8-spored;

spores bacciliform, curved, .5–.7 × 5–8 μ: paraphyses not observed.

On old wood and bark.

TYPE LOCALITY: Europe.

DISTRIBUTION: Toronto to Louisiana; also in Europe.

ILLUSTRATIONS: Massee, Brit. Fungus-Fl. **4**: 156. *f. 48–51.*

6. **Orbilia curvatispora** Boud. Bull. Soc. Myc. Fr. **4**: 80. 1888.

Apothecia reaching a diameter of 1 mm. but often less, transparent-white, becoming slightly yellowish, fleshy; hymenium convex, rarely depressed in the center; asci reaching a length of 40–45 μ and a diameter of 3–3.5 μ, 8-spored; spores slender, bacilliform, much curved, 1 × 10–15 μ; paraphyses filiform, clavate at their apices, 4–5 μ thick.

On rotten wood and bark.

TYPE LOCALITY: Europe.

DISTRIBUTION: Michigan to Indiana and Ontario; also in Europe.

ILLUSTRATIONS: Bull. Soc. Myc. Fr. **4**: *pl. 16, f. 6.*

The only American specimens seen consists of four collections made in Michigan and Indiana, determined by Dr. Geo. B. Cummins. Also one specimen from Ontario determined by Dr. H. S. Jackson.

7. **Orbilia Cucurbitae** (Ger.) Seaver, comb. nov.

Peziza Cucurbitae Gerard, Bull. Torrey Club **5**: 26. 1874.
Pezizella Cucurbitae Sacc. Syll. Fung. **8**: 285. 1889.

Apothecia sessile, when moist waxy, scutellate; hymenium pale tan-colored, when dry somewhat horny and dark purple-brown; spores simple, ellipsoid, 10 μ long.

On dead rinds of squash, *Cucurbita.*

TYPE LOCALITY: Poughkeepsie, New York.

DISTRIBUTION: Known only from the type locality.

8. **Orbilia epispora** (Nyl.) Karst. Not. Soc. Fauna Fl. Fenn. **11**: 248. 1871.

Peziza epispora Nyl. Not. Soc. Fauna Fl. Fenn. **10**: 58. 1869.

Apothecia, gregarious, whitish-translucent, .3–.5 mm. in diameter; asci clavate, reaching a length of 16–19 μ and a diameter 2–2.5 μ; spores bacilliform, .5 × 3–6 μ.

On *Polyporus igniarius.*

TYPE LOCALITY: Europe.

DISTRIBUTION: New Jersey to Wisconsin and Louisiana; also in Europe and South America.

EXSICCATI: C. L. Smith, Centr. Am. Fungi *50*.

The species is similar to *Orbilia leucostigma* but smaller.

9. **Orbilia vinosa** (Alb. & Schw.) Karst. Myc. Fenn. **1**: 101. 1871.

Peziza vinosa Alb. & Schw. Consp. Fung. 308.   1805.
Calloria vinosa Fries, Summa Veg. Scand. 359.   1849.
Mollisia vinosa Gill. Champ. Fr. Discom. 125.   1882.

Apothecia scattered, more rarely crowded, sessile, at first rounded and closed, expanding and becoming plane, or slightly concave, flesh-colored to dark-red, .2–1 mm. in diameter; asci cylindric-clavate, reaching a length of 40–50 μ and a diameter of 4–5 μ, 8-spored; spores bacilliform to fusoid, 1.5–2 $\times$ 12–17 μ; paraphyses filiform, subglobose above, 4 μ in diameter.

On rotten wood of various kinds; also on fabric.

TYPE LOCALITY: Europe.

DISTRIBUTION: Massachusetts and New Jersey to Michigan, Iowa and South Carolina; also in Europe.

ILLUSTRATIONS: Bull. Soc. Myc. Fr. **4**: *pl. 22, f. 1–6;* E. & P. Nat. Pfl. **1**¹: 217, *f. 169, D;* Rab. Krypt.-Fl. **1**³: 447, *f. 1;* Phill. Brit. Discom. *pl. 10, f. 63.*

EXSICCATI: Rav. Fungi Car. **4**: *19;* Ellis, N. Am. Fungi *142, 1313.*

A fine collection of this species was found on an old discarded bedtick in Woods Hole, Massachusetts during the summer of 1946.

10. **Orbilia phymatodes** (Phill.) Seaver, comb. nov.

Peziza phymatodes Phill. Grevillea **5**: 117.   1877.
Pezizella phymatodes Sacc. Syll. Fung. **8**: 285.   1889.

Apothecia scattered, at first spherical, then expanded and cup-shaped, reddish, or flesh-colored, smooth, faintly striate, the margin thin, paler; asci subclavate, 8-spored; spores elongate-fusoid, 3–4 $\times$ 13–16 μ; paraphyses filiform.

On reeds.

TYPE LOCALITY: Blue Canon, California.

DISTRIBUTION: Known only from the type locality.

ILLUSTRATIONS: Grevillea **5**: *pl. 88, f. 9.*

11. **Orbilia rubrococcinea** (Rehm) Sacc. Syll. Fung. **8**: 622. 1889.

Calloria rubrococcinea Rehm; Winter, Hedwigia **22**: 72. 1883.

Apothecia scattered, or gregarious, sessile, patelliform, scarlet-red, contracting when dry, expanding when moist, 1–3 mm. in diameter; asci clavate, reaching a length of 40–45 μ and a diameter of 5 μ, 8-spored; spores clavate, 3 × 12–15 μ; paraphyses filiform 3 μ in diameter above.

On bark of *Gleditsia triacanthos*.

TYPE LOCALITY: Kentucky.

DISTRIBUTION: Known only from the type locality.

12. **Orbilia coccinella** (Sommerf.) Fries, Summa Veg. Scand. 357. 1849.

Peziza coccinella Sommerf. in Wahlenb. Fl. Lapp. 296. 1826.
Mollisia coccinella Gill. Champ. Fr. Discom. 129. 1882.
Calloria coccinella Phill. Brit. Discom. 328. 1887.

Apothecia scattered, or gregarious, minute, subtremelloid, flesh-red, bright-red when dry, reaching a diameter of 1 mm. but often much less; asci cylindric-clavate, reaching a length of 30–50 μ, and a diameter of 4–5 μ, 8-spored; spores subglobose to ellipsoid, 2–2.5 × 3–5 μ; paraphyses very slender, with subglobose apices.

On wood and bark of various kinds, and occasionally on fungi *Polyporus* and *Diatrype*.

TYPE LOCALITY: Europe.

DISTRIBUTION: Delaware to Wisconsin, probably common; also in Europe.

ILLUSTRATIONS: Boud. Ic. Myc. *pl. 461;* Rab. Krypt.-Fl. **1**³: 447, *f. 6–8;* E. & P. Nat. Pfl. **1**¹: 217, *f. 169, A, B.*

13. **Orbilia rubinella** (Nyl.) Karst. Myc. Fenn. **1**: 97. 1871.

Peziza rubinella Nyl. Not. Soc. Fauna Fl. Fenn. **10**: 56. 1869.

Apothecia scattered, subgelatinous, flesh-red, or nearly scarlet, about .5 mm. in diameter; asci clavate to pyriform, 8–12-spored, reaching a length of 39–46 μ and a diameter of 8–11 μ; spores subglobose, about 2–3 μ in diameter; paraphyses filiform, about 1 μ thick, enlarged above to 1.5 μ.

On rotten wood and more rarely on woody fungi.

TYPE LOCALITY: Europe.

DISTRIBUTION: New Hampshire; also in Europe.

EXSICCATI: Reliq. Farlow. *133*.

The above cited specimen is the only one seen from North America.

14. Orbilia Eucalypti (Phill. & Hark.) Sacc. Syll. Fung. **8**: 628. 1889.

Calloria Eucalypti Phill. & Hark. Bull. Calif. Acad. Sci. **1**: 23. 1884.

Apothecia minute, scattered, patellate, flesh-red, the margin paler, smooth; asci cylindric, 8-spored; spores ovoid-ellipsoid, 3–5 × 7 μ; paraphyses filiform and apices pyriform.

On decorticated wood of *Eucalyptus;* also reported on *Salix.*

TYPE LOCALITY: California.

DISTRIBUTION: California and (Washington?).

15. Orbilia cruenta (Schw.) Seaver, comb. nov.

Peziza cruenta Schw. Trans. Am. Phil. Soc. II **4**: 177. 1832.
Peziza rufula Schw. Trans. Am. Phil. Soc. II **4**: 177. 1832.
Peziza fibriseda Berk. & Curt.; Berk. Grevillea **3**: 157. 1875.
Peziza saccharifera Berk. & Curt.; Berk. Grevillea **3**: 157. 1875.
Peziza regalis Cooke & Ellis, Grevillea **6**: 91. 1878.
Calloria occulta Rehm, Hedwigia **24**: 14. 1885.
Pezizella rufula Sacc. Syll. Fung. **8**: 283. 1889.
Pezizella regalis Sacc. Syll. Fung. **8**: 284. 1889.
Pezizella cruenta Sacc. Syll. Fung. **8**: 284. 1889.
Pseudohelotium fibrisedum Sacc. Syll. Fung. **8**: 298. 1889.
Orbilia occulta Sacc. Syll. Fung. **8**: 623. 1889.

Apothecia gregarious, sometimes confluent, sessile, at first closed, then becoming widely open, the margin often wavy, reaching a diameter of 1 mm., subgelatinous; hymenium slightly concave, crimson, or orange-red, brighter than the outside of the apothecium; asci narrowly clavate, with a long stem, 8-spored; spores irregularly 1-seriate, or sometime 2-seriate above, 1.5 × 4– 5 μ; paraphyses slender clavate, or lanceolate at their tips.

On bark of various trees.

TYPE LOCALITY: Bethlehem, Pennsylvania.

DISTRIBUTION: New Jersey to Montana and New Mexico.

EXSICCATI: Ellis, N. Am. Fungi *438* (as *Peziza regalis*), *848* (as *Peziza occulta*); Ellis & Ev. N. Am. Fungi *2326* (as *Peziza cruenta*).

The above synonymy is based largely on the conclusions of George Massee (Grevillea **22**: 99–100. 1894).

16. **Orbilia Caulophylli** Ellis & Ev. Proc. Acad. Sci. Phila. **1893**: 145. 1893.

Apothecia scattered, or gregarious, sessile, subgelatinous, pale rose-colored when fresh, orange when dry, about .5 mm. in diameter; hymenium plane, or slightly concave when fresh, more strongly concave when dry, the margin thin, laciniately toothed when fresh, clothed with short, hair-like glandular structures; asci clavate-cylindric, reaching a length of 40–50 μ and a diameter of 5–6 μ, 8-spored; spores 2-seriate, clavate-ovoid, hyaline, 1.5 × 5–6 μ; paraphyses filiform, not distinctly enlarged above.

On dead stems of *Caulophyllum thalictroides*.

TYPE LOCALITY: London, Canada.

DISTRIBUTION: Ontario to Michigan.

EXSICCATI: Ellis & Ev. N. Am. Fungi *2811*.

17. **Orbilia luteorubella** (Nyl.) Karst. Not. Soc. Fauna Fl. Fenn. **11**: 248. 1871.

Peziza luteorubella Nyl. Not. Soc. Fauna Fl. Fenn. **10**: 55. 1869.
Calloria luteorubella Phill. Brit. Discom. 333. 1887.

Apothecia scattered, or gregarious, subtremelloid, plane or nearly so, often depressed in the center, pallid, then pale yellowish-red, or yellowish-brown, .5–1.5 mm. in diameter; asci cylindric-clavate, reaching a length of 30–40 μ and a diameter of 4–4.5 μ; spores fusoid-filiform, 1–1.5 × 6–12 μ; paraphyses filiform, 1.5 μ thick, the apices clavate, or subsphaeroid, 2–2.5 μ thick.

On rotten wood of *Populus*, *Alnus*, *Salix*, etc.

TYPE LOCALITY: Europe.

DISTRIBUTION: New York to Michigan and Louisiana and the West Indies; also in Europe.

18. **Orbilia rubella** (Pers.) Karst. Not. Soc. Fauna Fl. Fenn. **11**: 248. 1871.

Peziza rubella Pers. Syn. Fung. 635. 1801.
Pezizella rubella Fuckel, Symb. Myc. 299. 1869.
Mollisia rubella Gill. Champ. Fr. Discom. 124. 1882.
Hyalinia rubella Nannf. Nova Acta Soc. Sci. Upsal. IV. **8**: 252. 1932.

Apothecia gregarious, sessile, at first closed and subglobose, expanding and becoming plane or nearly so, flesh-red to dark-red, waxy-gelatinous, .2–1.3 mm. in diameter; asci cylindric-clavate, reaching a length of 30–45 μ and a diameter of 4–5 μ,

8-spored; spores bacilliform, $1 \times 6-12\ \mu$, 2-seriate; paraphyses filiform, gradually enlarged above, $3\ \mu$ in diameter.

On rotten wood.

TYPE LOCALITY: Europe.

DISTRIBUTION: New Jersey to Pennsylvania, and West Virginia; also in Europe.

19. Orbilia pulviscula (Cooke) Seaver, comb. nov.

Peziza pulviscula Cooke, Hedwigia **14**: 84.   1875.
Peziza pulviscula Sacc. Syll. Fung. **8**: 278.   1889.

Apothecia gregarious, subtremelloid, at first globose, pallid to yellowish .4 mm. in diameter; asci cylindric, reaching a length of $30\ \mu$ and a diameter of $5\ \mu$; spores minute, spermatoid.

On stems of *Phytolacca*.

TYPE LOCALITY: New York.

DISTRIBUTION: New York and (New Jersey?).

While this species was placed in the subgenus *Mollisia* by its authors, the general description would indicate an *Orbilia*. This species was based on specimens collected by W. R. Gerard. Specimen in The New York Botanical Garden is evidently a part of the type.   A second specimen collected at Newfield, New Jersey has been referred to this species by J. B. Ellis.

20. Orbilia myriospora (Phill. & Hark.) Sacc. Syll. Fung. **8**: 631. 1889.

Calloria myriospora Phill. & Hark. Bull. Calif. Acad. Sci. **1**: 23.   1884.
Myridium myriosporum Clements, Gen. Fungi 174.   1909.

Apothecia minute, scattered, convex, pale rose-red; asci clavate; spores excessively minute, innumerable; paraphyses slender, abundant.

On dead stems of *Psoralea macrostachya*.

TYPE LOCALITY: California.

DISTRIBUTION: Known only from the type locality.

Saccardo (Syll. Fung. **8**: 631.   1889) founded a new subgenus *Myriella* on this species, but Clements had previously founded the genus *Myridium* on the same species.

DOUBTFUL AND EXCLUDED SPECIES

Orbilia assimilis (Cooke & Peck) Sacc. Syll. Fung. **8**: 629.   1889; *Peziza assimilis* Cooke & Peck, Grevillea **1**: 5.   1872.   Apothecia described as erumpent, dull-orange; spores $3 \times 39\ \mu$.   On stems of *Aster puniceus*, West Albany, New York.   This looks like *Calloria fusarioides*.

Pezizella carneorosea (Cooke & Hark.) Sacc. Syll. Fung. **8**: 284. 1889; *Peziza carneorosea* Cooke & Hark. Grevillea **9**: 130. 1881. Apothecia described as carneous-rosaceous, .3 mm. in diameter: spores ellipsoid, 3 × 5 μ. On twigs of *Eucalyptus* in California. Nothing more is known of this species. Apparently an *Orbilia*.

Pezizella citrinella (Schw.) Sacc. Syll. Fung. **8**: 287. 1889; *Peziza citrinella* Schw. Trans. Am. Phil. Soc. II. **4**: 177. 1832. This is probably an *Orbilia*. According to Schweinitz it is closely related to *Orbilia chrysocoma*.

Pezizella conchella (Schw.) Sacc. Syll. Fung. **8**: 284. 1889; *Peziza conchella* Schw. Trans. Am. Phil. Soc. II. **4**: 177. 1832. Apothecia minute, shell-shaped, sessile, subpellucid, red, confluent. On dead branches of *Morus alba*, Bethlehem, Pennsylvania. This may be an *Orbilia*. No material has been seen.

Orbilia paradoxa Vel. Monog. Discom. Bohem. **1**: 102. 1934. Reported by Dr. B. Kanouse from Washington, on log of *Alnus*. No material seen.

Orbilia rosella (Rehm) Sacc. Syll. Fung. **8**: 623. 1889; *Calloria rosella* Rehm, Hedwigia **23**: 56. 1884; *Laetinaevia rosella* Kanouse, Papers Mich. Acad. Sci. **21**: 100. 1936. Reported from Michigan by Dr. B. Kanouse. No material seen.

Orbilia Sarraziniana Boud. Rev. Myc. **7**: 221. 1885. This species has been reported from Michigan by Dr. B. Kanouse on decaying wood. No material has been seen.

26. **TRICHOBELONIUM** (Sacc.) Rehm in Rab. Krypt.-Fl. **1**³: 590. 1891.

Belonium subg. *Trichobelonium* Sacc. Syll. Fung. **8**: 495. 1889.

Apothecia sessile, gregarious, fleshy to subgelatinous, light- or dark-colored, seated on a white or colored mycelial subiculum; asci cylindric to clavate, usually 8-spored; spores ellipsoid to fusoid, becoming 3-septate.

Type species, *Peziza retincola* Rab.

When Saccardo established the subgenus he described the apothecia as byssoid or pilose. When Rehm raised the subgenus to generic rank he described the apothecia as seated on a spreading white, or colored mycelium. He later included in the genus *Trichobelonium albosuccineum* Rehm, which in our opinion is a synonym *Peziza leucorrhodina* Mont., the only North American species of the genus seen.

1. **Trichobelonium leucorrhodinum** (Mont.) Seaver, comb. nov.

Peziza leucorrhodina Mont. in Sagra, Hist. Cuba Pl. Cell. 360. 1842.
Peziza gelatinosa Ellis & Mart. Am. Nat. **17**: 1283. 1883.
Belonidium leucorrhodinum Sacc. Syll. Fung. **8**: 501. 1889.
Scutula leucorrhodina Speg. Anal. Soc. Ci. Arg. **26**: 58. 1888.
Orbilia gelatinosa Sacc. Syll. Fung. **8**: 624. 1889.

Mollisia gelatinosa Sacc. Syll. Fung. **8**: 624.   1889.
Trichobelonium albosuccineum Rehm, Hedwigia **39**: 89.   1900.

Apothecia gregarious, sessile, subgelatinous, pinkish, .25 mm. in diameter, seated on a white mycelial subiculum; hymenium becoming plane, or slightly convex; asci broad-clavate to ovate, reaching a length of 35–40 μ and a diameter of 15–20 μ, 8-spored; spores 2–3-seriate, fusoid, subhyaline, 3–3.5 × 12–16 μ; paraphyses thickened above, curved.

On the mycelium of species of Perisporiaceae (*Meliola*) on various hosts.

TYPE LOCALITY: Cuba.

DISTRIBUTION: Southern United States, the West Indies and tropical South America.

ILLUSTRATIONS: Hedwigia **39**: *pl. 5, f. 24.*

EXSICCATI: Rehm, Ascom. *1778.*

DOUBTFUL SPECIES

Trichobelonium hercynicum Lindau, Verh. Bot. Ver. Branden. **45**: 154. 1904; *Belonium hercynicum* Boud. Hist. Class. Discom. Eu. 118.   1907.   Reported from Michigan on old wood by Dr. B. Kanouse.   No material seen.

27. CALLORIA Fries, Summa Veg. Scand. 359.   1849.

Niptera Fries, Summa Veg. Scand. 359.   1849.
?Lanzia Sacc. Bot. Cent. **18**: 218.   1884.
Beloniella Sacc,; Rehm in Rab. Krypt.-Fl. **1**³: 638.   1892.
Eubelonis Clements, Gen. Fungi 175.   1909.
?Laetinaevia Nannf. Nova Acta Soc. Sci. Upsal. IV. **8**: 190.   1932.

Apothecia minute, usually less than 1 mm. in diameter, sessile, or subsessile, occasionally with a short, stem-like base, superficial, or suberumpent, dark-grayish, greenish, or more frequently bright-colored, red, yellow, or purplish, externally smooth or nearly so; asci clavate, typically 8-spored; spores ellipsoid, or fusoid, normally 1-septate (rarely 3-septate), hyaline; paraphyses filiform, simple, or branched, the ends either free, or agglutinated and forming an epithecium.

Type species, *Peziza fusarioides* Berk.

Most of the species originally included by Fries in this genus are now placed in *Orbilia*.   The species here accepted as the type was the last one mentioned by Fries but has come to be regarded as the type.   The spore characters were ignored by Fries.

Beloniella seems to have been established for species which are erumpent or a *Pyrenopeziza* with a septate spore.

On phanerogams, wood, stems etc.
　On dicotyledons.
　　On herbaceous stems.
　　　Apothecia orange or orange-red.
　　　　Spores 9–14 μ long.   1. *C. fusarioides.*
　　　　Spores 18–22 μ long.   2. *C. oleosa.*
　　　Apothecia not orange.
　　　　Apothecia vermilion.   3. *C. coccinea.*
　　　　Apothecia not vermilion.
　　　　　Spores 10–12 μ long, on *Solidago.* 4. *C. Solidaginis.*
　　　　　Spores 12–20 μ long, on *Lithospermum.* 5. *C. Lithospermi.*
　　On woody plants.
　　　Spores large, 12–18 μ long.
　　　　On cedar limbs and rubbish.   6. *C. aurea.*
　　　　On wood of *Arctostaphylos.*   7. *C. nitens.*
　　　Spores less than 12 μ long.
　　　　Apothecia purple.   8. *C. oregonensis.*
　　　　Apothecia yellow or orange.
　　　　　Spores 3–4 × 9–10 μ.   9. *C. Fairmani.*
　　　　　Spores 4–5 × 8–10 μ.   10. *C. kansensis.*
　　　　　Spores 4 × 12 μ.   11. *C. helotioides.*
　　On fallen leaf.   12. *C. glagosa.*
　On monocotyledon, *Carex.*   13. *C. caricinella.*
On cryptogams.
　On lichen, *Peltigera.*   14. *C. Mülleri.*
　On fern stipe, *Pteris.*   15. *C. cremea.*

1. **Calloria fusarioides** (Berk.) Fries, Summa Veg. Scand. 359. 1849.

Peziza fusarioides Berk. Mag. Zool. Bot. **1**: 46.　1837.
Mollisia fusarioides Gill. Champ. Fr. Discom. 120.　1882.

Apothecia scattered, or gregarious, originating beneath the cuticle, often collected in patches and confluent, at first subglobose, gradually expanding and becoming superficial, orbicular, or elongated, bright-orange, .5–1.5 mm. in diameter; hymenium slightly concave, or plane, orange-yellow; asci clavate, reaching a length of 70–95 μ and a diameter of 8–10 μ, 8-spored; spores long-ellipsoid, or fusiform, at first simple, finally becoming 1-septate (or occasionally 3-septate) 3.5–4 × 9–14 μ; paraphyses filiform, gradually enlarged above to 3.5 μ.

On stems of *Urtica* and possibly on other herbaceous stems.

TYPE LOCALITY: Europe.

DISTRIBUTION: New York and Delaware; probably widely distributed.

ILLUSTRATIONS: Berk. Mag. Zool. Bot. **1**: *pl. 2, f. 4;* Gill Champ. Fr. Discom. *pl. 81. f. 2;* E. & P. Nat. Pfl. **1¹**: 217, *f. 169* E. G.; Rab. Krypt.-Fl. **1³**: 448, *f. 1–3.*

2. **Calloria oleosa** (Ellis) Sacc. Syll. Fung. **8**: 639.   1889.

Peziza oleosa Ellis, Bull. Torrey Club **10**: 52.   1883.

Apothecia scattered, or gregarious, small, subglobose when fresh, orbicular and concave with a thick, obtuse margin when dry, oily-gelatinous; hymenium orange-red; asci fusoid, reaching a length of 75–90 μ and a diameter of 10–12 μ, 8-spored; spores 2-seriate, fusiform-navicular, 1-septate, hyaline, 3–3.5 × 18–22 μ; paraphyses absent or not observed.

On dead herbaceous stems.

TYPE LOCALITY: Pleasant Valley, Utah.

DISTRIBUTION: Michigan to Utah.

EXSICCATI: Clements, Crypt. Form. Colo. *75.*

Ellis states "Allied to *P. fusarioides.*"

3. **Calloria coccinea** (Earle) Seaver, comb. nov.

Niptera coccinea Earle in Green, Pl. Baker. **2**: 7.   1901.

Apothecia scattered, or gregarious, cup-shaped, .3–.5 mm. in diameter, soft, thin, subgelatinous, bright-vermilion throughout, or sometimes the margin bordered by a lighter, nearly white line; asci cylindric, reaching a length of 40–50 μ and a diameter of 4–6 μ; spores long-ellipsoid, becoming 1-septate, not constricted 3 × 8 μ; paraphyses filiform, indistinct.

On dead stems of *Corydalis.*

TYPE LOCALITY: Pagosa Peak, Colorado, elevation 10,000 ft.

DISTRIBUTION: Known only from the type locality.

4. **Calloria Solidaginis** Kanouse, Papers Mich. Acad. Sci. **20**: 66. 1935.

Apothecia erumpent through the epidermis of the host, solitary, or cespitose, often several appearing in rows, cup-shaped, not expanding, .5–1 mm. in diameter; hymenium dirty-white with violet tints, or light-brown when fresh, rose-lilac when dry; asci cylindric-clavate, 4-spored, reaching a length of 45–65 μ and a diameter of 6–8 μ; spores ellipsoid to subfusoid, straight, or slightly bent, with an oil-drop in each end, becoming 1-septate, 3–4 × 10–12 μ; paraphyses filiform, scarcely thickened, above and forming an epithecium.

On wet mats of dead stems of *Solidago* sp.

TYPE LOCALITY: Bank of Huron River, near Dexter, Michigan.

DISTRIBUTION: Known only from the type locality.

ILLUSTRATIONS: Papers Mich. Acad. Sci. **20**: *pl. 12, f. 2.*

5. Calloria Lithospermi (Ellis & Ev.) Seaver, comb. nov.

Niptera Lithospermi Ellis & Ev. Proc. Acad. Sci. Phila. **1893**: 147. 1893.

Apothecia erumpent-superficial, scattered, sessile, .75 mm. in diameter, at first closed, finally opening, blackish-brown on the outside, granular; hymenium concave, livid-white, becoming darker with age; asci cylindric, reaching a length of 45–55 μ and a diameter of 8–10 μ; spores 2-seriate, long-ellipsoid, becoming 1-septate and mostly constricted in the center, 3–3.5 × 12–20 μ; paraphyses obscure.

On dead stems of *Lithospermum canescens.*

TYPE LOCALITY: Mount Helena, Montana.

DISTRIBUTION: Known only from the type locality.

6. Calloria aurea (Ellis) Sacc. Syll. Fung. **8**: 640. 1889.

Ombrophila aurea Ellis, Bull. Torrey Club **8**: 74. 1881.

Apothecia at first obconic and concave above, at length becoming plane, or convex, with a subundulated margin, reaching a diameter of 1–3 mm., soft, golden-yellow; asci cylindric, narrowed at the base, reaching a length of 100 μ and a diameter of 10 μ; spores 1-seriate, acutely ellipsoid, finally 1-septate, 4 × 12–14 μ; paraphyses filiform, yellow.

In swamps, mostly on decaying cedar limbs lying partly in the water; also on old leaves and rubbish.

TYPE LOCALITY: New Jersey.

DISTRIBUTION: Known only from the type locality.

EXSICCATI: Ellis, N. Am. Fungi *395.*

7. Calloria nitens (Cash) Seaver, comb. nov.

Helotium nitens Cash, Mycologia **28**: 251. 1936.

Apothecia fleshy to subgelatinous, subsessile, pulvinate, convex, gregarious, sometimes confluent, reaching a diameter of 1.5 mm., 1 mm. high, pale yellow-orange to light salmon-orange; hymenium darker, even, smooth; asci clavate-cylindric, rounded with the wall slightly thickened at the apex, attenuated below into a long stem, reaching a length of 150–170 μ and a diameter of 7–8 μ; spores ellipsoid-clavate, at first simple, later becoming

1-septate, 4–4.5 × 13–18 μ; paraphyses filiform, freely branched above, 1 μ in diameter.

On wood of *Arctostaphylos Tracyi*.

TYPE LOCALITY: Spruce Cove, Trinidad, California.

DISTRIBUTION: Known only from the type locality.

8. **Calloria oregonensis** Kanouse, Papers Mich. Acad. Sci. **24**: 26. 1939.

Apothecia sessile, gregarious, 1 mm. in diameter, soft and subgelatinous, purplish; hymenium plane, purple; asci cylindric, 8-spored, reaching a length of 120–150 μ and a diameter of 7–8 μ, spores ellipsoid, becoming 1-septate and constricted at the septum, 5–6 × 10–12 μ; paraphyses filiform, stiff, abruptly enlarged at their apices, the capitate apex reaching a diameter of 4 μ.

On coniferous wood.

TYPE LOCALITY: Lake Tahkenitch, Oregon.

DISTRIBUTION: Known only from the type locality.

9. **Calloria Fairmani** Rehm, Ann. Myc. **9**: 366. 1911.

Belonium Fairmani Rehm, Ann. Myc. **9**: 367. 1911.

Apothecia scattered, globose, finally discoid, externally smooth, .1–.2 mm. in diameter, subgelatinous, yellowish; asci clavate, reaching a length of 60 μ and a diameter of 8–10 μ, 8-spored; spores ellipsoid, or subclavate, 2-seriate, becoming 1-septate, 3–4 × 9–10 μ; paraphyses filiform, yellowish above.

On decorticated wood.

TYPE LOCALITY: Lyndonville, New York.

DISTRIBUTION: Known only from the type locality.

10. **Calloria kansensis** Ellis & Ev. Bull. Torrey Club **25**: 507. 1898.

Apothecia gregarious, subgelatinous, pale-orange, globose, .3 mm. in diameter when fresh, of a deeper color and umbilicate when dry, with spreading, white hairs around the base; asci clavate-cylindric, reaching a length of 65–75 μ and a diameter of 8–10 μ, 8-spored; spores mostly 2-seriate, ellipsoid, with two large oil-drops, becoming 1-septate, 4–5 × 8–10 μ; paraphyses absent or not observed.

On rotten wood.

TYPE LOCALITY: Kansas.

DISTRIBUTION: Known only from the type locality.

11. **Calloria helotioides** (Rehm) Seaver, comb. nov.

Lanzia helotioides Rehm, Ann. Myc. **2**: 36. 1904.

Apothecia gregarious, or rarely solitary, sessile, or subsessile, at first cyathiform, then more or less patelliform, 2 mm. to 1 cm. in diameter; hymenium dilutely yellowish-red; asci cylindric, reaching a length of 100–120 μ and a diameter of 7–9 μ, 8-spored; spores ellipsoid, becoming 1-septate, scarcely constricted, 2-seriate, 4 × 12 μ; paraphyses filiform, hyaline, 2 μ thick.

On decaying wood.

TYPE LOCALITY: United States.

DISTRIBUTION: Known only from the type collection.

The specimens were sent to Dr. H. Rehm by G. F. Atkinson. No material has been seen.

12. **Calloria glagosa** (Ellis & Ev.) Seaver, comb. nov.

Peziza glagosa Ellis & Ev. Jour. Myc. **4**: 56. 1888.
Pezizella glagosa Sacc. Syll. Fung. **8**: 276. 1889.
Hymenoscyphus glagosus Kuntze, Rev. Gen. Pl. **3³**: 485. 1898.

Apothecia sessile, smooth, .25 mm. in diameter, milk-white when fresh, becoming subrufous, or amber-colored when dry; asci clavate, reaching a length of 70–75 μ and a diameter of 7–8 μ, gradually attenuated at the base, 8-spored; spores 2-seriate, or crowded above, long-ellipsoid, or clavate, 2.5–3 × 8–10 μ, becoming 1-septate; paraphyses abundant, filiform, not distinctly thickened above.

On a much-decayed, fallen leaf in swamp.

TYPE LOCALITY: Newfield, New Jersey.

DISTRIBUTION: Known only from the type locality.

Ellis seemed a little uncertain about the septation of the spores which leaves the species somewhat in doubt. W. L. White regards this as a *Helotium*.

13. **Calloria caricinella** (Peck) Seaver, comb. nov.

Helotium caricinellum Peck, Ann. Rep. N. Y. State Mus. **30**: 61. 1878.
Niptera caricinella Sacc. Syll. Fung. **8**: 484. 1889.

Apothecia scattered, sessile, .5–.7 mm. in diameter, reddish, or ochraceous-brown when moist, black, or blackish when dry; hymenium plane, or slightly concave; asci clavate, 8-spored; spores crowded, ellipsoid, 1-septate, 20–25 μ long.

On dead leaves of *Carex utriculata*.

TYPE LOCALITY: Adirondack Mountains.

Distribution: Known only from the type locality.
Illustrations: Ann. Rep. N. Y. State Mus. **30**: *pl. 1, f. 5–8.*

14. **Calloria Mülleri** (Willey) Seaver, comb. nov.

Phacopsis Mülleri Willey, Enum. Lich. Massachusetts 34.   1892.
Niptera Mülleri Vouaux, Bull. Soc. Myc. Fr. **30**: 182.   1914.

Apothecia adnate to the thallus of the lichen host, flat, or
slightly convex, rounded, at length confluent, flesh-colored,
becoming darker; asci cylindric, 8-spored; spores becoming
1-septate, 3.5–5.5 × 11–15 μ.

On the thallus of *Peltigera canina*.

Type locality: Massachusetts.

Distribution: Known only from the type locality.

No specimens seen.

15. **Calloria cremea** (Cash) Seaver, comb. nov.

Helotium cremeum Cash, Mycologia **28**: 249.   1936.

Apothecia subsessile, waxy, patellate, occasionally lobate,
reaching a diamater of 1 mm., exterior cream-colored to cream-
buff, margin whitish-furfuraceous; asci clavate, reaching a length
of 75–90 μ and a diameter of 10–12 μ, narrowed but rounded at
the apex, gradually attenuated toward the base, 8-spored;
spores 1–2-seriate, fusoid-clavate, 1-septate, slightly constricted
at the septum, 4–5 × 12–15 μ, upper cell slightly broader than
the lower, containing many oil-drops or granules; paraphyses
filiform, septate, unbranched reaching a diameter of 2 μ at the
apex.

On stipes of *Pteridium aquilinum pubescens*.

Type locality: Spruce Cove, Trinidad, California.

Distribution: Known only from the type locality.

Illustrations: Mycologia **28**: 250, *f. 4.*

Doubtful Species

Calloria citrina A. L. Smith, Jour. Linn. Soc. **35**: 15.   1901.   Described
from material collected in Dominica.   No material seen.

Niptera discolor (Mont. & Fries) Rehm in Rab. Krypt.-Fl. **1**[3]: 552.   1891;
Patellaria discolor Mont. & Fries, Ann. Sci. Nat. II. **5**: 290.   1836; *Mollisia
discolor* Phill. Brit. Discom. 175.   1887.   Reported from Michigan by Dr. B.
Kanouse.   No material seen.

Solenopezia grisea A. L. Smith, Jour. Linn. Soc. **35**: 14.   1901.   Spores
described as 1-septate, 10 × 15–17 μ.   On stalks of some monocotyledon in
Dominica.   No material seen.   Apparently a *Calloria*.

Peziza mycogena Ellis, Bull. Torrey Club **6**: 107. 1876. Apothecia minute, pale with a greenish tinge; spores ellipsoid, appearing to become 1-septate, 7.5 μ long. On old *Polyporus igniarius* buried in leaves, New Jersey. No material seen.

Niptera pella Clements, Crypt. Form. Colo. *88*. 1906. On *Streptopus amplexifolius*, Jack Brook, Colorado. No other publication seen.

Niptera tyrolensis (Sacc.) Rehm in Rab. Krypt.-Fl. **1³**: 554. 1891; *Mollisia tyolensis* Sacc. Syll. Fung. **8**: 333. 1889. This species has been reported from Michigan by Dr. B. Kanouse. No material has been seen.

Niptera uda (Pers.) Fuckel, Symb. Myc. 293. 1869; *Peziza uda* Pers. Syn. Fung. 634. 1801; *Mollisia uda* Gill. Champ. Fr. Discom. 127. 1882. Reported from Michigan by A. H. W. Povah. No material seen.

28. **BELONIUM** Sacc. Bot. Cent. **18**: 219. 1884.

Belonidium Rehm in Rab. Krypt.-Fl. **1³**: 561. 1891. Not Mont. & Dur. 1846.
Massea Sacc. Syll. Fung. **8**: 488. 1889.
Harknessiella Sacc. Syll. Fung. **8**: 845. 1889.

Apothecia sessile, or short-stipitate, cup-shaped to scutellate, black, brownish or occasionally yellowish, or reddish, usually smooth or not distinctly hairy; asci cylindric-clavate, 8-spored; spores fusiform to elongate, 2- to many-septate, hyaline; paraphyses filiform.

Type species, *Peziza graminis* Desm.

The presence of a short, stem-like base has been used to separate *Belonidium* from *Belonium*. The character is not regarded as a satisfactory one on which to separate the two genera.

On dicotyledonous plant tissues.
    Spores very large, 35–40 μ long.
        Spores 3-septate, on *Glyceria*.   1. *B. Glyceriae.*
        Spores more than 3-septate.
            Spores 4–8-septate.
                Spores 36–40 μ long.   2. *B. basitrichum.*
                Spores 40–50 μ long.   3. *B. introspectum.*
            Spores many-septate, 40–60 μ long.   4. *B. tympanoides.*
    Spores not over 35 μ long.
        Spores 30–33 μ long.
            Apothecia greenish.   5. *B. Parksi.*
            Apothecia not greenish.
                On decaying *Magnolia*.   6. *B. phlegmaceum.*
                On undetermined wood.   7. *B. quisquiliarum.*
        Spores 10–20 μ long.
            Spores 10–12 μ long.   8. *B. atrosanguineum.*
            Spores 12–20 μ long.

Spores 14–20 μ long, on *Aralia*.	9. *B. minimum.*
Spores 12–15 μ long.	
On twigs of *Abies*.	10. *B. inconspicuum.*
On *Populus* and *Salix*.	11. *B. aggregatum.*
On monocotyledonous tissues.	
On petioles of palms.	12. *B. sclerogenum.*
On grasses or sedges.	
Spores not over 20 μ long.	
On culms of *Spartina*.	13. *B. heteromorphum.*
On *Andropogon*.	14. *B. Andropogonis.*
Spores 20 μ or more long.	
On *Calamagrostis*, spores 24–30 μ long.	15. *B. intermedium.*
On *Carex*, spores 18–24 μ long.	16. *B. caricincolum.*
On *Arundinaria*, spores 20–24 μ long.	17. *B. eustegiaeforme.*
On various grasses, spores 21–40 μ long.	18. *B. culmicola.*

1. **Belonium Glyceriae** (Peck) Seaver, comb. nov.

Belonidium Glyceriae Peck, Bull. N. Y. State Mus. **139**: 19.   1909.

Apothecia 1–1.5 mm. in diameter, sessile, papillate on the under side; hymenium plane, or convex, pale-yellow; asci sub-clavate, or fusoid, reaching a length of 120–130 μ and a diameter of 14–18 μ, 8-spored; spores elongate-ellipsoid, straight, or slightly curved, 2-seriate, 4–5 × 35–40 μ, becoming 3-septate; paraphyses filiform.

On dead culms of *Glyceria nervata*.

TYPE LOCALITY: Lyndonville, New York.

DISTRIBUTION: Known only from the type locality.

2. **Belonium basitrichum** (Sacc.) Seaver, comb. nov.

Belonidium basitrichum Sacc. Atti Soc. Veneto Sci. Nat. Padova **4**[1]: 35.   1875.

Apothecia gregarious, sessile, patelliform, soft, entirely white, finally sordid-white, .25–.5 mm. in diameter; hymenium concave, or plane, becoming yellowish; asci clavate, reaching a length of 100–120 μ and a diameter of 14 μ, short-stipitate, 8-spored; spores 2-seriate, elongate-fusoid, curved, 3–4 × 36–40 μ, becoming 6–8-septate, hyaline; paraphyses filiform, clavate.

On rotten wood *Quercus*, and *Castanea*, seated on a black mycelial growth.

TYPE LOCALITY: Europe.

DISTRIBUTION: New York; also in Europe.

The only specimen seen of this species is one collected by C. L. Shear in Ringwood, New York and which seems to agree with the original description.

3. **Belonium introspectum** (Cooke) Sacc. Syll. Fung. **8**: 498. 1889.

Peziza introspecta Cooke, Hedwigia **14**: 84. 1875.

Apothecia gregarious, or scattered, sessile, translucent, at first hemispherical, finally expanded and nearly plane, .3–.4 mm. in diameter; asci clavate, 8-spored; spores fusiform, straight, or curved, with four or five oil-drops, finally 5-septate, 8 × 40–50 μ; paraphyses filiform.

On rotting wood.

TYPE LOCALITY: New Jersey.

DISTRIBUTION: Known only from the type locality.

4. **Belonium tympanoides** (Ellis & Ev.) Seaver, comb. nov.

Belonidium tympanoides Ellis & Ev. Proc. Acad. Sci. Phila. **1893**: 149. 1893.

Apothecia gregarious, fleshy, sessile, black, closed when dry, hemispherical when fresh, .3–.5 mm. in diameter; hymenium concave, pallid; asci clavate-cylindric, reaching a length of 80–100 μ and a diameter of 12 μ, 8-spored; spores ellipsoid-cylindric, becoming many-septate, 2.5 × 40–60 μ; paraphyses filiform, curved at their apices.

On rotten wood.

TYPE LOCALITY: London, Canada.

DISTRIBUTION: Known only from the type locality.

5. **Belonium Parksi** (Cash) Seaver, comb. nov.

Belonidium Parksi Cash, Mycologia **28**: 248. 1936.

Apothecia gregarious, cup-shaped, sessile, externally fuscous, furfuraceous, the margin inrolled, hysteroid, triangular, or irregularly folded; hymenium glaucous-green, drying chromium-green; asci clavate-cylindric, narrowed above, gradually tapering into a short stem, reaching a length of 65–80 μ and a diameter of 6–8 μ, 8-spored; spores fusoid, straight, or slightly curved, obliquely 1-seriate below, 2-seriate above, becoming 1–3-septate, about 3 × 13 μ; paraphyses filiform, slightly enlarged above, 1.5–2 μ thick.

On decorticated stems of *Vaccinium parvifolium*, *Rhamnus purshiana*, *Garrya elliptica*, *Physocarpus capitatus* and *Castanopsis chrysophylla*.

TYPE LOCALITY: Humboldt Co., California.

DISTRIBUTION: California.

ILLUSTRATIONS: Mycologia **28**: 250, *f. 2*.

6. **Belonium phlegmaceum** (Ellis) Seaver, comb. nov.

Peziza phlegmacea Ellis, Bull. Torrey Club **9**: 19.   1882.
Belonidium ? *phlegmaceum* Sacc. Syll. Fung. **8**: 500.   1889.

Apothecia seated on a subiculum of delicate, creeping, white, loosely matted threads, gregarious, sessile, circular, thin, white, soft; asci cylindric-clavate, reaching a length of 60 μ and a diameter of 12 μ; spores elongate-fusiform, attenuated below to a slender point, 2.5 $\times$ 30–33 μ; paraphyses filiform, scarcely thickened above.

On decaying *Magnolia*.

TYPE LOCALITY: Newfield, New Jersey.

DISTRIBUTION: Known only from the type locality.

7. **Belonium quisquiliarum** (Berk. & Curt.) Seaver, comb. nov.

Peziza quisquiliarum Berk. & Curt.; Berk. Jour. Linn. Soc. **10**: 366.   1868.
Massea quisquiliarum Sacc. Syll. Fung. **8**: 488.   1889.

Apothecia yellow, crateriform, expanding, sessile, the margin inflexed, 2–5 mm. in diameter; asci clavate, 8-spored; spores fusiform, curved, becoming 3-septate, 6–7 $\times$ 25 μ; paraphyses filiform.

On dead branches.

TYPE LOCALITY: Cuba.

DISTRIBUTION: Known only from the type locality.

ILLUSTRATIONS: Cooke, Mycographia *pl. 10, f. 37.*

8. **Belonium atrosanguineum** (Rehm) Seaver, comb. nov.

Calloria atrosanguinea Rehm, Ann. Myc. **5**: 518.   1907.

Apothecia parasitic in the mycelium of a *Stereum*, gregarious, black when dry, when moist blackish-red, punctiform, irregularly hemispherical, .15 mm. in diameter; asci clavate, reaching a length of 40–50 μ and a diameter of 10 μ, 8-spored; spores ellipsoid to clavate, becoming 1-septate, finally 3-septate, 3–3.5 $\times$ 10–12 μ; paraphyses filiform, 1 μ thick, globose at their apices, 4 μ in diameter.

On logs of conifers.

TYPE LOCALITY: Sumner, Washington.

DISTRIBUTION: Known only from the type locality.

9. **Belonium minimum** (Ellis & Ev.) Seaver, comb. nov.

Belonidium minimum Ellis & Ev. Proc. Acad. Sci. Phila. **1893**: 452.   1893.

Apothecia scattered, erumpent, sessile, pale rose-colored, .09–.11 mm. in diameter, subdiscoid, subgelatinous, furfuraceous, pilose, or nearly smooth; asci clavate-cylindric, reaching a length of 45–55 μ and a diameter of 6–7 μ, 8-spored; spores 2-seriate, fusoid, slightly curved, hyaline, 3-septate, 3–3.5 × 14–20 μ; paraphyses not observed.

On dead stems of *Aralia racemosa*.

TYPE LOCALITY: Granogue, Delaware.

DISTRIBUTION: Known only from the type locality.

10. **Belonium inconspicuum** Cash, Mycologia **28**: 304. 1936.

Apothecia superficial, sessile, occurring singly, smooth, patellate, thin-membranaceous, .3–.5 mm. in diameter, brownish-black; hymenium drab; asci clavate, tapering into a short stem, reaching a length of 50–65 μ and a diameter of 8–10 μ, 8-spored; spores ellipsoid to broad-clavate, obliquely 1-seriate to irregularly 2-seriate, hyaline, 3-septate, not constricted, 4–5 × 12–15 μ; paraphyses septate, branched with globose, olive-brown, glutinous tips 5 μ in diameter, forming an agglutinated layer.

On decorticated twigs of *Abies*.

TYPE LOCALITY: Grand Mesa, Colorado.

DISTRIBUTION: Known only from the type locality.

ILLUSTRATIONS: Mycologia **28**: 302, *f. 2*.

11. **Belonium aggregatum** Cash, Mycologia **28**: 303. 1936.

Apothecia sessile, urceolate to patellate, soft-fleshy, densely crowded on a more or less evident subiculum of pale-olivaceous hyphae, fuscous-black to olivaceous-black; hymenium of the same color with the margin paler and fimbriate, the entire fungus black when dry; asci cylindric-clavate, slightly narrowed at the apex, reaching a length of 50–65 μ and a diameter of 6–7 μ, 8-spored; spores 1–2-seriate, clavate to cylindric, 1–3-septate, not constricted, 2.5–3 × 13–15 μ; paraphyses filiform, unbranched, septate, hyaline, slightly enlarged at the apex, 2 μ in diameter.

On decorticated wood of *Populus tremuloides* and on *Salix*.

TYPE LOCALITY: Mesa Lakes, Colorado.

DISTRIBUTION: Known only from the type locality.

ILLUSTRATIONS: Mycologia **28**: 302, *f. 5*.

The subiculum would suggest a *Trichobelonium*.

12. **Belonium sclerogenum** (Berk. & Curt.) Seaver, comb. nov.

Peziza sclerogena Berk. & Curt.; Berk. Jour. Linn. Soc. **10**: 369. 1868.
Belonidium sclerogenum Sacc. Syll. Fung. **8**: 497. 1889.

Apothecia sessile, subhemispherical, yellowish, margin inflexed, finally applanate; asci clavate, 8-spored; spores 2-seriate, fusiform, curved, becoming 3-septate, 30–34 μ long.

On petioles of palms.

TYPE LOCALITY: Cuba.

DISTRIBUTION: Known only from the type locality.

13. **Belonium heteromorphum** (Ellis & Ev.) Seaver, comb. nov.

Peziza heteromorpha Ellis & Ev. Jour. Myc. **2**: 88.   1889.
Belonidium heteromorphum Sacc. Syll. Fung. **8**: 502.   1889.

Apothecia seated on a brownish-black, felt-like subiculum which extends for several inches along the culm, scattered, globose at first, with a small, round opening with a white margin, at length expanding to nearly plane, or even slightly convex, 2–3 mm. in diameter; hymenium pallid-white to flesh-colored; asci clavate-cylindric, reaching a length of 70 μ and a diameter of 6–7 μ, 8-spored; spores fusoid, slightly curved, finally 3-septate, or pseudoseptate, 2.5–3 × 20 μ.

On the base of culms of *Spartina polystachya*.

TYPE LOCALITY: Louisiana.

DISTRIBUTION: Known only from the type locality.

The subiculum suggests a *Trichobelonium*.

14. **Belonium Andropogonis** (Berk. & Curt.) Sacc. Syll. Fung. **8**: 493.   1889.

Peziza Andropogonis Berk. & Curt.; Berk. Grevillea **3**: 158.   1875.
Peziza aberrans Peck, Bull. Torrey Club **6**: 14.   1875.

Apothecia erumpent, at first elongated and hysteriform, gradually expanding and becoming elliptic in form, externally dark brownish-black, about .5 mm. long; hymenium concave, pale-yellowish; asci clavate, reaching a length of 60–70 μ and a diameter of 10–12 μ, 8-spored; spores irregularly 2-seriate, fusoid, 2–3 × 18 μ, 3-septate; paraphyses filiform, about 2 μ in diameter.

In *Andropogon*.

TYPE LOCALITY: South Carolina.

DISTRIBUTION: New Jersey to South Carolina.

EXSICCATI: Ellis, N. Am. Fungi *61;* Rab.-Winter, Fungi Eu. *3169* (from New Jersey).

15. **Belonium intermedium** (Rehm) Seaver, comb. nov.

Belonidium intermedium Rehm, Ann. Myc. **6**: 315. 1908.

Apothecia gregarious, sessile, at first closed, finally expanding and becoming patellate, brown, .2–.3 mm. in diameter; hymenium plane, yellow; asci clavate, reaching a length of 80 μ and a diameter of 12 μ, 8-spored; spores elongate-fusoid, 2-seriate, mostly straight, 3-septate, 5–5.5 × 24–30 μ; paraphyses filiform 2 μ in diameter, expanding above to 3.5.

On dead culms of *Calamagrostis canadensis*.

TYPE LOCALITY: Madison, Wisconsin.

DISTRIBUTION: Known only from the type locality.

16. **Belonium caricincolum** (Rehm) Seaver, comb. nov.

Belonidium caricincolum Rehm in Rab. Krypt.-Fl. **1³**: 564. 1891.

Apothecia gregarious, sessile, at first globose and closed, finally opening and becoming cup-shaped, externally brownish-black, .2–1 mm. in diameter; hymenium gray, or grayish-yellow to dark-olive; asci clavate, reaching a length of 60–70 μ and a diameter of 7–8 μ, 8-spored; spores fusiform, straight, or somewhat curved, finally 3-septate, 3 × 18–24 μ; paraphyses filiform, 3 μ thick.

On fallen leaves of *Carex*.

TYPE LOCALITY: Europe.

DISTRIBUTION: Colorado; also in Europe.

EXSICCATI: Clements, Crypt. Form. Colo. *284.* (as *Belonidium caricincolum*).

17. **Belonium eustegiaeforme** (Berk. & Curt.) Sacc. Syll. Fung. **8**: 494. 1889.

Peziza eustegiaeformis Berk. & Curt.; Berk. Grevillea **3**: 158. 1875.

Peziza Arundinariae Berk. & Curt.; Cooke, Bull. Buffalo Soc. Nat. Sci. **2**: 297. 1875.

Pyrenopeziza Arundinariae Sacc. Syll. Fung. **8**: 368. 1889.

Apothecia scattered, seated on a brownish spot, at first closed and rounded, or slightly elongated, expanding and becoming scutellate, externally dark brownish-black and furfuraceous; hymenium concave, lighter in color, pallid; asci clavate, reaching a length of 80 μ and a diameter of 10–15 μ, 8-spored; spores fusoid, hyaline, 5 × 20–24 μ, becoming 3-septate; paraphyses filiform.

On *Arundinaria macrosperma*.

TYPE LOCALITY: South Carolina.

DISTRIBUTION: South Carolina to Alabama and Georgia.
EXSICCATI: Ellis, N. Am. Fungi *668;* Rav. Fungi Am. *310.*

18. **Belonium culmicola** (Desm.) Seaver, comb. nov.

Peziza culmicola Desm. Ann. Sci. Nat. II. **6**: 244.   1836.
Belonium vexatum DeNot. Comm. Critt. Ital, **1**: 380.   1864.
Belonium Moliniae DeNot. Comm. Critt. Ital. **1**: 380.   1864.
Peziza vexata Karst. Not. Soc. Fauna. Fl. Fenn. **10**: 139.   1869.
Peziza subgibbosa Ellis, Bull. Torrey Club **6**: 108.   1876.
Phialea culmicola Gill. Champ. Fr. Discom. 103.   1882.
Belonium subgibbosum Sacc. Syll. Fung. **8**: 493.   1889.
Belonidium culmicolum Phill. Brit. Discom. 148.   1897.

Apothecia scattered, or gregarious, sessile, or with a very
short, stem-like base, at first rounded, expanding and becoming
cup-shaped, or saucer-shaped, .2–1 mm. in diameter, externally
yellowish, or slightly reddish; hymenium pallid, yellowish, or
rosy; asci clavate-cylindric, reaching a length of 120–150 μ and
a diameter of 12–15 μ, 8-spored; spores fusiform, straight, or
curved, becoming 3-septate, 4–5 × 21–40 μ; paraphyses filiform,
somewhat enlarged above, 6 μ in diameter, brown.

On dead stems of *Andropogon* and various other grasses.

TYPE LOCALITY: Europe.

DISTRIBUTION: New Jersey; also in Europe.

ILLUSTRATIONS: Phill. Brit. Discom. *pl. 5, f. 29.*

EXSICCATI: Ellis, N. Am. Fungi *850* (as *Peziza vexata* DeNot.)

DOUBTFUL AND EXCLUDED SPECIES

Belonium arabicolum Ellis & Ev. Proc. Acad. Sci. Phila. **1894**: 352.
1894. On stems of *Arabis furcata*, Mt. Paddo, Washington. This is not a
discomycete. It looks like a *Lophodermium*.

Belonidium aurantiacum Rehm in Rab. Krypt-Fl. **1**[3]: 564. 1891. Reported by A. H. W. Povah from Michigan on dead culms of *Calamagrostis*.
No material seen.

Belonium bicolor Ellis & Ev. Jour. Myc. **8**: 69. 1902. Apothecia sessile,
nearly black when dry, less than 1 mm. in diameter; spores 6–8 × 15–20 μ,
becoming 3-septate. On *Eupatorium*, Tuskegee, Alabama.

Belonium consanguineum Ellis & Ev. Jour. Myc. **8**: 70. 1902. On *Ilex*,
Tuskegee, Alabama. According to Ellis this is close to *B. bicolor*. Each is
known only from the type collection. Both need further study.

Belonium Delitschianum (Auersw.) Rehm in Rab. Krypt.-Fl. **1**[3]: 689.
1892; *Peziza Delitschiana* Auersw. in Rab. Fungi Eu. *912.* 1886; *Coronellaria
Delitschiana* Karst. Myc. Fenn. **1**: 183. 1871. This species has been reported
from Michigan by Dr. B. Kanouse on rotten logs. The type was recorded on
Scirpus lacustris. No American material has been seen. In spite of the
difference in habitat Dr. Kanouse regards the American and European specimens as identical. The spores are described as 3-septate and 6–7 × 25–40 μ.

Belonidium fuscopallidum, Bres. Verh. Zool.-Bot. Ges. Wien **52**: 434. 1902. This species has been reported from Michigan by Dr. B. Kanouse on wood of *Acer* sp. The apothecia are described as olive-green and the hyphae projecting beyond the margin of the apothecium; spores 4–6 × 18–21 μ. No material has been seen.

Belonium fuscum Phill. & Hark. Bull. Calif. Acad. Sci. **1**: 23. 1884. Apothecia short-stipitate, sooty-brown; spores 1–3-septate, 15–20 μ long. On dead stems of *Sanicula Menziesii*, California.

Belonidium hirtipes A. L. Smith, Jour. Linn. Soc. **35**: 14. 1901. Apothecia stipitate, 2 mm. in diameter with stem 2 mm. long; spores 3-septate, 5 × 20–25 μ. On wood in Dominica. No material seen.

Belonidium juncisedum (Karst.) Rehm in Rab. Krypt-Fl. **1³**: 568. 1891; *Mollisia junciseda* Karst. Myc. Fenn. **1**: 198. 1871. Reported from Michigan by Dr. B. Kanouse. No material seen.

Harknessiella purpurea (Phill. & Hark.) Sacc. Syll. Fung. **8**: 845. 1889; *Phillipsiella purpurea* Phill. & Hark. Bull. Calif. Acad. Sci. **1**: 23. 1884. On leaves of *Garrya elliptica*, California. The 3-septate spores would place this in the genus *Belonium*. Compare *Belonium Parksi* Cash which occurs on the same host.

Belonidium sclerotii A. L. Smith, Jour. Linn. Soc. **35**: 14. 1901. Spores described as 3-septate, 3 × 20 μ. On decorticated branch in Dominica.

29. **BELONIOSCYPHA** Rehm in Rab. Krypt.-Fl. **1³**: 743. 1893.

Apothecia scattered, stipitate, at first rounded, expanding and becoming campanulate, or turbinate, light-colored, externally smooth; asci clavate, or cylindric, 4–8-spored; spores elongated fusoid, becoming 3-septate, hyaline; paraphyses filiform.

Type species, *Peziza Campanula* Nees.

Asci 4-spored.   1. *B. lactea.*
Asci 8-spored.   2. *B. miniata.*

1. **Belonioscypha lactea** (Ellis & Ev.) Seaver, comb. nov. (PLATE 109.)

Helotium lacteum Ellis & Ev. Jour. Myc. **4**: 56. 1888.
Helotium lacteum Ellis & Ev. Proc. Acad. Sci. Phila. **1893**: 145. 1893.
Helotiella lactea Sacc. Syll. Fung. **11**: 415. 1895.

Apothecia thickly gregarious, or subconfluent, stipitate, less than 1 mm. in diameter, at first closed, opening and becoming turbinate, white with a tinge of yellow, amber when dry; hymenium concave, or nearly plane; stem short, stout, the length less than the diameter of the apothecium; asci cylindric-clavate, reaching a length of 75–80 μ and a diameter of 5 μ, tapering into a long, stem-like base, 4-spored; spores 1-seriate, or overlapping and becoming 2-seriate, fusoid, often curved, 2–3 × 10–15 μ; paraphyses filiform.

On decorticated wood.

TYPE LOCALITY: Cazenovia, New York.

DISTRIBUTION: New York and Pennsylvania.

Excellent specimens of this species were sent to the writer by L. O. Overholts Dec. 19, 1927 with the following note: "This was collected in abundance, and I can get enormous quantities of it, growing as thickly scattered as the sample, for 20 ft. or more along the side of an old *Liriodendron* log." His specimens agree perfectly with the type in the herbarium of The New York Botanical Garden.

This species was described twice as a new species both based on the same specimen. Ellis did not mention the 4-spored character of the asci but his material is apparently so.

2. **Belonioscypha miniata** Kanouse, Mycologia **39**: 640.   1947.

Apothecia stipitate, .2–.4 mm. in diameter, flat when moist with the margin slightly inrolled, delicately roughened by granules of an amorphous deposit outlining the cup, creamy-yellow when fresh, drying pale-yellow; stem very short; asci cylindric-clavate, reaching a length of 50–70 μ and a diameter of 6–9 μ, 8-spored; spores usually 2-seriate, fusoid, straight, or slightly curved, subhyaline, 3-septate, 2–3 × 10–12 μ; paraphyses filiform not enlarged above.

On old leaves of *Carex*.

TYPE LOCALITY: Lake Tahkenitch, Oregon.

DISTRIBUTION: Known only from the type locality·

ILLUSTRATIONS: Mycologia **39**: 649, *f. 4–6*.

30. **GORGONICEPS** Karst. Myc. Fenn. **1**: 15.   1871.

Belonopsis Sacc. Syll. Fung. **16**: 752.   1902.

Apothecia sessile, or short-stipitate, soft and waxy, yellowish, or ochraceous, turbinate to subdiscoid, or scutellate, seldom exceeding 2–3 mm. in diameter and often less than 1 mm.; asci cylindric to clavate, typically 8-spored; spores elongate, filiform, fusiform, or vermiform, usually becoming multiseptate, hyaline; paraphyses filiform, enlarged above and often branched.

Type species, *Gorgoniceps aridula* Karst.

The genus *Apostemidium* which is usually treated with the Geoglossaceae is close to the present genus and some regard them as

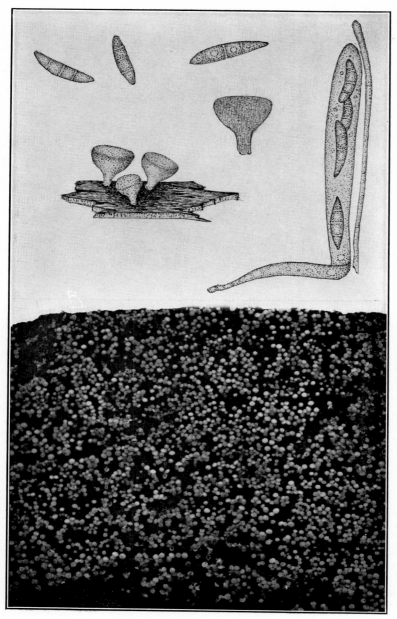

BELONIOSCYPHA LACTEA

synonymous.   Durand, however, treated them as distinct and retains the former with the Geoglossaceae because of its general resemblance to *Vibrissea*.

On coniferous plant tissues.
  Spores 35–60 μ long.
    On cones of conifers.   1. *G. Pumilionis.*
    On bark and scales of *Pinus*.   2. *G. aridula.*
  Spores 12–15 μ long, on pine needles.   3. *G. ontariensis.*
Not on coniferous plants.
  Spores very slender, 1.5–2 μ thick.   4. *G. montanensis.*
  Spores stout, 3–10 μ thick.
    Spores 3–4 × 30–37 μ, on rotten wood. 5. *G. iowensis.*
    Spores 5–7 × 40–45 μ, on wood and palms. 6. *G. confluens.*
    Spores 9–10 × 50–55 μ, on bamboo.   7. *G. jamaicensis.*

1. Gorgoniceps Pumilionis Rehm in Rab. Krypt-Fl. 1³: 692. 1892

Pezicula Pumilionis Rehm, Hedwigia **21**: 115.   1882.
Dermatella Pumilionis Sacc. Syll. Fung. **8**: 490.   1889.

Apothecia gregarious, sessile, or contracted into a short, stem-like base, at first rounded, expanding and becoming scutellate, pale-cinereous, becoming brownish-yellow, reaching a diameter of .1–.3 mm.; asci clavate, reaching a length of 75–80 μ and a diameter of 6–7 μ, attenuated above and gradually tapering below into a stem-like base; spores filiform, becoming septate (the number of septa usually 3 or 4), 2–2.5 × 35–40 μ; paraphyses filiform, about 1.5 μ in diameter.

On cones of conifers.

TYPE LOCALITY: Europe.

DISTRIBUTION: Colorado; also in Europe.

EXSICCATI: Clements, Crypt. Form. Colo. *290.*

The only American specimen of this species seen is the Clements specimen referred to above on scales of *Picea* sp.

2. Gorgoniceps aridula Karst. Myc. Fenn. 1: 185. 1871. (PLATE 110, FIG. 1.)

Apothecia gregarious, or scattered, sessile, or contracted into a very short, stem-like base, bluish-hyaline, when dry pale-brownish, reaching a diameter of .3–.8 mm.; hymenium bluish-hyaline, or pallid, plane, or convex; asci clavate, attenuated above and tapering below into a stem-like base, reaching a length of 100–125 μ and a diameter of 15 μ; spores fasciculate, filiform, straight, or curved, becoming septate (the number difficult to determine but apparently 16 to 20), reaching a length of 65 μ and a diameter of 2.5–3 μ; paraphyses filiform about 2 μ in diameter.

On coniferous bark and scales of *Pinus pungens*.

TYPE LOCALITY: Europe.

DISTRIBUTION: Pennsylvania; also in Europe.

ILLUSTRATIONS: E. & P. Nat. Pfl. 1¹: 208, *f. 163 A;* Rab. Krypt.-Fl. 1³: 652, *f. 1–5;* Mycologia **38**: 551 (upper figure).

The only American specimen of this species seen is one collected by Dr. L. O. Overholts and P. Spaulding (No. *10795*) in Pennsylvania. The plants are minute and the species is probably more common than indicated by the material at hand.

3. **Gorgoniceps ontariensis** (Rehm) Höhn. Mitt. Inst. Hochs. Wien **3**: 106. 1926.

Pezizella ontariensis Rehm, Ann. Myc. **11**: 167. 1913.

Apothecia scattered, sessile, at first globose, expanding becoming cup-shaped, finally discoid, contracted at the base, .5–1.5 mm. in diameter, pale yellowish-white, externally floccose; hymenium plane or nearly so, pale rose-colored; asci clavate, reaching a length of 45 μ and a diameter of 6–7 μ, 8-spored; spores filiform, overlapping in the ascus, 1 × 12–15 μ; paraphyses filiform, hyaline, 1.5 μ in diameter below, enlarged above to 3 μ.

On needles of *Pinus resinosa*.

TYPE LOCALITY: East Shore of Lake Huron, Ontario.

DISTRIBUTION: Known only from the type locality.

EXSICCATI: Rehm, Ascom. *2030* (apparently part of type material).

4. **Gorgoniceps montanensis** (Kanouse) Seaver, comb. nov.

Belonopsis montanensis Kanouse, Mycologia **33**: 461. 1941.

Apothecia superficial, 1.5–4.5 mm. in diameter, at first globose and closed, finally expanding and becoming plane, or slightly convex, the margin irregular with patches of the remains of the hymenial covering, broadly sessile, soft-waxy, solitary, gray-brown; hymenium orange when dry, slightly paler when moist; asci cylindric-clavate, long-stemmed, 8-spored, reaching a length of 130–155 μ and a diameter of 9–12 μ; spores needle-shaped, straight, or slightly curved, 1.5–2 × 38–50 μ, becoming about 7-septate; paraphyses filiform, twisted into coils or spirals at the apices, forming a thin epithecium.

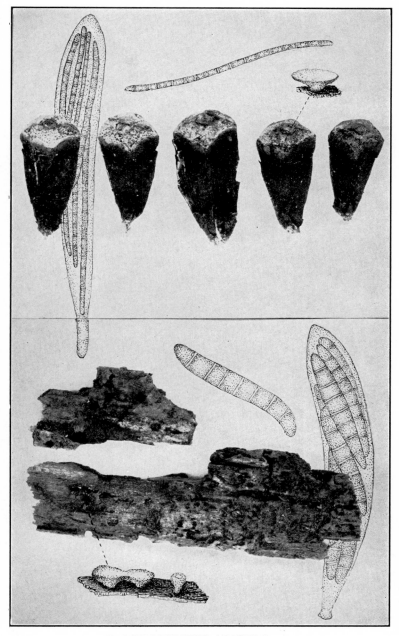

1. GORGONICEPS ARIDULA
2. GORGONICEPS CONFLUENS

On fallen leaves of beech and on fir needles.

TYPE LOCALITY: Echo Lake, Flathead National Forest, Montana.

DISTRIBUTION: Known only from the type locality.

5. **Gorgoniceps iowensis** Rehm, Ann. Myc. **4**: 338. 1906.

Apothecia scattered, at first subglobose, sessile, or contracted into a very short stem, expanding and becoming patellate, whitish, with a slight grayish-green tint when dry, pale-brownish, reaching a diameter of .2–.5 mm.; asci clavate, reaching a length of 80–100 μ and a diameter of 10–12 μ; spores subcylindric, or clavate, straight, or curved, becoming 7-septate, hyaline, 3–4 $\times$ 30–37 μ; paraphyses filiform, slightly enlarged above.

On rotten wood.

TYPE LOCALITY: Mt. Pleasant, Iowa.

DISTRIBUTION: New York and Iowa.

ILLUSTRATION: Bull. Lab. Nat. Hist. State Univ. Iowa **6**: *pl. 26, f. 2.*

6. **Gorgoniceps confluens** Seaver & Waterston, Mycologia **32**: 399. 1940. (PLATE 110, FIG. 2.)

Apothecia gregarious, occasionally crowded and several fusing together, sessile, or contracted into a very short, stem-like base, whitish or bluish-white, remaining light-colored, or becoming darker when dried, reaching a diameter of .5 mm., soft and waxy; hymenium plane, or slightly convex, similar in color to the outside of the apothecium; asci broad-clavate, with a very short, stem-like base, attenuated at the apex, reaching a length of 100 μ and a diameter of 14 μ, 8-spored; spores bunched together and overlapping, cylindric, fusoid, or subclavate, straight, or more often curved or double curved, becoming 7-septate, 5–7 $\times$ 40–45 μ; paraphyses filiform, about 2 μ in diameter.

On rotten wood and on palm stems, *Sabal.*

TYPE LOCALITY: Bermuda.

DISTRIBUTION: Known only from the type locality.

ILLUSTRATIONS: Mycologia **38**: 551 (lower figure).

Type collected in Bermuda by Stewardson Brown, N. L. Britton and Fred J. Seaver (No. *1487*) Nov. 29–Dec. 14, 1912. This is very similar to *G. iowensis* Rehm, which was described from material collected by the author in Iowa. The spores of the Bermuda specimens seem to be larger. Also collected in Paget Marsh on stems of native palm by Seaver & Waterston.

7. Gorgoniceps jamaicensis Seaver, Mycologia 38: 552.   1946.

Apothecia gregarious, or crowded, occasionally several coalescing, sessile or nearly so, becoming patellate, in dried specimens pale yellowish-amber, semitranslucent, reaching a diameter of .5 mm.; hymenium plane, or slightly concave; asci clavate, 8-spored, reaching a length of 140 μ and a diameter of 20 μ, tapering below into a short, stem-like base; spores fasciculate, cylindric with the ends attenuated, reaching a length of 50–55 μ and a diameter of 9–10 μ, becoming 7-septate; paraphyses filiform, about 2 μ in diameter.

On bamboo, *Bambusa vulgaris.*

Type collected by W. A. and Edna Murrill in Chester Vale, Jamaica, December 21–24, 1908, altitude 3000–4000 ft. (No. *311*).

This seems to differ from our Bermuda species in the much larger spores and asci.

Doubtful Species

Gorgoniceps dinemasporioides (Ellis & Ev.) Sacc. Syll. Fung. **8**: 506.   1889; *Peziza dinemasporioides* Ellis & Ev. Jour. Myc. **1**: 42.   1885.   This was described as a *Peziza* by Ellis and placed in the genus *Gorgoniceps* by Saccardo because of the filiform spores.   The spores are not filiform but fusoid and the asci appear to be borne in thin-walled perithecia, clothed with long *Chaetomium*-like hairs.   In the opinion of the author this is not a cup-fungus at all.

31. POCILLUM DeNot. Comm. Critt. Ital. 1: 361.   1864.

Apothecia stipitate, or substipitate, cup-shaped, or expanded; asci clavate-cylindric, 8-spored; spores filiform, hyaline.

Type species, *Helotium Cesatii* Mont.

This genus is close to *Gorgoniceps* apparently differing only in the stipitate, or substipitate character of the apothecia which ordinarily is not considered a very reliable one.   So far as observed only one species has been reported within our range.

1. Pocillum hypophyllum (Berk. & Curt.) Sacc. Syll. Fung. 8: 606.   1889.

Peziza hypophylla Berk. & Curt.; Berk. Jour. Linn. Soc. **10**: 369.   1868.

Apothecia hypophyllous, short-stipitate, externally brown, crateriform, the margin sinuate; hymenium livid; spores filiform.

On the under side of some leaf among the brown, velvety down.

Type locality: Cuba.

Distribution: Known only from the type locality.

No material has been seen and the above is all that is known of this species. It was originally placed in the subgenus *Mollisia*.

32. **APOSTEMIDIUM** Karst. Myc. Fenn. **1**: 15, 186.   1871.

Gorgoniceps Karst. Myc. Fenn. **1**: 15, 185. (in part).   1871.
Vibrissea Phill. Trans. Linn. Soc. II. **2**: 8.   1881.

Apothecia sessile, turbinate, or convex, soft-waxy, or subgelatinous; hymenium spread on the upper, convex, or plain surface, sterile below; asci narrowly cylindric, opening by a pore, 8-spored; spores in a parallel fascicle in the ascus, hyaline, filiform, many-septate, nearly as long as the ascus; paraphyses present.

Type species, *Peziza fiscella* Karst.

This genus is sometimes treated with the Geoglossaceae next to *Vibrissea* which it resembles in spore characters but in the opinion of the writer scarcely belongs there.

Paraphyses longer than the asci, repeatedly forked near
 the apices.   1. *A. Guernisaci.*
Paraphyses equaling or scarcely exceeding the asci, sim-
 ple or rarely forked.   2. *A. vibrisseoides.*

1. **Apostemidium Guernisaci** (Cr.) Durand, Ann. Myc. **6**: 456. 1908.   (PLATE 111.)

Vibrissea Guernisaci Crouan, Ann. Sci. Nat. IV. **7**: 176.   1857.
Gorgoniceps Guernisaci Sacc. Syll. Fung. **8**: 505.   1889.
Gorgoniceps turbinulata Rehm, Ann. Myc. **2**: 353.   1904.

Apothecia solitary, or gregarious, sessile by a rather broad base, turbinate, or convex; hymenium plain to convex, bluish-pallid to orange-ochraceous, 1–4 mm. in diameter, even, or sometimes wrinkled, often with a dimple in the center, dark-brownish below, somewhat furfuraceous; asci long-cylindric, apex rounded, reaching a length of 225–300 μ and a diameter of 5–6 μ; 8-spored; spores in a parallel fascicle in the ascus, hyaline, slenderly filiform, multiseptate, 1 × 200–250 μ; paraphyses numerous, filiform, hyaline, longer than the asci, the apical portion (about as much as projects beyond the asci) repeatedly forked or fasciculately branched, the tips abruptly pyriform-thickened, hyaline, 4–5 μ thick.

On dead sticks and wood in water.

TYPE LOCALITY: Europe.

DISTRIBUTION: New York and New England to Washington.

ILLUSTRATIONS: Ann. Sci. Nat. IV. **7**: *pl. 4, f. 24–27;* Pat. Tab. Fung. *f. 369;* Ann. Myc. **6**: *pl. 11, f. 115–118.*

EXSICCATI: Ellis & Ev. N. Am. Fungi *2738.*

The long, filiform spores often protrude at maturity, giving the hymenium a white, feathery appearance. This interesting species has been encountered by the writer but once in New York.

2. **Apostemidium vibrisseoides** (Peck) Boud. Ann. Myc. **4**: 240. 1906.

Helotium vibrisseoides Peck, Ann. Rep. N. Y. State Mus. **32**: 48.   1879.
?Vibrissea turbinata Phill. Trans. Linn. Soc. II. **2**: 8.   1881.
?Gorgoniceps turbinata Sacc. Bot. Cent. **18**: 219.   1884.
Gorgoniceps vibrisseoides Sacc. Syll. Fung. **8**: 505.   1889.
?Apostemidium obconicum Kanouse, Papers Mich. Acad. Sci. **21**: 97.   1936.

Apothecia gregarious, sessile, turbinate, 2–3 mm. in diameter; hymenium plain to convex, varying in color from bluish-pallid to yellowish, or ochraceous; asci reaching a length of 275–300 and a diameter of 5–6 μ, 8-spored; spores in a parallel fascicle in the ascus, long-filiform, hyaline, multiseptate, about 1 $\times$ 250 μ; paraphyses not numerous, slender, 1–1.5 μ thick below, simple, or occasionally forked, the apical portion clavate-thickened, colored, up to 6 μ thick, about as long as the asci.

On decaying sticks lying in the water.

TYPE LOCALITY: New York.

DISTRIBUTION: New Hampshire to New York.

ILLUSTRATIONS: Bull. N. Y. State Mus. **1**[2]: *pl. 2, f. 7–9;* Trans. Linn. Soc. II. **2**: *pl. 2, f. 14–18;* Ann. Myc. **6**: *pl. 11, f. 119–120.*

Apostemidium obconicum is said to differ in the color and more stipitate character of the apothecia. The writer doubts that the characters are dependable.

33. **PHAEOHELOTIUM** Kanouse, Papers Mich. Acad. Sci. **20**: 75. 1935.

Apothecia superficial, sessile to stipitate, at first turbinate, opening to a disk, finally convex and slightly umbilicate, soft-waxy, drying brittle to chalky, externally pulverulent, bright-colored; asci cylindric-clavate, 8-spored; spores ellipsoid to fusoid, straight, or slightly bent, 1-septate, brown; paraphyses filiform.

Type species, *Phaeohelotium flavum* Kanouse.

APOSTEMIDIUM VIBRISSEOIDES

1. **Phaeohelotium flavum** Kanouse, Papers Mich. Acad. Sci. **20**: 75. 1935.

Apothecia as above, 1–2 mm. broad and 1–1.5 mm. high; hymenium bright-yellow to orange-yellow, margin white to pale-yellow; asci reaching a length of 100–125 μ and a diameter of 8–9 μ; spores with two to four oil-drops, 4–5 × 14–16 μ, brown when mature; paraphyses 2–3 μ in diameter at their apices.

On decaying wood.

TYPE LOCALITY: Harbor Springs, Michigan.

DISTRIBUTION: Michigan.

ILLUSTRATIONS: Papers Mich. Acad. Sci. **20**: *pl. 15, f. 4.*

34. **PSEUDOPEZIZA** Fuckel, Symb. Myc. 290. 1869.

Drepanopeziza (Kleb.) Höhn. Ann. Myc. **15**: 332. 1917.
Pseudopeziza subg. *Drepanopeziza* Kleb. Zeitsch. Pflanzenkr. **16**: 76. 1906.

Apothecia erumpent, usually on living leaves, gregarious, minute, soft and fleshy, the margin often crenulate; asci cylindric-clavate, 4–8-spored; spores ovoid, simple, hyaline; paraphyses filiform and often much curved.

Type species, *Ascobolus Trifolii* Biv.-Bern.

On leguminose plants.
    On leaves of *Trifolium*.   1. *P. Trifolii.*
    On leaves of *Medicago*.   2. *P. Medicaginis.*
Not on leguminose plants.
    On *Ranunculus*.   3. *P. singularis.*
    On *Gentiana*.   4. *P. Holwayi.*
    On *Galium*.   5. *P. repanda.*
    On *Ribes*.   6. *P. Ribis.*
    On *Populus alba*.   7. *P. Populi-albae.*

1. **Pseudopeziza Trifolii** (Biv.-Bern.) Fuckel, Symb. Myc. 290. 1869.

Ascobolus Trifolii Bivona-Bernardi, Stirp. Rar. Sic. **4**: 27. 1816.
Peziza Trifoliorum Lib. Pl. Crypt. Ard. *324.* 1837.
Phyllachora Trifolii Sacc. Atti Soc. Veneto Sci. Nat. Padova **2**: 145. 1873.
Phacidium Trifolii Gill. Champ. Fr. Discom. 170. 1886.
Trochilia Trifolii DeNot.; Rehm in Rab. Krypt.-Fl. **1**[3]: 597. 1891.
Mollisia Trifolii Phill. Brit. Discom. 199. 1887.

Apothecia occurring on brown spots 1–3 mm. in diameter on living leaves, submerged, becoming erumpent, at first closed, opening and becoming nearly flat, .3–.5 mm. in diameter, brownish, waxy; asci clavate, reaching a length of 60–80 μ and a diameter of 10–14 μ, 8-spored; spores ellipsoid, or ovoid, simple, usually with one oil-drop, 5–7 × 10–15 μ; paraphyses filiform, enlarged above, 3–4 μ thick.

On living leaves of species of *Trifolium*.

TYPE LOCALITY: Europe.

DISTRIBUTION: Probably throughout North America where clover grows; also in Europe.

ILLUSTRATIONS: Bivona-Bernardi, Stirp. Rar. Sic. **4**: *pl. 6, f. 3;* Rab. Krypt.-Fl. **1**³: 594, *f. 1–6;* E. & P. Nat. Pfl. **1**¹: 216. *f. 168. A–C;* Briosi & Cavara, Fung. Par. *68.*

EXSICCATI: Ellis & Ev. N. Am. Fungi *2626;* Rab.-Winter-Paz. Fungi Eu. *3872* (from California); Barth. Fungi Columb. *1755, 3164;* Clements, Crypt. Form. Colo. *285;* Seym. & Earle, Econ. Fungi *511;* Rehm, Ascom. *614c* (from London, Canada).

Nannfeldt (Nova Acta Soc. Sci. Upsal. IV. **8**: 179.) regards the form on *Trifolium arvense* as distinct and lists it under the name *Pseudopeziza Trifolii-arvensis*.

2. **Pseudopeziza Medicaginis** (Lib.) Sacc. Syll. Fung. **8**: 724. 1889.

Phacidium Medicaginis Lib. Pl. Crypt. Ard. *176.* 1832.
?Pyrenopeziza Medicaginis Fuckel, Symb. Myc. 295. 1869.
Phyllachora Medicaginis Sacc. Malpighia **1**: 455. 1888.

Apothecia minute, expanded and subdiscoid, ochraceous-brown, seated on yellowish, suborbicular spots, opening with a laciniate aperture through the epidermis, .5–1 mm. in diameter; asci stipitate, subcylindric, attenuated above, 8-spored, reaching a length of 75–80 μ and a diameter of 10 μ; spores 2-seriate, ovoid, simple, hyaline, 4–6 × 8–11 μ; paraphyses filiform-clavate.

On living or dying leaves of species, of *Medicago*, *Trigonella* and *Melilotus*.

TYPE LOCALITY: Europe.

DISTRIBUTION: Throughout North America or wherever the hosts are grown.

ILLUSTRATIONS: Briosi & Cavara, Fungi, Par. *262.*

EXSICCATI: Ellis, N. Am. Fung. *1319;* Ellis & Ev. N. Am. Fungi *1319, 2627, 3059,* Fungi Columb. *533;* Barth. Fungi Columb. *2554, 3059, 3927, 4250, 4345, 4551,* Seaver, N. Dak. Fungi *18;* Griff. West Am. Fungi *337;* Seym. & Earle, Econ. Fungi *506;* Solh. Myc. Sax. Exsicc. 8.

Nannfeldt (Nova Acta Soc. Sci. Upsla. IV. **8**: 179) regards Fuckel's material as distinct and assigns the new name *Pseudopeziza Jonesii* to it.

Sydow (Ann. Myc. **34**: 396) regards the form on *Melilotus alba* as distinct and describes the species *Pseudopeziza Meliloti*.

3. **Pseudopeziza singularis** (Peck) Davis, Trans. Wisc. Acad. Sci. **17**: 863. 1914.

Peziza singularis Peck, Ann. Rep. N. Y. State Mus. **35**: 142. 1884.
Mollisia singularis Sacc. Syll. Fung. **8**: 328. 1889.

Apothecia minute but often forming irregular, confluent patches 1 mm. or more in diameter, seated on pallid spots, dingy-gray, or pale amber-colored; hymenium plane, or convex; asci clavate, reaching a length of 50–70 μ and a diameter of 7–10 μ; spores 1-seriate or partially 2-seriate, elongate-ellipsoid, 4–5 $\times$ 10–15 μ; paraphyses filiform.

On living leaves of species of *Ranunculus*.
TYPE LOCALITY: East Berne, New York.
DISTRIBUTION: New York and Wisconsin.
EXSICCATI: Davis, Fungi Wisc. *35*.

4. **Pseudopeziza Holwayi** P. Henn. Hedwigia **41**: Beibl. (64). 1902.

Apothecia hypophyllous, erumpent, at first subglobose, finally cup-shaped to discoid, sessile, the margin brownish-black, .3 to .9 mm. in diameter; hymenium plane or nearly so, cinereous-violet; asci clavate, or subfusoid, 8-spored, reaching a length of 50–80 μ and a diameter of 10–14 μ; spores 2-seriate, ellipsoid, or fusoid, 4–5 $\times$ 15–21 μ; paraphyses simple, filiform, thickened above.

On living leaves of *Gentiana*.
TYPE LOCALITY: Mexico.
DISTRIBUTION: Known only from the type locality.

P. Hennings states that it is not unlikely that at maturity the spores become 1-septate which, if true, would place it in the genus *Fabraea*.

5. **Pseudopeziza repanda** (Fries) Karst. Acta Soc. Fauna Fl. Fenn. **2**[6]: 161. 1885.

Phacidium repandum Fries, Vet. Akad Handl. **1819**: 108. 1819.
Leptotrochila repanda Karst. Myc. Fenn. **1**: 246. 1871.

Apothecia at first lightly submerged, soon becoming superficial but deep seated in the tissues of the host, opening with a laciniate margin, .3 mm. in diameter, externally greenish-black;

hymenium concave, dingy-brown; asci clavate, reaching a length of 50–60 μ and a diameter of 6–7 μ, 8-spored; spores 2-seriate, or partially so, fusoid, 3–4 × 16–12 μ; paraphyses filiform.

On living leaves of *Galium* and occasionally on other herbaceous plants.

TYPE LOCALITY: Europe.

DISTRIBUTION: Oregon; also in Europe.

ILLUSTRATIONS: Alb. & Schw. Consp. Fung. *pl. 4, f. 6* (as *Xyloma herbarum.*)

The only American specimens seen are two collected in Oregon on *Galium boreale* and *Galium triflorum*. The fungus seems to agree well with European specimens.

6. **Pseudopeziza Ribis** Rehm; Kleb. Cent. Bakt. II. **15**: 336. 1905.

Gloeosporium Ribis Kleb. Zeit. Pflanzenk. **16**: 76. 1906.
Drepanopeziza Ribis Höhn. Ann. Myc. **15**: 332. 1917.

Apothecia erumpent, with a short, stout, stem-like base, becoming nearly plane, .25 mm. in diameter; asci clavate, reaching a length of 80–100 μ and a diameter of 18–20 μ, 8-spored; spores ovoid, 7–8 × 12–17 μ; paraphyses filiform, branched, scarcely enlarged above.

On leaves of species of *Ribes*.

TYPE LOCALITY: Europe.

DISTRIBUTION: Ontario and Oregon; also in Europe.

7. **Pseudopeziza Populi-albae** Kleb. Haupt.-Nebenfr. Ascom. 344. 1918.

Drepanopeziza Populi-albae Nannf. Nova Acta Soc. Sci. Upsal. IV. **8**: 170. 1932.
Trochilia populorum Desm. Bull. Soc. Bot. Fr. **4**: 858. 1857. Sensu Edgerton, Mycologia **2**: 169. 1910.

Apothecia at first somewhat globose, expanding and becoming plane, or slightly concave, surrounded by the upturned fragments of the ruptured epidermis, .1–.2 mm. in diameter; asci clavate, reaching a length of 60–80 μ and a diameter of 12–14 μ, 8-spored; spores 1–2-seriate, ellipsoid, usually with two large oil-drops, hyaline, 5–7.2 × 12–16 μ; paraphyses filiform, somewhat enlarged above.

On leaves of *Populus alba*.

TYPE LOCALITY: Europe.

DISTRIBUTION: New York and Georgia; also in Europe.

ILLUSTRATIONS: Mycologia **2**: 170, *f. 1–5;* 172, *f. 6, 7.*
The conidial stage *Marsonia Castagnei* (Desm. & Mont.)
Magn. is said to be widely distributed in the United States but
the perfect stage rarely collected.

DOUBTFUL SPECIES

Pseudopeziza Bistortae (Lib.) Fuckel, Symb. Myc. 290. 1869. This
species has been reported from Alaska in Harriman Alaska Exped. **5**: 23, on
Polygonum viviparum. No material seen.

Pseudopeziza Cerastiorum Arenariae Sacc.; Sacc. Peck, & Trelease in
Harriman Alaska Exped. **5**: 23. 1904. This variety was described by Saccardo from material collected on Yakutat Bay, Alaska on dead leaves of
Arenaria lateriflora. Said to differ from the species in its larger spores,
4–4.5 × 14–16 μ.

35. **FABRAEA** Sacc. Michelia **2**: 331. 1881.

Diplocarpon Wolf, Bot. Gaz. **54**: 231. 1912.

Apothecia at first submerged, becoming superficial, sessile,
becoming expanded and subdiscoid, soft and fleshy; asci cylindric
to clavate, usually 8-spored; spores ellipsoid, or ovoid, becoming
1-septate; paraphyses filiform.

Type species, *Phacidium congener* Ces.

Occurring on herbaceous plants.
On *Ranunculus.*	1. *F. Ranunculi.*
On *Fragaria.*	2. *F. Earliana.*
On woody plants.	
On *Rosa.*	3. *F. Rosae.*
On *Pyrus, Amelanchier,* and *Cydonia.*	4. *F. maculata.*

1. **Fabraea Ranunculi** (Fries) Karst. Acta Soc. Fauna Fl. Fenn. **2**[6]: 161. 1885.

Dothidea Ranunculi Fries, Syst. Myc. **2**: 562. 1823.
Phlyctidium Ranunculi Wallr. Fl. Crypt. Germ. **2**: 420. 1833.
Excipula Ranunculi Rab. Deutsch. Krypt.-Fl. **1**: 153. 1844.
Phacidium Ranunculi Lib. Pl. Crypt. Ard. *69.* 1830.
Pseudopeziza Ranunculi Fuckel, Symb. Myc. 290. 1869.
Mollisia Ranunculi Phill. Brit. Discom. 200. 1887.
Phacidium litigiosum Rob. & Desm. Ann. Sci Nat. III. **8**: 181. 1847.
Fabraea litigiosa Sacc. Syll. Fung. **8**: 735. 1889.

Apothecia occurring for the most part on discolored spots on
the under side of the leaves, at first submerged, becoming
erumpent and finally superficial, at first rounded, later cup-
shaped, then nearly plane, externally brown, or brownish-black;
hymenium concave to plane, dull-gray, or brownish; asci clavate,

reaching a length of 60–85 μ and a diameter of 12–14 μ, 8-spored; spores ellipsoid to ovoid, at first simple, becoming 1-septate, 5–6 × 12–15 μ; paraphyses filiform 3 μ in diameter above.

On the under side of living leaves of *Ranunculus*.

TYPE LOCALITY: Europe.

DISTRIBUTION: Maine to Toronto and Nebraska.

EXSICCATI: Barth. Fungi Columb. *2625;* Reliq. Farlow. *638.*

2. **Fabraea Earliana** (Ellis & Ev.) Seaver, comb. nov.

Leptothyrium Fragariae Lib. Pl. Crypt. Ard. *162.* 1832.
Ascochyta colorata Peck, Ann. Rep. N. Y. State Mus. **38**: 94. 1884.
Peziza Earliana Ellis & Ev. Bull. Torrey Club **11**: 74. 1884.
Marsonia Fragariae Sacc. Malpighia **10**: 276. 1896.
Diplocarpon Earliana Wolf, Jour. Elisha Mitch. Sci. Soc. **39**: 158. 1924.
Mollisia Earliana Sacc. Syll. Fung. **8**: 328. 1889.

Conidial stage forming reddish, or purplish spots on the leaves of the host; acervuli black, usually epiphyllous; conidia unsymmetrical, 1-septate, the upper cell larger, constricted at the septum, 5–7 × 18–30 μ.

Apothecia discoid, mostly hypophyllous, dark-brown to black, reaching .3 mm. in diameter, at first submerged, becoming erumpent at maturity, splitting with an irregular margin, closing when dry; hymenium yellow to brown; asci clavate, reaching a length of 55–70 μ and a diameter of 15–20 μ; 8-spored; spores ellipsoid, curved, hyaline, unequally 2-celled, the upper larger, constricted at the septum, 4–6 × 18–28 μ; paraphyses capitate.

On fallen leaves of *Fragaria* sp.

TYPE LOCALITY: Anna, Illinois.

DISTRIBUTION: Widely distributed in eastern United States and Canada.

ILLUSTRATIONS: Ann. Rep. N. Y. State Mus. **38**: *pl. 2, f. 9, 10;* Jour. Elisha Mitch. Sci. Soc. **39**: *pl. 9–15;* Tech. Bull. N. C. Agr. Exp. Sta. **28**: *pl. 7, f. 3–5;* Ber. Deutsch. Bot. Ges. **42**: 193, *f. 9–13.*

EXSICCATI: Ellis, N. Am. Fungi *2043.*

3. **Fabraea Rosae** (Wolf) Seaver, comb. nov.

Asteroma Rosae Lib. Mem. Soc. Linn. Paris **5**: 405. 1827.
Actinonema Rosae Fries, Summa Veg. Scand. 424. 1849.
Dicoccum Rosae Bonord. Bot. Zeit. **11**: 282. 1853.
Marsonia Rosae Trail, Trans. Crypt. Soc. Scotland **14**: 46. 1889.
Diplocarpon Rosae Wolf, Bot. Gaz. **54**: 231. 1912.

Conidial stage forming large, dark-brown, or blackish spots, sometimes becoming confluent and involving the entire leaf;

conidia formed on erumpent acervuli, 1-septate and constricted at the septum, 5–6 × 18–25 μ.

Apothecia epiphyllous, sphaerical to discoid, .25 mm. in diameter, at first closed, opening with a stellate aperture; asci subclavate, reaching a length of 70–80 μ and a diameter 15 μ, 8-spored; spores ellipsoid, hyaline, unequally 2-celled, constricted at the septum, 5–6 × 20–25 μ; paraphyses filiform, enlarged at their apices.

On leaves of *Rosa* sp.

TYPE LOCALITY: Auburn, Alabama.

DISTRIBUTION: The conidial stage, black-spot of rose, is widely distributed wherever the rose is cultivated.

ILLUSTRATIONS: Mem. Soc. Linn. Paris **5**: *pl. 5, f. 3;* Bot. Gaz. **54**: *pl. 13.*

4. **Fabraea maculata** (Lév.) Atkinson, Science N. S. **30**: 452. 1909.

Entomosporium maculatum Lév. in Moug. & Nestl. Stirp. Crypt. *1457.* 1860.

Apothecia minute, inconspicuous in dried specimens, when moist swelling up and making their appearance through the ruptured epidermis, often ellipsoid in form; asci 8-spored; spores hyaline, 1-septate.

On leaves of *Pyrus, Amelanchier* and *Cydonia.*

TYPE LOCALITY: New York.

DISTRIBUTION: Widely distributed in North America on cultivated pear and quince.

When Atkinson published a preliminary report of this he promised a more complete description later but this apparently was never done.

DOUBTFUL SPECIES

Fabraea cincta Sacc. & Scalia; Sacc. Peck, & Trelease in Harriman Alaska Exped. **5**: 23. 1904. On suborbicular or irregular spots 1–2 mm. in diameter; apothecia erumpent, 300–400 μ in diameter; asci 18–20 × 70–90 μ, 8-spored; spores elongate-ellipsoid, 5–6.5 × 15–20 μ; paraphyses filiform with capitate apices. On fading leaves of *Rubus,* Yakutat Bay, Alaska. No material has been seen.

36. **PESTALOPEZIA** Seaver, Mycologia **34**: 300. 1942.

Apothecia superficial, sessile, or subsessile, at first subglobose, becoming expanded and subdiscoid, externally pruinose, or tomentose, light-colored; hymenium becoming nearly plane, dark-colored, almost black; asci 8-spored, subcylindric; paraphyses filiform and rather strongly enlarged.

With a *Pestalotia* as its conidial stage.

Type species, *Dermatea brunneopruinosa* Zeller.

On leaves of *Gaultheria Shallon* in the far west.   1. *P. brunneopruinosa*.
On leaves of *Rhododendron maximum* in the east.   2. *P. Rhododendri*.

1. **Pestalopezia brunneopruinosa** (Zeller) Seaver, Mycologia 34: 300. 1942.

Pestalotia gibbosa Hark. Bull. Calif. Acad. Sci. 2: 439. 1887.
Dermatea brunneopruinosa Zeller, Mycologia 26: 291. 1934.

Apothecia superficial, .6–2 mm. in diameter, closed at first, then cup-shaped, sessile to short-stipitate, externally powdery-villose, fawn-colored to wood-brown; hymenium fuscous-black due to the brownish tips of the paraphyses which form an epithecium; asci clavate-cylindric, reaching a length of 120–140 μ and a diameter of 10–14 μ; spores ellipsoid, hyaline, 1-seriate, containing one or two oil-drops, 7–10 × 14–20 μ; paraphyses slender, clavate, light-brown above, 4–6 μ in diameter at their tips.

On large dead spots on leaves of *Gaultheria Shallon* associated with *Pestalotia gibbosa* which represents its conidial stage.

TYPE LOCALITY: Coos County, Oregon.

DISTRIBUTION: Washington to northern California.

2. **Pestalopezia Rhododendri** Seaver, Mycologia 34: 300. 1942. (PLATE 112.)

Apothecia sparingly scattered near the center of circular or subcircular dead spots apparently caused by the conidial or associated stage of the fungus, *Pestalotia*, the spots becoming brown and bordered with concentric rings of variegated colors from red to brownish the apothecia not exceeding 1 mm. in diameter, appearing as minute, light-colored balls, gradually expanding and exposing the dark-colored discs; asci subcylindric to clavate, tapering into a short, stem-like base, reaching a length of 150 μ and a diameter of 14 μ, 8-spored; spores 1-seriate, ellipsoid, hyaline, 8 × 16 μ; paraphyses slightly enlarged above, pale-brown, reaching a diameter of 6 μ at their apices.

Associated with what appears to be the *Pestalotia* stage on *Rhododendron*, the spores with three brown cells, 10 × 20 μ,

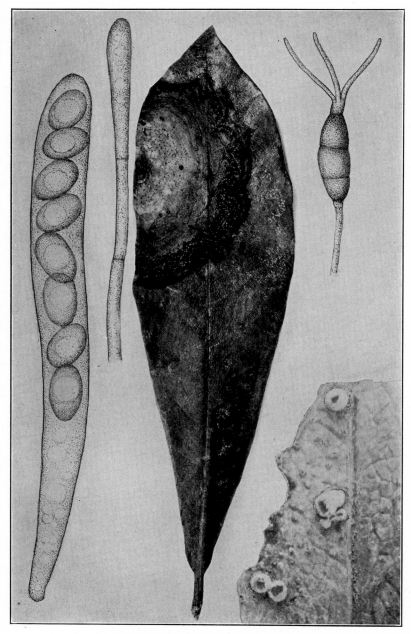

PESTALOPZIA RHODODENDRI

exclusive of the basal cell and bearing three appendages at the opposite end.

TYPE LOCALITY: Pineola, North Carolina.

DISTRIBUTION: Known only from the type locality.

ILLUSTRATIONS: Mycologia **34**: 299, *f. 1.*

EXSICCATI: Ellis & Ev. Fungi Columb. *331* (as *Dermatea lobata*).

DOUBTFUL SPECIES

Cenangella Ericae (Fries) Rehm in Rab. Krypt.-Fl. **1**[3]: 232. 1889; *Cenangium Ericae* Fries, Syst. Myc. **2**: 188. 1822. This species was described as the perfect stage of *Pestalotia Callunae* Ces. on *Calluna*. One specimen on *Ledum groehlandicum* seems to agree with the one reported on *Calluna*. This is the only North American specimen seen. If the reported connection is correct this species should be transferred to the genus *Pestalopezia*.

37. **TAPESIA** (Pers.) Rehm in Rab. Krypt.-Fl. **1**[3]: 574. 1891.

Peziza [subg.] *Tapesia* Pers. Myc. Eu. **1**: 270. 1822.

Apothecia superficial, seated on a conspicuous web of white, or dark-colored mycelium, minute, usually less than 1 mm. in diameter, soft and waxy, scutellate, or discoid, the excipulum dark; hymenium often at first light-colored, becoming dark; asci cylindric, or clavate, 4–8-spored; spores ellipsoid, fusoid, or elongated, never or rarely septate, hyaline; paraphyses filiform.

Type species, *Peziza fusca* Pers.

This genus differs from *Mollisia* only in the presence of the subiculum.

Apothecia seated on a brown or black subiculum.
  On wood or branches of various kinds.
    Spores relatively large, 12–14 μ long.   1. *T. fusca.*
    Spores mostly less than 12 μ long.
      On swollen branches of *Bigelovia*.   2. *T. tumefaciens.*
      Not on swollen branches.
        On old decaying wood.
          Hymenium white or yellow.   3. *T. culcitella.*
          Hymenium brown.   4. *T. mollisioides.*
        On recently killed branches.
          On stems of *Rosa*.   5. *T. Rosae.*
          On twigs of *Ribes*.   6. *T. ribicola.*
          On branches of *Sambucus*.   7. *T. coloradensis.*
          On branches of *Carpinus* and *Betula*.   8. *T. lividofusca.*
  On leaves of *Abies balsamea*.   9. *T. balsamicola.*
Apothecia on a white mycelium.
  Spores 2–3 × 7–10 μ, on *Betula*.   10. *T. secamenti.*
  Spores 7 × 15 μ, on grass and old wood.   11. *T. earina.*

1. **Tapesia fusca** (Pers.) Fuckel, Symb. Myc. 302.   1869.

Peziza fusca Pers. Obs. Myc. **1**: 29.   1796.
Peziza subiculata Schw. Trans. Am. Phil. Soc. II. **4**: 174.   1832.
Peziza griseopulveracea Schw. Trans. Am. Phil. Soc. II. **4**: 174.   1832.
Mollisia fusca Karst. Myc. Fenn. **1**: 207.   1871.
?Peziza vincta Cooke & Peck, Grevillea **1**: 6.   1872.
Tapesia cinerella Rehm, Hedwigia **21**: 102.   1882.
Tapesia vincta Sacc. Syll. Fung. **8**: 372.   1889.
Tapesia subiculata Sacc. Syll. Fung. **8**: 380.   1889.
Tapesia griseopulveracea Sacc. Syll. Fung. **8**: 380.   1889.
Tapesia Rhois Fairman, Proc. Rochester Acad. Sci. **3**: 216.   1900.

Apothecia seated on a widely, densely effused subiculum of brown filaments, scattered, or gregarious, at first urn-shaped, black, finally expanding and becoming cinereous-brown, or cinereous-yellow, reaching a diameter of 1 mm.; hymenium pallid-cinereous to pale-yellowish; asci fusoid-clavate, reaching a length of 55–68 μ and a diameter of 6–8 μ, 8-spored; spores fusoid, usually simple, occasionally doubtfully 1-septate, 2–3 × 12–14 μ; paraphyses filiform, 2–4 μ in diameter.

On wood and bark of various kinds.

TYPE LOCALITY: Europe.

DISTRIBUTION: New York to Oregon and North Carolina; probably widely distributed.

ILLUSTRATIONS: Rab. Krypt.-Fl. **1**[3]: 573. *f. 1–5:* E. & P. Nat. Pfl. **1**[1]: 211. *f. 165, A. B.;* Greville, Scot. Crypt. Fl. **4**: *pl. 192.*

No specimen of *Peziza vincta* Cooke & Peck has been seen but the description suggests the above.

2. **Tapesia tumefaciens** Ellis & Ev. Bull. Torrey Club **24**: 281. 1897.

Apothecia densely gregarious, or in broad strips appearing through cracks in the bark, seated on a thick subiculum, when fresh about 2 mm. in diameter, mouse-colored, soft and fleshy, the margin fringed with olivaceous, straight hairs; hymenium slightly concave; asci clavate-cylindric, reaching a length of 70–75 μ and a diameter of 8–10 μ, 8-spored; spores obliquely 1-seriate, ellipsoid, hyaline, 5–6 × 8–11 μ; paraphyses stout, about 2 μ thick.

On swollen dead stems of *Bigelovia graveolens*.

TYPE LOCALITY: Colorado.

DISTRIBUTION: Known only from the type locality.

Ellis states: "When dry the opposite sides of the ascomata are rolled together in a hysteriform manner. The habit is that of *Angelina rufescens* Duby." "The mycelium penetrates the wood and causes the stems to swell in the same manner as *Montagnella tumefaciens* E. & M."

3. **Tapesia culcitella** (Cooke & Ellis) Sacc. Syll. Fung. **8**: 374. 1889.

Peziza culcitella Cooke & Ellis, Grevillea **6**: 7.   1877.

Apothecia gregarious, about 1 mm. broad, at first depressed urn-shaped, expanding to plane, seated on a sparse, brown, byssoid stratum, at first milk-white, pallid, or yellowish when dry, externally quite smooth; asci narrowly clavate-cylindric; spores narrow-ellipsoid, $2 \times 8 \mu$; paraphyses filiform.

Inside an old oak stump, *Quercus.*

TYPE LOCALITY: Newfield, New Jersey.

DISTRIBUTION: Newfoundland and New Jersey.

EXSICCATI: Ellis, N. Am. Fungi *388.*

This looks like *Mollisia melaleuca* except for the subiculum.

4. **Tapesia mollisioides** (Schw.) Sacc. Michelia **2**: 140.   1880.

Peziza mollisiaeoides Schw. Trans. Am. Phil. Soc. II. **4**: 174.   1832.

Apothecia seated on a brownish-black subiculum, minute, about .12 mm. in diameter, becoming plane, or slightly concave, sessile, brown; hymenium paler; asci clavate, reaching a length of 50–55 μ and a diameter of 10 μ, 8-spored; spores ellipsoid to subfusoid, 3×8–9 μ, slightly curved; paraphyses filiform.

On old wood.

TYPE LOCALITY: Pennsylvania.

DISTRIBUTION: New England to Ohio and South Carolina.

5. **Tapesia Rosae** (Pers.) Fuckel, Symb. Myc. 301.   1869.

Peziza Rosae Pers. Syst. Myc. 656.   1801.
Mollisia Rosae Karst. Myc. Fenn. **1**: 208.   1871.
Lachnea Rosae Gill. Champ. Fr. Discom. 92.   1882.

Apothecia sparse, or gregarious, seated on a broadly effused subiculum of brown hyphae, subtomentose, or smooth, brown, slightly concave, coriaceous, contracted and deformed when dry, 1 mm. in diameter; hymenium grayish-white; asci cylindric-clavate, reaching a length of 45–60 μ and a diameter of 6–8 μ, 8-spored; spores, fusoid, 2–2.5 $\times$ 7–9 μ; paraphyses filiform.

On stems of species of *Rosa.*

TYPE LOCALITY: Europe.
DISTRIBUTION: New York to Pennsylvania and Colorado;
also in Europe.
ILLUSTRATIONS: Boud. Ic. Myc. *pl. 539.*
EXSICCATI: Clements, Crypt. Form. Colo. *524.*

6. **Tapesia ribicola** Cash, Mycologia **28**: 300.   1936.

Apothecia sessile, superficial, depressed-globose, then patellate, irregularly contorted when dry, often hysteroid, or triangular, gregarious, soft-fleshy, reaching a diameter of 1.5 mm.,
externally brownish-black to black and smooth near the base,
dull-gray to white and fimbriate at the margin; hymenium light-
gray; asci cylindric-clavate, short-stipitate, narrowed at the apex,
8-spored, reaching a length of 35–50 μ and a diameter of 4.5–5 μ;
spores 2-seriate, elongate-ellipsoid to clavate, hyaline, simple,
straight, or slightly curved, 1.5–2 × 7–11 μ; paraphyses filiform,
hyaline, septate, unbranched, not enlarged above.
On twigs of *Ribes montigenum.*
TYPE LOCALITY: Mesa Lakes, Colorado.
DISTRIBUTION: Colorado.
ILLUSTRATIONS: Mycologia **28**: 302, *f. 3.*

7. **Tapesia coloradensis** Ellis & Ev. Proc. Acad. Sci. Phila.
1894: 350.   1894.

Apothecia scattered, or gregarious, sessile, seated on the
epidermis or on the surface of the inner bark, surrounded by a
thin, brown, subiculum composed of slender, pale-brown,
sparingly septate hyphae, reaching a diameter of 1–2 mm., externally nearly black; hymenium glaucous-white, or livid-white
at first, becoming nearly black when dry; asci clavate-cylindric,
reaching a length of 45–55 μ and a diameter of 8–10 μ, 8-spored;
spores ellipsoid, 2-seriate, smoky-hyaline, often slightly curved,
2.5–3.5 × 8–11 μ; paraphyses stout, evanescent.
On bark of *Sambucus melanocarpa.*
TYPE LOCALITY: Cameron Pass, Colorado, elevation 10,000 ft.
DISTRIBUTION: Known only from the type locality.

8. **Tapesia lividofusca** (Fries) Rehm in Rab. Krypt.-Fl. **1**[3]: 576.
1891.

Peziza lividofusca Fries, Syst. Myc. **2**: 147.   1822.
Niptera lividofusca Fuckel, Symb. Myc. Nacht. **2**: 58.   1873.
Mollisia lividofusca Gill. Champ. Fr. Discom. 127.   1882.

Apothecia for the most part gregarious, occasionally confluent, seated on a thick, brown, mycelial subiculum, at first globose and closed, opening and becoming cup-shaped, later expanded, shallow cup-shaped, externally brown, reaching a diameter of .2–2 mm., waxy; hymenium, gray, or yellowish-white; asci clavate, reaching a length of 50–60 μ and a diameter of 5–6 μ, 8-spored; spores fusiform, straight, or slightly curved, 2.5–3 $\times$ 9–12 μ; paraphyses filiform 2 μ in diameter.

On dead branches of *Carpinus, Betula* etc.

TYPE LOCALITY: Europe.

DISTRIBUTION: Reported from Illinois and Nebraska: also in Europe.

Dr. B. Kanouse (Papers Mich. Acad. Sci. **20**: 69. 1935.) reports *Tapesia lividofusca* var. *fallax* (Desm.) Rehm on cones of *Picea mariana* from Michigan. No material has been seen.

9. **Tapesia balsamicola** (Peck) Sacc. Syll. Fung. **8**: 376. 1889.

Peziza balsamicola Peck, Ann. Rep. N. Y. State Mus. **34**: 51. 1883.

Apothecia seated on a subiculum consisting of two kinds of mycelium one black, coarse and producing large, three- or four-lobed, spore-like bodies, the other hyaline, delicate, producing narrow-fusiform conidia, minute, less than .5 mm. in diameter, sessile, smooth, waxy, whitish; asci clavate, reaching a length of 40–50 μ; spores, subfusoid, with three or four oil-drops, 5–7 $\times$ 15–20 μ; paraphyses filiform.

On dead or dying leaves of *Abies balsamea*.

TYPE LOCALITY: Stony Clove, Catskill Mountains, New York.

DISTRIBUTION: Known only from the type locality.

ILLUSTRATIONS: Ann. Rep. N. Y. State Mus. **34**: *pl. 1, f. 14–21.*

Part of the type collection is in the herbarium of The New York Botanical Garden.

10. **Tapesia secamenti** Fairman, Ann. Myc. **8**: 329. 1910.

Apothecia seated on a white mycelium which is arachnoid and radiating from the cup bottom, later becoming extended and intricately interwoven and composed of simple, or sparingly branched, continuous, smooth, hyaline hyphae 3 μ thick, globose, then expanded, brownish-olivaceous, sessile, about 1 mm. in diameter; hymenium white at first, often becoming pale straw-colored; asci fusoid-cylindric, reaching a length of 57 μ and a

diameter of 6 μ; spores 1-seriate, ellipsoid, or fusoid, 2–3 ×
7–10 μ; paraphyses indistinct.

On old chips of *Betula*.

TYPE LOCALITY: Lyndonville, New York.

DISTRIBUTION: New York and Michigan.

11. **Tapesia earina** (Ellis) Sacc. Syll. Fung. **8**: 377.   1889.

Peziza earina Ellis, Bull. Torrey Club **8**: 124.   1881.

Apothecia thin, soft, seated on a loose mycelium of dirty-
white, matted, cottony hairs, circular about 1 mm. in diameter;
hymenium brick-red when dry; asci cylindric, reaching a length of
100–125 μ and a diameter of 12 μ, 8-spored; spores 1-seriate, ellipsoid, usually with one oil-drop, 7 × 15 μ; paraphyses rather stout,
thickened above.

On dead stems and leaves of grasses and on weather-beaten
wood, still partly covered with snow.

TYPE LOCALITY: Pleasant Valley, Utah, elevation 6,000 feet.

DISTRIBUTION: Known only from the type locality.

DOUBTFUL AND EXCLUDED SPECIES

Tapesia anomala (Pers.) Fuckel, Symb. Myc. 300.   1869; *Peziza anomala*
Pers. Obs. Myc. **1**: 29.   1796; *Solenia anomala* Fuckel, Symb. Myc. Nacht. **1**:
290.   1871.   This is now regarded as a *Solenia*.   Widely distributed.

Tapesia arachnoidea (Schw.) Sacc. Syll. Fung. **8**: 384.   1889; *Peziza
arachnoidea* Schw. Trans. Am. Phil. Soc. II. **4**: 174.   1832.   The type of this
species was examined by the writer on May 2, 1916, and no apothecia could
be found.   The description is brief and incomplete and the species must remain in doubt.

Tapesia atrofusca (Berk. & Curt.) Sacc. Syll. Fung. **8**: 373.   1889; *Peziza
atrofusca* Berk. & Curt.; Berk. Grevillea **3**: 156.   1875.   The original description states "Cups dark brown, with an inflexed crenulated granulated margin
springing from a pale ochre membranaceous stratum."   This is all that is
known of the species.   No specimens seen.   Probably *Tapesia fusca*.

Tapesia Bloxami (Berk. & Curt.) Sacc. Syll. Fung. **8**: 380.   1889; *Peziza
Bloxami* Berk. & Curt. Ann. Mag. Nat. Hist. II. **7**: 181.   1851.   This species
has been rejected by Phillips (Brit. Discom. 408) since no fruit could be found.

Tapesia cinerella Rehm, Hedwigia **21**: 102.   1882.   Reported on stems of
Gaultheria Shallon from Washington by Dr. B. Kanouse.   No material seen.

Tapesia daedalea (Schw.) Sacc. Syll. Fung. **8**: 379.   1889; *Peziza daedalea*
Schw. Trans. Am. Phil. Soc. II. **4**: 174.   1832.   Specimens distributed under
this name (Rav. Fungi Car. **1**: *37*) are referred apparently by Ellis to *Solenia
poriaeformis*.   The type in the Schweinitz was examined by the writer May 2,
1916.

Tapesia derelicta Morgan, Jour. Myc. **8**: 186.   1902.   No specimen has
been seen but the white subiculum and septate spores strongly suggest *Arachnopeziza aurelia*.

Tapesia discincola (Schw.) Sacc. Syll. Fung. **8**: 384. 1889; *Peziza discincola* Schw. Trans. Am. Phil. Soc. II. **4**: 174. 1832. The type of this species in the Schweinitz Collection has been examined and found very unsatisfactory.

Tapesia megaloma (Schw.) Sacc. Syll. Fung. **8**: 378. 1889; *Peziza megaloma* Schw. Trans. Am. Phil. Soc. II. **4**: 175. 1832. This was reported on rotten wood. No specimens have been seen and details of the fruits are unknown.

Tapesia melaleucoides var. *Vaccinii* Rehm in Rab. Krypt.-Fl. **1³**: 578. 1891. This species which was distributed by Rehm (Ascom. *153 b*) as *Tapesia fusca* has been reported from Michigan by Dr. B. Kanouse. No American material has been seen.

Tapesia pruinata (Schw.) Sacc. Syll. Fung. **8**: 379. 1889; *Peziza pruinata* Schw. Schr. Nat. Ges. Leipzig **1**: 120. 1822. This is *Solenia poriaeformis* (Pers.) Fries.

Tapesia sanguinea (Pers.) Fuckel, Symb. Myc. 303. 1869; *Peziza sanguinea* Pers. Tent. Disp. Fung. 34. 1797; *Patellaria sanguinea* Karst. Myc. Fenn. **1**: 231. 1871; *Patellea sanguinea* Rehm in Rab. Krypt.-Fl. **1³**: 284. 1889. This species belongs with the Patellariaceae which are not treated in this work.

Tapesia scariosa (Berk. & Curt.) Sacc. Syll. Fung. **8**: 375. 1889; *Peziza scariosa* Berk. & Curt.; Berk. Grevillea **3**: 156. 1875. Original description reads: "Cups collapsing, confluent, black with a whitish margin springing from a black subiculum; sporidia slender, subfusiform. Allied to *P. fusca*." A small specimen and drawing by Massee are in the herbarium of the New York Botanical Garden. Nothing more is known of it.

Tapesia tela (Berk. & Curt.) Sacc. Syll. Fung. **8**: 373. 1889; *Peziza tela* Berk. & Curt.; Berk. Grevillea **3**: 156. 1875. This is *Cyphella tela* (Berk. & Curt.) Massee, Jour. Myc. **6**: 179. 1891.

Tapesia tenebrosa (Karst.) Nannf. Nova Acta Soc. Sci. Upsal. IV. **8**: 165. 1932. Reported from Michigan by Dr. B. Kanouse on oak bark. Characterized by the stroma-like base of coarse hyphae on which the apothecia are seated. Asci 5–6 × 30–40 μ; spores 1.5–2 × 5–5.5 μ.

38. **MOLLISIA** Karst. Myc. Fenn. **1**: 187. 1871.

Mollisiopsis Rehm, Ann. Myc. **6**: 315. 1908.
Dibelonis Clements, Gen. Fungi 87, 175. 1909.

Apothecia usually dark-colored, sordid, the hymenium occasionally light-colored, externally smooth, or occasionally slightly roughened; asci clavate to cylindric, or curved, sometimes short and subglobose to ellipsoid, fusiform, or baciliform; paraphyses usually filiform, more rarely lanceolate, hyaline, or colored.

Type species, *Peziza cinerea* Batsch.

The genus *Mollisiopsis* was established for species with lanceolate paraphyses.

This is a large and difficult genus, new species having been described on the flimsiest pretext and often based on a single specimen in many cases none too good. It is hoped that the following enumeration may be of some assistance to the collector.

Occurring on phanerogams.
  On dicotyledonous plant tissues.
    On rotten wood.
      Spores 2–5 μ thick.
        Spores not exceeding 10 μ long.
          Hymenium cinereous or livid.
            Spores 6–10 μ long.   1. *M. cinerea.*
            Spores 3.5–4.5 μ long.   2. *M. Fairmani.*
          Hymenium whitish.   3. *M. melaleuca.*
          Hymenium pale-brown.   4. *M. microcarpa.*
        Spores 10 μ or more long.
          Hymenium orange-yellow.   5. *M. aureofulva.*
          Hymenium bluish-gray.   6. *M. caesia.*
          Hymenium horn-colored.   7. *M. subcornea.*
          Hymenium sooty-black.   8. *M. fumigata.*
          Hymenium whitish.   9. *M. glenospora.*
      Spores slender, 1.5–2 μ thick.
        Hymenium whitish.
          Spores 6–9 μ long.   10. *M. vulgaris.*
          Spores 4–6 μ long.   11. *M. caespiticia.*
        Hymenium leather-colored.   12. *M. encoelioides.*
        Hymenium lilac-colored.   13. *M. lilacina.*
    On dead herbaceous stems.
      Spores 16–18 μ long.   14. *M. apiophila.*
      Spores not exceeding 10 μ long.
        On dead stems of *Polygonum.*   15. *M. Polygoni.*
        On dead stems of *Asclepias.*   16. *M. Asclepiadis.*
        On dead stems of *Thalictrum.*   17. *M. subcinerea.*
        On compositae.
          Spores scarcely 5 μ long.   18. *M. exigua.*
          Spores 10 μ long.   19. *M. erigeronata.*
        On stems of various kinds.   20. *M. atrata.*
    On living or recently killed leaves.
      On living *Potentilla.*   21. *M. Dehnii.*
      On leaves of *Amophila.*   22. *M. clavigera.*
      On leaves of *Angelica.*   23. *M. Angelicae.*
      On leaves of *Antennaria.*   24. *M. lanaria.*
      On leaves of *Gaultheria.*   25. *M. Gaultheriae.*
      On petioles of *Juglans.*   26. *M. abdita.*
      On a variety of herbaceous plants.   27. *M. Oenotherae.*
  On monocotyledonous plant tissues.
    Spores 25–30 μ long.
      On *Andropogon.*   28. *M. atriella.*
      On *Scirpus.*   29. *M. scirpina.*

Spores less than 25 μ long.
  Spores 15–21 μ long.

On *Juncus*.	30. *M. stictoidea*.
On leaves of *Iris*.	31. *M. Iridis*.

  Spores 9–14 μ long.

On grasses of various kinds.	32. *M. hydrophila*.
On *Typha*.	33. *M. epitypha*.
On *Juncus alpinus*.	34. *M. alpina*.

  Spores 4–6 μ long.

On petioles of *Sabal*.	35. *M. Sabalidis*.
On dead stems of *Smilax*.	36. *M. nipteroides*

Occurring on cryptogams.
  On fern stems, *Osmunda*.   37. *M. tenella*.
  On fungi.

On *Polyporus*.	38. *M. incrustata*.
On *Scoleconectria*.	39. *M. Scoleconectriae*.

1. **Mollisia cinerea** (Batsch) Karst. Myc. Fenn. **1**: 189.   1871.

Peziza cinerea Batsch, Elench. Fung. Cont. **1**: 197.   1786.
Niptera cinerea Fuckel, Symb. Myc. 292.   1869.

Apothecia gregarious, or scattered, at first subglobose, then expanding and becoming nearly plane, or occasionally repand, cinereous, with a whitish margin; hymenium undulate, cinereous to yellowish, often becoming nearly black; asci cylindric-clavate, 8-spored, reaching a length of 45–60 μ and a diameter of 5–7 μ, 8-spored; spores ellipsoid to fusoid, 2-seriate, straight, or curved, 2–3 × 6–10 μ; paraphyses filiform, enlarged above to 3–5 μ in diameter.

On decaying wood of many kinds.

TYPE LOCALITY: Europe.

DISTRIBUTION: Throughout North America; also in Europe.

ILLUSTRATIONS: Batsch, Elench. Fung. *pl. 26, f. 137;* Boud. Ic. Myc. *pl. 541;* Phill. Brit. Discom. *pl. 6, f. 32;* E. & P. Nat. Pfl. **1**[1]: 213, *f. 166, A, B;* Rab. Krypt.-Fl. **1**[3]: 505, *f. 7, 8.*

EXSICCATI: Clements, Crypt. Form. Colo. *87;* Wilson & Seaver, Ascom. *39.*

Numerous varieties of this species have been recorded by Saccardo, among them *culmina* and *caricina*.   *Mollisia culmina* var. *caricina* (Sacc.) Rehm has been recorded from Michigan by Dr. B. Kanouse on *Carex*.   No material has been seen.

2. **Mollisia Fairmani** (Ellis & Ev.) Sacc. Syll. Fung. **8**: 302. 1889.

Peziza Fairmani Ellis & Ev. Jour. Myc. **4**: 56.   1888.
Pseudohelotium Fairmani Sacc. Syll. Fung. **8**: 302.   1889.

Apothecia sessile, gregarious, concave with the margin sub-incurved, then expanding to plane, or slightly convex, reaching a diameter of .75 mm., externally clothed with glandular hair-like structures; hymenium livid, paler toward the margin; asci clavate-cylindric, reaching a length of 35–40 μ and a diameter of 4–5 μ; spores 2-seriate or partially so, ellipsoid, the lower end often more acute, 2–2.5 × 3.5–4.5 μ.

On inner bark lying on the ground.

TYPE LOCALITY: Lyndonville, New York.

DISTRIBUTION: Known only from the type locality.

3. **Mollisia melaleuca** (Fries) Sacc. Syll. Fung. **8**: 337.   1889.

Peziza melaleuca Fries, Syst. Myc. **2**: 150.   1822.
Niptera melaleuca Fuckel, Symb. Myc. 292.   1869.
Patellaria melaleuca Quél. Ench. Fung. 325.   1886.
Mollisia lignicola Phill. Grevillea **15**: 113.   1887.
Pyrenopeziza lignicola Sacc. Syll. Fung. **8**: 366.   1889.

Apothecia sessile, waxy, becoming patellate, externally rough, becoming black; hymenium whitish; asci cylindric-clavate, acuminate, 8-spored; spores 2-seriate, cylindric, slightly curved, 2 × 8 μ; paraphyses filiform, 2.5 μ in diameter.

On old wood.

TYPE LOCALITY: Europe.

DISTRIBUTION: Maine to Ontario and Michigan; also in Europe.

4. **Mollisia microcarpa** (Fuckel) Sacc. Syll. Fung. **8**: 340.   1889.

Niptera (?) *microcarpa* Fuckel, Symb. Myc. Nacht. **1**: 46.   1871.

Apothecia gregarious, or scattered, sessile, 1 mm. in diameter, watery-transparent, pale-brown, externally lightly strigose, at first closed, then opening; hymenium concave, flexuous, same color; asci clavate-cylindric, 8-spored, reaching a length of 30 μ and a diameter of 8 μ; spores 2-seriate, fusoid, 2.5 × 8 μ; paraphyses filiform.

On rotten wood.

TYPE LOCALITY: Europe.

DISTRIBUTION: Michigan; also in Europe.

Reported from Michigan by Dr. B. Kanouse but no specimen seen.

5. **Mollisia aureofulva** (Cooke) Sacc. Syll. Fung. **8**: 339. 1889.

Peziza aureofulva Cooke, Hedwigia **14**: 83. 1875.

Apothecia subgregarious, soft-waxy, minute, at first hemispherical, then expanded, .5 mm. in diameter, contracted when dry; externally brown; hymenium convex, orange-yellow; asci clavate-cylindric; spores ellipsoid, with two oil-drops, 5 × 8–15 μ; paraphyses enlarged above.

On old wood.

TYPE LOCALITY: New Jersey.

DISTRIBUTION: Known only from the type locality.

6. **Mollisia caesia** (Fuckel) Sacc. Syll. Fung. **8**: 340. 1889.

Niptera caesia Fuckel, Symb. Myc. Nacht. **1**: 47. 1871.

Apothecia gregarious, sessile, reaching a diameter of 2 mm., becoming plane, finally convex, externally smooth; hymenium bluish-gray; asci clavate-cylindric, reaching a length of 55 μ and a diameter of 8 μ; spores fusiform, with two oil-drops, 3 × 14–16 μ; paraphyses filiform, enlarged above to 3 μ in diameter.

On rotting wood of *Fagus*.

TYPE LOCALITY: Europe.

DISTRIBUTION: North Dakota; also in Europe.

EXSICCATI: Brenckle, Fungi Dak. *533*.

7. **Mollisia subcornea** Phill. & Hark. Bull. Calif. Acad. Sci. **1**: 22. 1884.

Apothecia scattered, or crowded, occasionally coalescing, sessile, patellate, reddish horn-color, .5–1 mm. in diameter; asci clavate, 8-spored; spores 2-seriate, ellipsoid-fusoid, occasionally 1-septate, 3 × 10–12 μ; paraphyses filiform.

On branches of *Eucalyptus globulus*.

TYPE LOCALITY: California.

DISTRIBUTION: Known only from the type locality.

No specimen has been seen.

8. **Mollisia fumigata** (Ellis & Ev.) Sacc. Syll. Fung. **8**: 343. 1889.

Peziza fumigata Ellis & Ev. Bull. Torrey Club **11**: 41. 1884.

Apothecia densely gregarious, minute, .13–.16 mm. in diameter, sooty-black; hymenium nearly plane, minutely papillose; asci subcylindric, reaching a length of 50 μ and a diameter of 7 μ;

spores obliquely arranged, or 2-seriate, with two or three oil-drops, slightly curved, 3 × 9–12 μ.

On rotten wood of *Magnolia*.

TYPE LOCALITY: New Jersey.

DISTRIBUTION: Known only from the type locality.

No specimen found in New York Botanical Garden Herbarium.

9. **Mollisia glenospora** (Ellis & Ev.) Sacc. Syll. Fung. **8**: 343. 1889.

Peziza glenospora Ellis & Ev. Bull. Torrey Club **11**: 18.   1884.

Apothecia sessile, orbicular, .25–.5 mm. in diameter, dull-white, becoming darker, the margin fringed with a row of erect, subcylindric, continuous, hair-like cells, 2 × 8–10 μ; asci cylindric-clavate, reaching a length of 75–80 μ and a diameter of 9–10 μ; spores ellipsoid, or more acute at one end, with a single oil-drop, 7–8 × 12–15 μ; paraphyses curved and swollen at their apices.

On rotten *Magnolia*.

TYPE LOCALITY: New Jersey.

DISTRIBUTION: Known only from the type locality.

10. **Mollisia vulgaris** (Fries) Gill. Champ. Fr. Discom. 119. 1882.

Peziza vulgaris Fries, Syst. Myc. **2**: 146.   1822.
Pezizella vulgaris Sacc. Syll. Fung. **8**: 278.   1889.
Phialea vulgaris Rehm in Rab. Krypt.-Fl. **1³**: 709.   1892.

Apothecia subcespitose, membranaceous, sessile, 1 mm. in diameter; hymenium concave, white, or pallid-white; asci cylindric-clavate, reaching a length of 35–45 μ and a diameter of 5–6 μ, 8-spored; spores elongate-ellipsoid, straight, or curved, 1.5–2 × 6–9 μ; paraphyses filiform.

On rotten wood.

TYPE LOCALITY: Europe.

DISTRIBUTION: New Jersey and South Carolina; also in Europe.

EXSICCATI: Rav. Fungi Car. **4**: *18*.

Berkeley (Grevillea **3**: 159.   1875) records two varieties of this species: var. *myceticola* on decaying *Polyporus*, and on oak Santee River, South California; *sanguinella* with a pale-orange disc, on *Liquidambar*, from South Carolina and New England.

Peziza albella With. (Arrang. Brit. Pl. **4**: 350. 1796) has been cited by various authors including Fries as a synonym. As pointed out by Phillips (Rev. Myc. **12**: 140. 1890) this was an error, apparently referring to *Peziza albida*. This species is described as having apothecia .25–.75 of an inch in diameter while the species listed under this name, *Pezizella albella*, is only 1 mm. in diameter or less.

11. **Mollisia caespiticia** Karst. Myc. Fenn. **1**: 188. 1871.

Peziza caespiticia Karst. Not. Soc. Fauna. Fl. Fenn. **10**: 159. 1869.

Apothecia for the most part occurring in cespitose clusters, at first globose and closed, finally opening and becoming cup-shaped, at length nearly plane, externally brownish, .2–1.5 mm. in diameter, waxy; hymenium grayish-white; asci clavate, reaching a length of 30–45 μ and a diameter of 3–4 μ, 8-spored; spores elongated and somewhat fusoid, straight, or curved, 2-seriate, 1–1.5 $\times$ 4–6 μ; paraphyses filiform, 3 μ in diameter.

On old wood, *Salix*, *Ulmus* etc.

TYPE LOCALITY: Europe.

DISTRIBUTION: Michigan; also in Europe.

12. **Mollisia encoelioides** Rehm in Rab. Krypt.-Fl. **1**³: 522. 1891.

Apothecia thickly gregarious, sessile but narrowed below, at first globose and closed, finally opening and becoming cup-shaped, .3–1.2 mm. in diameter, externally brownish and somewhat rough; hymenium leather-colored; asci clavate, reaching a length of 30–35 μ and a diameter of 5 μ, 8-spored; spores cylindric, straight, or somewhat curved, 2-seriate, 1.5 $\times$ 5–6 μ; paraphyses filiform, 2.5 μ in diameter, above irregular, 5 μ in diameter and brown.

On wood of *Salix;* also reported on spruce branches.

TYPE LOCALITY: Europe.

DISTRIBUTION: Michigan; also in Europe.

13. **Mollisia lilacina** Clements, Bot. Surv. Nebr. **4**: 15. 1896.

Apothecia gregarious, sessile, discoid, lilac, or pale-livid, testaceous when dry, paler beneath, margin elevated, white-crenulate, reaching a diameter of 1 mm.; asci small, clavate, reaching a length of 30–40 μ and a diameter of 3–4 μ; spores cylindric, straight, or curved, with one to three oil-drops and pseudoseptate, 1-seriate, or becoming 2-seriate, 1.5–2 × 7–10 μ; paraphyses cylindric, granular, 2 μ in diameter.

On bark of *Ulmus americana*, and on stems of *Helianthus annuus*.

TYPE LOCALITY: Nebraska.

DISTRIBUTION: Known only from the type locality.

EXSICCATI: Rehm, Ascom. *1222* (from Lincoln, Nebraska).

14. **Mollisia apiophila** Dearness, Mycologia **8**: 98.   1916.

Apothecia minute, .3–1 mm. in diameter when fresh, hardly visible to the naked eye when dry, flesh-colored, subsessile, turbinate, about as high as wide; asci reaching a length of 45–52 μ and a diameter of 10–12 μ; spores 2-seriate, with two to four oildrops, 4–6 × 16–18 μ; paraphyses forming a yellowish layer.

On old stems of *Apios tuberosa*.

TYPE LOCALITY: London, Ontario, Canada.

DISTRIBUTION: Known only from the type locality.

While this was described as a *Mollisia*, the color and form would suggest that it might be a *Helotium*.

15. **Mollisia Polygoni** (Lasch.) Gill. Champ. Fr. Discom. 120.   1882.

Peziza Polygoni Lasch, in Rab. Herb. Myc. *1127*.   1866.
Peziza lactuosa Cooke, Hedwigia **14**: 83.   1875.
Niptera Polygoni Rehm, Ber. Naturh. Ver. Augsburg **26**: 21.   1881.

Apothecia gregarious, sessile, at first hemispherical, the margin incurved, crenulate, then expanding and becoming saucer-shaped, externally brownish, reaching a diameter of 1 mm. though often less; hymenium dull-grayish, or slightly yellowish; asci clavate, reaching a length of 30–40 μ and a diameter of 5–6 μ, 8-spored; spores ellipsoid to clavate, straight, or curved, 2 × 7–9 μ; paraphyses filiform.

On stems of species of *Polygonum*.

TYPE LOCALITY: Europe.

DISTRIBUTION: New Jersey to Iowa and probably widely distributed in North America; also in Europe.

ILLUSTRATIONS: Bull. Lab. Nat. Hist. State Univ. Iowa **5**: *pl. 23, f. 2.*

EXSICCATI: Ellis, N. Am. Fungi *442*.

This was commonly collected by the writer in Iowa on old *Polygonum* stems.

16. **Mollisia Asclepiadis** Ellis & Ev. Proc. Acad. Sci. Phila. **1895**: 428. 1895.

Apothecia at first covered by the epidermis, then erumpent-superficial, circular, .75 mm. in diameter, pale slate-colored, the margin coarsely toothed, or subentire; hymenium plane to convex; asci cylindric, 8-spored, reaching a length of 55–60 μ and a diameter of 6 μ; spores for the most part 2-seriate, ellipsoid, or subinequilateral, with usually two oil-drops, 3 × 8–10 μ; paraphyses slender.

On dead stems of *Asclepias*.

TYPE LOCALITY: London, Canada.

DISTRIBUTION: Known only from the type locality.

EXSICCATI: Ellis & Ev. N. Am. Fungi *3334*.

17. **Mollisia subcinerea** (Rehm) Seaver, comb. nov.

Mollisiopsis subcinerea Rehm, Ann. Myc. **6**: 315. 1908.

Apothecia gregarious, sessile, at first globose-closed, then expanding, .3–1 mm. in diameter, externally brownish, when dry margin inrolled; hymenium hyaline, yellowish when dry; asci clavate, reaching a length of 45–50 μ and a diameter of 6–7 μ, 8-spored; spores subclavate, straight, 2-seriate, 2 × 8–9 μ; paraphyses lanceolate, 4–5 μ in diameter, prominent.

On dead stems of *Thalictrum;* also reported on aspen stems.

TYPE LOCALITY: Lyndonville, New York.

DISTRIBUTION: New York and Michigan.

Distinguished from *Mollisia cinerea* by its lanceolate paraphyses.

18. **Mollisia exigua** (Cooke) Seaver, comb. nov.

Peziza exigua Cooke, Hedwigia **14**: 83. 1875.
Pezizella exigua Sacc. Syll. Fung. **8**: 283. 1889.

Apothecia scattered, minute, scarcely visible to the unaided eye, sessile, subtremellose, cup-shaped, finally plane, or convex, the margin scarcely elevated; asci minute, lanceolate; spores linear scarcely 5 μ long.

On stems of *Erigeron*.

TYPE LOCALITY: New Jersey.

DISTRIBUTION: Known only from the type locality.

Cooke in describing the species states: "Oculo nudo inconspicua." Specimens in The New York Botanical Garden are apparently part of the type material. Several collections were made by Ellis.

19. Mollisia erigeronata (Cooke) Sacc. Syll. Fung. **8**: 323.   1889.

Peziza erigeronata Cooke, Hedwigia **14**: 83.   1875.
?Peziza astericola Cooke & Ellis, Grevillea **6**: 90.   1878.
?Mollisia astericola Sacc. Syll. Fung. **8**: 323.   1889.

Apothecia gregarious, sessile, at first subglobose, expanding and becoming nearly plane, brownish-black, becoming black, .5 mm. in diameter; hymenium livid-cinereous, the margin slightly elevated; asci short-clavate; spores irregularly 2-seriate, long-ellipsoid, $2 \times 10 \mu$; paraphyses not observed.

On stems of *Erigeron* and *Solidago.*

TYPE LOCALITY: New Jersey.

DISTRIBUTION: New Jersey to London, Canada.

EXSICCATI: Ellis & Ev. N. Am. Fungi *? 3335, 3040;* Ellis & Ev. Fungi Columb. *245, 1310.*

20. Mollisia atrata (Pers.) Karst. Myc. Fenn. **1**: 200.   1871.

Peziza atrata Pers. Syn. Fung. 669.   1801.
Pyrenopeziza atrata Fuckel, Symb. Myc. 294.   1869.
Peziza atrocinerea Cooke; Phill. Brit. Discom. 176.   1887.
Urceola atrata Quél. Ench. Fung. 322.   1886.
Mollisia atrocinerea Phill. Brit. Discom. 176.   1887.

Apothecia gregarious, or scattered, at first globose, expanding and becoming subdiscoid, cinereous, or brownish-black when dry, the margin white and more or less undulate, less than 1 mm. in diameter; asci clavate-fusoid, reaching a length of $25–40 \mu$ and a diameter of $5–6 \mu$; spores elongate-ellipsoid, $1.5–2 \times 5–8 \mu$; paraphyses filiform.

On dead herbaceous stems of many kinds.

TYPE LOCALITY: Europe.

DISTRIBUTION: New York to Ontario and California and south to Alabama; also in Europe.

EXSICCATI: Ellis, N. Am. Fungi *443;* Ellis & Ev. N. Am. Fungi *2625;* Fungi Columb. *409;* Seaver, N. Dak. Fungi *14.*

21. Mollisia Dehnii (Rab.) Karst. Myc. Fenn. **1**: 206.   1871.

Peziza Dehnii Rab. Bot. Zeit. **1**: 12.   1843.
Beloniella Dehnii Rehm in Rab. Krypt.-Fl. **1³**: 639.   1892.
Dibelonis Dehnii Clements, Gen. Fungi 175.   1909.

Apothecia gregarious, innate, sessile, finally plane, brown, becoming black when dry, .5–1 mm. in diameter; hymenium livid-cinereous, asci cylindric-clavate, reaching a length of 50–60 μ and a diameter of 6–7.5 μ; spores 2-seriate, fusoid-elongate, simple, slightly curved, or nearly straight, 3 × 10–16 μ; paraphyses filiform, 2 μ thick.

On living leaves and stems of species of *Potentilla*.

TYPE LOCALITY: Europe.

DISTRIBUTION: New York to Montana and Iowa; also in Europe.

ILLUSTRATIONS: Alb. & Schw. Consp. Fung. *pl. 4, f. 6;* Bull. Lab. Nat. Hist. State Univ. Iowa **5**: *pl. 23, f. 1.*

EXSICCATI: Ellis & Ev. N. Am. Fungi *2039;* Fungi Columb. *531;* Barth. Fungi Columb. *2237;* Kellerm. Ohio Fungi *103;* Seaver, N. Dak. Fungi *38;* Brenckle, Fungi Dak. *5;* Griff. West Am. Fungi *193;* Wilson & Seaver, Ascom. *15.*

22. **Mollisia clavigera** (Ellis & Ev.) Sacc. Syll. Fung. **8**: 354. 1889.

Peziza clavigera Ellis & Ev. Jour. Myc. **4**: 100. 1888.

Apothecia protruding when moist, contracted and hysteriform when dry, .25–.5 mm. wide and .5–1 mm. long, the margin fimbriate; hymenium cinereous, or livid-white; asci cylindric-clavate, narrowed below into a stem-like base; spores 2-seriate, 4–4.5 × 12–15 μ with two or three oil-drops; paraphyses slender, enlarged above to 3–4 μ.

On leaves of *Ammophila longifolia.*

TYPE LOCALITY: Sand Coulee, Montana.

DISTRIBUTION: Known only from the type locality.

No specimen seen.

23. **Mollisia Angelicae** Dearness, Mycologia **16**: 145. 1924.

Apothecia dark-brown, superficial, mostly hypophyllous, occasionally epiphyllous, sessile, rugose-costate, urceolate, reaching a diameter of .125 mm.; asci cylindric to fusoid, 8-spored, reaching a length of 36–40 μ and a diameter of 9 μ; spores 2-seriate, or obliquely 1-seriate, narrowed at one end, 2.5–3 × 9–15 μ; paraphyses scant.

On dead parts of living leaves of *Angelica atropurpurea.*

TYPE LOCALITY: London, Canada.

DISTRIBUTION: Known only from the type locality.

Dearness states that more than 100 leaves were collected bearing this *Mollisia*.

24. Mollisia lanaria Fairman, Ann. Myc. **9**: 150. 1911.

Apothecia hypophyllous, nestling in the woolly coating of the leaves, sessile, minute, .25 mm. in diameter, circular in outline, numerous, gregarious, brown; hymenium lighter, light-brown; asci clavate-cylindric, sessile or short-stipitate, narrowed slightly above, reaching a length of 40–45 μ and a diameter of 6–7 μ; spores long-cylindric, 2–3 × 14–24 μ; paraphyses subcylindric.

On under side of leaves of *Antennaria plantaginifolia*.

TYPE LOCALITY: Orchard Creek Road, Ridgway, New York.

DISTRIBUTION: Known only from the type locality.

ILLUSTRATIONS: Ann. Myc. **9**: 150. *f. 1–2.*

No specimens seen. Fariman states that on the same leaves is occasionally found a *Septoria* which might be connected with *Mollisia lanaria*.

25. Mollisia Gaultheriae Ellis & Ev. Proc. Acad. Sci. Phila. **1894**: 349. 1894.

Apothecia occurring on large (1–2 cm. broad) orbicular, dirty-brown spots which are sometimes irregular in shape and occupying the greater part of the leaf, amphigenous but more numerous on the under side, .5 mm. in diameter, dark liver-colored, nearly black when dry, the margin slightly dentate; hymenium concave; asci cylindric-clavate, reaching a length of 45–55 μ and a diameter of 10–12 μ; spores 2-seriate, ellipsoid, 3–3.5 × 10–12 μ; paraphyses filiform, branched.

On living leaves of *Gaultheria Shallon*.

TYPE LOCALITY: Seattle, Washington,

DISTRIBUTION: Known only from the type locality.

26. Mollisia abdita (Ellis) Sacc. Syll. Fung. **8**: 353. 1889.

Peziza abdita Ellis, Bull. Torrey Club **9**: 19. 1882.

Apothecia orbicular, pale, .6–.7 mm. in diameter, with a narrow, jagged, membranaceous margin, protruding when fresh through a narrow slit in the epidermis from which it is again withdrawn and entirely disappears when dry; asci cylindric-clavate, reaching a length of 40–45 μ and a diameter of 4–5 μ; spores 2-seriate, clavate, 2–2.5 × 10 μ; paraphyses clavate.

On fallen petioles of *Juglans regia.*
TYPE LOCALITY: New Jersey.
DISTRIBUTION: Known only from the type locality.
EXSICCATI: Ellis, N. Am. Fungi *849.*
When dry its presence is indicated only by a narrow slit in the epidermis, hence the name.

27. **Mollisia Oenotherae** (Cooke & Ellis) Seaver, comb. nov.
Dacryomyces Lythri Desm. Pl. Crypt. Fr. *1545.* 1846.
Peziza Oenotherae Cooke & Ellis, Grevillea **6**: 90. 1878.
Sphaeronema corneum Cooke & Ellis, Grevillea **6**: 84. 1878.
Gloeosporium ? *tremellinum* Sacc. Michelia **2**: 168. 1880.
Gloeosporium ? *rhoinum* Sacc. Fungi Ital. *f. 1035.* 1881.
Hainesia rhoina Ellis & Sacc. in Sacc. Syll. Fung. **3**: 699. 1884.
Pezizella Oenotherae Sacc. Syll. Fung. **8**: 278. 1889.
Tubercularia Rhois Halsted, Bull. Torrey Club **20**: 251. 1893.
Patellina Fragariae Stevens & Peterson, Phytopathology **6**: 264. 1916.
Sporonema pulvinatum Shear, Bull. Torrey Club **34**: 308. 1907.
Pezizella Lythri Shear & Dodge, Mycologia **13**: 161. 1921.

Apothecia scattered, with a short, stem-like base, externally light-brown to amber, .5–1 mm. in diameter; hymenium nearly plane, whitish especially when moist; asci cylindric, or clavate, reaching a length of 55–70 μ and a diameter of 7–8 μ, 8-spored; spores straight, or slightly curved, 1-seriate, or often becoming 2-seriate, often enlarged at one end, 2 $\times$ 8 μ; paraphyses filiform, branched, scarcely enlarged above, 1–1.5 μ in diameter.

On dead spots on living or dead leaves, petioles and stems of *Oenothera* and a larger variety of other hosts.
TYPE LOCALITY: New Jersey.
DISTRIBUTION: Widely distributed in North America; also in Europe.
ILLUSTRATIONS: Mycologia **13**: 146, *f. 1.;* 148, *f. 2.;* 149, *f. 3;* 150, *f. 4;* 151, *f. 5; pl. 8–10;* Phytopathology **6**: 267, *f. 19–26.*
EXSICCATI: Ellis, N. Am. Fungi *733, 846;* Ellis & Ev. N. Am. Fungi *2074, 2278;* Fungi Columb. *244, 287;* Seym. & Earle, Ecom. Fungi *273.*
For a complete discussion of this species, its synonymy, morphology and hosts see Mycologia **13**: 135–170.

28. **Mollisia atriella** (Cooke) Sacc. Syll. Fung. **8**: 352. 1889.
Peziza atriella Cooke, Hedwigia **14**: 83. 1875.

Apothecia gregarious, sessile, soft-waxy, .2 mm. in diameter, at first hemispherical, then expanded, smooth, black; hymenium cinereous-black; asci clavate-cylindric; spores narrowly fusiform, 2.5 × 30 μ; paraphyses filiform.

On culms of *Andropogon* and *Spartina*.

TYPE LOCALITY: New Jersey.

DISTRIBUTION: New Jersey and Louisiana.

EXSICCATI: Ellis & Ev. N. Am. Fungi *1779* (on *Spartina polystachya*).

29. **Mollisia scirpina** (Peck) Sacc. Syll. Fung. **8**: 348. 1889.

Peziza scirpina Peck, Ann. Rep. N. Y. State Mus. **28**: 67. 1876.

Apothecia minute, scattered, erumpent, smooth, externally black, grayish within; asci cylindric, 8-spored; spores crowded, fusiform, straight, or slightly curved, with two oil-drops, 20–25 μ long.

On dead stems of *Scirpus caespitosus*.

TYPE LOCALITY: Adirondack Mountain, N. Y.

DISTRIBUTION: Known only from the type locality.

30. **Mollisia stictoidea** (Cooke & Ellis) Sacc. Syll. Fung. **8**: 348. 1889.

Peziza stictoidea Cooke & Ellis, Grevillea **6**: 8. 1877.

Apothecia erumpent, sparse, .1–.12 mm. in diameter, becoming stictoid when dry, brownish-black; hymenium pallid; asci clavate; spores fusoid, 3 × 16 μ.

On *Juncus*.

TYPE LOCALITY: New Jersey.

DISTRIBUTION: Known only from the type locality.

ILLUSTRATIONS: Grevillea **6**: *pl. 96, f. 37*.

The erumpent character of the apothecia would indicate a *Pyrenopeziza*. Cooke states: "Margin longitudinally connivent when dry, and retracted, hence it resembles a minute *Stictis*."

31. **Mollisia Iridis** (Rehm) Sacc. Syll. Fung. **8**: 331. 1889.

Micropeziza Iridis Rehm, Ascom. Lojk. 16. 1882.

Apothecia when dry oblongate, or rotundate, more or less hemispheric, finally expanded, sessile, gregarious, subfuscous, margin yellowish-green, .3 mm. in diameter; hymenium brownish: asci clavate, reaching a length of 50 μ and a diameter of 9 μ, 4–8-spored; spores ellipsoid, 2-seriate, with one to four oil-drops, 4 × 18–21 μ; paraphyses filiform, enlarged above to 3.5 μ.

On stems of *Iris*.

Type locality: Europe.

Distribution: New York and Ontario; also in Europe.

Several collections of this species from Canada determined by H. S. Jackson are in The New York Botanical Garden. The apothecia are seated on large reddish discolored areas on the leaves.

32. **Mollisia hydrophila** Karst. Myc. Fenn. **1**: 196.   1871.

Peziza hydrophila Karst. Not. Soc. Fauna Fl. Fenn. **10**: 163.   1869.

Apothecia gregarious, entirely white, or pallid, finally sordid-white to cinereous, becoming brown when dry, .7–1 mm. in diameter; hymenium convex, concave when dry; asci fusoid-cylindric, reaching a length of 50–70 μ and a diameter of 5–7 μ; spores acicular, 1.5–2.5 × 9–12 μ; paraphyses filiform, enlarged above.

On culms of grasses, *Phragmites, Andropogon, Sorghum* etc.

Type locality: Europe.

Distribution: New Jersey; also in Europe.

Exsiccati: Ellis, N. Am. Fungi *847*.

33. **Mollisia epitypha** Karst. Myc. Fenn. **1**: 197.   1871.

Niptera epitypha Karst. Not. Soc. Fauna. Fl. Fenn. **11**: 247.   1871.

Apothecia gregarious, sessile, at first globose, opening and becoming hemispherical, finally convex, at first entirely brownish-cinereous, then cinereous-livid, .6–1 mm. in diameter, externally black when dry; hymenium grayish-pallid; asci fusoid-clavate, reaching a length of 50–60 μ and a diameter of 5–6 μ; spores 2-seriate, straight, 2 × 10–14 μ; paraphyses filiform.

On stems of *Typha latifolia;* also reported on *Carex*.

Type locality: Europe.

Distribution: Michigan; also in Europe.

The only American specimen of this species seen is one from Michigan collected and determined by Geo. B. Cummins.

34. **Mollisia alpina** Rostr. Medd. Grønl. **3**: 609.   1891.

Apothecia gregarious, sessile, at first globose, then concave, brownish-black, reaching a diameter of .25 μ; asci cylindric-clavate, reaching a length of 40 μ and a diameter of 6–8 μ; spores ellipsoid, 4 × 10–21 μ, with two or three oil-drops.

On dead stems of *Juncus alpinus.*

TYPE LOCALITY: Greenland.

DISTRIBUTION: Known only from the type locality in Greenland.

35. **Mollisia Sabalidis** (Ellis & Mart.) Seaver, comb. nov.

Dermatea Sabalidis Ellis & Mart. Am. Nat. **18**: 1147.   1884.
Cenangium Sabalidis Sacc. Syll. Fung. **8**: 562.   1889.

Apothecia gregarious, or scattered, shallow cup-shaped, or scutellate, scarcely exceeding .3 mm. in diameter, externally tobacco-brown and clothed with very poorly developed hair-like structures which give the apothecium a powdery appearance; hymenium slightly concave, dark-brown; asci clavate, reaching a length of 35 μ and a diameter of 6–7 μ, 8-spored; spores 2-seriate or irregularly crowded, ellipsoid, 1–1.3 $\times$ 4–6 μ; paraphyses filiform, slender, scarcely enlarged above.

On dead petioles of *Sabal serrulata.*

TYPE LOCALITY: Green Cove Springs, Florida.

DISTRIBUTION: Known only from the type locality.

The apothecia do not appear to the writer to be erumpent. It is certainly not a *Dermea.*

36. **Mollisia nipteroides** Ellis & Ev. Proc. Acad. Sci. Phila. **1893**: 147.   1893.

Apothecia cup-shaped, substipitate, or almost sessile, reaching a diameter of 1–1.5 mm., umber-colored but whitened outside by a pruinose coat and a short, erect, glandular pubescence which is more abundant toward the margin; asci cylindric, reaching a length of 35–40 μ and a diameter of 5–6 μ; spores 2-seriate, with two or three oil-drops, straight, or slightly curved, 1–1.5 $\times$ 4–5 μ; paraphyses branched above and bearing minute, globose conidia.

On dead stems of *Smilax.*

TYPE LOCALITY: St. Martinville, Louisiana.

DISTRIBUTION: Known only from the type locality.

The only specimen seen is the type collection in The New York Botanical Garden.

37. **Mollisia tenella** (Cooke & Ellis) Sacc. Syll. Fung. **8**: 349. 1889.

Peziza tenella Cooke & Ellis, Grevillea **7**: 40.   1878.

Apothecia hypophyllous, scattered, at first hemispherical, finally expanded, .15–.2 mm. in diameter; hymenium pallid; asci clavate; spores linear, straight, 5 μ long.

On fronds of *Osmunda*.

TYPE LOCALITY: Newfield, New Jersey.

DISTRIBUTION: Known only from the type locality.

EXSICCATI: Ellis, N. Am. Fungi *669*.

A part of the type collection is in The New York Botanical Garden.

38. Mollisia incrustata (Ellis) Seaver, comb. nov.

Peziza incrustata Ellis, Am. Nat. **17**: 192. 1883.
Mollisia trametis Ellis & Ev. Proc. Acad. Sci. Phila. **1893**: 147. 1893.
Pseudohelotium incrustatum Sacc. Syll. Fung. **8**: 301. 1889.

Apothecia growing on the margin and inner surface of the pores of the host, obconic, .11–.15 mm. in diameter, honey-colored, of fibrous structure, the ends of the fibers projecting so that the outer surface and margin appear granulose-pubescent, or as if covered with sharp-pointed granules, at first convex-hemispheric, resembling a *Nectria*, soon becoming concave with a distinct subfimbriate margin; asci clavate, reaching a length of 20–23 μ and a diameter of 4–5 μ; spores 2-seriate, or obliquely 1-seriate, narrow-ellipsoid, or clavate, 1.5–2 × 3.5–4.5 μ; paraphyses not observed.

Parasitic on *Polyporus Stevensii* and on a resupinate *Polyporus*.

TYPE LOCALITY: Newfield, New Jersey.

DISTRIBUTION: Known only from the type locality.

The color of this species would suggest a *Helotium* or *Orbilia* rather than a *Mollisia*. Should be more carefully studied.

39. Mollisia Scoleconectriae Cash & Davidson, Mycologia **32**: 733. 1940.

Apothecia cespitose on and sometimes completely covering stromata and perithecia of *Scoleconectria scolecospora*, .2–.4 mm. in diameter, brownish-black, sessile, soft-fleshy, globose with small circular opening, then cupulate with the margin fimbriate, incurved when dry; hymenium dark olive-gray; asci cylindric-clavate, reaching a length of 25–35 μ and a diameter of 3–4 μ, 8-spored; spores 2-seriate, hyaline, simple, narrow-cylindric, straight, or slightly curved, 1 × 5–7 μ; paraphyses filiform, hyaline, unbranched, 1 μ in diameter.

On stromata of *Scoleconectria scolecospora* on twigs of *Pinus Strobus*.
TYPE LOCALITY: Huntingdon County, Pennsylvania.
DISTRIBUTION: Pennsylvania and New York.
ILLUSTRATIONS: Mycologia **32**: 729, *f. 1, D*.

DOUBTFUL SPECIES

Mollisia alabamensis Ellis & Ev. Jour. Myc. **8**: 69.   1902.   Apothecia erumpent, pale rose-colored, becoming almost black; spores 6–7 × 12–13 μ; paraphyses forming a brown epithecium.   On canes of *Rubus villosus*, Alabama. May be a *Cenangium*.

Pezizella aquifoliae (Cooke & Ellis) Sacc. Syll. Fung. **8**: 288.   1889; *Peziza aquifoliae* Cooke & Ellis, Grevillea **6**: 91.   1878.   Apothecia subaurantiaceous, attached by white, radiating mycelium; spores ovoid.   On leaves of *Ilex opaca*. The type in The New York Botanical Garden is too scant to permit of critical study.

Mollisia benesuada (Tul.) Phill. Brit. Discom. 174.   1887; *Peziza benesuada* Tul. Ann. Sci. Nat. III. **20**: 169.   1853.   This species has been reported from Michigan and Washington by Dr. B. Kanouse.   No American specimens have been seen except one doubtfully determined from Pennsylvania.   Also var. *polyspora* Kauff. has been recorded.

Pezizella brassicaecola (Berk.) Sacc. Syll. Fung. **8**: 283.   1889; *Peziza brassicaecola* Berk. Grevillea **3**: 157.   1875.   Apothecia brownish; spores 10 μ long.   On cabbage stems, *Brassica*, New England.   No material seen.

Peziza conorum Ellis, Bull. Torrey Club **6**: 133.   1877.   On cones of *Pinus inops* lying on the ground in New Jersey.   Apothecia yellowish-brown sessile, convex; spores 7.5–10 × 15–20 μ.   No material seen.

Mollisia complicatula Rehm in Rab. Krypt.-Fl. **1³**: 520.   1891.   Apothecia erumpent, brownish-black, .3–1 mm. in diameter; spores 2.5 × 9–12 μ. On wood of *Lonicera* and *Populus*.   Reported from Michigan and Washington. No material seen.

Mollisia euparaphysata Schröt. Krypt.-Fl. Schles. **3²**: 107.   1908.   This species is reported from Michigan on *Carex* by Dr. B. Kanouse under the name *Mollisiopsis euparaphysata* (Schröt.) Rehm.   No material has been seen.

Pezizella exidiella (Berk. & Curt.) Sacc. Syll. Fung. **8**: 288.   1889; *Peziza exidiella* Berk. & Br.; Berk. Grevillea **3**: 158.   1875.   Berkeley places this in the subgenus *Mollisia*.   The description is so brief as to mean little.

Pseudohelotium fibrisedum (Berk. & Curt.) Sacc. Syll. Fung. **8**: 298.   1889; *Peziza fibriseda* Berk. & Curt.; Berk. Grevillea **3**: 157.   1875.   Apothecia irregular, orange, externally clothed with sugar-like granules, the margin broken, laciniate; hymenium concave.   On *Ulmus americana*, Virginian Mountains.   No specimen available and the description is too meager to enable it to be identified.   See *Orbilia cruenta*.

Pezizella floriformis (Peck) Sacc. Syll. Fung. **8**: 287.   1889; *Peziza floriformis* Peck, Ann. Rep. N. Y. State Mus. **33**: 31.   1880.   Apothecia at first subcyathiform, then floriform, with the margin wavy, hymenium cream-colored; spores spermatoid.   On decaying wood of *Acer*, Verona, New York. No material seen.

Peziza heterocarpa Ellis, Bull. Torrey Club **6**: 134. 1877. Apothecia minute, stipitate, about 1 mm. in diameter and high, pale-greenish; spores fusoid-ellipsoid, 20 μ long. On dead stems of *Bidens*, in New Jersey. No material seen. This was placed in the subgenus *Mollisia* by Ellis.

Peziza hypnicola Ellis, Bull. Torrey Club **6**: 134. 1877. Apothecia sessile, pallid-orange, concave; asci subcylindric; spores 1-septate, 2.5 $\times$ 9–10 μ. On *Hypnum sylvaticum* in Pennsylvania. This was placed in the subgenus *Mollisia* by Ellis.

Mollisia introviridis (Cooke & Ellis) Sacc. Syll. Fung. **8**: 339. 1889; *Peziza introviridis* Cooke & Ellis, Grevillea **7**: 7. 1878; Ellis, N. Am. Fungi *566*. The type of this species was first labeled *Peziza sanguinea*. What induced Ellis to change his mind is not apparent.

Mollisia leucostigma (Fuckel) Rehm in Rab. Krypt.-Fl. **1**³: 516. 1891; *Niptera leucostigma* Fuckel, Symb. Myc. Nacht. **2**: 59. 1873. This represents Fuckel's conception of *Peziza leucostigma* Fries but does not agree with the usual interpretation of the species. See *Orbilia leucostigma*.

Peziza melichroa Cooke, Grevillea **7**: 47. 1878; *Pseudohelotium melichroum* Sacc. Syll. Fung. **8**: 301. 1889. Apothecia described as minute, saccharine-granulated; spores ellipsoid, 5 μ long.

Mollisia miltophthalma (Berk. & Curt.) Sacc. Syll. Fung. **8**: 334. 1889; *Peziza miltophthalma* Berk. & Curt.; Berk. Grevillea **3**: 158. 1875. Apothecia black with yellow hymenium; spores minute. On branches of *Cornus florida*, South Carolina. No material seen.

Mollisia papillata Earle, Bull. N. Y. Bot. Gard. **3**: 290. 1905. Apothecia black, clothed with clavate papillae; spores 2 $\times$ 8–10 μ, On old weathered chips, Stanford University, California. The papillate hairs would suggest a *Lachnella*. No material seen.

Mollisia paullopuncta (Cooke & Ellis) Sacc. Syll. Fung. **8**: 342. 1889; *Peziza paulupuncta* Cooke & Ellis, Grevillea **7**: 7. 1878. Apothecia brownish-black; hymenium cinereous; spores 1 $\times$ 5 μ. On bark of *Acer*, New Jersey. No material seen. The spelling of the specific name was changed by Saccardo as indicated above.

Peziza regalis Cooke & Ellis, Grevillea **6**: 91. 1878. Apothecia .1 mm. in diameter, white; asci clavate; spores 7 μ long. On apple bark, New Jersey. No material seen. This was originally placed in the subgenus *Mollisia*.

Pseudohelotium sacchariferum (Berk.) Sacc. Syll. Fung. **8**: 298. 1889; *Peziza saccharifera* Berk. Grevillea **3**: 157. 1875. Apothecia gregarious, pallid-orange, irregular, externally saccharine; disc concave. On *Liquidambar*, Alabama. Originally placed in the subgenus *Mollisia*. Its identity is uncertain. See *Orbilia cruenta*.

Mollisia stenostoma (Berk. & Curt.) Sacc. Syll. Fung. **8**: 354. 1889; *Peziza stenostoma* Berk. & Curt.; Berk. Grevillea **3**: 159. 1875. Apothecia elongate; spores narrow-ellipsoid. Looks like a *Hysterium*. No specimen seen.

Mollisia Teucrii (Fuckel) Rehm in Rab. Krypt.-Fl. **1**³: 524. 1891; *Niptera Teucrii* Fuckel, Symb. Myc. Nacht. **1**: 47. 1871; *Pseudohelotium Teucrii* Sacc. Syll. Fung. **8**: 294. 1889; *Pezizella Teucrii* Rehm in Rab. Krypt.-Fl. **1**³: 1264. 1896. Originally reported on stems of *Teucrium*. Reported from Michigan by Dr. B. Kanouse on wood of *Alnus*. No specimens seen,

39. **MOLLISIELLA** (Phill.) Massee, Brit. Fungus-Fl. **4**:
221. 1895.

Mollisia subg. *Mollisiella* Phill. Brit. Discom. 193.   1887.
Unguiculariopsis Rehm, Ann. Myc. **7**: 400.   1909.

Apothecia small cupulate, becoming expanded, usually occurring on other fungi, externally dark-colored, brownish,
tomentose, or clothed with poorly developed hairs; asci clavate,
or cylindric, usually 8-spored; spores at maturity 1-seriate,
globose; paraphyses filiform, slightly enlarged above.

Type species, *Peziza ilicincola* Berk. & Br.

Mollisiella ilicincola (Berk. & Br.) Massee, Brit. Fungus-Fl. **4**:
222. 1895. (PLATE 113.)

Peziza ilicincola Berk. & Br. Ann. Mag. Nat. Hist. III. **7**: 450.   1859.
Peziza hysterigena Berk. & Br. Jour. Linn. Soc. **14**: 106.   1873.
Peziza Ravenelii Berk. & Curt.; Berk. Grevillea **3**: 152.   1875.
Pseudohelotium ilicincolum Sacc. Syll. Fung. **8**: 304.   1889.
Lachnellula hysterigena Sacc. Syll. Fung. **8**: 391.   1889.
Cenangium Ravenelii Sacc. Syll. Fung. **8**: 568.   1889.
Mollisia ilicincola Phill. Brit. Discom. 193.   1887.
Unguiculariopsis ilicincola Rehm, Ann. Myc. **7**: 400.   1909.

Apothecia occurring in fasciculate clusters 1–2 mm. in diameter, the individual apothecia irregularly cupulate often compressed from mutual pressure with the margins strongly incurved,
externally furfuraceous, whitish, or brownish, with poorly developed hairs; hymenium concave, pallid-brown, or purplish to
rosy; asci cylindric to clavate, 8-spored, reaching a length of 40–50 μ
and a diameter of 5–6 μ, 8-spored; spores usually 1-seriate,
4–5 μ in diameter; paraphyses filiform, slightly enlarged above.

Usually on other fungi, *Myriangium, Patellaria, Hysterium*
and *Tryblidiella*.

TYPE LOCALITY: Europe.

DISTRIBUTION: North Carolina to Florida; also in Europe.

ILLUSTRATIONS: Ann. Mag. Nat. Hist. III. **7**: *pl. 16, f. 17.*
Jour. Linn. Soc. **31**: *pl. 18, f. 15–18;* Mycologia **31**: 94, *f. 1.*

EXSICCATI: Rav. Fungi Car. **2**: *46.*

40. **CATINELLA** Boud. Hist. Class. Discom. Eu. 150.   1907.

?Bulgariella Karst. Acta Soc. Fauna Fl. Fenn. **2**⁶: 142.   1885.

Apothecia patellate or nearly so, dark-greenish, subgelatinous;
asci cylindric or subcylindric, 8-spored; spores simple, greenish;
paraphyses filiform granular.

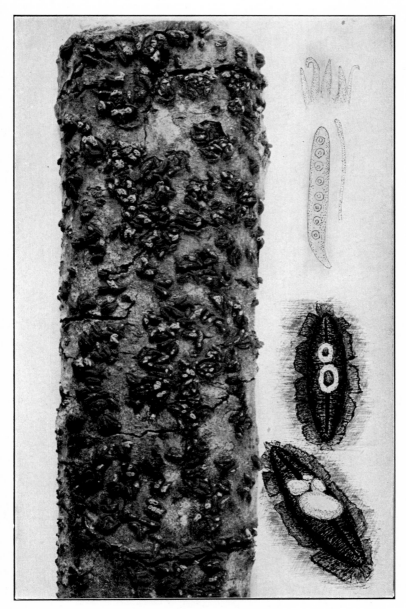

MOLLISIELLA ILICINCOLA

Type species, *Peziza olivacea* Batsch.

1. **Catinella nigroolivacea** (Schw.) Durand, Bull. Torrey Club **49**: 16. 1922. (PLATE 114.)

?Peziza olivacea Batsch, Elench. Fung. 127. 1783.
Peziza nigroolivacea Schw. Schr. Nat. Ges. Leipzig **1**: 121. 1822.
Patellaria pulla nigroolivacea Fries, Syst. Myc. **2**: 160. 1822.
Bulgaria nigrita Fries, Elench. Fung. **2**: 16. 1828.
Lemalis rufoolivacea Schw. Trans. Am. Phil. Soc. II. **4**: 184. 1832.
Rhizina nigroolivacea Curr. Trans. Linn. Soc. **24**: 493. 1864.
Peziza viridiatra Berk. & Curt. Jour. Linn. Soc. **10**: 369. 1868.
Patellaria violacea Berk. & Br. Jour. Linn. Soc. **14**: 108. 1875.
Patellaria hirneola Berk. & Br. Jour. Linn. Soc. **14**: 108. 1875.
Patellaria applanata Berk. & Br. Jour. Linn. Soc. **14**: 108. 1875.
Peziza fuscocarpa Ellis & Holw. Jour. Myc. **1**: 5. 1885.
Patellaria olivacea Phill. Brit. Discom. 361. 1887.
?Humaria olivacea Sacc. Syll. Fung. **8**: 148. 1889.
Pezicula viridiatra Sacc. Syll. Fung. **8**: 315. 1889.
Phaeopezia fuscocarpa Sacc. Syll. Fung. **8**: 474. 1889.
Bulgariella pulla nigroolivacea Sacc. Syll. Fung. **8**: 638. 1889.
Bulgariella nigrita Sacc. Syll. Fung. **8**: 638. 1889.
Patinella violacea Sacc. Syll. Fung. **8**: 770. 1889.
Patinella olivacea Sacc. Syll. Fung. **8**: 770. 1889.
Patinella hirneola Sacc. Syll. Fung. **8**: 771. 1889.
Patinella applanata Sacc. Syll. Fung. **8**: 771. 1889.
Humaria marchica Rehm in Rab. Krypt.-Fl. **1³**: 952. 1894.
Phaeopezia marchica Sacc. Syll. Fung. **11**: 415. 1895.
Aleuria marchica Sacc. & Syd. in Sacc. Syll. Fung. **16**: 739. 1902.
Humaria fuscocarpa Morgan, Jour. Myc. **8**: 189. 1902.
Aleuria fuscocarpa Sacc. & Syd. in Sacc. Syll. Fung. **16**: 739. 1902.
Catinella olivacea Boud. Hist. Class. Discom. Eu. 150. 1907.

Apothecia sessile, solitary, or several crowded together, attached to the substratum by numerous radiating, dark-brown fibers more conspicuous in young plants, at first subglobose and closed, then expanding with a permanently upturned margin, at first entirely greenish-yellow, becoming dark-green, finally blackish with an olive tint, when old the exterior brownish and furfuraceous and vertically striate, fleshy and somewhat gelatinous when fresh, brittle when dry; reaching a diameter of 1 cm. but usually much smaller, mycelial fibers about the base very coarse, straight, or strongly kinked, septate, dark-brown, reaching a diameter of 10 μ, radiating 2–3 mm. beyond the base of the apothecium; asci narrowly cylindric-clavate, 8-spored, reaching a length of 75–90 μ and a diameter of 5–6 μ; spores 1-seriate, irregularly ellipsoid, often slightly constricted near the center so as to appear slipper-shaped, containing one or two oil-drops, pale-olive, becoming brown, 4–5 × 7–10 μ; paraphyses cylindric, simple or rarely branched.

On rotten wood of various kinds.

TYPE LOCALITY: North Carolina.

DISTRIBUTION: Throughout eastern North America, the West Indies and Ceylon; also in Europe.

ILLUSTRATIONS: Batsch, Elench. Fung. *pl. 12, f. 51;* Boud. Ic. Myc. *pl. 452;* Trans. Linn. Soc. **24**: *pl. 51, f. 10–12.*

EXSICCATI: N. Am. Fungi *2325:* Seaver, N. Dak. Fungi *28.*

The species is easily recognized by its greenish apothecia and peculiar shaped, greenish spores.

41. **PYRENOPEZIZA** Fuckel, Symb. Myc. 293. 1869.

?Phillipsiella Cooke & Ellis; Cooke, Grevillea **7**: 48. 1878.

Apothecia at first submerged in the plant tissues, then erumpent, sessile, at first rounded and closed, opening and becoming cup-shaped, or more or less expanded, externally dark-colored, usually brown, or brownish-black, and often roughened; hymenium usually concave, lighter than the outside of the apothecium; asci clavate, or subclavate, 4–8-spored; spores ellipsoid to fusiform, or elongate-ellipsoid, straight, or more rarely curved, simple; paraphyses filiform, usually enlarged above.

Type species, *Peziza rugulosa* Fuckel.

Distinguished from *Mollisia* by the erumpent character of the apothecia a character which is very difficult to recognize in mature plants. The two genera are, to say the least, very close together.

On dicotyledonous plant tissues.
  On woody plants, trees or shrubs.
    On canes of *Rubus.*

Spores 7–9 μ long.	1. *P. Rubi.*
Spores 12–13 μ long.	2. *P. lacerata.*

    On leaves.

Spores 10–15 μ long, on *Quercus Prinus.*	3. *P. prinicola.*
Spores 7–8 μ long, on *Magnolia.*	4. *P. protrusa.*
Spores 5–7 μ long, on *Acer.*	5. *P. leucodermis.*
Spores 4–5 μ long, on *Tilia.*	6. *P. minuta.*

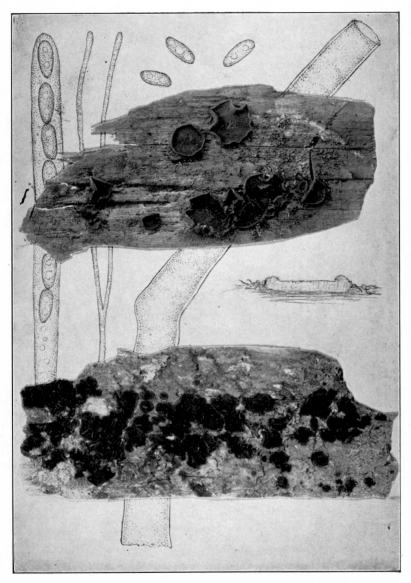

CATINELLA NIGROOLIVACEA

On herbaceous plant tissues.

Spores 25–30 μ long, on *Thalictrum*.   7. *P. Thalictri*.

Spores less than 25 μ long.

  Spores mostly 10–12 μ long.

    Spores 14–16 μ long.

      On *Festuca*.   8. *P. Ellisii*.

      On *Solidago*.   9. *P. subatra*.

      On *Potentilla*.   10. *P. coloradensis*.

    Spores 12–13 μ long, on *Cephalanthus*.   11. *P. Cephalanthi*.

    Spores 8–12 μ long.

      On stems of *Linum*.   12. *P. californica*.

      On stems of various kinds.

        Hymenium gray or rose-colored.   13. *P. compressula*.

        Hymenium reddish-yellow.   14. *P. Absinthii*.

    Spores 7–9 μ long.

      On *Artemisia*.   15. *P. Artemisiae*.

      On *Apocynum*.   16. *P. Dearnessii*.

      On *Smilax herbacea*.   17. *P. smilacicola*.

On monocotyledonous stems, *Carex*.

  Spores 12–20 μ long.

    Spores 18–20 μ long.   18. *P. caricina*.

    Spores 12–15 μ long.

      Spores 4–5 × 13–15 μ.   19. *P. yogoensis*.

      Spores 2 × 12–14 μ.   20. *P. cervinula*.

    Spores 10–12 μ long.   21. *P. Caricis*.

    Spores 5–7 μ long.   22. *P. multipunctoidea*.

1. **Pyrenopeziza Rubi** (Fries) Rehm, Ascom. *416*. 1878.

Excipula Rubi Fries, Syst. Myc. **2**: 190. 1822.
Mollisia Rubi Karst. Acta Soc. Fauna Fl. Fenn. **2**⁶: 136. 1885.

Apothecia gregarious, or occasionally crowded, at first immersed, rounded, finally erumpent through the ruptured epidermis, sessile, expanding and becoming cup-shaped, .3–1 mm. in diameter, the margin inrolled when dry; brownish-black; asci clavate, reaching a length of 45–55 μ and a diameter of 6–8 μ, 8-spored, spores elongate-clavate, or almost cylindric, 1.5–2.5 × 7–9 μ; paraphyses filiform, 1.5 μ in diameter.

On dead canes of *Rubus*.

TYPE LOCALITY: Europe.

DISTRIBUTION: Delaware to Oregon and Washington; also in Europe.

2. **Pyrenopeziza lacerata** (Cooke & Peck) Sacc. Syll. Fung. **8**: 361. 1889.

Peziza lacerata Cooke & Peck, Grevillea **1**: 6. 1872.

Apothecia gregarious, globose, then expanded, dark-brown, the margin coarsely lacerated into subtriangular, irregular teeth; hymenium cinereous, becoming blackish when dry; asci sub-cylindric; spores cylindric-clavate, 12–13 μ long.

On *Rubus odoratus.*

TYPE LOCALITY: Adirondack Mountains, New York.

DISTRIBUTION: New York and West Virginia.

EXSICCATI: Ellis & Ev. N. Am. Fungi *2044.*

3. **Pyrenopeziza prinicola** (Ellis & Ev.) Cash, Jour. Wash. Acad. Sci. **30**: 302. 1940.

Peziza prinicola Ellis & Ev. Jour. Myc. **4**: 99. 1888.
Mollisia prinicola Sacc. Syll. Fung. **8**: 330. 1889.

Apothecia erumpent, becoming superficial, scattered rather thickly and evenly over the lower surface of the leaf, minute, .15–.27 mm. in diameter, subturbinate, then patellate, the entire apothecium black when dry; the margin slightly crenulate; hymenium alutaceous to dark-brown when fresh; asci clavate, narrowed below, reaching a length of 33–40 μ and a diameter of 6–8 μ, 8-spored; spores irregularly 2–3-seriate, clavate, becoming pseudoseptate, 1–1.5 × 10–15 μ; paraphyses filiform.

On dead leaves of *Quercus Prinus* and other species of *Quercus.*

TYPE LOCALITY: Louisiana.

DISTRIBUTION: Louisiana and Georgia.

4. **Pyrenopeziza protrusa** (Berk. & Curt.) Sacc. Syll. Fung. **8**: 364. 1889.

Peziza protrusa Berk. & Curt.; Berk. Grevillea **3**: 159. 1875.
Pseudopeziza protrusa Sacc. Syll. Fung. **8**: 726. 1889.

Apothecia erumpent, dot-like, surrounded by the ruptured cuticle, externally granulated, chestnut; hymenium concave, white; asci clavate, reaching a length of 45–50 μ and a diameter of 5 μ; spores ellipsoid-clavate, 1.5 × 7–8 μ; paraphyses filiform-clavate.

On the under side of the leaves of *Magnolia glauca.*

TYPE LOCALITY: South Carolina.

DISTRIBUTION: New Jersey and South Carolina.

EXSICCATI: Ellis, N. Am. Fungi *143;* Rehm, Ascom. *310* (from New Jersey); Thüm. Myc. Univ. *519* (from New Jersey).

5. **Pyrenopeziza leucodermis** Cash, Jour. Wash. Acad. Sci. **29:** 49. 1939.

Apothecia hypophyllous, subepidermal, then superficial, sessile, scattered thickly over the leaf, at first nearly globose, then patellate, membranous-fleshy, pale-brown, .15–.25 mm. in diameter, externally furfuraceous, the margin even or slightly fimbriate, inrolled when dry; hymenium subhyaline; asci clavate, or narrow-ellipsoid, reaching a length of 28–35 μ and a diameter of 5–6 μ; spores obliquely 1–2-seriate, straight, cylindric to clavate, .7–1 $\times$ 5–7 μ; paraphyses filiform, slightly enlarged above.

On leaves of *Acer leucoderme* and *Acer floridanum*.

TYPE LOCALITY: Athens, Georgia.

DISTRIBUTION: Georgia.

6. **Pyrenopeziza minuta** Cash, Jour. Wash. Acad. Sci. **30:** 302. 1940.

Apothecia hypophyllous, subcuticular, becoming superficial, closely aggregated in pale spots on withering leaves, patellate, pale-brown when moist, entire fungus black when dry, .15–.17 mm. in diameter, soft, fleshy to membranous; asci clavate to fusoid, rather abruptly narrowed above and below, reaching a length of 28–33 μ and a diameter of 4–5 μ, 8-spored; spores 2-seriate, or irregularly 3-seriate, clavate, .7–1 $\times$ 4–5 μ; paraphyses filiform, enlarged above, 2–3 μ in diameter.

On withering leaves of *Tilia heterophylla* var. *Michauxii*.

TYPE LOCALITY: Princeton, Georgia.

DISTRIBUTION: Known only from the type locality.

7. **Pyrenopeziza Thalictri** (Peck) Sacc. Syll. Fung. **8:** 360. 1889.

Peziza Thalictri Peck, Ann. Rep. N. Y. State Mus. **29:** 55. 1878.

Apothecia abundant, sessile, bursting through the epidermis, small, punctiform when dry, externally black, the margin usually whitish, or cinereous and subfimbriate; hymenium cinereous; asci cylindric; spores long-ellipsoid, 5 $\times$ 25–30 μ paraphyses not observed.

On dead stems of *Thalictrum*.

TYPE LOCALITY: Center, New York.

DISTRIBUTION: New York to Ontario and South Dakota.

8. **Pyrenopeziza Ellisii** (Rehm) Massee, Jour. Bot. **34**: 149. 1896.

Niptera Ellisii Rehm; Ellis & Ev. N. Am. Fungi *2329;* Sacc. Syll. Fung. **11**: 416. 1895. nom. nud.

Apothecia sparse, or gregarious, erumpent, at first subglobose and closed, then expanding, .5–.7 mm. in diameter, grayish-black, externally brownish-black; asci clavate, reaching a length of 90–100 μ and a diameter of 10–12 μ, 8-spored; spores 2-seriate, elongate-cylindric, usually curved, 3.5–4 × 14–16 μ; paraphyses filiform, 3 μ in diameter.

On dead stems of *Festuca tenella;* also reported on *Ammophila longifolia.*

TYPE LOCALITY: Newfield, New Jersey.

DISTRIBUTION: New Jersey and Montana.

ILLUSTRATIONS: Jour. Bot. **34**: *pl. 357, f. 6, 7.*

EXSICCATI: Ellis, N. Am. Fungi *565* (as *Peziza denigrata* Kunze); Ellis & Ev. N. Am. Fungi *2329.*

9. **Pyrenopeziza subatra** (Cooke & Peck) Sacc. Syll. Fung. **8**: 359. 1889.

Peziza subatra Cooke & Peck; Peck, Ann. Rep. N. Y. State Mus. **28**: 66. 1876.

Apothecia erumpent, black, soft, or waxy, at first hemispheri-cal, finally expanding, slightly rugose; hymenium fuliginous, the margin paler; asci subcylindric, 8-spored; spores narrow-ellipsoid, or fusoid, 15–16 μ long; paraphyses filiform.

On dead herbaceous stems.

TYPE LOCALITY: North Greenbush, New York.

DISTRIBUTION: New York and New Jersey; also reported from Michigan.

EXSICCATI: Ellis, N. Am. Fungi *445;* Ellis & Ev. Fungi Columb. *21.*

A portion of what appears to be the type of this species is in the collection of The New York Botanical Garden. A number of specimens have been doubtfully referred to this species. All should be more carefully studied.

10. **Pyrenopeziza coloradensis** Ellis & Ev. Bull. Torrey Club **27**: 60. 1900.

Apothecia scattered, superficial, black, with a narrow, slightly incurved margin, reaching a diameter of 1 mm.; hymen-ium concave, or plane; asci clavate, reaching a length of 100–110 μ and a diameter of 15 μ; spores 2-seriate, ellipsoid, with two oil-drops, 10 $\times$ 20 μ; paraphyses stout.

On dead stems of *Potentilla*.

TYPE LOCALITY: Red Mountain, Colorado, elevation 12,000 feet.

DISTRIBUTION: Known only from the type locality.

11. **Pyrenopeziza Cephalanthi** Fairman, Jour. Myc. **10**: 229. 1904.

Apothecia scattered, small, sessile, black outside, the margin occasionally white; hymenium white at first, growing yellowish, then black with age; asci clavate-cylindric, reaching a length of 60 μ and a diameter of 6–7 μ; spores cylindric, or elongate-fusoid, 2–3 $\times$ 10–13 μ with several oil-drops; paraphyses filiform.

On dead limb of *Cephalanthus occidentalis* lying in a ditch.

TYPE LOCALITY: Ridgway, New York.

DISTRIBUTION: Known only from the type locality.

12. **Pyrenopeziza californica** Sacc. Ann. Myc. **2**: 15. 1904.

Pezizella californica Rehm, Ann. Myc. **5**: 520. 1907.

Apothecia superficial, gregarious, cyathiform, base contracted, black outside and inside, smooth, when dry mouth contracted, externally striate, .5–1 mm. in diameter, the margin conspicuously laciniate; asci clavate, reaching a length of 50–68 μ and a diameter of 5.5–6.5 μ, 8-spored; spores 1-seriate, ellipsoid, 3.5–4.5 $\times$ 9–12 μ; paraphyses filiform, enlarged above.

On dead stems of *Linum Lewisii*.

TYPE LOCALITY: Mt. Eddy, Siskiyou Co., California.

DISTRIBUTION: Known only from the type locality.

Saccardo suggests that this might be the type of a new genus (*Geasterina*).

13. **Pyrenopeziza compressula** Rehm in Rab. Krypt.-Fl. **1**[3]: 618. 1892.

Apothecia gregarious, at first globose and immersed, becoming erumpent and sessile, cup-shaped, externally brown to brownish-black, .1–.35 mm. in diameter; hymenium concave, gray, or slightly rose-colored; asci clavate, reaching a length of 45–60 μ and a diameter of 5–6 μ, 8-spored; spores fusoid, occasionally slightly curved, 1.5–3 $\times$ 8–12 μ; paraphyses filiform, 2 μ in diameter.

On dead stems of various plants.

TYPE LOCALITY: Europe.

DISTRIBUTION: Colorado, California and (Quebec?); also in Europe.

EXSICCATI: Clements, Crypt. Form. Colo. *287.*

14. **Pyrenopeziza Absinthii** (Lasch) Rehm in Rab. Krypt.-Fl. 1³: 625. 1892.

Peziza Absinthii Lasch; Rab. in Klotzsch, Herb. Viv. Myc. **14**: *17;* Bot. Zeit. **8**: 438. 1850.
Pyrenopeziza fuscorubra Rehm; Winter, Flora **55**: 526. 1872.
Niptera fuscorubra Rehm, Ber. Naturh. Ver. Augsburg **26**: 21.   1881.
Pseudohelotium Absinthii Sacc. Syll. Fung. **8**: 297.   1889.
Mollisia fuscorubra Sacc. Syll. Fung. **8**: 322.   1889.

Apothecia thickly gregarious, at first globose and closed, submerged, becoming erumpent and sessile, cup-shaped, finally expanded and saucer-shaped, externally reddish-brown, smooth, .5–1.2 mm. in diameter; hymenium reddish-yellow; asci cylindric-clavate, reaching a length of 60–70 μ and a diameter of 6–7 μ, 8-spored; spores elongate-ellipsoid, 2.5 × 8–10 μ; paraphyses sparse, filiform, 2 μ in diameter.

On dead stems of various plants, *Iva xanthiifolia, Erigeron, Bidens, Helianthus, Artemisia* etc.

TYPE LOCALITY: Europe.

DISTRIBUTION: New Jersey to Kansas and North Dakota.

EXSICCATI: Ellis, N. Am. Fungi *60;* Brenckle, Fungi Dak. *215.*

15. **Pyrenopeziza Artemisiae** (Lasch) Rehm, Ascom. *66.* 1872.

Peziza Artemisiae Lasch in Klotzsch, Herb. Viv. Myc. *335*; Grevillea **1**: 131. 1873.
Mollisia Artemisiae Phill. Brit. Discom. 188.   1887.

Apothecia erumpent, turbinate, substipitate, expanding and becoming flattened, brownish-black; hymenium concave to plane, whitish; asci cylindric, reaching a length of 50 μ and a diameter of 5 μ; spores elongate-cylindric, 1.5–2 × 7–9 μ; paraphyses filiform, 2 μ thick, enlarged above to 3 μ.

On dead stems of *Artemisia vulgaris.*

TYPE LOCALITY: Europe.

DISTRIBUTION: New York and New Jersey to Michigan and Ontario.

EXSICCATI: Barth. Fungi Columb. *3866.*

A variety, *Solidaginis*, on *Solidago juncea*, has been recognized by Rehm, Ascom. *1957* from material collected in Ontario, Canada by Dr. John Dearness.

16. **Pyrenopeziza Dearnessii** Rehm, Ann. Myc. **9**: 286.   1911.

Apothecia gregarious, at first submerged, becoming erumpent through the laciniately splitting epidermis, at first globose, sessile, brownish, expanding, .2–.5 mm. in diameter, when dry more or less covered by the laciniate epidermis; asci clavate, reaching a length of 50 μ and a diameter of 7–8 μ; spores elongate-ellipsoid, 2-seriate, 3 × 6–8 μ; paraphyses filiform 2 μ thick, expanding to 3–3.5 μ at their apices.

On dead stems of *Apocynum androsaemifolium*.
Type locality: London, Canada.
Distribution: Ontario, Canada.

17. **Pyrenopeziza smilacicola** Dearn. & House, N. Y. State Mus. Circ. **24**: 28.   1940.

Apothecia thickly scattered, subcorticular at first, then erumpent, and finally sessile-superficial, the margin inrolled when dry, .2–.35 mm. in diameter, externally brown; asci clavate, reaching a length of 48–55 μ and a diameter of 6.5–9 μ; spores 2-seriate, subclavate, 2 × 7–8.5 μ; paraphyses filiform, longer than the asci.

On the dead stems of *Smilax herbacea*.
Type locality: Essex Co., New York.
Distribution: Known only from the type locality.

18. **Pyrenopeziza caricina** (Lib.) Rehm in Rab. Krypt.-Fl. **1³**: 634.   1892.

Peziza caricina Lib. Pl. Crypt. Ard. 230.   1834.
Pseudopeziza caricina Sacc. Syll. Fung. **8**: 727.   1889.

Apothecia scattered, at first submerged, then erumpent, externally dark-brown, later black, waxy; hymenium grayish-white; asci clavate, reaching a length of 100–110 μ and a diameter of 15–20 μ, 8-spored; spores elongated, slightly curved, 5–6 × 18–20 μ; paraphyses filiform, slightly enlarged above.

On dead stems of *Carex*.
Type locality: Europe.
Distribution: Colorado; also in Europe.
Exsiccati: Clements, Crypt. Form. Colo. *286*.

19. **Pyrenopeziza yogoensis** (Ellis & Gall.) Sacc. Syll. Fung. **10**: 17. 1892.

Peziza yogoensis Ellis & Gall. Jour. Myc. **5**: 65.   1889.

Apothecia erumpent, .2 mm. in diameter, with an incurved, fimbriate margin, olivaceous within; hymenium pale; asci cylindric, reaching a length of 55–60 μ and a diameter of 15–18 μ; spores obliquely 1-seriate, or 2-seriate, ellipsoid, a little narrower at one end, 4–5 × 13–15 μ; paraphyses stout, not abundant.

On dead leaves of *Carex*.

TYPE LOCALITY: Yogo, Belt Mountains, Montana.

DISTRIBUTION: Known only from the type locality.

20. **Pyrenopeziza cervinula** (Cooke) Sacc. Syll. Fung. **8**: 369. 1889.

Peziza cervinula Cooke, Hedwigia **14**: 84.   1875.
Peziza multipuncta Peck; Thüm. Myc. Univ. *1412*.   1879.

Apothecia thickly scattered, erumpent, minute, .1–.2 mm. in diameter, at first subglobose, opening and becoming subhemispheric, externally deer-colored; hymenium white; asci clavate, reaching a length of 40 μ; spores cylindric, straight, or curved, 2 × 12–14 μ; paraphyses not observed.

On culms of *Carex*.

TYPE LOCALITY: New Jersey.

DISTRIBUTION: New Jersey and New York.

EXSICCATI: Ellis, N. Am. Fungi *440;* Thüm. Myc. Univ. *1412;* Roum. Fungi Select. Exsicc. *4628* (as *Peziza multipuncta*).

21. **Pyrenopeziza Caricis** Rehm in Rab. Krypt.-Fl. **1**³: 633. 1892.

Mollisia Karstenii var. *Caricis* Rehm, Hedwigia **23**: 53.   1884.

Apothecia gregarious, at first globose and closed, immersed, then erumpent through the epidermis, expanding and becoming cup-shaped, or saucer-shaped, externally dark-colored, brownish-black, .2–.5 mm. in diameter; hymenium yellowish; asci clavate, reaching a length of 40–50 μ and a diameter of 8–9 μ, 8-spored; spores, fusoid, straight, or slightly bent, 2–2.5 × 10–12 μ; paraphyses filiform, 2 μ in diameter above.

On dead stems of *Carex*.

TYPE LOCALITY: Europe.

DISTRIBUTION: New York; also in Europe.

22. **Pyrenopeziza multipunctoidea** Dearn. & House, N. Y. State Mus. Circ. **24**: 28. 1940.

Apothecia at first immersed, minute, thickly scattered, often in rows, emerging through a usually elliptic rupture of the cuticle, .15–.22 mm. in diameter; asci cylindric, or fusoid, reaching a length of 33–48 μ and a diameter of 6–7.5 μ, 8-spored; spores partially 2-seriate, ellipsoid, 2.5–2.75 $\times$ 5–7 μ; paraphyses numerous, filiform.

On dead leaves of *Carex lacustris.*

TYPE LOCALITY: Newcomb, Essex Co., New York.

DISTRIBUTION: Known only from the type locality.

Part of the type material is in The New York Botanical Garden.

DOUBTFUL AND EXCLUDED SPECIES

Pyrenopeziza cariosa (Peck) Sacc. Syll. Fung. **8**: 367.   1889; *Peziza cariosa* Peck, Ann. Rep. N. Y. State Mus. **24**: 95.   1872.   On rotten wood, Catskill Mountains.   Description incomplete.

Pyrenopeziza doryphora Clements, Crypt. Form. Colo. *288.* 1907.   On dead stems of *Carex*, Colorado.   A variety, *Heleocharidis* Clements, Crypt. Form. Colo. *525.* 1908, has been listed.   So far as known these have not been officially published.

Pyrenopeziza foliicola (Karst.) Sacc. Michelia **1**: 65.   1877; *Mollisia foliicola* Karst. Myc. Fenn. **1**: 201.   1871.   This species has been reported from Georgia but on a doubtful determination.

Phillipsiella nigrella Cooke & Hark. Bull. Calif. Acad. Sci. **1**: 23.   1884. On leaves of *Quercus agrifolia*, California.

Pyrenopeziza nigrella Fuckel, Symb. Myc. Nacht. **3**: 30.   1875.   Reported from Michigan by Dr. B. Kanouse.   On decaying stems.   No material seen.

Pyrenopeziza nigritella (Phill. & Hark.) Sacc. Syll. Fung. **8**: 357.   1889; *Peziza nigritella* Phill. & Hark. Bull. Calif. Acad. Sci. **1**: 22.   1884.   Apothecia blackish; spores 6 $\times$ 15 μ.   On dead stems of *Galium.*

42. **STAMNARIA** Fuckel, Symb. Myc. 309. 1869.

Apothecia erumpent-superficial, occurring singly, or more often in cespitose clusters, more or less gelatinous, horny when dry, sessile, or short-stipitate; asci clavate, usually 8-spored; spores simple, ellipsoid, hyaline; paraphyses filiform, slightly enlarged above.

Type species, *Peziza Persoonii* Moug.

On species of *Equisetum.*   1. *S. americana.*
On foliage of *Thuja.*   2. *S. Thujae.*

1. **Stamnaria americana** Massee & Morgan, Jour. Myc. **8**: 183. 1902.   (PLATE 115.)

Apothecia erumpent in clusters of three or four each, or in rows 6–7 mm. long, the individual apothecia sessile, or subsessile, at first rounded, gradually expanding and becoming turbinate, reaching a diameter of .5–.7 mm., pale-orange; hymenium plane, or slightly concave, similar in color to the outside of the apothecium, whitish-pruinose from the ends of the protruding asci and paraphyses; asci clavate, 8-spored, reaching a length of 150–200 μ and a diameter of 16 μ; spores 1-seriate, or partially 2-seriate above, ellipsoid, straight, or curved, usually with one or two large oil-drops surrounded with a granular contents, hyaline, 7–9 × 24–32 μ; paraphyses filiform, rather strongly enlarged above, pale-orange in mass.

On species of *Equisetum*, especially *Equisetum robustum* and *Equisetum hyemale.*

TYPE LOCALITY: Preston, Ohio.

DISTRIBUTION: New York and New Jersey to Virginia, Indiana and (Oregon?).

ILLUSTRATIONS: Papers Mich. Acad. Sci. **19**: *pl. 8.*

EXSICCATI: Ellis, N. Am. Fungi *1274* (as *Peziza Personii*); Ellis & Ev. Fungi Columb. *333* (as *Stamnaria Equiseti*); Kellerm. Ohio Fungi **2**: *18.*

2. **Stamnaria Thujae** Seaver, Mycologia **28**: 186.   1936. (PLATE 116.)

Apothecia occurring singly, or more often in congested masses, erumpent through the epidermis on the under side of the foliage of the host, translucent with a slight yellowish or pinkish tint, exceedingly soft and gelatinous, shrinking much in drying, the individual apothecia small, not usually exceeding .2 mm. in diameter; asci clavate, reaching a length of 55 μ and a diameter of 15 μ, 8-spored; spores irregularly disposed in the ascus, ellipsoid, hyaline, granular, 6–7 × 10–12 μ; paraphyses very slender branched.

On foliage of *Thuja occidentalis.*

TYPE LOCALITY: Baileys Harbor, Wisconsin.

DISTRIBUTION: Known only from the type locality.

43. **OMBROPHILA** Fries, Summa Veg. Scand. 357.   1849.

Neobulgaria Petrak, Ann. Myc. **19**: 44.   1921.

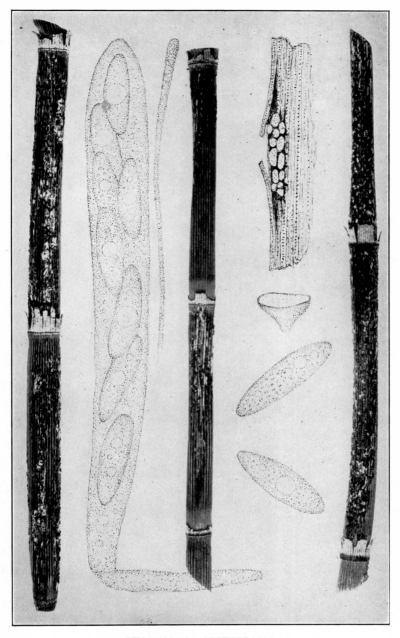

STAMNARIA AMERICANA

Apothecia fleshy to gelatinous, stipitate but the stem often much reduced in length, or 1 cm. or more in length; asci clavate to cylindric, usually 8-spored; spores ellipsoid to fusoid, hyaline, simple; paraphyses variable.

Type species, *Peziza Clavus* Alb. & Schw.

The genus has been used as a "waste basket" and most of the species placed here have been referred to other genera. However, the type species is very characteristic, widely distributed, and well known, so the genus is allowed to stand.

1. **Ombrophila Clavus** (Alb. & Schw.) Cooke, Grevillea **8**: 84. 1879. (PLATE 76.)

Peziza Clavus Alb. & Schw. Consp. Fung. 306. 1805.
Peziza Clavus violascens Alb. & Schw. Consp. Fung. 306. 1805.
Helotium pileatum Peck, Ann. Rep. N. Y. State Mus. **28**: 67. 1876.
?Helotium palustre Peck, Ann. Rep. N. Y. State Mus. **32**: 48. 1879.
Leotia aquatica Lib. Rev. Myc. **2**: 18. 1880.
Helotium Clavus Gill. Champ. Fr. Discom. 153. 1883.
Cudonia aquatica Quél. Ench. Fung. 267. 1886.
?Ombrophila albiceps Peck, Ann. Rep. N. Y. State Mus. **42**: 34. 1889.
Cudoniella fructigena Rostr. Medd. Grønl. **3**: 605. 1891.
Cudoniella aquatica Sacc. Syll. Fung. **8**: 42. 1889.

Apothecia scattered, fleshy, to subgelatinous, stipitate, or subsessile, pallid, or purplish; reaching a diameter of 1 cm.; hymenium plane, or convex, same color; stem variable, short, or very long, gradually expanding into the apothecium, the whole subturbinate; asci cylindric-clavate, reaching a length of 70–90 μ and a diameter of 9–10 μ, 8-spored; spores elongate-ellipsoid, or subfusoid, often unequal-sided, 4–5 × 10–15 μ; paraphyses filiform.

On leaves and wood submerged in water or in very wet places.

TYPE LOCALITY: Europe.

DISTRIBUTION: New England to Washington; probably widely distributed.

ILLUSTRATIONS: Alb. & Schw. Consp. Fung. *pl. 11, f. 5;* Phill. Brit. Discom. *pl. 10, f. 62;* Boud. Ic. Myc. *pl. 434.*

When Albertini and Schweinitz described *Peziza Clavus* they listed two varieties: a. *pallens* and b. *violascens.* When Fries established the genus *Ombrophila*, he raised these forms to specific rank calling the first *Ombrophila, purpurascens* and the second *Ombrophila violacea.* Just why *violascens* was changed to *violacea* is not apparent.

Ellis distributed specimens under these names (N. Am. Fungi *392* as *Ombrophila violacea* Fr. and *393* as *Ombrophila purpurascens* Fr.) He indicates that the two were collected together and so far as the writer can see they are identical. Whether these are identical with either of the forms listed by Fries is a question. Ellis and Everhart distributed this again (Fungi Columb. *22*) as *Ombrophila purpurascens* Fr. One specimen in the Ellis collection is labeled *Ombrophila purpurascens* Fr. (stipitate var.). This is certainly *Ombrophila Clavus* and indicates that Ellis regarded the two as having a varietal relationship.

Dr. B. Kanouse lists for Michigan one collection of *Ombrophila violacea* (Hedw.) Fr. The small specimen examined seems to be entirely sessile although *Octospora violacea* Hedw. is illustrated (Hedw. Descr. *2*: *pl. 8, f. a.*) as being at first cylindric, expanding above and at maturity with a stout stem, closely resembling Tulasne's (Fung. Carp. *3*: *pl. 17, f. 1–8*) figures of *Coryne sarcoides*. There is nothing to indicate that *Ombrophila violacea* Fries was based on *Octospora violacea* Hedw., although some have so regarded it. So it will be seen that there is much confusion over the identity of these forms.

Phillips (Brit. Discom. 325) states referring to *Ombrophila Clavus:* "A most variable species both in size and shape, the stem at one time absent, at another elongated to twice the diameter of the disc; sometimes very thick, at other times slender." The variability of this species doubtless is responsible for the confusion which has arisen concerning it.

DOUBTFUL AND EXCLUDED SPECIES

Ombrophila albofusca Ellis, Bull. Torrey Club **9**: 73. 1882. The type of this species has been examined but no apothecia could be found.

Ombrophila aurata (Berk. & Rav.) Phill. Grevillea **19**: 74. 1891; *Peziza aurata* Berk. & Rav. in Rav. Fung. Car. **3**: 37. 1855. No specimen of this species is available but the description suggests *Chlorociboria*.

Ombrophila flavens Feltg. Vorst. Pilz-Fl. Luxenb. **1**[3]: 76. 1903. This species has been reported on wood of *Salix* from Washington by Dr. B. Kanouse. No material seen. Description suggests a *Helotium*.

Ombrophila hirtella Rehm, Ann. Myc. **6**: 314. 1908. No material has been seen but from the description it would appear to be a *Lachnella*. It was described from material sent from Glenco, Illinois by E. T. and S. A. Harper.

Ombrophila lilacina (Wulf.) Karst. Myc. Fenn. **1**: 90. 1871; *Elvela lilacea* Wulf. in Jacq. Coll. Bot. **1**: 347. 1786; *Peziza lilacina* Fries, Syst. Myc. **2**: 140. 1822. This species has been reported from Dominica by A. L. Smith, Jour. Linn. Soc. **35**: 15. 1901. We have no other knowledge of the species.

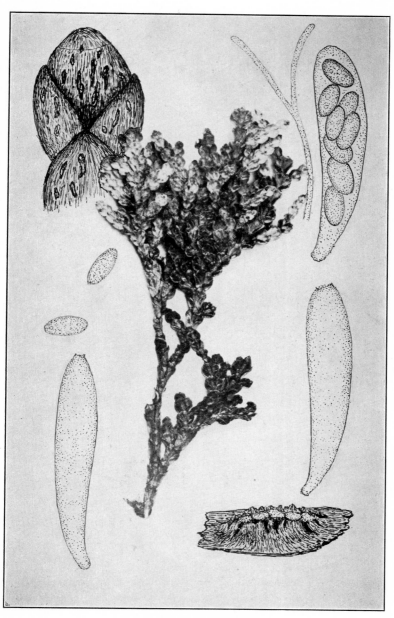

STAMNARIA THUJAE

Ombrophila limosa Rehm, Ann. Myc. **11**: 154. 1913. Described from material on dead leaves of *Carex limosa* from London, Canada. No specimens have been seen.

Ombrophila Lysichitonis Kanouse, Mycologia **39**: 672. 1947. Reported on *Lysichiton camtschatcense* from Washington by Dr. B. Kanouse. The description strongly suggests *Ombrophila Clavus.*

Ombrophila limosella (Karst.) Rehm, Ascom. *508;* Ber. Naturh. Ver. Augsburg **26**: 122. 1881; *Peziza limosella.* Karst. Not. Soc. Fauna Fl. Fenn. **10**: 152. 1869; *Ombrophila violacea* var. *limosella* Karst. Myc. Fenn. **1**: 88. 1871. This species has been reported from Washington by Dr. B. Kanouse. As will be seen it was originally regarded as a variety of *Ombrophila violacea,* the identity of which is uncertain as indicated above.

Ombrophila pellucida A. L. Smith, Jour. Linn. Soc. **35**: 14. 1901. Described from material collected in Dominica. No material seen.

Ombrophila similis (Berk. & Curt.) Sacc. Syll. Fung. **8**: 614. 1889; *Bulgaria similis* Berk. & Curt.; Berk. Jour. Linn. Soc. **10**: 370. 1868. On dead wood in Cuba. This is said to differ from *Coryne sarcoides* in its much smaller spores. Its identity is uncertain.

Ombrophila subsqualida Rehm in Rab. Krypt.-Fl. **1³**: 1226. 1896. This species has been reported from Michigan by Dr. B. Kanouse. No material has been seen.

Ombrophila subaurea Cooke, Bull. Buffalo Soc. Nat. Sci. **2**: 300. 1875. Saccardo suggests that this is close to *Ombrophila enterochroma* (*Kriegeria enterochroma*) and the size of the spores, 5 × 30 μ, would so indicate. Specimens distributed by Ellis (N. Am. Fungi *394*) do not agree with Peck's description and are apparently small specimens of *Coryne sarcoides.*

44. PHAEOBULGARIA Seaver, Mycologia **24**: 253. 1932.

Apothecia medium large, decidedly gelatinous within, distinguished from *Bulgaria* by its inoperculate asci and brown spores; asci clavate, 8-spored; paraphyses filiform.

Type species, *Peziza inquinans* Pers.

Only one species of this genus known at the present time. Fries included this with *Bulgaria* since at that time no distinction was made between the operculates and the inoperculates.

Phaeobulgaria inquinans (Pers.) Nannf., Nova Acta Soc. Sci. Upsal. IV. **8**: 311. 1932. (PLATE 117, FIG. 2.)

Peziza polymorpha Oeder, Fl. Dan. **8**: 8. 1769.
Tremella turbinata Hudson, Fl. Angl. **2**: 563. 1778.
Peziza nigra Bull. Hist. Champ. Fr. 238. 1791.
Peziza brunnea Batsch, Elench. Fung. **1**: 125. 1783.
Octospora elastica Hedw. Descr. **2**: 28. 1789.
Burcardia turbinata Schmidel, Ic. Pl. 263. 1793.
Tremella agaricoides Retz. Fl. Scand. 295. 1795.
Peziza inquinans Pers. Tent. Disp. Fung. 33. 1797.
Ascobolus inquinans Nees, Syst. Pilze Schw. 268. 1817.

Bulgaria inquinans Fries, Syst. Myc. **2**: 167.   1822.
Coryne turbinata Bonord. Handh. Myk. 149.   1851.
Bulgaria polymorpha Wettst. Verh. Zool.-Bot. Ges. Wien **35**: 595.   1886.

Apothecia occurring singly, or more often several from a common base in cespitose clumps, at first rounded, or short cylindric, finally expanding and becoming turbinate with the hymenium concave, becoming plane and finally convex with the margin repand, reaching a diameter of 1–4 cm., externally brownish; hymenium black, or bluish-black and shining; asci clavate, reaching a length of 150–200 μ and a diameter of 9–10 μ, 8-spored; spores ellipsoid, unequal-sided and often narrower at one end, becoming pale-brown, 6–7 $\times$ 12–14 μ; paraphyses slender yellow, or violet brown, 1 μ in diameter.

On bark of trees, especially the various species of oak, *Quercus;* occasionally reported on other trees.

TYPE LOCALITY: Europe.

DISTRIBUTION: New York to Alabama, Washington and California; also in Europe.

ILLUSTRATIONS: Fl. Dan. *pl. 464;* Bull. Herb. Fr. *pl. 116;* Batsch. Elench. Fung. **1**: *pl. 11, f. 50;* Schmidel, Ic. Pl. *pl. 70;* Schaeff. Fung. Bavar. *pl. 158;* Hedw. Descr. **2**: *pl. 6, E.;* Bull. Lab. Nat. Hist. State Univ. Iowa **6**: *pl. 37, f. 2.*

EXSICCATI: Rav. Fungi Car. **5**: *43;* Ellis, N. Am. Fungi *448;* Reliq. Farlow. *100.*

45. **ASCOTREMELLA** Seaver, Mycologia **22**: 53.   1930.

Haematomyces Authors (in part) not Berk. & Br. Jour. Linn. Soc. **14**: 108. 1875.

Apothecia densely crowded, or cespitose, tremelloid, sessile, or substipitate; asci cylindric but often much swollen so that the spores appear relatively small, 8-spored; spores ellipsoid, or more or less irregular in form, usually containing two small oil-drops, hyaline; paraphyses slender, simple, or branched.

Type species, *Haematomyces fagineus* Peck.

Apothecia forming cerebriform masses.   1. *A. faginea.*
Apothecia cespitose, turbinate.   2. *A. turbinata.*

1. **Ascotremella faginea** (Peck) Seaver, Mycologia **22**: 53.   1930. (PLATE 118.)

Haematomyces fagineus Peck, Ann. Rep. N. Y. State Mus. **43**: 33.   1890.

Apothecia tremelloid, cerebriform, reaching a diameter of 2–4 cm., or forming a continuous mass 8–10 cm. in extent, gyrose-

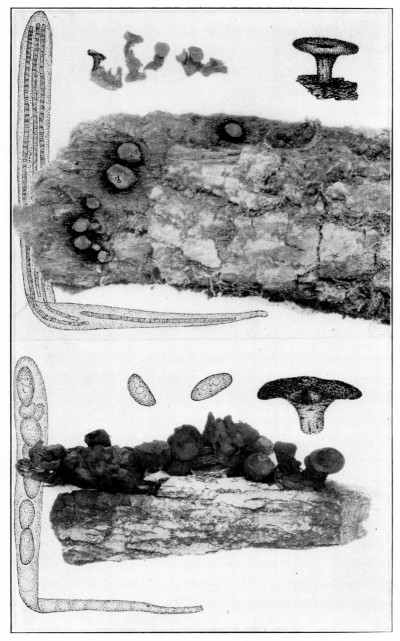

1. HOLWAYA GIGANTEA
2. PHEUDOBULGARIA INQUINANS

lobate, smooth, shining, raisin-colored without and within, the substance gelatinous, becoming horny when dry; asci subcylindric, reaching a length of 50 μ and a diameter of 6–7 μ; spores usually 1-seriate, narrow-ellipsoid, hyaline, 4–5 × 7 μ; paraphyses slender, slightly enlarged above.

On trunks of beach, *Fagus americana;* also reported on *Tilia.*

Type locality: Rainbow, Franklin Co., New York.

Distribution: New York to Ontario and Michigan; also reported from England.

Illustrations: Ann. Rep. N. Y. State Mus. **43**: *pl. 4, f. 5–7;* Mycologia **22**: *pl. 11;* Trans. Brit. Myc. Soc. **29**: 152, *f. 13.*

2. **Ascotremella turbinata** Seaver, Mycologia **22**: 53. 1930. (Plate 119.)

Apothecia extremely gelatinous, closely crowded when young, giving rise to cespitose clusters as they mature, the individual apothecia at first rounded, becoming turbinate, or subturbinate, externally light-colored, reaching a diameter of about 2 cm. and of about the same height, the substance shrinking to a thin film when dry; hymenium much darker than the outside of the apothecium, brownish, nearly circular in form, convex, plane, or very slightly concave, even or nearly so, occasionally with a few folds about the margin; asci cylindric, or subcylindric, often strongly swollen; spores small, ellipsoid, 3–4 × 6–7 μ; paraphyses slender, often branched.

On rotten wood.

Type locality: Ithaca, New York.

Illustration: Mycologia **22**: *pl. 12.*

Distribution: New York.

Excluded Species

Peziza concrescens Schw. Schr. Nat. Ges. Leipzig **1**: 118. 1822. This species which was described as a cartilaginous discomycete is a *Tremella* according to Burt.

46. **CORYNE** Tul. Fung. Carp. **3**: 190. 1865.

Producing both conidial and ascigerous fruiting bodies from the same base, the structures occurring in cespitose clumps of similar purple color and both gelatinous, the conidial bodies club-shaped or tongue-shaped; apothecia turbinate, becoming expanded, sessile, or substipitate; asci clavate, 8-spored; spores fusoid, hyaline, becoming several-septate; paraphyses filiform.

Type species, *Peziza sarcoides* Pers.

The name *Coryne* was first used by Nees (Syst. Pilze 157. 1817) for the conidial stage of this fungus. Later it was adopted by Tulasne for the ascigerous stage and would seem to be the tenable name for the genus.

Spores large, 10–30 μ long.
    Spores 4–6 × 10–18 μ.   1. *C. sarcoides.*
    Spores 6–7 × 24–30 μ.   2. *C. urnalis.*
Spores small, 2–3 × 5–6 μ.   3. *C. microspora.*

1. **Coryne sarcoides** (Pers.) Bonord. Handb. Myk. 149. 1851.

?Octospora violacea Hedw. Descr. **2**: 32. 1789. Not *Ombrophila violacea* Fries. 1849.
Peziza sarcoides Pers. Syn. Fung. 633. 1801.
Peziza janthina Fries, Syst. Myc. **2**: 130. 1822.
Bulgaria sarcoides Fries, Syst. Myc. **2**: 168. 1822.
Tremella sarcoides Fries, Syst. Myc. **2**: 217. 1822.
Sarcodea sarcoides Karst. Not. Soc. Fauna Fl. Fenn. **11**: 232. 1871.
Bulgaria striata Ellis & Ev. Jour. Myc. **1**: 90. 1885.
?Coryne solitaria Rehm in Rab. Krypt.-Fl. **1³**: 488. 1891.
Ombrophila sarcoides Phill. Brit. Discom. 323. 1887.
Coryne striata Sacc. Syll. Fung. **8**: 643. 1889.
?Orbilia atropurpurea Clements, Bot. Surv. Nebr. **4**: 16. 1896.

Apothecia sessile, or substipitate, gelatinous, flesh-red to dark-purple, externally veined, turbinate, later expanding, 2 mm. to 1 cm. in diameter; hymenium at first concave, becoming plane, or repand, often irregularly deformed; asci clavate, reaching a length of 100–135 μ and a diameter of 7–8 μ, 8-spored; spores ellipsoid-fusoid, 4–6 × 10–18 μ, rarely larger, becoming several-septate, the septa often indistinct; paraphyses filiform, often adhering.

On rotten wood.

TYPE LOCALITY: Europe.

DISTRIBUTION: Maine to Oregon and probably throughout North America; also in Europe, and the West Indies.

ILLUSTRATIONS: Hedw. Descr. **2**: *pl. 8, f. 1–7;* Bull. Herb. Fr. *pl. 410, f. 1;* (as *Peziza tremelloidea*) Gill. Champ. Fr. Discom. *pl. 49* (as *Aleuria purpurascens*); E. & P. Nat. Pfl. **1¹**: 209, *f. 164 E, F;* Bull. Lab. Nat. Hist. State Univ. Iowa **5**: *pl. 22, f. 1;* Tul. Fung. Carp. **3**: *pl. 17, f. 1–8.*

EXSICCATI: Ellis, N. Am. Fungi *1280* (as *Bulgaria purpurea*); Ellis & Ev. N. Am. Fungi *2606* (as *Tremella sarcoides*).

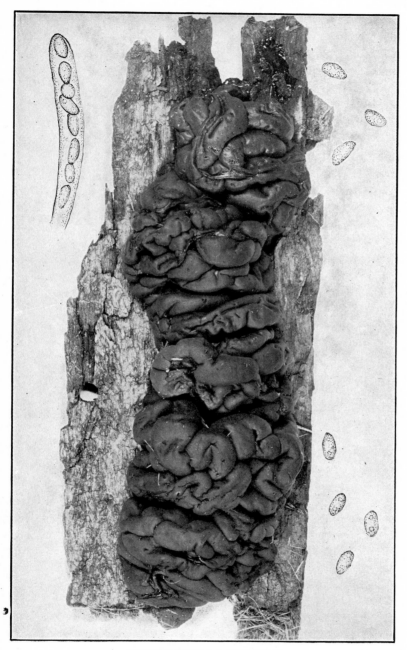

ASCOTREMELLA FAGINEA

2. **Coryne urnalis** (Nyl.) Sacc. Atti Soc. Veneto Sci. Nat. Padova
4: 131. 1875.

Bulgaria urnalis Nyl. Not. Soc. Fauna Fl. Fenn. **10**: 73. 1869.
Ombroplila urnalis Karst. Myc. Fenn. **1**: 87. 1871.
Coryne purpurea Fuckel, Symb. Myc. 284. 1869.

Apothecia similar to those of *Coryne sarcoides* but much larger, reaching a diameter 2–3 cm.; asci clavate reaching a length of 160–190 μ and a diameter of 10–15 μ, 8-spored; spores 2-seriate, elongate-fusoid, 6–7 × 24–30 μ, granular within, becoming 1–9-septate; paraphyses filiform, slightly enlarged above, the apices adhering.

On rotten wood of various kinds.

TYPE LOCALITY: Europe.

DISTRIBUTION: Same as that of *Coryne sarcoides.*

3. **Coryne microspora** Ellis & Ev. Bull. Torrey Club **24**: 282.
1897.

Ombrophila microspora Sacc. & Syd. in Sacc. Syll. Fung. **14**: 802. 1899.

Apothecia cespitose, fleshy-gelatinous, light liver-colored, reaching a diameter of 1 cm., contracted below into a short, thick, stem, wrinkled and folded; hymenium concave with a depression in the center, similar in color to the outside of the apothecium; asci cylindric-clavate, reaching a length of 70–75 μ and a diameter of 5–6 μ, 8-spored; spores 1-seriate, fusoid, 2–3 × 5–6 μ, 2-nucleate (becoming septate?); paraphyses filiform.

On decaying logs.

TYPE LOCALITY: Canada.

DISTRIBUTION: Canada; also reported from Washington.

The type has been examined and as stated by Ellis it looks exactly like *Coryne urnalis* except for the minute spores.

DOUBTFUL SPECIES

Coryne vinosa Berk. & Curt.; Berk. Jour. Linn. Soc. **10**: 341. 1868. Described as vinose, fusiform, compressed, rugose with a small head of similar color, about 4 mm. high, single, or cespitose; conidia minute, abundant. On rotten wood in woods, Cuba. Very doubtful.

Coryne unicolor Berk. & Curt.; Massee (as synonym) Jour. Myc. **6**: 182. 1891. This is *Dacryopsis unicolor* (Berk. & Curt.) Massee, Jour. Myc. **6**: 181.

47. **HOLWAYA** Sacc. Syll. Fung. **8**: 646. 1889.

Apothecia stipitate, reaching a diameter of 1.5 cm., dark-colored, brownish-black, or greenish; asci 8-spored; spores septate, filiform, or vermiform, hyaline.

Type species, *Bulgaria Ophiobolus* Ellis.

The position of the genus is uncertain. Durand (Bull. Torrey Club **28**: 354) places it in the Patellariaceae because the paraphyses cling together above the asci forming an epithecium. This would hardly seem to the writer to be sufficient reason for placing it there since the consistency of the apothecia does not suggest such a relationship.

1. **Holwaya gigantea** (Peck) Durand, Bull. Torrey Club **28**: 354. 1901. (PLATE 117, FIG. 1.)

Stilbum giganteum Peck, Ann. Rep. N. Y. State Mus. **24**: 93.   1872.
Coryne Ellisii Berk. Grevillea **2**: 33.   1873.
Patellaria leptosperma Peck, Ann. Rep. N. Y. State Mus. **30**: 62.   1878.
Bulgaria Ophiobolus Ellis, Am. Nat. **17**: 193.   1883.
Graphium giganteum Sacc. Syll. Fung. **4**: 611.   1886.
Holwaya Ophiobolus Sacc. Syll. Fung. **8**: 646.   1889.
Lecanidion leptospermum Sacc. Syll. Fung. **8**: 800.   1889.
Dacryopsis Ellisiana Massee, Jour. Myc. **6**: 181.   1891.
Chlorosplenium canadense Ellis & Ev. Proc. Acad. Sci. Phila. **1893**: 146.   1893.
Holwaya tiliacea Ellis & Ev. Am. Nat. **31**: 427.   1897.

Conidial structures gregarious, or single, fleshy-gelatinous, consisting of a slender or tapering stem with a length of 3–10 mm. and a diameter of 2 mm. and an ellipsoid, soft, viscid head, 2–6 × 2–4 mm.; conidiophores very slender, branched; conidia hyaline, ellipsoid, 1 × 3 μ.

Apothecia cespitose, or single, scattered, stipitate, cup-shaped, becoming expanded and plane, or with the margin reflexed and umbilicate, orbicular, or irregular from mutual pressure, reaching a diameter of 1.5 cm., greenish-black, fleshy-gelatinous, shrinking much in drying; hymenium similar in color to the outside of the apothecium; stem tapering upward, covered with a greenish-brown tomentum which often disappears with age; asci narrowly clavate, reaching a length of 120–200 μ and a diameter of 10–12 μ; spores more or less fasciculate, filiform-cylindric, or very narrowly clavate-cylindric with ends rounded or one end occasionally acute, straight, or curved, hyaline, 14–20-septate, 3–4 × 30–75 μ; paraphyses filiform, slender, longer than the asci, globose at their apices and clinging together.

On rotten logs of *Tilia*, *Acer*, *Quercus*, and *Magnolia*, in the crevices of the bark or on the bare wood.

TYPE LOCALITY: Buffalo, New York.

DISTRIBUTION: New York to Ontario, Iowa and West Virginia.

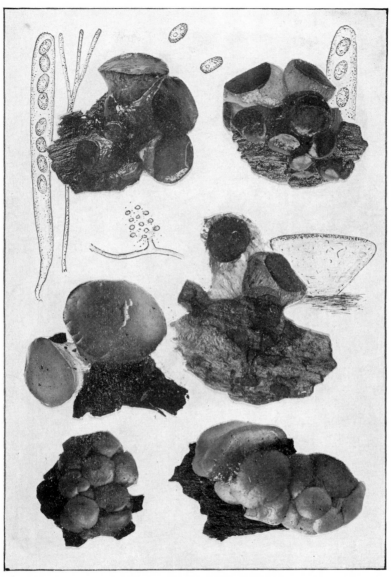

ASCOTREMELLA TURBINATA

ILLUSTRATIONS: Ann. Rep. N. Y. State Mus. **24**: *pl. 3, f. 7–9;*
Bull. Torrey Club **28**: *pl. 26.*
EXSICCATI: Ellis, N. Am. Fungi *1383;* Reliq. Farlow. *126 a,*
126 b.
According to Nannfeldt (Nova Acta Soc. Sci. Upsal. IV. **8**:
306. 1932) *Crinula caliciiformis* Fries is identical with this
species and would replace *Graphium giganteum* (Peck) Sacc. as
had been pointed out by Von Höhnel.

48. **ACERVUS** Kanouse, Papers Mich. Acad. Sci. **23**: 149. 1938.

Apothecia sessile, densely cespitose in clusters varying from
few to many, much contorted from mutual pressure, arising from
a black, tough, rubbery sclerotiform base, reaching a diameter of
4 cm., externally slightly verrucose, soft-leathery and pliant
when fresh, somewhat friable when dry, orange-colored.

Type species, *Acervus aurantiacus* Kanouse.

1. **Acervus aurantiacus** Kanouse, Papers Mich. Acad. Sci. **23**:
    140. 1938.

Apothecia as above; hymenium orange-colored; asci cylindric,
reaching a length of 100–140 μ and a diameter of 7–10 μ, 8-
spored; spores ellipsoid, simple, usually with one oil-drop,
smooth, 3.5–4 $\times$ 6–8 μ; paraphyses clavate, curved, orange-
colored, 9 μ in diameter.

On soil at the base of an elm tree.

TYPE LOCALITY: South Lyons, Michigan.

DISTRIBUTION: Known only from the type locality.

ILLUSTRATIONS: Papers Mich. Acad. Sci. **23**: 150, *f. 1.*

49. **LACHNELLA** Fries, Corpus Fl. Prov. Suec. **1**: 343.   1835;
        Summa Veg. Scand. 365.   1849.

Lachnum Retz. Fl. Scand. Prodr. 256. 1779.
Dasyscyphus S. F. Gray, Nat. Arrang. Brit. Pl. **1**: 670. 1821.
Peziza tribe *Dasyscyphae* Fries, Syst. Myc. **2**: 89. 1822.
Trichopeziza Fuckel, Symb. Myc. 295. 1869.
Hyalopeziza Fuckel, Symb. Myc. 297. 1869.
Dasycypha Fuckel, Symb. Myc. 304. 1869.
Coronellaria Karst. Myc. Fenn. **1**: 14. 1871.
Hyaloscypha Boud. Bull. Soc. Myc. Fr. **1**: 118. 1885.
Erinella Quél. Ench. Fung. 301. 1886.
Cistella Quél. Ench. Fung. 319. 1886.
Solenopezia Sacc. Syll. Fung. **8**: 477. 1889.
Atractobolus Tode; Kuntze, Rev. Gen. Pl. **3³**: 445. 1898.

Unguicularia Höhn. Ann. Myc. **3**: 404. 1905. Not *Unguicularia* D. C.
1825.
Helolachnum Torrend, Broteria **9**: 53. 1910.

Apothecia sessile, or stipitate, externally densely clothed with
hairs; hairs usually flexuose, smooth, or more often delicately
roughened; asci cylindric to clavate, usually 8-spored; spores
ellipsoid to fusoid, simple, or rarely pseudoseptate, paraphyses
filiform to lanceolate.

Type species, *Peziza corticalis* Pers.

The genus *Lachnella* was founded by Fries (Fl. Corpus Prov.
Sueciae **1**: 343. 1835) on *Peziza alboviolascens*, a *Cyphella*,
which he apparently mistook for a discomycete as indicated by
the fact that a few years later (Summa Veg. Scand. 365. 1849)
he used the name for a genus of Discomycetes. Six species were
included, one of which *Peziza flammea* Alb. & Schw. is an operculate and later became the type of the genus *Perrotia* (Boud. Bull.
Soc. Myc. Fr. **17**: 24. 1901) and is included in our volume on
the North American Cup-fungi (operculates). It is proposed
that the name *Lachnella* be conserved for a genus of Discomycetes
as originally intended by Fries, and that *Peziza corticalis* Pers. be
adopted as the type since this is probably the most widely distributed and best known inoperculate species included by Fries
in his genus. Although not mentioned by Fries, the spores of
this species are much elongated, relatively large and occasionally
become septate at maturity although this character is by no
means a constant one.

Since Fries' time various interpretations have been placed on
the genus by different workers. In 1887, William Phillips took
up the genus and used it in the same sense as Fries (1849) including four of his six species, one species, *Lachnella alboviolacens*
(Alb. & Schw.) Fries, being ruled out as a *Cyphella* and one other
Lachnella rhabarbarina (Berk.) Fries being transferred to *Dermatea*. Many other species were included in the genus by
Phillips. No particular stress was placed on the spores but the
genus was described as follows:

"*Cups small, stipitate or sessile; flesh thin, firm, waxy; externally
pilose or villous; asci cylindrical or subclavate; sporidia 8, colour-
less; paraphyses filiform or acerose.*"

In 1884, Saccardo (Bot. Cent. **18**: 216) recognized the genus
as follows: "(*Lachnella* et *Helotium* e. p. *Erinella* Quél. e. p.

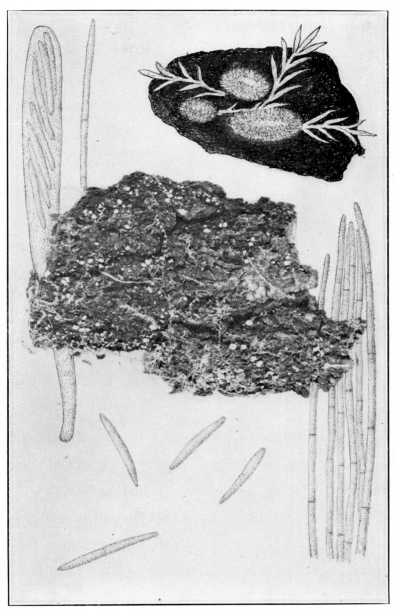

LACHNELLA CORTICALIS

Velutaria Fuckel e. p.)" and treated *Dasyscypha*, *Trichopeziza* and *Hyalopeziza* as subgenera. In 1889 he used the genus *Lachnella* in a more restricted sense keeping it for those forms with filiform paraphyses. Those species with what he calls fusoid paraphyses, which is apparently another name for acerose are placed in two different genera *Trichopeziza* Fuckel (1849) with subsessile apothecia and *Dasycypha* with stipitate apothecia. Whether the form of the paraphyses is a good character on which to base a genus is a question. The presence or absence of a stem is scarcely more reliable here than in *Helotium* and *Phialea*. This is so variable in a given species that it is not regarded by us as a sound character on which to found a genus.

In 1897, Lindau, (Engler & Prantl, Nat. Pfl. 1¹: 201. 1897.) also took up the genus *Lachnum* Retz (1770) treating *Trichopeziza* Fuckel (1849) as a synonym. If these two genera are synonyms, *Trichopeziza* would have priority as a post-Friesian genus. *Lachnum* (*Trichopeziza*) is distinguished by Lindau from *Lachnella* and *Dasyscypha* by its lanceolate paraphyses.

In 1931, Clements and Shear (Genera Fungi 327) used *Lachnella* with the *Peziza flammea* Alb. & Schw. as the type placing it in the Helotiaceae which belong to the inoperculates and ignoring the fact that it had been made the type of the genus *Perrotia* founded by Boudier and placed in the operculates.

It will be seen from the above that no two authors have had exactly the same conception of this genus resulting in almost hopeless confusion. The writer is not presumptuous enough to think that he can straighten out this hopeless tangle, but he can at least present his own views. For the time being we are inclined to use *Lachnella* in much the same sense used by early workers, and include a number of the recent segregates.

Hairs hyaline, appearing white, or cinereous.
  On coniderous plant tissues.
    Spores ellipsoid to fusoid, not over 20 μ long.
      On species of *Larix*.

Spores 6–8 × 15–20 μ.	1. *L. Willkommii.*
Spores 4–7 × 10–18 μ.	2. *L. occidentalis.*
Not restricted to *Larix*.	
Spores 14–20 μ long.	3. *L. Hahniana.*
Spores 11–15 μ long.	4. *L. oblongospora.*
Spores usually less than 11 μ long.	
On *Pseudotsuga taxifolia*.	
Forming cankers on branches.	5. *L. Pseudotsugae.*
Not forming cankers.	6. *L. ciliata.*

Not restricted to *Pseudotsuga.*
  On resin on *Abies balsamea.*   7. *L. resinaria.*
  On bark of various conifers.   8. *L Agassizii.*
Spores fusiform, sharp-pointed, 20–25 μ
long.   9. *L. Ellisiana.*
Not restricted to coniferous plants.
  On phanerogams.
    Apothecia becoming shallow cup-shaped.
      On dicotyledonous hosts.
        On twigs and branches of woody
plants.
          Spores 15–20 μ long, on bark.   10. *L. corticalis.*
          Spores less than 15 μ long.
            H y m e n i u m   orange-col-
ored.   11. *L. bicolor.*
            Hymenium ochraceous, on
Rubus.   12. *L. subochracea.*
        On rotten wood of various kinds.
          Apothecia fasciculate, on olive.   13. *L. fasciculata.*
          Apothecia not fasciculate.
            Paraphyses filiform.   14. *L. hyalina.*
            Paraphyses lanceolate.
              Hairs capped with
crystals.   15. *L. crucifera.*
              Hairs not capped with
crystals.   16. *L. virginea.*
        On leaves, living or dead.
          On leaves of *Quercus.*
            Hairs capitate, spores 5–
7 μ long.   17. *L. capitata.*
            Hairs not capitate.
              Spores 2.5–3 × 15–20 μ.   18. *L. ciliaris.*
              Spores 1.5 × 7–8 μ.   19. *L. pollinaria.*
           Not on leaves of *Quercus.*
            On leaves of *Garraya.*   20. *L. tautilla.*
            On leaves of *Vaccinium.*   21. *L. virginella.*
            On leaves of *Gaultheria.*   22. *L. Gaultheriae.*
            On leaves of *Halesia.*   23. *L. Halesiae.*
        On exposed roots and debris.   24. *L. pygmaea.*
      On monocotyledons.
        On culms of *Juncus.*   25. *L. diminuta.*
        On stems of *Calamagrostis.*   26. *L. agrostina.*
        On leaves of *Arundinaria.*   27. *L. Arundinariae.*
    Apothecia cylindric, soleniform.   28. *L. Solenia.*
  On cryptogams.
    On rootstocks of *Equisetum.*   29. *L. inquilina.*
    On ferns.
      On stems of *Pteris.*   30. *L. pteridicola.*
      On stems of *Aspidium* and *Adiantum.*   31. *L. aspidicola.*
      On fronds of *Dicranopteris.*   32. *L. Dicranopteridis.*

On fungi.
  On *Rhytisma* on *Acer*.   33. *L. Rhytismatis.*
  On stromata of *Acanthonitschkia*.   34. *L. Acanthonitschkeae.*
Hairs pale-yellow to sulphur-yellow.
  On phanerogams.
    On dead wood.
      Spores 3–5 × 10–14 μ.   35. *L. succina.*
      Spores 2 × 7–8 μ.   36. *L. albolutea.*
    On herbaceous stems.
      Paraphyses lanceolate.   37. *L. sulphurea.*
      Paraphyses filiform.
        On stems of *Arctium;* spores 1.5–2 ×
          8–12 μ.   38. *L. canadensis.*
        On stems of *Iva*, spores; spores 3 ×
          12–14 μ.   39. *L. Ivae.*
    On fallen pine needles.   40. *L. pulverulenta.*
  On cryptogams, tree fern, *Gleichenia*. 41. *L. Gleicheniae.*
Hairs greenish or purple.
  Hairs pale-purple.   42. *L. atropurpurea.*
  Hairs green or greenish.
    Spores large, 5–6 × 12–15 μ.   43. *L. viridicoma.*
    Spores small, less than 10 μ long.
      Spores 1–1.5 × 5–5.5 μ.   44. *L. microspora.*
      Spores 2 × 7–9 μ.   45. *L. pulveracea.*
Hairs brown to blackish.
  On phanerogams.
    On dicotyledons.
      On coniferous plants of various kinds.
        Spores 4 × 7 μ.   46. *L. arida.*
        Spores 5 × 12–15 μ.   47. *L. fuscosanguinea.*
      Not on conifers.
        On woody plant tissues.
          On bark of *Populus tremuloides*.
            Paraphyses lanceolate.   48. *L. populicola.*
            Paraphyses clavate.   49. *L. populina.*
          Not restricted to *Populus*.
            Paraphyses lanceolate on
              leaves of *Cornus*.   50. *L. Corni.*
            Paraphyses clavate, on rotten
              wood.   51. *L. cerina.*
        On herbaceous stems; *Eupatorium*. 52. *L. Eupatorii.*
      On monocotyledons, *Fragmites, Andropogon*
        etc.   53. *L. albotestacea.*
    On cryptogams, *Pteris aquilina*.   54. *L. Pteridis.*

1. **Lachnella Willkommii** Hartig, Wicht. Krankh. Waldbäume 98. 1874. (PLATE 123.)

?Peziza calycina Schum. Enum. Pl. Saell. **2**: 424. 1803.
Peziza calycina Laricis Chaillet; Fries, Elench. Fung. **2**: 8. 1828.

Peziza Laricis Rehm, Grevillea **4**: 169.   1876.
Lachnea calycina Gill. Champ. Fr. Discom. 71.   1880.
Helotium Willkommii Wettestein, Bot. Cent. **31**: 285.   1887.
Trichoscypha Willkommii Boud. Hist. Class. Discom. Eu. 125.   1907.
Trichoscyphella Willkommii Nannf. Nova Acta Soc. Sci. Upsal. IV. **8**: 300.   1932.

Conidial stage which precedes the apothecial stage consists of waxy, whitish stromata with irregular labyrinthiform cavities in which the microconidia are abstricted from the tips of slender, subulate, simple, or verticillately-branched sporophores; microconidia hyaline, simple, ellipsoid, or allantoid, 1–2 × 2–8 μ.

Apothecia sessile, or short-stipitate, scattered, or gregarious in small groups and occasionally several closely crowded together, at first closed and rounded, becoming expanded and finally scutellate, externally white and clothed with a dense covering of white hairs, reaching a diameter of 1–3 mm., hymenium becoming nearly plane, yellow to orange; hairs flexuous, of nearly uniform thickness throughout their length, minutely roughened; asci cylindric-clavate, reaching a length of 100–120 μ and a diameter of 8–10 μ, 8-spored; spores ellipsoid, often with the ends narrowed, hyaline, granular within, 6–8 × 15–20 μ, or rarely as long as 25 μ; paraphyses slender, clavate, reaching a diameter of about 3 μ.

On branches of *Larix europaea*.

TYPE LOCALITY: Europe.

DISTRIBUTION: Massachusetts to Idaho; also in Europe.

ILLUSTRATIONS: Hartig, Text-book Dis. Trees (trans.) 119, *f. 58, 59;* Boud. Ic. Myc. *pl. 518* (in part); Mycologia **26**: *pl. 9, f. 1–5, pl. 10.*

2. **Lachnella occidentalis** (Hahn & Ayers) Seaver, comb. nov.

Dasyscypha occidentalis Hahn & Ayers, Mycologia **26**: 90.   1934.

Conidial stage consisting of a fleshy stroma with irregular labyrinthiform cavities in which the microconidia are abstricted from the tips of slender, pointed, verticillately branched sporophores; microconidia ellipsoid, or allantoid, simple, 1–1.5 × 2–5 μ.

Apothecia fleshy, abundant, scattered, or grouped, at first globose and closed, opening in a roundish form, the margin incurved, later expanding and becoming saucer-shaped under moist conditions, laterally compressed and closed when dry, short-stipitate, externally densely clothed with white hairs, reaching a diameter of 1–3 mm., salmon-orange; hairs long, flexuous, roughened on the outside, 3–4 μ in diameter; asci clavate, reaching a length of 130 μ and a diameter 12 μ, 8-spored; spores obliquely 1-seriate, becoming 1-septate on germination, ellipsoid, 4–7 × 10–18 μ; paraphyses filiform, swollen at their apices, 1–2 μ in diameter.

On species of *Larix*, *L. europaea*, *L. Laricina*, *L. leptolepis*, and *L. occidentalis*.

TYPE LOCALITY: Hills, British Columbia.

DISTRIBUTION: Vermont to Pennsylvania and British Columbia.

ILLUSTRATIONS: Mycologia **26**: *pl. 12, f. 6–10; pl. 13.*

3. **Lachnella Hahniana** Seaver, nom. nov.

Dasyscypha calycina Fuckel, Symb. Myc. 305. 1869. Not *Peziza calycina* Schum. 1803.

Conidial stage consisting of erumpent, fleshy, or waxy stromata containing simple, or labyrinthiform loculi in which microconidia are abstricted from the tips of short, subulate, simple, or verticillately branched sporophores; microconidia continuous, hyaline; ellipsoid, or allantoid, 1–2 × 2–5 μ.

Apothecia abundant, solitary, or grouped, short-stipitate, at first globose and closed, opening with a rounded form and expanding to a saucer-shaped structure in moist weather, reaching a diameter of 3 mm., externally densely clothed with white hairs; hymenium, concave, ochraceous to salmon-orange; hairs cylindric with slightly swollen extremities, 3–4 μ thick, minutely roughened; asci clavate, reaching a length of 100–165 μ and a diameter of 8–10 μ, 8-spored; spores 1-seriate, smooth, ellipsoid, simple, often becoming 1-septate on germination, 5–8 × 14–20 μ; paraphyses filiform, intermixed with broader filaments with swollen extremities, rounded, or with subacute apices, occasionally spathulate, 1–4 μ thick.

On *Larix europaea*, *Larix leptolepis*, and *Pseudotsuga taxifolia*.

TYPE LOCALITY: Europe.

DISTRIBUTION: Massachusetts; also in Europe.

ILLUSTRATIONS: Mycologia **26**: *pl. 8; pl. 9, f. 6–10.*

4. **Lachnella oblongospora** (Hahn & Ayers) Seaver, comb. nov.

Dasyscypha oblongospora Hahn & Ayers, Mycologia **26**: 88. 1934.

Conidial stage consisting of waxy, erumpent, stromata with irregular, labyrinthiform cavities in which the microconidia are abstricted from the tips of the slender, subulate, simple, or verticillately branched sporophores; microconidia hyaline, simple, ellipsoid, or allantoid, 1–1.5 × 2–5 μ.

Apothecia erumpent, waxy, scattered, or grouped, at first globose, closed, opening and becoming urn-shaped, expanding under moist conditions, laterally compressed and closed when dry, short-stipitate, externally clothed with white, or grayish-white hairs; reaching a diameter of 2 mm.; hymenium salmon-orange to orange-buff; hairs long, flexuous and minutely roughened, septate, 3–4 μ thick; asci clavate, reaching a length of 70–100 μ and a diameter of 7–10 μ, 8-spored; spores 1-seriate, or irregularly disposed, ellipsoid, hyaline, smooth, at first simple, often becoming 1-septate before germination, 4–6 × 11–15 μ; paraphyses filiform, scarcely swollen at their apices, 1–2 μ in diameter.

On coniferous branches of *Larix laricens, L. europaea, L. leptolepis, Picea pungens, Pinus pungens, Pinus virginiana* and *Pseudotsuga taxifolia*.

TYPE LOCALITY: Bethel, Vermont (On *Larix laricina*).

DISTRIBUTION: Vermont to Pennsylvania and Michigan.

ILLUSTRATIONS: Mycologia **26**: *pl. 11; pl. 12, f. 1–5.*

5. **Lachnella Pseudotsugae** (Hahn) Seaver, comb. nov.

Dasyscypha Pseudotsugae Hahn, Mycologia **32**: 138.   1940.

Conidial stage consisting of waxy-fleshy, light-buff stromata with labyrinthiform cavities in which the conidia are borne and from which they exude in a droplet or tendril; conidia produced on the ends of simple, or branched conidiophores; hyaline ellipsoid, 1.8–3 × 2.4–4 μ.

Apothecia waxy-fleshy, scattered, or closely grouped, at first globose and closed, opening in a roundish form, the margin incurved, urn-shaped, becoming widely expanded and discoid under moist conditions, laterally compressed when dry, 1–3.5 mm. in diameter; hymenium light orange-yellow to orange; hairs minutely roughened, cylindrical with obtuse ends, 3–3.5 μ in diameter, hyaline-white; asci clavate, reaching a length of 50–60 μ and a diameter of 4–5.4 μ, 8-spored; spores 1-seriate, ellipsoid to fusoid, 2–4 × 4–7 μ; paraphyses filiform, slightly swollen above.

Forming cankers on living branches of *Pseudotsuga taxifolia*.

TYPE LOCALITY: Lokoya, Napa County, California.

DISTRIBUTION: California to British Columbia.

ILLUSTRATIONS: Mycologia **32**: 139. *f. 1–6.*

6. **Lachnella ciliata** (Hahn) Seaver, comb. nov.

Dasyscypha ciliata Hahn, Mycologia **32**: 141. 1940.

Conidial stage not observed.

Apothecia waxy-fleshy, short-stipitate, usually scattered, occasionally grouped, at first globose, expanding and becoming discoid, externally whitish, 1–2 mm. in diameter; hymenium concave, or plane, orange; hairs cylindric with acute ends, minutely roughened, giving rise to a fringe-like margin, 3 μ thick; asci clavate, reaching a length of 70–80 μ and a diameter of 7–10 μ, 8-spored; spores obliquely 1-seriate, ovoid to ellipsoid, 4–6 $\times$ 8–12 μ; paraphyses filiform, slightly swollen at their apices.

On dead branches of *Pseudotsuga taxifolia*.

TYPE LOCALITY: Portland Heights, Oregon.

DISTRIBUTION: Oregon and British Columbia.

ILLUSTRATIONS: Mycologia **32**: 143. *f. 2*.

7. **Lachnella resinaria** Phill. Brit. Discom. 242. 1887.

Peziza resinaria Cooke & Phill.; Cooke, Grevillea **3**: 185. 1875.
Dasyscypha resinaria Rehm, Ascom. Lojk. 11. 1882.
Lachnellula resinaria Rehm. in Rab. Krypt.-Fl. **1**³: 864. 1893.

Apothecia short-stipitate, at first globose and closed, expanding and becoming shallow cup-shaped, externally densely clothed with white hairs, reaching a diameter of .5–1.5 mm.; hymenium concave, pale-orange; hairs hyaline-white, cylindric, flexuous, septate, reaching a diameter of 3–3.5 μ; asci cylindric above, tapering below, 8-spored; spores ellipsoid, 2.5 $\times$ 5 μ; paraphyses filiform, slightly enlarged above.

On resin from the branches of balsam fir, *Abies balsamea*.

TYPE LOCALITY: Europe.

DISTRIBUTION: Minnesota.

ILLUSTRATIONS: Bull. Torrey Club **29**: *pl. 1*.

Reported from Minnesota, as causing canker growth on *Abies balsamea* (Bull. Torrey Club **29**: 23. 1902.) While frequently reported from Europe, surprisingly little material has been encountered from this country.

8. **Lachnella Agassizii** (Berk. & Curt.) Seaver, comb. nov.
(PLATE 121.)

?Elvela calyciformis Batsch, Elench. Fung. Cont. **1**: 195. 1786.
?Peziza calyciformis Willd. Fl. Berol. 404. 1787.
?Octospora calyciformis Hedw. Descr. **2**: 78. 1789.

Peziza Agassizii Berk. & Curt.; Cooke & Peck, Grevillea **1**: 5.   1872.
Peziza subtilissima Cooke, Grevillea **3**: 121.   1875.
Helotium calyciforme Wettstein, Bot. Cent. **31**: 319.   1887.
Lachnella subtilissima Phill. Brit. Discom. 244.   1887.
Dasyscypha subtilissima Sacc. Syll. Fung. **8**: 438.   1889.
Dasyscypha Agassizii Sacc. Syll. Fung. **8**: 438.   1889.
Dasyscypha incarnata Clements, Bull. Torrey Club **30**: 88.   1903.

Apothecia gregarious, short-stipitate, at first subglobose, becoming expanded and shallow cup-shaped, or nearly plane, externally clothed with flexuose, white hairs; hymenium bright orange-yellow, fringed about with marginal hairs, reaching a diameter of 4 mm., but often much smaller; hairs hyaline, white to the unaided eye, blunt, or attenuated, externally covered with coarse granules, reaching a length of 100 μ and a diameter of 2–3 μ; asci cylindric or subcylindric, reaching a length of 60–95 μ and a diameter of 3–4 μ; spores 1- or 2-seriate, narrow-ellipsoid, hyaline, 3–4 × 6–10 μ; paraphyses clavate, reaching a diameter of 2–5 μ at their apices.

On bark and wood of conifers, *Abies balsamea*, *Picea Mariana*, *Picea rubra*, *Pinus monticola*, *Pinus Strobus*, *Pseudotsuga taxifolia* and *Tsuga canadensis*.

TYPE LOCALITY: Lake Superior region (Calumet).

DISTRIBUTION: Labrador to Pennsylvania, Montana and Colorado.   Probably widely distributed in subboreal regions; ? also in Europe.

ILLUSTRATIONS: Hedw. Descr. **2**: *pl. 22, B;* Grevillea **3**: *pl. 40, f. 169;* Mycologia **21**: *pl. 20;* **35**: 107, *f. 2.*

EXSICCATI: Ellis, N. Am. Fungi *1311;* Barth. Fungi Columb. *4530;* Clements, Crypt. Form. Colo. *82;* W. B. Cooke, Mycobiota N. Am. *67.*   Rehm, Ascom. *1854* (from Wisconsin).

Bingham and Ehrlich after an exhaustive study (Mycologia **35**: 95–111.   1943) consider *Dasyscypha Agassizii* as distinct from *Peziza calyciformis* Willd. of Europe, the differences based largely on spore measurements.   The differences are so slight that it seems to the writer a question whether or not they should be separated.

9. **Lachnella Ellisiana** (Rehm) Seaver, comb. nov.

Peziza Ellisiana Rehm, Grevillea **4**: 169.   1876.
Dasyscypha Ellisiana Sacc. Syll. Fung. **8**: 459.   1889.

Imperfect stage abundant, yellowish-green, minute, consisting of an erumpent stroma, 106–132 μ diam., at first closed, then opening up with a single exposed chamber, or compound, with

LACHNELLA AGASSIZII

more than one locule, 243–433 μ diam., microconidia fusiform, .9–1.2 $\times$ 5–5.8 μ abstricted from the tips of short sporophores, subulate, acute, simple, or verticillately branched.

Apothecia scattered, or gregarious, sessile, or very short-stipitate, at first subglobose, expanding and becoming shallow cup-shaped, reaching a diameter of 1–2 mm., externally densely clothed with white or whitish hairs; hymenium plane, or slightly concave, bright-orange; hairs cylindric or subcylindric, septate, hyaline, minutely roughened on the outside, reaching a diameter of 4–6 μ and of variable length up to 200 μ; asci clavate, reaching a length of 65 μ and a diameter of 6–7 μ, 8-spored; spores 2-seriate or partially so, fusiform and very sharp-pointed, 2–3 $\times$ 20–25 μ; paraphyses lanceolate, or sublanceolate, rather slender, about half the diameter of the ascus.

On bark of species of *Pinus*, *Picea* and *Larix*.

TYPE LOCALITY: New Jersey.

DISTRIBUTION: Maine to Alabama and Texas.

ILLUSTRATIONS: Mycologia **26**: *pl. 21; pl. 22, f. 1–7.*

EXSICCATI: Rav. Fungi Am. *175;* Ellis & Ev. N. Am. Fungi *2324, 3231;* Fungi Columb. *641, 1224;* Rehm, Ascom. *303;* Thüm. Myc. Univ. *716.*

Resembling *Lachnella bicolor* but differing in the very long, sharply pointed spores. *Dasyscypha lachnoderma* (Berk.) Rehm, which was thought by Ellis to be identical with this species is quite distinct according to Ayers and Hahn (Mycologia **26**: 157–180). Some of the above exsiccati were distributed under the latter name.

10. **Lachnella corticalis** (Pers.) Fries, Summa Veg. Scand. 365. 1849. (PLATE 120.)

Peziza corticalis Pers. Tent. Disp. Fung. 33. 1797.
Helotium corticale Karst. Myc. Fenn. **1**: 159. 1871.
Lachnea corticalis Gill. Champ. Fr. Discom. 84. 1880.
Peziza borealis Ellis & Holw.; Arth. Bull. Geol. Nat. Hist. Surv. Minn. **3**: 35. 1887.
Lachnella canescens Phill. Brit. Discom. 259. 1887.
Dasyscypha borealis Sacc. Syll. Fung. **8**: 457. 1889.
Lachnella rhizophila Ellis & Ev. Proc. Acad. Phila. **1894**: 348. 1894.
Dasyscypha canescens Massee, Brit. Fungus-Fl. **4**: 346. 1895.
Lachnum corticale Clements, Crypt. Form. Colo. *86.* 1906.

Apothecia thickly gregarious, sessile, or short-stipitate, at first closed and subglobose, expanding and becoming shallow cup-shaped with the mouth constricted, reaching a diameter of 1 mm., externally densely clothed with grayish hairs which may change to yellowish, or brownish with age, the hairs often tipped with masses of minute, rod-like bodies which give them a crystaline appearance; hymenium concave, cream-colored, or yellowish; hairs gradually attenuated, semiacute, septate, pale-yellow below, hyaline toward the tip, reaching a length of 100–150 μ and a diameter of 3–4 μ at the base; asci clavate, reaching a length of 100 μ and a diameter of 10 μ, tapering below; spores obliquely 2-seriate, fusoid, slightly curved, about 4 × 15–20 μ or rarely as long as 30 μ, often with a row of oil-drops, giving the spore a pseudoseptate appearance, or often 1-septate; paraphyses protruding far beyond the asci, cylindric below, semiacute at their apices, 2–3 μ in diameter.

On bark of various deciduous trees.

TYPE LOCALITY: Europe.

DISTRIBUTION: New York to Winnipeg, Washington and Colorado; also in Europe.

ILLUSTRATIONS: Boud. Ic. Myc. *pl. 517;* E. & P. Nat. Pfl. 1^1: 202, *f. 159 H,*

EXSICCATI: Clements, Crypt. Form. Colo. *86.*

11. **Lachnella bicolor** (Bull.) Phill. Brit. Discom. 249.　1887. (PLATE 123, FIG. 2.)

Peziza bicolor Bull. Hist. Champ. Fr. 243.　1791.
Dasyscypha bicolor Fuckel, Symb. Myc. 305.　1869.
Lachnum bicolor Karst. Myc. Fenn. **1:** 172.　1871.
Lachnea bicolor Gill. Champ. Fr. Discom. 70.　1880.
Erinella bicolor Quél. Ench. Fung. 303.　1886.

Apothecia closely gregarious, or scattered, sessile, or short-stipitate, at first closed, becoming expanded, externally clothed with a dense covering of white hairs, reaching a diameter of 1–2 mm.; hymenium nearly plane, orange-yellow; hairs hyaline, cylindric, septate, externally roughened with granules, reaching a length of 200 μ, and a diameter of 4 μ; often with crystaline caps, asci cylindric-clavate, reaching a length of 40–50 μ and a diameter of 5–6 μ; spores fusiform, straight, or slightly curved, 1.5–2 × 6–12 μ; paraphyses lanceolate, very sharp pointed.

On small twigs of various trees and shrubs.

TYPE LOCALITY: Europe.

DISTRIBUTION: New York to Washington and Alaska to Mexico; also in Europe.　Probably widely distributed.

ILLUSTRATIONS: Bull. Herb. Fr. *pl. 410, f. 3;* Phill. Brit. Discom. *pl. 8, f. 46;* Gill. Champ. Discom. *pl. 61, f. 3;* Rab. Krypt.-Fl. **1**³: 865, *f. 1–4.*

EXSICCATI: Ellis & Ev. Fungi Columb. *1222;* Barth. Fungi Columb. *3117.*

During the summer of 1929 a fine collection of the species was made near the Alpine Laboratory on Pikes Peak, Colorado on dead stems of *Rubus* (Seaver & Shope *443*).

12. Lachnella subochracea (Cooke & Peck) Seaver, comb. nov.

Peziza subochracea Cooke & Peck, Grevillea **1**: 6. 1872.
Trichopeziza subochracea Sacc. Syll. Fung. **8**: 408. 1889.

Apothecia sessile, or substipitate, scattered, at first subglobose, then expanded and becoming discoid with the margin slightly elevated, yellowish, clothed with white hairs; hymenium ochraceous, nearly plane, or slightly concave with the margin slightly elevated; hairs poorly developed, hyaline; asci clavate, reaching a length of 80 μ and a diameter of 8 μ; spores partially 2-seriate, or irregularly disposed, fusiform, hyaline, 3 $\times$ 12–13 μ; paraphyses filiform, or slightly enlarged above.

On dead stems of *Rubus.*

TYPE LOCALITY: Adirondack Mountains.

DISTRIBUTIONS: New York and Newfoundland.

ILLUSTRATIONS: Grevillea **1**: *pl. 1, f. 4.*

Although this has been placed in the genus *Trichopeziza* by Saccardo, the apothecia are not conspicuously hairy as in most of the species of the genus. The apothecia are delicately pubescent but the hairs poorly developed.

13. Lachnella fasciculata (Seaver & Waterston) Seaver, comb. nov.

Dasyscypha fasciculata Seaver & Waterston, Mycolgia **32**: 397. 1940.

Apothecia thickly gregarious, occurring singly, or more often in dense, fasciculate clumps, several apparently springing from the same base and so closely compact that they appear to be one compound fruit body, short-stipitate, externally clothed with a dense covering of white hairs, the clumps scarcely exceeding 1 mm. in diameter, the individual apothecia much less; hymenium concave, pale-orange; hairs flexuous, hyaline, externally roughened, about 2 μ in diameter; asci clavate, reaching a length of 35–40 μ and a diameter of 4 μ, 8-spored; spores minute, fusoid, hyaline, 1.5 × 6 μ; paraphyses filiform semiacute but scarcely lanceolate.

Type collected on rotten stumps of olive tree, *Olea europaea,* Walsingham, Bermuda, Nov. 30, 1938.

DISTRIBUTION: Known only from the type locality.

14. Lachnella hyalina (Pers.) Phill. Brit. Discom. 267.   1887.

Peziza hyalina Pers. Syn. Fung. 655.   1801.
Helotium hyalinum Karst. Not. Soc. Fauna Fl. Fenn. **11:** 240.   1871.
Lachnea hyalina Gill. Champ. Fr. Discom. 79.   1880.
Urceola hyalina Quél. Ench. Fung. 321.   1886.
Pseudohelotium hyalinum Fuckel, Symb. Myc. 298.   1869.
Hyaloscypha hyalina Boud. Ic. Myc. **4:** 308.   1911.

Apothecia gregarious, sessile, at first globose and closed, opening and becoming cup-shaped, then expanded and saucer-shaped, soft and externally slightly downy, .1–.5 mm. in diameter; hymenium hyaline, occasionally slightly yellowish or rosy; hairs delicate, hyaline, or white; asci fusoid, reaching a length of 35–45 μ and a diameter of 7–10 μ, 8-spored; spores elongated, or slightly fusiform, 2-seriate, straight, or slightly curved, 2–2.5 × 6–10 μ; paraphyses filiform, 2 μ thick.

On dead wood and bark of various kinds.

TYPE LOCALITY: Europe.

DISTRIBUTION: New Jersey to Washington, California and Louisiana; also in Europe.

ILLUSTRATIONS: Phill. Brit. Discom. *pl. 8, f. 48;* Boud. Ic. Myc. *pl. 525.*

EXSICCATI: Ellis & Ev. N. Am. Fungi *2810.*

15. Lachnella crucifera Phill. Brit. Discom. 250.   1887.
(PLATE 122.)

Peziza crucifera Phill. Gard. Chron. II. **10:** 397.   1878.
Peziza sulphurella Peck, Ann. Rep. N. Y. State Mus. **30:** 59.   1878.
Dasyscypha crucifera Sacc. Syll. Fung. **8:** 440.   1889.
Dasyscypha sulphurella Sacc. Syll. Fung. **8:** 459.   1889.

Apothecia stipitate, reaching a diameter of .5–1 mm., shallow cup-shaped with hymenium pale-yellowish, externally clothed with hairs intermixed with crystals of calcium oxalate giving them a granular appearance; stem reaching a length of .5–1 mm. and also clothed with hairs; hairs hyaline, clavate, septate, reaching a length of 80 μ and a diameter of 4 μ; asci clavate, reaching a length of 40 μ and a diameter of 5–6 μ, 8-spored; spores minute, about 2 × 5–6 μ; paraphyses lanceolate.

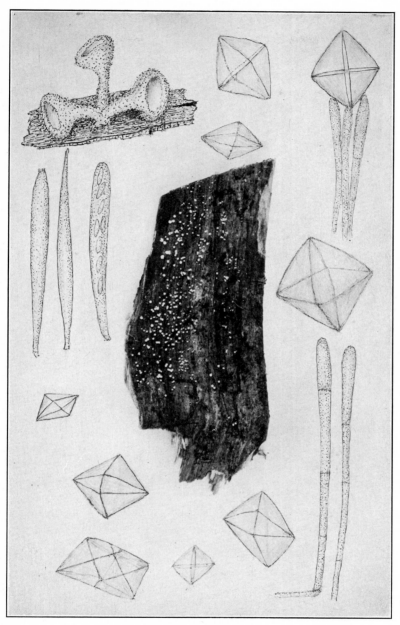

LACHNELLA CRUCIFERA

On rotten wood.

TYPE LOCALITY: Europe.

DISTRIBUTION: New York to Michigan; also in Europe.

ILLUSTRATIONS: Gard. Chron. II. **10**: *f. 71;* Mycologia **28**: 5, *f. 1.*

In Papers of the Michigan Academy of Sciences (**23**: 153. 1938) Dr. B. Kanouse calls attention to the fact that *Peziza sulphurella* Peck and *Peziza crucifera* Phill. are identical. It is an interesting coincidence that both were described in the same month of the same year. Which name actually has priority it is difficult to say. The above synonymy is based on the conclusions of Dr. Kanouse.

16. **Lachnella virginea** (Batsch) Phill. Brit. Discom. 248. 1887.

Lachnum agaricinum Retz. Fl. Scand. Prodr. 256. 1779.
Peziza virginea Batsch, Elench. Fung. 125. 1783.
Octospora nivea Hedw. (fil.) Obs. Bot. 13. 1802.
Trichopeziza nivea Fuckel, Symb. Myc. 296. 1869.
Dasyscypha virginea Fuckel, Symb. Myc. 305. 1869.
Lachnum niveum Karst. Myc Fenn. **1**: 168. 1971.
Lachnum virgineum Karst. Myc. Fenn. **1**: 169. 1871.
Lachnea virginea Gill. Champ. Fr. Discom. 68. 1880.
?Peziza latebrosa Ellis, Bull. Torrey Club **9**: 18. 1882.
Erinella virginea Quél. Ench. Fung. 304. 1886.
Lachnella nivea Phill. Brit. Discom. 245. 1887.
?Dasyscypha latebrosa Sacc. Syll. Fung. **8**: 434. 1889.
Dasyscypha nivea Sacc. Syll. Fung. **8**: 437. 1889.
Lachnum niveum var. *Fairmani* Rehm, Ann. Myc. **6**: 316. 1908.

Apothecia gregarious, or scattered, at first globose and closed, becoming expanded and shallow cup-shaped, reaching a diameter of 1 mm., contracted below into a short stem which reaches a length of 1–2 mm., clothed with white hairs; hymenium nearly plane, white, or becoming slightly yellowish; hairs hyaline, septate, covered with minute granules, mostly clavate, reaching a diameter of 3 μ below and about 6 μ above, the length variable but up to 80 μ; asci clavate-cylindric, about 4–5 × 45–60 μ, 8-spored; spores fusoid, 1.5–2.5 × 6–10 μ; paraphyses lanceolate, extending far beyond the asci, hyaline, 4–6 μ in diameter at the thickest point.

On wood of various kinds and on woody stems.

TYPE LOCALITY: Europe.

DISTRIBUTION: Widely distributed in North America; also in Europe.

ILLUSTRATIONS: Hedw. (fil.) Obs. Bot. *pl. 8, f. B;* Sow. Engl. Fungi, *pl. 65;* Gill Champ. Fr. Discom. *pl. 59, f. 1–2;* Bull. Lab. Nat. Hist. State Univ. Iowa **6**: *pl. 25, f. 3* (except paraphysis).

EXSICCATI: Ellis, N. Am. Fungi *387;* Ellis & Ev. N. Am. Fungi, *2912;* Fungi Columb. *19, 2015.*

This is a species concerning which there has been much confusion. Phillips in his British Discomycetes regards *L. niveum* and *L. virgineum* as distinct species and states that the latter can be distinguished from the former by its lanceolate paraphyses. Rehm treats them as distinct but assigns lanceolate paraphyses to both and seemed to know nothing of *L. niveum* as diagnosed by Phillips. Gillet in his Discomycetes of France treats them as distinct and attempts to draw a slight difference in the form of the cups or length of the hairs.

In his early work on the Discomycetes the writer encountered this species and figured it with clavate paraphyses which may have been due to faulty observation, for in going over all of the specimens in the herbarium of The New York Botanical Garden listed under these two names, he finds that all of them have lanceolate paraphyses. Since none of the diagnostic characters used by European authors, in separating these two species appear to have any particular value, they are here combined. One specimen examined seems to show two kinds paraphyses, lanceolate and filiform.

17. Lachnella capitata (Peck) Seaver, comb. nov.

?Peziza crystallina Fuckel, Symb. Myc. 306. 1869.
Peziza capitata Peck, Ann. Rep. N. Y. State Mus. **30**: 60. 1878.
Trichopeziza capitata Sacc. Syll. Fung. **8**: 417. 1889.
?Dasyscypha crystallina Sacc. Syll. Fung. **8**: 440. 1889.
?Dasyscypha scintillans Massee, Brit. Fungus-Fl. **4**: 328. 1895.
Dasyscypha capitata Kanouse, Mycologia **39**: 646. 1947.

Apothecia very minute, sessile, at first globose and closed, finally expanding, externally clothed with a dense coat of white hairs; hymenium white, or yellowish; hairs hyaline with transmitted light, often terminated with a crystalline cap; asci cylindric, or subcylindric, reaching a length of 30 μ; spores cylindric, slender 5–7 μ long; paraphyses thick, longer than the asci, lanceolate.

On leaves of *Quercus.*

TYPE LOCALITY: Albany, New York, type on *Quercus alba.*

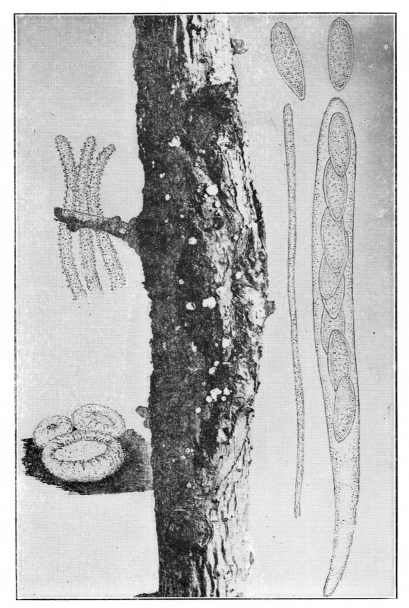

LACHNELLA WILLKOMMII

DISTRIBUTION: New York to Missouri, and Pennsylvania; also in Europe?

EXSICCATI: Rab.-Winter, Fungi Eu. *3168* (from Ohio); Thüm. Myc. Univ. *813* (from New York).

18. **Lachnella ciliaris** (Schrad.) Phill. Brit. Discom. 251.   1887.

Peziza ciliaris Schrad.; Gmel. Syst. Nat. **2**: 1453.   1791.
Peziza echinulata Auersw. Hedwigia **7**: 136.   1868.
Hyalopeziza ciliaris Fuckel, Symb. Myc. 298.   1869.
Lachnea ciliaris Gill. Champ. Fr. Discom. 68.   1880.
Erinella ciliaris Quél. Ench. Fung. 304.   1886.
Trichopeziza ciliaris Rehm, Ber. Naturh. Ver. Augsburg **26**: 64.   1881.
Lachnella echinulata Phill. Brit. Discom. 249.   1887.
Dasyscypha ciliaris Sacc. Syll. Fung. **8**: 443.   1889.
Lachnum echinulatum Rehm in Rab. Krypt.-Fl. **1**³: 876.   1893.

Apothecia minute, sessile, or very short-stipitate, subglobose and usually closed when dry, open when moist, white, densely clothed with white hairs; hymenium white; hairs hyaline with transmitted light, reaching a diameter of 6–7 μ and a length of 40–60 μ; asci clavate, reaching a length of 50–60 μ and a diameter of 8–9 μ, 8-spored; spores fusoid, partially 2-seriate, 2.5–3 $\times$ 15–20 μ; paraphyses sparse, filiform, reaching a diameter of 3 μ, lanceolate above.

On fallen leaves of *Quercus*.

TYPE LOCALITY: Europe.

DISTRIBUTION: Ohio; also in Europe.

EXSICCATI: Ellis, N. Am. Fungi *1312* (as *Peziza echinulata* Awd.).

19. **Lachnella pollinaria** (Cooke.) Seaver, comb. nov.

Peziza pollinaria Cooke, Bull. Buffalo Soc. Nat. Sci. **2**: 292.   1875.
Trichopeziza pollinaria Sacc. Syll. Fung. **8**: 416.   1889.
Lachnum pollinarium Cash, Jour. Wash. Acad. Nat. Sci. **29**: 48.   1939.

Apothecia subgregarious, minute, sessile, at first globose, then expanding, clothed with short, white hairs; asci subcylindric, reaching a length of 35–40 μ and a diameter of 5 μ; spores irregularly 2-seriate, fusoid, 1.5 $\times$ 7–8 μ; paraphyses stout, lanceolate, 3–4 μ in diameter.

On old oak leaves.

TYPE LOCALITY: New Jersey.

DISTRIBUTION: New Jersey to Georgia.

The original description of this species is brief and unsatisfactory. The above description is based on material in the Ellis collection. The species will bear more investigation.

20. **Lachnella tautilla** (Phill. & Hark.) Seaver, comb. nov.

Peziza tautilla Phill. & Hark. Bull. Calif. Acad. Sci. **1**: 21.   1884.
Dasyscypha tautilla Sacc. Syll. Fung. **8**: 445.   1889.

Apothecia scattered, stipitate, minute, white, cup-shaped, clothed to the base with slender, hyaline-white hairs; asci broad-clavate, 8-spored; spores ovoid, 1×4 μ; paraphyses filiform.

Among the hairs on the under side of living leaves of *Garrya elliptica*.

TYPE LOCALITY: Tamalpais, California.

DISTRIBUTION: Known only from the type locality.

EXSICCATI: Ellis & Ev. N. Am. Fungi *2041;* Rab.-Winter, Fungi Eu. *3468* (from California).

21. **Lachnella virginella** (Cooke) Seaver, comb. nov.

Peziza virginella Cooke; Cooke & Ellis, Grevillea **4**: 178.   1876.
Dasyscypha virginella Sacc. Syll. Fung. **8**: 444.   1889.
Lachnum virginellum Zeller, Mycologia **26**: 293.   1934.

Apothecia scattered, long-stipitate, externally white-tomentose; hymenium whitish; hairs hyaline, reaching a length of 40–50 μ; asci clavate, reaching a length of 40 μ and a diameter of 4–5 μ, short-stipitate, 8-spored; spores fusoid, 1×6–7 μ; paraphyses lanceolate.

On leaves of species of *Vaccinium*.

TYPE LOCALITY: New Jersey.

DISTRIBUTION: New Jersey to Pennsylvania and Louisiana.

EXSICCATI: Ellis & Ev. N. Am. Fungi *2144*.

22. **Lachnella Gaultheriae** (Ellis & Ev.) Seaver, comb. nov.

Dasyscypha Gaultheriae Ellis & Ev. Erythea **1**: 199.   1893.
Lachnum Gaultheriae Zeller, Mycologia **26**: 292.   1934.

Apothecia seated on large, semicircular, grayish to brownish spots with a dark-purple border, stipitate, nearly closed at first, then opening and becoming cup-shaped, pale-orange, clothed externally with a dense coat of short, white hairs; hymenium pale-orange, reaching a diameter of 1 mm.; stem short, rather less than the diameter of the apothecium; asci clavate-cylindric, reaching a length of 40 μ and a diameter of 3–4 μ, 8-spored; spores 2-seriate, ellipsoid, or clavate-ellipsoid, 1.25×4–5 μ.

On leaves of *Gaultheria Shallon*.

TYPE LOCALITY: Seattle, Washington.

DISTRIBUTION: Washington to Oregon and California.

According to Zeller (Mycologia **26**: 293) this species is very closely related to *Trichopeziza echinulata* Rehm, Ascom. *259*.

23. **Lachnella Halesiae** Cash, Jour. Wash. Acad. Sci. **29**: 48. 1939.

Apothecia scattered over the lower surface of the host, sessile, or substipitate, at first subglobose with the margin inrolled and the hymenium hidden, then expanding and becoming subdiscoid, .2–.6 mm. in diameter, externally clothed with white hairs; hymenium translucent, apricot-colored; hairs hyaline, septate, delicately roughened, not enlarged above; asci broadcylindric, 8-spored, reaching a length of 47–50 μ and a diameter of 7–8 μ; spores obliquely 1-seriate, fusoid, 1.5–2 $\times$ 12–15 μ; paraphyses lanceolate, longer than the asci 2.5–4 μ thick.

On leaves of *Halesia carolina*.

TYPE LOCALITY: Athens, Georgia.

DISTRIBUTION: Known only from the type locality.

ILLUSTRATIONS: Jour. Wash. Acad. Sci. **29**: 47, *f. 2*.

24. **Lachnella pygmaea** (Fries) Phill. Brit. Discom. 242. 1887. (PLATE 125, FIG. 2.)

Peziza pygmaea Fries, Syst. Myc. **2**: 79. 1822.
Helotium luteolum Currey, Trans. Linn. Soc. **24**: 153. 1864.
Helotium rhizophilum Fuckel, Fungi Rhen. *1598*. 1865.
Ciboria rhizophila Fuckel, Symb. Myc. 312. 1869.
Helotium pygmaeum Karst. Not. Soc. Fauna Fl. Fenn. **11**: 214. 1871.
Lachnea pygmaea Gill. Champ. Fr. Discom. 71. 1880.
Peziza nuda Phill. & Plow. Grevillea **8**: 101. 1880.
Helotium affinissimum Peck, Ann. Rep. N. Y. State Mus. **33**: 32. 1883.
Erinella pygmaea Quél. Ench. Fung. 303. 1886.
Hymenoscypha Hedwigii Phill. Brit. Discom. 130. 1887.
Hymenoscypha rhizophila Phill. Brit. Discom. 144. 1887.
Lachnella nuda Phill. Brit. Discom. 247. 1887.
Lachnella luteola Phill. Brit. Discom. 247. 1887.
Helotium rhizogenum Ellis & Ev. Jour. Myc. **4**: 100. 1888.
Helotium Phillipsii Sacc. Syll. Fung. **8**: 220. 1889.
Phialea Hedwigii Sacc. Syll. Fung. **8**: 260. 1889.
Phialea affinissima Sacc. Syll. Fung. **8**: 272. 1889.
Dasyscypha pygmaea Sacc. Syll. Fung. **8**: 436. 1889.
Dasyscypha luteola Sacc. Syll. Fung. **8**: 440. 1889.
Hymenoscypha flexipes Cooke & Phill. Grevillea **19**: 106. 1891.
Phialea flexipes Sacc. Syll. Fung. **10**: 9. 1892.

Ciboria pygmaea Rehm in Rab. Krypt.-Fl. **1**³: 760.   1893.
Helotium Hedwigii Massee, Brit. Fungus-Fl. **4**: 243.   1895.
Helotium nudum Massee, Brit. Fungus-Fl. **4**: 498.   1895.
Calycina rhizogena Kuntze, Rev. Gen. Pl. **3**³: 449.   1898.
Hymenoscyphus affinissimus Kuntze, Rev. Gen. Pl. **3**³: 485.   1898.
Lachnum Hedwigii Bres. Ann. Myc. **1**: 121.   1903.
Lachnum pygmaeum Bres. Ann. Myc. **1**: 121.   1903.
Ciboria carbonaria Feltg. Vorst. Pilz-Fl. Luxenb. **1**³: 44.   1903.
Helotium carbonarium Boud. Hist. Class. Discom. Eu. 113.   1907.
Helotium flexipes Boud. Hist. Class. Discom. Eu. 114.   1907.
Hymenoscypha nuda Boud. Hist. Class. Discom. Eu. 122.   1907.
Helotium subrubescens Rehm, Ann. Myc. **7**: 524.   1909.
Helolachnum aurantiacum Torrend, Broteria **9**: 53.   1910.
Ciboria subrubescens Dodge, Trans. Wis. Acad. Sci. **17**: 1033.   1914.
Lachnum rhizophilum Vel. Monog. Discom. Bohem. **1**: 258.   1934.

Apothecia gregarious or cespitose, typically occurring in a few cespitose clusters of five to ten each, surrounded by a few growing singly, stipitate, 2–7 mm. high, 2–4 mm. in diameter, at first infundibuliform, then spreading and plane; stem slender, flexuous, usually somewhat thickened just below the disc, pale-yellow to flesh-color, or dull-orange, more or less whitish-puberulent, often appearing smooth in the dried condition, the margin rather obtuse, when dried becoming somewhat elevated above the hymenium, finely and obscurely puberulent; hymenium pale-yellow to deep-yellow, often varying toward orange, or apricot, retaining the color on drying, or becoming more ochraceous; hairs 20–50 μ long, clavate, rough, hyaline-white, 4–6 μ in diameter, 1–2-septate; asci small, cylindric, 60–75 × 4.5–6 μ; spores 2-seriate, 1-celled, narrow, broadest just above the middle, slightly tapering toward a point, at the lower end, round, or only slightly pointed above, straight, 1.9–2.4 × 7–11 μ; paraphyses more or less lance-pointed, protruding above the asci, septate, 3–4.5 μ in diameter.

Apothecia arising at ground level on partly buried plant debris of all sorts, especially on roots and rhizomes of grasses and other herbaceous plants and on limbs of both frondose and coniferous trees.

TYPE LOCALITY: Europe.

DISTRIBUTION: Throughout northern United States and southern Canada; also in Europe.

ILLUSTRATIONS: Trans. Linn. Soc. **24**: *pl. 25, f. 11–12, 18;* **25**: *pl. 55, f. 7–18;* Bull. Lab. Nat. Hist. State Univ. Iowa **6**: *pl. 24, f. 3 a–d.*

EXSICCATI: Rehm, Ascom. *1852* (from Wisconsin).

The synonymy is based largely on the studies of Dr. W. L. White.

25. **Lachnella diminuta** (Rob.) Phill. Brit. Discom. 253. 1887.

Peziza diminuta Rob.; Desm. Pl. Crypt. Fr. *1538;* Ann. Sci. Nat. III. **8**: 185. 1847.
Lachnea diminuta Gill. Champ. Fr. Discom. 71. 1880.
Dasyscypha diminuta Sacc. Syll. Fung. **8**: 449. 1889.

Apothecia scattered, or crowded, short-stipitate, at first globose, then expanded and hemispherical, minute, .5 mm. in diameter, whitish-tomentose; hymenium concave, yellowish, or orange; asci cylindric-clavate, 8-spored; spores long-ellipsoid, $2 \times 12 \ \mu$; paraphyses filiform.

On dry culms of *Juncus* and grass stems.

TYPE LOCALITY: Europe.

DISTRIBUTION: Michigan and North Dakota; also in Europe.

EXSICCATI: Brenckle, Fungi Dak. *527.*

26. **Lachnella agrostina** (Peck) Seaver, comb. nov.

Peziza agrostina Peck, Ann. Rep. N. Y. State Mus. **29**: 55. 1878.
Trichopeziza agrostina Sacc. Syll. Fung. **8**: 421. 1889.

Apothecia scattered, minute, less than 1 mm. in diameter, subsessile, subglobose, externally hairy, of a dull-pinkish color; hymenium pallid, or cream-colored; hairs about the margin bent inward when moist, usually with longer, subulate, whitish points, the others not subulate, often rough and septate; asci cylindric, 8-spored; spores elongated, $6–8 \ \mu$ in length; paraphyses longer than the asci, lanceolate.

On dead stems of *Calamagrostis canadensis*.

TYPE LOCALITY: West Albany, New York.

DISTRIBUTION: Known only from the type locality.

Part of the type collection is in The New York Botanical Garden.

27. **Lachnella Arundinariae** (Cash) Seaver, comb. nov.

Lachnum Arundinariae Cash, Jour. Wash. Acad. Sci. **30**: 301. 1940.

Apothecia sessile, scattered, minute, .1–.2 mm. in diameter, translucent when moist, clothed with white hairs, margin fimbriate, at first subglobose, then becoming discoid; hymenium plane or nearly so, hyaline to slightly pink; hairs slightly roughened,

3 μ in diameter; asci cylindric, 8-spored, reaching a length of
27–33 μ and a diameter of 3–3.5 μ; spores narrow-clavate,
2-seriate, .7–1 × 5–6.5 μ; paraphyses lanceolate, about 3 μ in
diameter.

On leaves of *Arundinaria tecta*.

TYPE LOCALITY: Athens, Georgia.

DISTRIBUTION: Known only from the type locality.

ILLUSTRATIONS: Jour. Wash. Acad. Sci. **30**: 303. *f. 2*.

28. **Lachnella Solenia** (Peck) Seaver, comb. nov.   (PLATE 124.)

Peziza Solenia Peck, Ann. Rep. N. Y. State Mus. **25**: 99.   1873.
Peziza soleniaeformis Ellis & Ev. Jour. Myc. **4**: 55.   1888.   Not Berk. &
    Curt.   1875.
Dasyscypha soleniiformis Sacc. Syll. Fung. **8**: 436.   1889.
Solenopezia Solenia Sacc. Syll. Fung. **8**: 477.   1889.

Apothecia gregarious, sessile, minute, not exceeding .3 mm.
in diameter, short-cylindric, a little longer than broad, constricted
at the mouth, externally clothed with brown hairs but with a
white margin around the mouth; hymenium not much exposed;
hairs consisting of two kinds, those about the side of the apothecium dark-brown, clavate, septate, slightly roughened and
knotted reaching a diameter of 5–6 μ and a length of 60–80 μ,
the marginal hairs similar in size and form but hyaline tipped
and covered with minute granules; asci clavate, reaching a
length of 65 μ and a diameter of 10 μ; spores 2-seriate, fusoid,
hyaline, becoming 1-septate, usually with four small oil-drops,
about 3–4 × 12–13 μ; paraphyses filiform slightly enlarged
above.

On dead stems of *Eupatorium ageratoides*, and on rotten wood.

TYPE LOCALITY: Watkins Glen, New York.

DISTRIBUTION: New York.

EXSICCATI: Ellis, N. Am. Fungi *384;* Thüm. Myc. Univ. *1114*.

The above description and accompanying illustrations were
drawn from material in the herbarium of The New York Botanical Garden which is apparently part of the type collection.

29. **Lachnella inquilina** Karst. Acta Soc. Fauna Fl. Fenn. **2**⁶: 132.
    1885.

Helotium inquilinum Karst. Myc. Fenn. **1**: 147.   1871.
Trichopeziza inquilina Sacc. Syll. Fung. **8**: 424.   1889.
Pezizella inquilina Rehm in Rab. Krypt.-Fl. **1**³: 675.   1892.
Lachnum inquilinum Schröt. Krypt.-Fl. Schles. **3**²: 96.   1893.

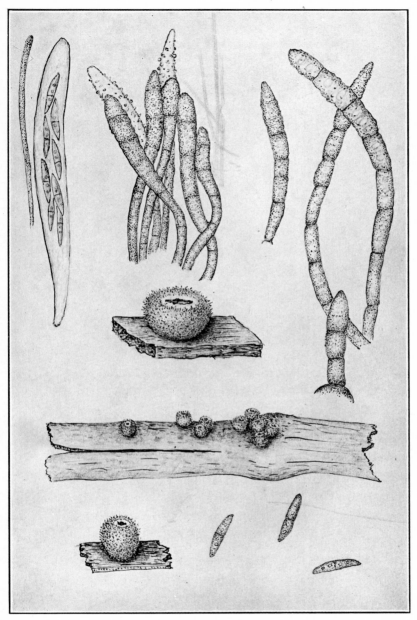

LACHNELLA SOLENIA

Apothecia scattered, or gregarious, short-stipitate, at first globose and closed, later becoming expanded, reaching a diameter of .3–.5 mm., entirely white, or becoming brownish with age, clothed externally with white hairs; hymenium plane, or slightly concave; hairs short, not exceeding 50 μ in length and reaching a diameter of 3–4 μ, hyaline, rarely septate, more or less irregularly curved, or flexuous, the tips obtuse, smooth, or granular; asci clavate, narrowed at the tips, reaching a length of 35–45 μ and a diameter of 5–6 μ, 8-spored; spores 2-seriate, hyaline, clavate-cylindrical, 2×8–10 μ; paraphyses lanceolate-acute at their tips, 3 μ thick.

On decaying rootstock of *Equisetum hyemale*, on the ground in wet places.

TYPE LOCALITY: Europe.

DISTRIBUTIONS: New York and Ontario; also in Europe.

Our only knowledge of this is from Durand's report in Bull. Torrey Club **29**: 464. 1902.

30. Lachnella pteridicola Seaver, nom. nov.

Hyalopeziza Pteridis Kanouse, Mycologia **39**: 660. 1947. Not *Lachnella Pteridis* Phill. 1887.

Apothecia solitary, sessile to substipitate, about .2 mm. in diameter, cup-shaped, translucent-white; hairs hyaline-white, rough, reaching a length of 75 μ and a diameter of 6 μ; asci cylindric-clavate, reaching a length of 30–35 μ and a diameter of 4.5 μ, 8-spored; spores bacilloid, .75–1 $\times$ 4–5.5 μ; paraphyses filiform.

On stems of *Pteris* sp.

TYPE LOCALITY: Lake Quinault, Washington.

DISTRIBUTION: Known only from the type locality.

31. Lachnella aspidicola (Berk. & Br.) Phill. Brit. Discom. 245· 1887.

Peziza aspidicola Berk. & Br. Ann. Mag. Nat. Hist. II. **13**: 465. 1854.
Mollisia aspidicola Quél. Bull. Soc. Bot. Fr. **26**: 234. 1879.
Dasyscypha aspidicola Sacc. Syll. Fung. **8**: 451. 1889.
Helotium aspidicolum Rehm, Hedwigia **20**: 35. 1881.

Apothecia scattered, or gregarious, sessile, or short-stipitate, at first rounded and closed, then expanded and cup-shaped, reaching a diameter of .4 mm. and about the same in height, white-floccose; hymenium white, or yellowish-white; asci clavate, reaching a length of 25–35 μ and a diameter of 5–6 μ, 8-spored;

spores long-ellipsoid, or subclavate, 1–2 × 5–8 μ; paraphyses filiform, 3 μ in diameter.

On dead stems of ferns, *Aspidium* and *Adiantum* and undetermined species.

TYPE LOCALITY: Europe.

DISTRIBUTION: Reported from Michigan and New Jersey; also in Europe.

32. **Lachnella Dicranopteridis** (Seaver & Whetzel) Seaver, comb. nov.

Dasyscypha Dicranopteridis Seaver & Whetzel, Sci. Sur. Porto Rico. **8**: 74. 1926.

Apothecia scattered, shortly stipitate, or subsessile, externally reddish, densely clothed with white hairs, reaching a diameter of .5 mm.; hairs cylindric, roughened on the outside; asci clavate, 8-spored, reaching a length of 40–50 μ and a diameter of 5 μ, tapering into a rather long, stem-like base; spores fusiform, the ends acute and almost bristle-like, not distinctly septate but often with suggestions of septa, 2 × 8–10 μ; paraphyses not indicated.

On fronds of *Dicranopteris pectinata*.

TYPE LOCALITY: Porto Rico.

DISTRIBUTION: Known only from the type locality.

33. **Lachnella Rhytismatis** Phill. Grevillea **8**: 101. 1880.

Dasyscypha Rhytismatis Sacc. Syll. Fung. **8**: 453. 1889.

Apothecia minute, gregarious, white, at first globose and closed, then expanding, clothed with white hairs; hymenium white; hairs short, hyaline with transmitted light, bearing at their tips globular or crystalline heads; stem rather long, hairy to the base; asci cylindric, 8-spored; spores fusoid, 1 × 3–5 μ; paraphyses as broad as the asci but longer, lanceolate.

On *Rhytisma acerinum*, on leaves of *Acer*.

TYPE LOCALITY: Europe.

DISTRIBUTION: Maryland; also in Europe.

EXSICCATI: Ellis & Ev. Fungi Columb. *1819*.

The original spelling of the specific name was "*Rhytismae*" but was corrected by Saccardo as indicated above.

34. **Lachnella Acanthonitschkeae** (Cash & Davidson) Seaver, comb. nov.

Dasyscypha Acanthonitschkeae Cash & Davidson, Mycologia **32**: 730. 1940.

Apothecia scattered, one to eight on a single stroma of the host fungus, nearly globose at first, then cup-shaped to subglobose, with a circular opening, white, .1–.2 mm. in diameter and height, externally white-tomentose, the hymenium translucent-white; asci cylindric, 8-spored, reaching a length of 22–24 μ and a diameter of 3–3.5 μ; spores 1-seriate, ellipsoid, 1.5–2 × 3.5–4 μ; paraphyses filiform, hyaline, simple, not enlarged at their tips.

On stromata of *Acanthonitschkea coloradensis*.

TYPE LOCALITY: Grand Mesa, Colorado.

DISTRIBUTION: Known only from the type locality.

ILLUSTRATIONS: Mycologia **32**: 729, *f. 1, E, F.*

35. **Lachnella succina** (Phill.) Seaver, comb. nov.

Peziza succina Phill. Grevillea **5**: 116. 1887.
Dasyscypha succina Sacc. Syll. Fung. **8**: 458. 1889.

Apothecia stipitate, scattered, or crowded, at first globose, then expanded, concave, or nearly plane, reaching a diameter of 5 mm., the stem short and brownish, externally clothed with a mass of rather poorly developed, sulphur-yellow hairs giving the whole exterior of the apothecium a decidedly yellow color, the individual hairs being yellow, or subhyaline; hymenium flesh-colored; asci clavate, reaching a length of 80 μ and a diameter of 10–12 μ; spores ellipsoid, 3–5 × 10–14 μ; paraphyses cylindric, slightly enlarged above.

On dead wood of *Quercus* and (*Celtis* ?).

TYPE LOCALITY: Blue Cannon, Sierra Nevada Mountains, California.

DISTRIBUTION: California and Utah.

ILLUSTRATIONS: Grevillea **5**: *pl. 89, f. 12.*

EXSICCATI: Ellis, N. Am. Fungi. *839.*

36. **Lachnella albolutea** (Pers.) Karst. Acta Soc. Fauna Fl. Fenn. **2**[6]: 132. 1885.

Peziza sulphurea albolutea Pers. Syn. Fung. 649. 1801.
Peziza variecolor Fries, Syst. Myc. **2**: 100. 1822.
?Peziza turbinulata Schw. Trans. Am. Phil. Soc. II. **4**: 173. 1832.
Tapesia variecolor Fuckel, Symb. Myc. Nacht. **1**: 336. 1871.
Helotium aboluteum Karst. Myc. Fenn. **1**: 160. 1871.
Lachnea variecolor Gill. Champ. Fr. Discom. 83. 1880.
Lachnella variecolor Phill. Brit. Discom. 259. 1887.
Dasyscypha turbinulata Sacc. Syll. Fung. **8**: 456. 1889.
?Lachnella citrina Peck, Ann. Rep. N. Y. State Mus. **46**: 35. 1893.
Dasyscypha sulphuricolor Peck, Bull. N. Y. State Mus. **157**: 25. 1911.

Apothecia scattered, sessile, at first rounded, expanding and becoming subdiscoid, reaching a diameter of 2 mm., externally clothed with sulphur-yellow hairs; hymenium slightly concave, whitish, or slightly yellowish; hairs when fresh sulphur-yellow, becoming darker with age, slightly roughened on the outside, often a little enlarged above; asci cylindric to clavate, reaching a length of 40–50 μ and a diameter of 6 μ, 8-spored; spores ellipsoid to slightly clavate, straight, or curved, about 2 × 7–8 μ; paraphyses filiform.

On old wood.

TYPE LOCALITY: Europe.

DISTRIBUTION: Iowa to Missouri and Cuba; also in Europe.

ILLUSTRATIONS: Pers. Ic. & Descr. *pl. 8, f. 4–5.;* Gill. Champ. Fr. Discom. *pl. 67, f. 7.*

EXSICCATI: Ellis, N. Am. Fungi *364* (as *Peziza turbinulata* Schw.).

37. **Lachnella sulphurea** (Pers.) Quél. Ench. Fung. 315.   1886.
(PLATE 125, FIG. 1.)

Peziza sulphurea Pers. Tent. Disp. Fung. 33.   1797.
Trichopeziza sulphurea Fuckel, Symb. Myc. 296.   1869.
Peziza sulphurea leucophaea Pers. Myc. Eu. **1**: 250.   1871.
Lachnum leucophaeum Karst. Myc. Fenn. **1**: 177.   1871.
Trichopeziza leucophaea Rehm, Ascom. *65.*   1872.
Lachnea sulfurea Gill. Champ. Fr. Discom. 81.   1880.
Peziza cenangioides Ellis, Bull. Torrey Club **8**: 123.   1881.
Lachnella cenangioides Sacc. Syll. Fung. **8**: 396.   1889.
Lachnum sulfureum Rehm in Rab. Krypt-Fl. **1**³: 891.   1893.

Apothecia scattered, sessile but narrowed below approaching substipitate, at first closed, becoming expanded and plane, margin slightly raised, reaching a diameter of 1–1.5 mm., externally densely clothed with sulphur-yellow hairs; hymenium concave, or plane, whitish, or creamy-white; hairs straight, or slightly curved, septate, cylindric, either blunt, or gradually tapering with semiacute apices, minutely roughened, pale-yellowish, or occasionally yellowish-brown; asci clavate, reaching a length of 50–75 μ and a diameter of 5 μ, 8-spored; spores slender-ellipsoid, straight, or slightly curved, or fusoid, 2-seriate, 1.5–2 × 8–10 μ or rarely 15–16 μ long; paraphyses lanceolate, about as thick as the ascus.

On dead herbaceous stems.

TYPE LOCALITY: Europe.

1. LACHNELLA SULPHUREA
2. LACHNELLA PYGMAEA

DISTRIBUTION: New York to Iowa; also in Europe. Probably widely distributed.

ILLUSTRATIONS: Boud. Ic. Myc. *pl. 513;* Bull. Torrey Club **27**: *pl. 31, f. 12.*

EXSICCATI: Ellis & Ev. N. Am. Fungi *2631;* Brenckle, Fungi Dak. *664.*

Phillips (Brit. Discom. 265.) treats *leucophaea* as a synonym of the above and states: "certainly not worthy of specific rank."

38. **Lachnella canadensis** (Ellis & Dearn.) Seaver, comb. nov.

Pseudohelotium canadense Ellis & Dearn. Proc. Canad. Inst. **14**: 89. 1897.

Apothecia scattered, short-stipitate, or subsessile, whitish, with a yellowish tinge, about 1 mm. in diameter, pubescent, the margin fringed with tufts of short, pale hairs and when dry the opposite sides rolled together so as to become elliptical, or triangular in outline; hymenium concave, yellowish; asci cylindric-clavate, reaching a length of 40 μ and a diameter of 4 μ; spores 2-seriate, long-ellipsoid, 1.5–2 $\times$ 8–12 μ; paraphyses filiform.

On dead stems of *Arctium Lappa.*

TYPE LOCALITY: London, Canada.

DISTRIBUTION: Known only from the type locality.

39. **Lachnella Ivae** (Rehm) Seaver, comb. nov.

Dasyscypha Ivae Rehm, Ann. Myc. **11**: 397. 1913.

Apothecia gregarious, sessile, at first globose and closed, then cup-shaped, finally expanded, constricted at the base, externally yellow, then brownish, .5–1.3 mm. in diameter, the margin clothed with straight, septate, hyaline, then yellowish hairs 4–5 $\times$ 150 μ; asci cylindric-clavate, reaching a length of 70 μ and a diameter of 9 μ, 8-spored; spores ellipsoid, 2-seriate, 3 $\times$ 12–14 μ; paraphyses filiform, slightly enlarged above 4 μ in diameter.

On dead stems of *Iva xanthifolia.*

TYPE LOCALITY: Kulm, N. Dakota.

DISTRIBUTION: Known only from the type locality.

40. **Lachnella pulverulenta** (Lib.) Quél. Ench. Fung. 316. 1886.

Peziza pulverulenta Lib. Pl. Crypt. Ard. *125.* 1832.
Trichopeziza pulverulenta Fuckel, Symb. Myc. 297. 1869.
Peziza solfatera Cooke & Ellis, Grevillea **7**: 7. 1878.
Lachnella solfatera Phill. Brit. Discom. 246. 1887.
Dasyscypha pulverulenta Sacc. Syll. Fung. **8**: 462. 1889.
Dasyscypha solfatera Sacc. Syll. Fung. **8**: 463. 1889.

Apothecia scattered, sessile, or very short-stipitate, lemon-yellow, at first globose, opening and becoming cup-shaped, .2–1 mm. in diameter, externally clothed with hairs; hymenium concave, whitish to yellow; hairs reaching a length of 30–90 μ and a diameter of 4–5 μ, rough, yellow, often with yellowish-brown masses at their apices; asci cylindric-clavate, reaching a length of 30–50 μ and a diameter of 3–4 μ, 8-spored; spores fusoid, 1.5–2 × 5–8 μ; paraphyses filiform, 2 μ in diameter.

On fallen pine needles.

TYPE LOCALITY: Europe.

DISTRIBUTION: New Jersey and Michigan; also in Europe.

EXSICCATI: Ellis, N. Am. Fungi *439*.

The variety *purpurascens* has been reported by Dr. B. Kanouse on fir needles from Olympic National Park (Mycologia **39**: 647. 1947).

41. Lachnella Gleicheniae (Cash) Seaver, comb. nov.

Lachnum Gleicheniae Cash, Mycologia. **30**: 105. 1938.

Apothecia developing in the sunken areas of the stems, sulphur-yellow to pale-orange, fading with age, stipitate, at first globose, then cup-shaped, .2–.7 mm. in diameter, clothed with yellow, strongly verrucose hairs 3–5 μ in diameter; stem smooth, .1 mm. in diameter and .1–.4 mm. high; asci cylindric, or sub-cylindric, reaching a length of 40–50 μ and a diameter of 3–5 μ, 8-spored; spores irregularly 2-seriate, fusiform, or fusoid, straight, or slightly unequal sided, 1–2 × 9–11 μ; paraphyses extending beyond the asci, lanceolate, 3–3.5 μ thick.

On stipes of *Gleichenia*, and stems of tree fern.

TYPE LOCALITY: Hawaii.

DISTRIBUTION: Known only from the type locality.

ILLUSTRATIONS: Mycologia **30**: 99, *f. 2*.

This is very close to *L. Dicranopteridis* (Seaver & Whetzel) Seaver.

42. Lachnella atropurpurea (Durand) Seaver, comb. nov.

Lachnum atropurpureum Durand, Jour. Myc. **10**: 100. 1904.

Apothecia gregarious, or scattered, stipitate, occasionally several apothecia at the summit of a common stem, pale-purple, reaching 1 mm. in diameter, externally clothed with pale-purple hairs; hymenium concave, pale-purple; stem slender, the length equaling the diameter of the apothecium; hairs cylindrical, smooth, closely septate, reaching a length of 80 μ and a diameter of 5 μ, pale-purple with transmitted light; asci clavate-cylindric, reaching a length of 40–50 μ and a diameter of 5–6 μ, 8-spored; spores 1-seriate, hyaline, smooth, ellipsoid, 2–3 × 6–8 μ; paraphyses narrowly lanceolate above, 3–4 μ thick.

On *Eucalyptus*.

TYPE LOCALITY: Stanford University, California.

DISTRIBUTION- Known only from the type locality.

43. **Lachnella viridicoma** (Peck) Seaver, comb. nov.

Peziza viridicoma Peck, Ann. Rep. N. Y. State Mus. **31**: 46. 1877.
Trichopeziza viridicoma Sacc. Syll. Fung. **8**: 414. 1889.

Apothecia minute, reaching a diameter of .3 to .5 mm. sessile, or tapering below into a short, stem-like base, externally pale-green, clothed with a dense covering of hairs; hymenium becoming plane or nearly so, whitish, or slightly yellowish; hairs slender, reaching a length of 40 μ and a diameter of 1.5–2 μ at the base, gradually tapering above, subhyaline but more or less agglutinated in a greenish-yellow matrix; asci clavate, reaching a length of 80 μ and a diameter of 10 μ; spores 2-seriate, ellipsoid, usually slightly curved, with two distinct oil-drops, 5–6 × 12–15 μ; paraphyses very slender, hyaline.

On rotten wood.

TYPE LOCALITY: Sandlake, New York.

DISTRIBUTION: New York and Toronto.

44. **Lachnella microspora** (Kanouse) Seaver, sp. nov.

Lachnella tricolor var. *microspora* Kanouse, Papers Mich. Acad. Sci. **20**: 73. 1935.

Apothecia gregarious, stipitate, externally pale-blue, or bluish-green, densely hairy, .5 mm. in diameter; hymenium concave, yellowish; hairs slender, nearly hyaline with transmitted light, bluish-green to the unaided eye; asci clavate, 8-spored; spores 1–1.5 × 5–5.5 μ; paraphyses filiform.

On decaying wood.

TYPE LOCALITY: Cascade Glen, Ann Arbor, Michigan.

DISTRIBUTION: Known only from the type locality.

Dr. B. Kanouse recorded this variety as indicated above. The difference in size of the spores would, in the opinion of the writer, give it specific rank. It is one of the few species which has greenish hairs and in this is unique.

45. **Lachnella pulveracea** (Alb. & Schw.) Seaver, comb. nov.

Peziza pulveracea Alb. & Schw. Consp. Fung. 342.   1805.
Cenangium pulveraceum Fries, Syst. Myc. **2**: 181.   1822.
Tympanis seriata Schw. Trans. Am. Phil. Soc. II. **4**: 237.   1832.
Cenangium microspermum Sacc. & Ellis; Sacc. Michelia **2**: 571.   1882.
Lachnum viridulum Massee & Morgan; Morgan, Jour. Myc. **8**: 187.   1902.
    Not *Dasyscypha viridula* (Schrad.) Sacc. Syll. Fung. **8**: 437.   1889.
Dasyscypha chlorella Seaver, Mycologia **3**: 63.   1911.

Apothecia gregarious, occasionally cespitose, at first rounded and closed, gradually expanding and becoming cup-shaped (mouth constricted in dried specimens), reaching a diameter of .5 mm., clothed with greenish hairs, narrowed into a short, stout stem the base of which is surrounded by a radiating mass of white mycelium, or nearly sessile, pale-greenish; hymenium concave, dark-green; hairs irregularly tapering above into a semi-acute apex, reaching a length of 60–80 μ and a diameter of 3–4 μ; asci cylindric or subcylindric, 8-spored, reaching a length of 45–60 and a diameter of 6 μ; spores 2-seriate, smooth, hyaline, cylindric, or slightly curved, 2 × 7–9 μ; paraphyses filiform.

On *Ilex opaca* and old wood of various kinds.

TYPE LOCALITY: Europe.

DISTRIBUTION: New Jersey to Alabama and Colorado; also in Europe.

ILLUSTRATIONS: Alb. & Schw. Consp. Fung. *pl. 8, f. 2.*

While the hairs are greenish they fade in old specimens and become cinereous.   It seems to us that our plants are identical with *Peziza pulveracea* although authentic material is lacking.

46. **Lachnella arida** (Phill.) Seaver, comb. nov.   (PLATE 126.)

Peziza arida Phill. Grevillea **5**: 117.   1877.
Dasyscypha flavovirens Bres. Fungi Trid. **1**: 92.   1887.
Dasyscypha arida Sacc. Syll. Fung. **8**: 455.   1889.
Lachnum Engelmani Earle in Greene, Pl. Baker. **1**: 25.   1901.

Apothecia scattered, very short-stipitate, or apparently sessile, at first globose, then expanded but often irregularly angular, or subhysteriform, especially when dry, excipular cells dark-brown, giving rise to a dense covering of hairs, reaching a diameter of 4–8 mm.; hymenium concave, yellowish; hairs cylindrical, brown, septate, minutely roughened on the outside, the ends blunt or semiacute, reaching a length of 200–300 μ and a diameter of 5 μ; asci clavate, 8-spored, reaching a length of 50 μ

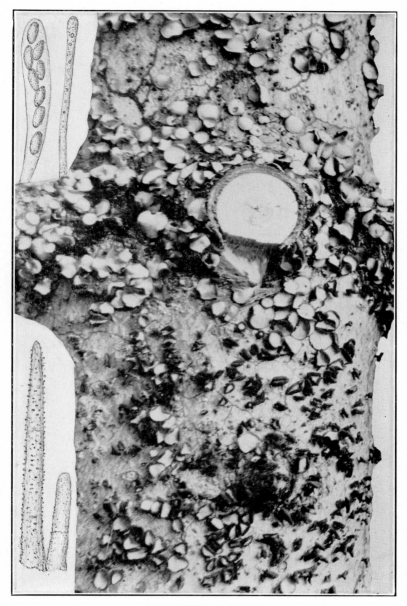

LACHNELLA ARIDA

and a diameter of 5 μ; spores partially 2-seriate, ellipsoid, 4 × 7 μ; paraphyses filiform.

On bark of conifers (*Abies, Larix, Picea*), or more rarely on the leaves.

TYPE LOCALITY: Blue Canon, Sierra Nevada Mountains, California.

DISTRIBUTION: Newfoundland to Alberta and south to Colorado and California; also in Europe.

ILLUSTRATIONS: Grevillea 5: *pl. 89, f. 13;* Bres. Fungi. Trid. *pl. 104, f. 1.*

EXSICCATI: Ellis & Ev. N. Am. Fungi *2146* (as *Peziza fusco-sanguinea*); Fungi Columb. *1223, 1311; 1414;* Clements, Crypt. Form. Colo. *81* (as *Dasyscypha cerina* (Pers.) Fuckel); W. B. Cooke, Mycobiota N. Am. *66.*

One of the commonest species in the Rocky Mountains on spruces and firs, and conspicuous by reason of the bright-yellow color of the hymenium and distinguished from most other conifer-inhabiting species, by the dark exteriors of the apothecia.

47. **Lachnella fuscosanguinea** (Rehm) Karst. Medd. Soc. Fauna Fl. Fenn. **16**: 16.   1888.

Dasyscypha fuscosanguinea Rehm, Ascom. *112;* Ber. Naturh. Ver. Augsburg **26**: 30.   1881.
Lachnella Pini Brunch. Bergens Mus. Aarb. **1892**[8]: 8.   1892.
Dasyscypha Pini Hahn & Ayers, Mycologia **26**: 487.   1934.

Apothecia scattered, or occasionally several in close contact, short-stipitate, at first rounded, becoming shallow cup-shaped with the margin slightly upturned, reaching a diameter of 2–3 mm., becoming hysteriform, or angular when dry, excipular cells dark-brown, giving rise to a dense covering of brown hairs; hymenium bright-yellowish; hairs cylindrical, brown, septate, externally minutely roughened, reaching a diameter of 4–5 μ and a length of 150–200 μ; asci cylindric reaching a length of 90–100 μ and a diameter of 9 μ; spores 1–2-seriate, ellipsoid with the ends narrowed, 5 × 12–15 μ; paraphyses clavate, reaching a diameter of 3 μ at their apices.

On bark of *Pinus Murrayana*, and other species of *Pinus*.

TYPE LOCALITY: Europe.

DISTRIBUTION: Colorado and Michigan; also in Europe.

ILLUSTRATIONS: Rab. Krypt.-Fl. **1**[3]: 827, *f. 1–4;* Mycologia **26**: *pl. 53.*

Our material agrees with Rehm's description except the color which is bright-yellow rather than rose-red. It differs from *Lachnella arida* with which it is thought by some to be identical, in its larger spores and asci. The two forms are easily confused.

48. Lachnella populicola Seaver, sp. nov.

Apothecia thickly gregarious, sessile, or subsessile, at first globose, becoming expanded with the margin incurved, reaching a diameter of 1–1.5 mm., externally clothed with brown hairs; hymenium pale-yellow; hairs blunt, septate, pale-brown, externally minutely roughened, reaching a length of 80–90 μ and a diameter of 5–6 μ; asci clavate reaching a length of 50–55 μ and a diameter of 5–6 μ, 8-spored; spores 2-seriate, fusoid, about 2 × 8–9 μ; paraphyses lanceolate, reaching a diameter of 4 μ.

Apotheciis sessilis vel subsessilis, 1–1.5 mm. diam., extus pilis brunneis vestitis; ascis clavatis, 5–6 × 80–90 μ, 8-sporis; sporis fusoideis, 2 × 8–9 μ; paraphysibus lanceolatis.

On old wood of *Populus tremuloides*.
TYPE LOCALITY: Tolland, Colorado.
DISTRIBUTION: Colorado.

This species closely resembles *L. populina* but differs in its smaller spores and asci and in having lanceolate paraphyses. From external characters the two can scarcely be distinguished.

49. Lachnella populina Seaver, sp. nov.

Apothecia thickly gregarious, sessile, or subsessile, at first globose, becoming expanded but with the margin incurved, when dry either rounded or hysteriform, reaching a diameter of 1–1.5 mm., externally densely clothed with brown hairs; hymenium yellowish-brown in dried specimens (the color in living material was not indicated); hairs cylindric, flexuous, reaching a diameter of 5–6 μ and a length of 150 μ, brown, roughened with minute granules; asci rather short and stubby, cylindric, scarcely narrowed at the base, reaching a length of 65–80 μ and a diameter of 9–10 μ; spores obliquely 1-seriate, or becoming 2-seriate to irregularly disposed, straight, or slightly curved, irregularly fusoid, 4 × 13–15 μ, or occasionally as long as 18 μ; paraphyses clavate, rather strongly enlarged above where they reach a diameter of 4 μ.

Apotheciis dense gregariis, breve stipitatis vel subsessilis, pallide flavidis 1–2 mm. diam.; ascis clavatis, 7–10 × 90–100 μ, 8-sporis; sporis fusoideis, 3–4 × 7–10 μ; paraphysibus filiformibus, 2–3 μ diam.

On bark and decorticated wood of *Populus tremuloides*.
TYPE LOCALITY: University of Colorado Summer Camp.
DISTRIBUTION: Colorado and Montana to Oregon.

This species resembles in general appearance *Lachnella fusco-sanguinea* but differs in habitat and in the longer more narrowly fusoid spores. Two collections were obtained by the writer and the late Ellsworth Bethel in 1910 and two collections during the summer of 1929 (Seaver and Shope *107* and *140*). Also one specimen was received from Dr. J. R. Weir from Montana in 1917. All are on the same substratum *Populus tremuloides*.

50. **Lachnella Corni** (Cash.) Seaver, comb. nov.

Lachnum Corni Cash, Jour. Wash. Acad. Sci. **29**: 47. 1939.

Apothecia stipitate, scattered, cup-shaped, brown, less than 1 mm. in diameter, the margin inrolled when dry, often becoming triangular; externally clothed with hairs; hymenium whitish to pale olive-buff; stem cylindrical .5 mm. long and 1–.7 mm. thick; hairs brown, non-septate, delicately roughened; asci narrow-cylindric, 8-spored, reaching a length of 33–37 μ and a diameter of 3–3.5 μ; spores 2-seriate, ellipsoid, .5–1 $\times$ 4–6 μ; paraphyses lanceolate, 3–3.5 μ thick.

On leaves of *Cornus Amomum*.
TYPE LOCALITY: Clark County, Georgia.
DISTRIBUTION: Known only from the type locality.
ILLUSTRATIONS: Jour. Wash. Acad. Sci. **29**: 47, *f. 1*.

51. **Lachnella cerina** Karst. Acta Soc. Fauna Fl. Fenn. **2**[6]: 131. 1885.

Peziza cerina Pers. Obs. Myc. **1**: 43. 1796.
Dasyscypha cerina Fuckel, Symb. Myc. 305. 1869.
Helotium cerinum Karst. Myc. Fenn. **1**: 156. 1877.
Lachnea cerina Gill. Champ. Fr. Discom. 70. 1880.
Erinella cerina Quél. Ench. Fung. 303. 1886.
Lachnella virginica Ellis & Ev. Proc. Acad. Sci. Phila. **1894**: 349. 1894.

Apothecia gregarious, at first closed and subglobose, expanding and becoming shallow cup-shaped, with a short, dark stem, clothed externally with brown hairs, reaching a diameter of 1 mm.; hymenium becoming nearly plane, yellow; hairs brown, septate, roughened on the outside, reaching a length of 120 μ and a diameter of 4–5 μ; asci cylindric, or subcylindric, reaching a length of 40–50 μ and a diameter of 4–5 μ, 8-spored;

spores ellipsoid, 1- or 2-seriate, 2.5–3 × 5–6 μ; paraphyses filiform, extending far beyond the asci, reaching a diameter of 2 μ.
On wood of deciduous trees.

TYPE LOCALITY: Europe.

DISTRIBUTION: West Virginia, Tennessee, Newfoundland,
and Washington; also in Europe.

ILLUSTRATIONS: Boud. Ic. Myc. *pl. 509;* Gill. Champ. Fr.
Discom. *pl. 60, f. 1;* Rab. Krypt.-Fl. 1³: 826. *f. 1–4;* Phill. Brit.
Discom. *pl. 7, f. 44.*

52. **Lachnella Eupatorii** (Schw.) Seaver, comb. nov.

Peziza Eupatorii Schw. Trans. Am. Phil. Soc. II. **4**: 174.   1832.
Trichopeziza Eupatorii Sacc. Syll. Fung. **8**: 426.   1889.
Dasyscypha Eupatorii Massee, Jour. Bot. **34**: 146.   1896.

Apothecia scattered, sessile, at first closed, then expanding
and becoming plane, somewhat contorted when dry, 2–5 mm. in
diameter, externally clothed with hairs; hairs reaching a length
of 70–120 μ and a diameter of 5–7 μ, 3–7-septate, externally
sometimes roughened with particles of lime, dark-brown and
translucent except near the apex which is paler, almost hyaline, or
purple with potassium hydrate, often arranged in fascicles; asci
clavate, reaching a length of 100 μ and a diameter of 8–9 μ,
8-spored; spores 2-seriate, straight, or slightly curved, finally
becoming 1-septate, 3 × 12–20 μ; paraphyses 2–3 μ thick,
almost cylindric.

On dead stems of *Eupatorium purpureum.*

TYPE LOCALITY: Bethlehem, Pennsylvania.

DISTRIBUTION: Known only from the type locality.

53. **Lachnella albotestacea** (Desm.) Phill. Brit. Discom. 273.
1887.

Peziza albotestacea Desm. Pl. Crypt. Fr. *1415;* Ann. Sci. Nat. II.   **19**: 368·
    1843.
Lachnum albotestaceum Karst. Myc. Fenn. **1**: 175.   1871.
Trichopeziza albotestacea Sacc. Atti Ist. Veneto VI. **2**: 491.   1884.

Apothecia scattered, sessile, or very short-stipitate, at first
rounded, then expanding, externally brick-red, or reddish-brown,
fading with age, .5 mm. in diameter; hymenium flesh-colored;
hairs slender, flexuous, reddish, hyaline at their tips; asci cylindric-clavate, 8-spored; spores fusoid, 1–2 × 7–10 μ; paraphyses
lanceolate.

On dead grasses, *Phragmites*, *Andropogon* and undetermined grasses.

TYPE LOCALITY: Europe.

DISTRIBUTION: New Jersey to Washington; also in Europe.

EXSICCATI: Ellis & Ev. N. Am. Fungi *2038, 2630.*

54. **Lachnella Pteridis** (Alb. & Schw.) Phill. Brit. Discom. 256. 1887.

Peziza Pteridis Alb. & Schw. Consp. Fung. 338.   1805.
Urceola Pteridis Quél. Ench. Fung. 321.   1886.
Mollisia Pteridis Gill. Champ. Fr. Discom. 121.   1882.
Trichopeziza Pteridis Rehm, Ber. Naturh. Ver. Augsburg **26**: 100.   1881.

Apothecia scattered, or gregarious, sessile, at first closed and rounded, finally expanding and becoming cup-shaped, externally dark olive-brown and clothed with adpressed, brown hairs; hymenium concave, dull-yellowish; hairs 30–50 μ long, 4 μ thick, septate; asci clavate, 8-spored, reaching a length of 35–55 μ and a diameter of 6–7 μ; spores fusiform simple, 1.5–2.5 $\times$ 6–9 μ; paraphyses filiform.

On dead stems of *Pteridium aquilinum.*

TYPE LOCALITY: Europe.

DISTRIBUTION: New York and New Jersey to Ontario, Canada; also in Europe.

ILLUSTRATIONS: Alb. & Schw. Consp. Fung. *pl. 12, f. 7;* Phill. Brit. Discom. *pl. 8, f. 47.*

EXSICCATI: Ellis, N. Am. Fungi *141.*

DOUBTFUL AND EXCLUDED SPECIES

Trichopeziza acerina (Cooke & Ellis) Sacc. Syll. Fung. **8**: 417.   1889; *Peziza acerina* Cooke & Ellis, Grevillea **7**: 40.   1878.   On leaves of *Acer* sp· Ellis, N. Am. Fungi *666.*   Material unsatisfactory.

Dasyscypha acuum (Alb. & Schw.) Sacc. Syll. Fung. **8**: 443.   1889; *Peziza acuum* Alb. & Schw. Consp. Fung. 330.   1805; *Helotium acuum* Karst. Myc. Fenn. **1**: 147.   1871; *Lachnella acuum* Phill. Brit. Discom. 246.   1887.   This is recorded from California on leaves of *Sequoia.*

Dasyscypha albocitrina (Cooke) Sacc. Syll. Fung. **8**: 446.   1889; *Peziza albocitrina* Cooke, Grevillea **7**: 47.   1878.   On *Vaccinium* leaves.   Cooke states "closely allied to *P. virginella.*"

Lachnella albolabra Ellis & Ev. Bull. Torrey Club **24**: 467.   1897.   On dead shoots of *Ribes prostratum.*   Type specimen in The New York Botanical Garden needs further study.

Dasyscypha albopileata (Cooke) Sacc. Syll. Fung. **8**: 445.   1889; *Peziza albopileata* Cooke, Hedwigia **14**: 82.   1875.   On fallen leaves of *Magnolia.* The description suggests *Lachnella virginea.*

Lachnella albopileata var. *subaurata* Ellis, Grevillea **19**: 107.   1891.   On leaves of *Clethra alnifolia*.

Trichopeziza alboviridis (Cooke) Sacc. Syll. Fung. **8**: 415.   1889; *Peziza alboviridis* Cooke, Grevillea **7**: 47.   1878.   On decorticated *Myrica* or in fissures of the bark.   Described as aeruginous-villose.

Lachnum alneum Vel. Monog. Discom. Bohem. **1**: 247.   1934.   This species has been reported from Washington on log of *Alnus* by Dr. B. Kanouse. It appears to be *Lachnella virginea*.

Hyaloscypha alniseda Vel. Monog. Discom. Bohem. **1**: 283.   1934.   Apothecia described as .1–.3 mm. in diameter, sessile, watery-gray, fringed with a single row of simple hairs with bulbose bases; asci 30–40 × 5 μ; spores 2 × 5–7 μ; paraphyses filiform.   Reported on log of *Alnus*, from Washington by Dr. B. Kanouse.   No material has been seen.

Dasyscypha Arundinariae (Berk.) Sacc. Syll. Fung. **8**: 448.   1889; *Peziza Arundinariae* Berk. Grevillea **3**: 155.   1875.   Reported on stems of *Arundinaria* from South Carolina.   No material seen.   *Peziza cannea* Cooke is said to be identical.   Compare *Pyrenopeziza Arundinariae*.

Lachnella ascoboloidea (Schw.) Sacc. Syll. Fung. **8**: 400.   1889; *Peziza ascoboloidea* Schw. Trans. Am. Phil. Soc. II. **4**: 175.   1832.   Reported on *Vitis labrusca*.   No material seen.

Trichopeziza Aspidii (Lib.) Fuckel, Symb. Myc. 297.   1869; *Peziza Aspidii* Lib. Pl. Crypt. Ard. *226*.   1834; *Lachnum Aspidii* Karst. Medd. Soc. Fauna Fl. Fenn. **16**: 27.   1888.   This species has been reported from Michigan, material determined by George B. Cummins.

Dasyscypha Bakeri Earle in Greene, Pl. Baker. **2**: 6.   1901.   On dead stems of *Corydalis Brandegei*.   Clothed with brown hairs.   Spores, 3 × 12 μ.

Trichopeziza brevipila (Rob.) Sacc. Syll. Fung. **8**: 404.   1889; *Peziza brevipila* Rob. in Desm. Pl. Crypt. Fr. *1742;* Ann. Sci. Nat. III. **11**: 362.   1849; *Lachnella brevipila* Quél. Ench. Fung. 313.   1886; *Beloniella brevipila* Rehm in Rab. Krypt.-Fl. **1³**: 641.   1892.   Reported from Michigan on overwintering stems of *Aster* sp.   The spores are described as 27–32 μ long and finally many-septate which would place it in the genus *Belonium* as treated here.

Dasyscypha brunneola (Desm.) Sacc. Syll. Fung. **8**: 460.   1889; *Peziza brunneola* Desm. Pl. Crypt. Fr. *1156*; Ann. Sci. Nat. II. **17**: 96.   1842; *Lachnum brunneolum* Karst. Myc. Fenn. **1**: 180.   1871.   This species has been reported from America but no American material has been seen.

Dasyscypha callochaetes (Ellis & Ev.) Sacc. Syll. Fung. **8**: 462.   1889; *Lachnella callochaetes* Ellis & Ev. Jour. Myc. **4**: 99.   1888.   On fallen leaves of *Myrica cerifera*.   About 1 mm. in diameter, sparingly clothed with erect, black, bristle-like hairs.   Spores 2–3 × 12–14 μ.

Lachnum calyculaeforme (Schum.) Karst. Myc. Fenn. **1**: 178.   1871; *Peziza calyculaeformis* Schum. Enum. Pl. Saell. **2**: 425.   1803; *Dasyscypha calyculaeformis* Sacc. Syll. Fung. **8**: 454.   1889.   This has been reported from Colorado (Mycologia **28**: 304) and Washington but there seems to be some doubt as to its determination.

Peziza Campanula Ellis, Bull. Torrey Club **8**: 73.   1881.   This is apparently a *Cyphella*.

Trichopeziza candida Clements, Bot. Surv. Nebr. **4**: 15.   1896.   On bark and twigs of *Tilia americana*.   Clements states "possibly *Trichopeziza Tiliae* (Peck) Sacc." which is a *Cyphella*.

Lachnum carneolum (Sacc.) Rehm in Rab. Krypt.-Fl. **1**[3]: 881. 1893; *Hyalopeziza carneola* Sacc. Michelia **1**: 253. 1878. *Dasyscypha carneola* Sacc. Syll. Fung. **8**: 447. 1889. Reported by Dr. B. Kanouse from Michigan on *Eriophorum callitrix*.

Trichopeziza carneorubra (Ellis) Sacc. Syll. Fung. **8**: 405. 1889; *Peziza carneorubra* Ellis; Cooke, Bull. Buffalo Soc. Nat. Sci. **3**: 22. 1877. On stems of *Erigeron*. "Sessile, rosy flesh-color, 1 mm. broad; sporidia linear binucleate." No material seen.

Coronellaria Castanopsidis Kanouse, Mycologia **33**: 464. 1941. On leaves of *Castanopsis chrysophylla*, Mt. Hood, Oregon. The spores are described as 4×14–16μ and spuriously septate. The apothecia are clothed with adpressed, hair-like hyphae. No material has been seen.

Dasyscypha chamaeleontina (Peck) Sacc. Syll. Fung. **8**: 433. 1889; *Peziza chamaeleontina* Peck, Ann. Rep. N. Y. State Mus. **30**: 60. 1878. On under surface of hemlock (*Tsuga*) wood lying on the ground. Described as white changing to yellow when bruised. Peck states "There is scarcely any appearance of hairiness on the cups." It is however placed in *Dasyscypha* by Saccardo.

Dasyscypha caulicola (Fries) Sacc. Syll. Fung. **8**: 463. 1889; *Peziza caulicola* Fries, Syst. Myc. **2**: 94. 1822; *Helotium caulicola* Karst. Myc. Fenn. **1**: 134. 1871; *Lachnella caulicola* Phill. Brit. Discom. 236. 1887. Reported on *Arundinaria* in North America and palm stems in Cuba. Also on stems of *Urtica* from Michigan.

Hyalopeziza ciliata Fuckel, Symb. Myc. 298. 1869. This species is reported by Dr. B. Kanouse from Washington on *Delphinium*. No material has been seen.

Lachnella cinereofusca (Schw.) Sacc. Syll. Fung. **8**: 399. 1889; *Peziza cinereofusca* Schw. Schr. Nat. Ges. Leipzig **1**: 119. 1822; *Cyphella cinereofusca* Sacc. Syll. Fung. **6**: 674. 1888; *Velutaria cinereofusca* Bres.; Rehm in Rab. Krypt.-Fl. **1**[3]: 645. 1892. As will be seen, Saccardo treated this first as a Basidiomycete and later as a Discomycete. It is a *Cyphella*.

Dasyscypha clandestina (Bull.) Fuckel, Symb. Myc. 305. 1869; *Peziza clandestina* Bull. Hist. Champ. Fr. 251, *pl. 416, f. 5.* 1791; *Lachnum clandestinum* Karst. Myc. Fenn. **1**: 178. 1871. This species has been recorded for America. Characterized by its crystaline hairs.

Trichopeziza coarctata Ellis & Ev. Am. Nat. **31**: 427. 1897. On dead branches of *Vaccinium myrtilloides*. Type in The New York Botanical not very satisfactory.

Trichopeziza comata (Schw.) Sacc. Syll. Fung. **8**: 431. 1889; *Peziza comata* Schw. Trans. Am. Phil. Soc. II. **4**: 173. 1832. On old leaves of *Quercus*. Minute, white-tomentose. Spores not described.

Dasyscypha crinella (Ellis & Ev.) Sacc. Syll. Fung. **8**: 450. 1889; *Peziza crinella* Ellis & Ev. Bull. Torrey Club **10**: 76. 1883. On dead stems of *Carex crinita* lying in the water. Described as white, thin, and delicate. Spores 2×9–12μ. Ellis, N. Am. Fungi *1273.*

Trichopeziza crossota (Ellis) Sacc. Syll. Fung. **8**: 413. 1889; *Peziza crossota* Ellis, Bull. Torrey Club **8**: 124. 1881. According to Ellis this is closely related to *Peziza Meleagris* which is an old specimen of *Perrotia flammea*. Material very poor.

Peziza digitalis Alb. & Schw. Consp. Fung. 315. 1803. Reported from North America by Schweinitz. This is *Cyphella pendula*.

Trichopeziza distincta (Peck) Sacc. Syll. Fung. **8**: 421. 1889; *Peziza distincta* Peck, Ann. Rep. N. Y. State Mus. **30**: 60. 1878. On dead stems of *Andropogon furcatus*. Externally blackish, the margin tomentose-hairy, tawny, or olivaceous, the disk pink when moist. Spores fusiform, 20–25 µ long.

Trichopeziza earoleuca (Berk. & Br.) Sacc. Syll. Fung. **8**: 409. 1889; *Peziza earoleuca* Berk. & Br. Jour. Linn. Soc. **14**: 105. 1875. On herbaceous stems, sticks, and wood. Externally clothed with white hairs. Spores 4 µ long. Reported from Ceylon. Also recorded from America (Rav. Fungi. Am. *631*) and one doubtful specimen from Bermuda collected by the writer. Should be studied.

Dasyscypha epixantha (Cooke) Sacc. Syll. Fung. **8**: 458. 1898; *Peziza epixantha* Cooke, Grevillea **7**: 3. 1878. On twigs of *Quercus*. Spores linear, 10 µ long.

Dasyscypha eryngiicola Ellis & Ev. Bull. Torrey Club **25**: 506. 1898. On dead stems of *Eryngium*. White-tomentose, 1 mm. in diameter. Spores 2.5–3 × 12–15 µ.

Lachnella extricata (Berk & Curt.) Sacc. Syll. Fung. **8**: 401. 1889; *Peziza extricata* Berk. & Curt.; Berk. Grevillea **3**: 152. 1875. On dead stems of some umbelliferous plant. Specimens as stated by Berkeley not in very good condition.

Dasyscypha Fairmani Rehm, Ann. Myc. **7**: 535. 1909. On decaying wood. Said to differ from *D. albolutea* in the larger spores which are 3 × 12–14 µ.

Solenopezia fimbriata Ellis & Barth. Jour. Myc. **8**: 174. 1902. Type material of this species loaned the writer by Dr. Elam Bartholomew of Hays, Kansas shows it to be one of the Phacidiaceae and apparently a *Diplonaevia*. It is close to *Diplonaevia melaleuca* Ellis & Ev. described from material occurring on decorticated wood of *Populus tremuloides* from Montana while the above species was described from material occurring on the same substratum in Colorado. The two species are very similar but appear to differ in spore measurements. Whether this difference will be found to be of specific importance when more material has been studied remains to be seen.

Dasyscypha fimbriifera (Berk. & Curt.) Sacc. Syll. Fung. **8**: 452. 1889; *Peziza fimbriifera* Berk. & Curt.; Berk. Jour. Linn. Soc. **10**: 367. 1868. Apothecia white, short-stipitate, externally farinose-tomentose, toward the margin clothed with long, straight hairs. On stems of ferns in Cuba. Fruit not described and no specimen seen.

Pseudohelotium fibrisedum Sacc. Syll. Fung. **8**: 298. 1889; *Peziza fibriseda* Berk. & Curt.; Berk. Grevillea **3**: 157. 1875; *Peziza saccharifera* Berk. & Curt.; Berk. Grevillea **3**: 157. 1875; *Pseudohelotium sacchariferum* Sacc. Syll. Fung. **8**: 298. 1889. Illustrations in the herbarium of The New York Botanical Garden, drawn by Geo. Massee show the exterior of the apothecia to be clothed with club-shaped hairs. *Orbilia cruenta*.

Dasyscypha flavidula Rehm, Ann. Myc. **7**: 542. 1909. Doubtfully reported on fern stems from Porto Rico by the writer.

Trichopeziza flavofuliginea (Alb. & Schw.) Sacc. Syll. Fung. **8**: 413. 1889; *Peziza flavofuliginea* Alb. & Schw. Consp. Fung. 319, *pl. 11, f. 7*. 1805; *Dasyscypha flavofuliginea* Fuckel, Symb. Myc. Nacht. **1**: 337. 1871. Reported from America but material doubtful.

Peziza frondicola Ellis & Ev. Jour. Myc. **4**: 99. 1888; *Pirottaea frondicola* Sacc. Syll. Fung. **8**: 388. 1889. On fallen leaves of *Osmunda*. Type specimen is very poor. Nannfeldt says scarcely a *Pirottaea*.

Trichopeziza fulvocana (Schw.) Sacc. Syll. Fung. **8**: 430. 1889; *Peziza fulvocana* Schw. Schr. Nat. Ges. Leipzig **1**: 120. 1822. About 2 mm. in diameter, yellowish-brown-tomentose. Spores not described.

Dasyscypha fuscescens (Pers.) Rehm, Ascom. *457*. 1878; *Peziza fuscescens* Pers. Syn. Fung. 654. 1801. On fallen leaves of *Fagus*. This species has been recorded for North America. American material has been reported on leaves of *Andromeda*.

Dasyscypha fuscidula (Cooke) Sacc. Syll. Fung. **8**: 462. 1889; *Peziza fuscidula* Cooke, Bull. Buffalo Soc. Nat. Sci. **3**: 22. 1877. On leaves of *Andromeda*.

Lachnella fuscobarbata (Schw.) Sacc. Syll. Fung. **8**: 400. 1889; *Peziza fuscobarbata* Schw. Trans. Am. Phil. Soc. II. **4**: 173. 1832. No material seen and description inadequate.

Lachnum fuscofloccosum Rehm, Ann. Myc. **5**: 520. 1907. Reported on sticks from Sumner, Washington. No material seen.

Lachnum hyalinellum Rehm in Rab. Krypt.-Fl. **1**³: 874. 1893. This is reported from Michigan by Dr. B. Kanouse on cones of *Picea Mariana*. Also reported from Washington on *Alnus*. No material has been seen.

Pirottaea ? Hydrangeae (Schw.) Sacc. Syll. Fung. **8**: 390. 1889; *Peziza Hydrangeae* Schw. Schr. Nat. Ges. Leipzig. **1**: 121. 1822. Apothecia described as cinereous-white, subglobose, clothed with long, black hairs. No specimen seen.

Dasyscypha hystricula (Ellis & Ev.) Sacc. Syll. Fung. **8**: 445. 1889; *Lachnella hystricula* Ellis & Ev. Jour. Myc. **4**: 99. 1888. On the under side of leaves of *Magnolia grandiflora*. Minute, white, clothed with rough hairs. Spores subfusoid, 3-septate, $2 \times 12 \mu$. Description suggests an *Erinellina*.

Dasyscypha illota (Berk. & Curt.) Sacc. Syll. Fung. **8**: 457. 1889; *Peziza illota* Berk. & Curt.; Berk. Jour. Linn. Soc. **10**: 368. 1868. Short-stipitate, solid, crateriform, fawn-colored, tomentose; hymenium brown. On bark. Cuba. No material seen.

Dasyscypha inspersa (Berk. & Curt.) Sacc. Syll. Fung. **8**: 437. 1889; *Peziza inspersa* Berk. & Curt.; Berk. Jour. Linn. Soc. **10**: 368. 1868. Cups subglobose, externally densely white-farinaceous; hymenium flesh-red. On rotten wood in Cuba. No material seen.

Lachnella incarnescens (Schw.) Sacc. Syll. Fung. **8**: 399. 1889; *Peziza incarnescens* Schw. Trans. Am. Phil. Soc. II. **4**: 173. 1832. Sessile, 2–5 mm. in diameter, externally brownish-black; hymenium flesh-colored. No fruit described. On rotten wood, Bethlehem, Pennsylvania. No material seen.

Trichopeziza Kalmiae (Peck) Sacc. Syll. Fung. **8**: 411. 1889; *Peziza Kalmiae* Peck, Ann. Rep. N. Y. State Mus. **25**: 99. 1873. On dead stems of *Kalmia angustifolia.* Externally dull-gray, the hymenium pinkish-brown. Spores 5 × 10 μ.

Trichopeziza labrosa (Phill. & Hark.) Sacc. Syll. Fung. **8**: 419. 1889; *Lachnella labrosa* Phill. & Hark. Bull. Calif. Acad. Sci. **1**: 21. 1884. On dead leaves of *Arctostaphylos pungens.* Olive-brown, clothed with brown hairs. Spores 3–6 × 10–14 μ.

Dasyscypha Lentaginis (Schw.) Sacc. Syll. Fung. **8**: 458. 1889; *Peziza Lentaginis* Schw. Trans. Am. Phil. Soc. II. **4**: 175. 1832. On branches of *Viburnum Lentago.* Fruit not indicated.

Trichopeziza leonina (Schw.) Sacc. Syll. Fung. **8**: 430. 1889; *Peziza leonina* Schw. Schr. Nat. Ges. Leipzig **1**: 119. 1822. On wood of *Ulmus.*

Dasyscypha longipila (Peck) Sacc. Syll. Fung. **8**: 463. 1889; *Peziza longipila* Peck, Ann. Rep. N. Y. State Mus. **32**: 20. 1879. On dead stems of *Eupatorium maculatum.* Clothed with long, tawny-brown hairs. Spores subfusoid, 2–3 × 7–10 μ.

Dasyscypha luteoalba (Schw.) Sacc. Syll. Fung. **8**: 457. 1889; *Peziza luteoalba* Schw. Trans. Am. Phil. Soc. II. **4**: 173. 1832. On bark. Hairs yellow; hymenium white.

Dasyscypha luteodisca (Peck) Sacc. Syll. Fung. **8**: 449. 1889; *Lachnella luteodisca* Peck, Ann. Rep. N. Y. State Mus. **33**: 31. 1880. On dead *Scirpus validus.* Hairs white; hymenium yellow.

Trichopeziza marginata (Cooke) Sacc. Syll. Fung. **8**: 416. 1889; *Peziza marginata* Cooke, Hedwigia **14**: 82. 1875. On fallen leaves of *Quercus* and *Andromeda.* Clothed with brown hairs. Spores minute, linear, spermatoid. See Ellis N. Am. Fungi *386.* Should be studied.

Trichopeziza melaxantha (Fries) Sacc. Syll. Fung. **8**: 428. 1889; *Peziza melaxantha* Fries, Syst. Myc. **2**: 150. 1822. On old wood. Externally black; hymenium yellowish.

Trichopeziza myricacea (Peck) Sacc. Syll. Fung. **8**: 409. 1889; *Peziza myricacea* Peck, Ann. Rep. N. Y. State Mus. **30**: 59. 1878. On dead stems of *Myrica Gale.* The type of this species in the herbarium is very scant and unsatisfactory.

Dasyscypha membranata (Schw.) Sacc. Syll. Fung. **8**: 457. 1889; *Peziza membranata* Schw. Trans. Am. Phil. Soc. II. **4**: 175. 1832. On old wood. Clothed with adpressed hairs, entirely black. Fruit not indicated.

Lachnum Nardi Rehm in Rab. Krypt.-Fl. **1**³: 883. 1893. This species originally reported on *Nardus* has been recorded by Dr. B. Kanouse on some sedge from Washington. No material seen.

Trichopeziza nigrocincta (Berk. & Curt.) Sacc. Syll. Fung. **8**: 421. 1889; *Peziza nigrocincta* Berk. & Curt.; Berk. Grevillea **3**: 155. 1875. On sheaths of grasses. "Minute dot-like, plane, scarlet, rough with short, black hairs externally; asci-clavate, sporidia oblong curved."

Trichopeziza obscura (Cooke) Sacc. Syll. Fung. **8**: 410. 1889; *Peziza obscura* Cooke, Grevillea **7**: 3. 1878. On twigs of *Quercus.* Sessile, blackish-brown. Spores linear, 10 μ long.

Dasyscypha ochracea (Schw.) Sacc. Syll. Fung. **8**: 455. 1889; *Peziza ochracea* Schw. Trans. Am. Phil. Soc. II. **4**: 172. 1832. No specimen was found in the Schweinitz collection when examined May 22, 1931.

Trichopeziza Opulifoliae (Schw.) Sacc. Syll. Fung. **8**: 429. 1889; *Peziza Opulifoliae* Schw. Trans. Am. Phil. Soc. II. **4**: 175. 1832. On dead branches of *Spiraea opulifolia*.

Unguicularia oregonensis Kanouse, Mycologia **33**: 467. 1941. The apothecia are described as less than 1 mm. in diameter. The asci are 7–8 × 35–50 μ and the spores 3–3.5 × 7–8 μ; paraphyses curved. No material seen.

Trichopeziza Osmundae (Cooke & Ellis) Sacc. Syll. Fung. **8**: 423. 1889; *Peziza Osmundae* Cooke & Ellis, Grevillea **6**: 7. 1877; Ellis, N. Am. Fungi *136*. On fronds of *Osmunda*. Minute, white. Spores 5 μ long.

Lachnum pallideroseum (Saut.) Rehm in Rab. Krypt.-Fl. **1**³: 885. 1893; *Peziza palliderosea* Saut. Pilze Salzb. **2**: 14. ? ; *Phialea palliderosea* Sacc. Syll. Fung. **8**: 264. 1889. Reported on grasses from Washington. No material seen. Said to become rose-red when dry. Spores 1.5 × 7–10 μ.

Lachnella papillaris (Bull.) Karst. Acta Soc. Fauna Fl. Fenn. **2**⁶: 132. 1885; *Peziza papillaris* Bull. Hist. Champ. Fr. 244. 1791; *Helotium papillare* Karst. Myc. Fenn. **1**: 160. 1871; *Lachnea papillaris* Gill. Champ. Fr. Discom. 80. 1880. This species has been reported from Michigan by Dr. B. Kanouse. She states that the crystals at the ends of the hairs are a help in determining the species.

Dasyscypha patula (Pers.) Sacc. Syll. Fung. **8**: 443. 1889; *Peziza patula* Pers. Syn. Fung. 654. 1801; *Lachnella patula* Phill. Brit. Discom. 251. 1887; *Hyalopeziza patula* Fuckel, Symb. Myc. 298. 1869. This species has been reported from America but no American material has been seen.

Trichopeziza penicillata (Schw.) Sacc. Syll. Fung. **8**: 429. 1889; *Peziza penicillata* Schw. Schr. Nat. Ges. Leipzig. **1**: 120. 1822. On bark of *Vitis*. No fruit described.

Trichopeziza pomicolor (Berk. & Rav.) Sacc. Syll. Fung. **8**: 429. 1889; *Peziza pomicolor* Berk & Rav.; Cooke, Bull. Buffalo Soc. Nat. Sci. **2**: 294. 1875; Berk. Grevillea **3**: 157. 1875. On bark of *Taxodium distichum*. Externally pomicolor, the hymenium olivaceous. Spores not described. This species is listed by Saccardo twice; once under *Trichopeziza* and once as a *Pseudohelotium*. See *Chlorosplenium chlora*.

Dasyscypha prolificans (Schw.) Sacc. Syll. Fung. **8**: 456. 1889; *Peziza prolificans* Schw. Trans. Am. Phil. Soc. II. **4**: 172. 1832. On old trunks. No fruit described.

Dasyscypha puberula (Berk. & Curt.) Sacc. Syll. Fung. **8**: 461. 1889; *Peziza puberula* Berk. & Curt.; Berk. Grevillea **3**: 155. 1875. On fallen leaves of *Fraxinus*. "Cups, fawn-coloured, globose, furfuraceo-tomentose; stem short, pallid as is the hymenium, which becomes brighter with age." No fruit described.

Trichopeziza punctiformis (Fries) Fuckel, Symb. Myc. 296. 1869; *Peziza punctiformis* Fries, Syst. Myc. **2**: 105. 1822. On leaves of *Myrica, Alnus*, and *Quercus*. Recorded for North America.

Trichopeziza relicina (Fries) Fuckel, Symb. Myc. 296. 1869; *Peziza relicina* Fries, Syst. Myc. **2**: 103. 1822; *Lachnum relicinum* Karst. Myc. Fenn. **1**: 182. 1871. Recorded from Alaska.

Trichopeziza roseoalba (Schw.) Sacc. Syll. Fung. **8**: 428. 1889; *Peziza roseoalba* Schw. Schr. Nat. Ges. Leipzig **1**: 122. 1822. On bark of *Cornus florida*. Type examined in 1931 and found unsatisfactory for study.

Trichopeziza rufiberbis (Schw.) Sacc. Syll. Fung. **8**: 426. 1889; *Peziza rufiberbis* Schw. Trans. Am. Phil. Soc. II. **4**: 173. 1832. On unidentified stems.

Dasyscypha scabrovillosa (Phill.) Sacc. Syll. Fung. **8**: 458. 1889; *Peziza scabrovillosa* Phill. Grevillea **7**: 22. 1878. On *Rubus nutkanus*. This suggests *Lachnella bicolor*.

Trichopeziza setigera (Phill.) Sacc. Syll. Fung. **8**: 407. 1889; *Peziza setigera* Phill. Grevillea **7**: 22. 1878; *Lachnum setigerum* Rehm, Ann. Myc. **3**: 518. 1905. On dead stems of species of *Aralia*. Clothed with brown hairs; hymenium brown. Spores 3.5 × 16 μ.

Pezizella soleniiformis (Berk. & Curt.) Sacc. Syll. Fung. **8**: 280. 1889; *Peziza soleniiformis* Berk. & Curt.; Berk. Grevillea **3**: 160. 1875. This is probably a *Solenia*.

Lachnella solitaria (Schw.) Sacc. Syll. Fung. **8**: 401. 1889; *Peziza solitaria* Schw. Trans. Am. Phil. Soc. II. **4**: 175. 1832. Solitary, 6 mm. in diameter, scantily clothed with white hairs. No fruit described.

Peziza sphaerincola Schw. Trans. Am. Phil. Soc. II. **4**: 172. 1832. Described as minute, stipitate, externally brown, hairy. On some *Sphaeria*.

Dasyscypha spiraeicola (Karst.) Sacc. Syll. Fung. **8**: 442. 1889; *Peziza spiraeaecola* Karst. Not. Soc. Fauna Fl. Fenn. **10**: 192. 1869; *Lachnum spiraeaecolum* Rehm in Rab. Krypt.-Fl. **1**³: 880. 1893. Reported from Michigan from material collected by C. H. Kauffman and identified by Geo. B. Cummins.

Dasyscypha stipiticola (Schw.) Sacc. Syll. Fung. **8**: 465. 1889; *Peziza stipiticola* Schw. Trans. Am. Phil. Soc. II. **4**: 172. 1832. On old stems. No fruit indicated.

Lasiobelonium subflavidum Ellis & Ev. Bull. Torrey Club **24**: 136. 1897. The description of this species suggests *Perrotia flammea*.

Dasyscypha subhirta (Schw.) Sacc. Syll. Fung. **8**: 462. 1889; *Peziza subhirta* Schw. Trans. Am. Phil. Soc. II. **4**: 173. 1832. On leaves. Described as yellow, minutely hirsute. Spores not described.

Dasyscypha subtilissima (Cooke) Sacc. Syll. Fung. **8**: 438. 1889; *Peziza subtilissima* Cooke, Grevillea **3**: 121. 1875. On bark of firs, *Abies*. Cooke states that this can scarcely be distinguished from *Peziza calycina* Schum. but since its identity is in doubt it does not help much. See *Lachnella Agassizii*.

Dasyscypha sulphurella (Peck) Sacc. Syll. Fung. **8**: 459. 1889; *Peziza sulphurella* Peck, Ann. Rep. N. Y. State Mus. **30**: 59. 1878. The species is described as stipitate, pale-yellow, the hairs capitate; spores 7.5 μ long; paraphyses lanceolate. On stems of *Myrica Gale*, Adirondack Mountains. The type in the herbarium of The New York Botanical Garden is too scant to permit of critical study. See *Lachnella crucifera*.

Solenopezia Symphoricarpi Ellis & Ev. Jour. Myc. **9**: 165. 1903. Described from material collected on *Symphoricarpos* at Steamboat Springs, Colorado. The apothecia are described as 1 mm. in diameter and covered with a coat of dark-colored hairs; spores fusoid, or subclavate, septate. No material could be found in the Ellis Collection. Compare *Lachnella Symphoricarpi* Ellis & Ev. Bull. Torrey Club **24**: 467. 1897. The two may be the same. Material not satisfactory.

Trichopeziza Tiliae (Peck) Sacc. Syll. Fung. **8**: 428. 1889; *Peziza Tiliae* Peck, Ann. Rep. N. Y. State Mus. **24**: 96. 1872; *Cyphella Tiliae* Cooke, Grevillea **20**: 9. 1891. Cooke was undoubtedly correct in referring this species to *Cyphella*. Burt (Ann. Missouri Bot. Gard. **1**: 364) recognizes this combination. Frequently collected on *Tilia*.

Dasyscypha translucida (Berk. & Curt.) Sacc. Syll. Fung. **8**: 439. 1889; *Peziza translucida* Berk. & Curt.; Berk, Grevillea **3**: 155. 1875. On twigs of *Castanea*. Description very inadequate.

Dasyscypha tuberculiformis Ellis & Ev. Bull. Torrey Club **27**: 60. 1900. On dead stems of *Aquilegia coerulea*. Ellis states: "The specimens were not well matured so that the form and size of the sporidia will have to be more accurately ascertained hereafter."

Dasyscypha uncinata (Phill.) Sacc. Syll. Fung. **8**: 456. 1889; *Peziza uncinata* Phill. Grevillea **5**: 117. 1877. On pine needles, *Pinus* and oak wood, *Quercus*. Clothed with cinereous hairs. Spores 1 × 6 μ. Specimens in The New York Botanical Garden meager.

Trichopeziza urticina (Peck) Sacc. Syll. Fung. **8**: 403. 1889; *Peziza urticina* Peck, Ann. Rep. N. Y. State Mus. **32**: 46. 1879. On dead stems of nettle, *Urtica canadensis*. Peck states: "The plants are so small that to the naked eye they appear like mere white grains." A part of the type is in The New York Botanical Garden. The species should be investigated.

Trichopeziza venturioides (Ellis & Ev.) Sacc. Syll. Fung. **8**: 419. 1889; *Lachnella venturioides* Ellis & Ev. Jour. Myc. **4**: 99. 1888. On fallen leaves of *Gaylussacia dumosa*. Minute, clothed with black bristles. Spores 1.5–2 × 10–12 μ. See Ellis & Ev. N. Am. Fungi *2145*.

Peziza villosa Pers. Syn. Fung. 655. 1801. This is *Cyphella villosa* (Pers.) Karst.

Trichopeziza Vitis (Schw.) Sacc. Syll. Fung. **8**: 429. 1889; *Peziza Vitis* Schw. Trans. Am. Phil. Soc. II. **4**: 173. 1832. Schweinitz's type examined May, 1931 shows only lichen apothecia.

Dasyscypha vixvisibilis (Schw.) Sacc. Syll. Fung. **8**: 456. 1889; *Peziza vixvisibilis* Schw. Trans. Am. Phil. Soc. II. **4**: 175. 1832. On the bark of *Castanea*. Fruit not described.

Solenopezia vulpina (Cooke) Sacc. Syll. Fung. **8**: 478. 1889; *Peziza vulpina* Cooke, Hedwigia **14**: 82. 1875. As pointed out by the writer (Bull. Torrey Club **36**: 202–203. 1909), this is a synonym of *Nectria Peziza* (Tode) Fries. Because of its resemblance to a cup-fungus, Cooke described it as a *Peziza* and Saccardo on the basis of his description placed it in the genus *Solenopezia*.

Pirottaea yakutatiana Sacc.; Sacc. Peck, & Trelease in Harriman Alaska Exped. **5**: 25. 1904. Apothecia 1 mm. broad and high, externally hairy, black; asci 5.5–6 × 40–45 μ, 8-spored; spores 1-seriate, fusoid, 2–2.5 × 9 μ.

50. DIPLOCARPA Massee, Brit. Fungus-Fl. **4**: 307. 1895.

Apothecia small, at first closed, finally expanding and becoming shallow cup-shaped, attenuated below into a short, stemlike base, densely clothed with short, septate, brown hairs giving the entire exterior a brown color; hymenium concave, olive-

green; asci clavate, 8-spored; spores fusoid, containing several
oil-drops and finally becoming septate; paraphyses filiform and
surmounted with a fusiform conidium-like body.

Type species, *Diplocarpa Curreyana* Massee.

Diplocarpa Curreyana Massee, Brit. Fungus-Fl. **4**: 307.   1895.
(PLATE 127.)

Peziza diplocarpa Currey, Trans. Linn. Soc. **24**: 153.   1864.
Lachnella diplocarpa Phill. Brit. Discom. 232.   1887.

Apothecia gregarious, or closely congested, appearing sessile
but actually short-stipitate, at first closed, then expanding and
becoming shallow cup-shaped, reaching a diameter of 1–2 mm.
externally dark-brown, decidedly rough, tomentose, the rough-
ening often vertically striated near the margin; hairs short,
septate, brown, with the tips often sharp-pointed; stem 1–1.5
mm. long, relatively thick, about half as thick as long; hymenium
concave, olive-green, becoming brownish with age; asci clavate,
reaching a length of 70 μ and a diameter of 7 μ; spores partially
2-seriate, ellipsoid, with two or three small oil-drops, finally
becoming 1- or 2-septate about 3 $\times$ 9 μ; paraphyses slender,
surmounted with fusoid, septate, spore-like tips which reach a
length of 28–32 μ and a diameter of 6–7 μ.

On much rotted wood, Wychwood, Wisconsin.

TYPE LOCALITY: England.

DISTRIBUTION: Wisconsin; also in Europe.

ILLUSTRATIONS: Trans. Linn. Soc. **24**: *pl. 25, f. 30, 32, 33;*
Phill. Brit. Discom. *pl. 7, f. 43;* Mycologia **29**: 176, *f. 1.*

For a full discussion of this species see Mycologia **29**: 174–177.
1937.

51. **LACHNELLULA** Karst. Medd. Soc. Fauna Fl. Fenn. **11**:
138.   1884.

Apothecia stipitate, or subsessile, bright-colored, yellowish,
or orange, externally whitish and densely clothed with hairs
which are white to the unaided eye, or in one species slightly
rufous; asci cylindric, or clavate, 8-spored; spores globose, hyaline; paraphyses filiform, or subclavate.

Type species, *Peziza chrysophthalma* Pers.

Distinguished from *Lachnella* by its globose spores.

Spores reaching a diameter of 5–6 μ.	1. *L. chrysophthalma.*
Spores reaching a diameter of 3.5 μ.	2. *L. microspora.*

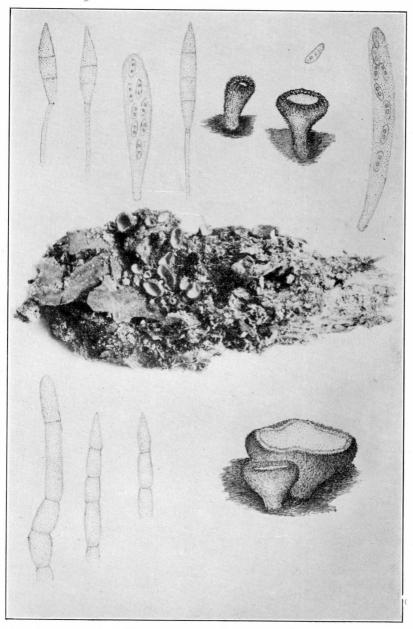

DIPLOCARPA CURREYANA

1. **Lachnellula chrysophthalma** (Pers.) Karst. Medd. Soc. Fauna
Fl. Fenn. **11**: 138.   1884.   (PLATE 128, FIG. 1.)

Peziza chrysophthalma Pers. Myc. Eu. **1**: 259.   1822.
Peziza calycina Abietis Fries, Syst. Myc. **2**: 91.   1822.
Helotium chrysophthalmum Karst. Fungi Fenn. *832*.   1869.
Pithya suecica Fuckel, Symb. Myc. Nacht. **3**: 32.   1875.

Apothecia gregarious, erumpent through the outer bark,
short-stipitate, at first subglobose, expanding and becoming
shallow cup-shaped and apparently sessile, reaching a diameter
of 5–6 mm., clothed externally with a dense covering of white
hairs; hymenium concave, or nearly plane, orange-red; hairs
cylindric, reaching a diameter of 4 μ and a length of 50–60 μ,
densely covered with minute granules, hyaline; asci cylindric,
tapering below into a rather slender, stem-like base which is often
forked, reaching a length of 80 μ and a diameter of 5–6 μ; spores
1-seriate, globose, reaching a diameter of 5–6 μ; paraphyses
filiform, slightly enlarged above, reaching a diameter of 2–2.5 μ,
filled with oil-drops.

On bark of conifers.

TYPE LOCALITY: Europe.

DISTRIBUTION: Ontario to Idaho, Montana and Colorado;
also in Europe.

ILLUSTRATIONS: Rab. Krypt.-Fl. **1**³: 828, *f. 1–5; E. &. P.*
Nat. Pfl. **1**¹: 200, *f. 138, A.*

Frequently collected in the Rocky Mountains on bark of
Abies and *Picea* and occasionally on *Pinus*. Also reported by
L. O. Overholts from Ontario on *Pinus Banksiana* (Mycologia
25: 420.   1933).

2. **Lachnellula microspora** Ellis & Ev. Proc. Acad. Sci. Phila.
1893: 451.   1893.

Apothecia sessile, or very short-stipitate, at first nearly
closed, expanding and reaching a diameter of 2–3 mm., the
margin fringed and the outside of the apothecium clothed with
rufo-cinereous hairs; hymenium pale-orange when fresh, subrufous when dry; hairs stout, simple, or branched, subfasciculate,
reaching a length of 70–80 μ and a diameter of 3.5 μ; asci clavate-
cylindric, subsessile, reaching a length of 55–60 μ and a diameter
of 6–7 μ; spores 1-seriate, globose, hyaline, reaching a diameter
of 3.5 μ; paraphyses present.

On bark of spruce trees,

TYPE LOCALITY: New Harbor, Newfoundland.

DISTRIBUTION: Known only from the type locality.

The type of this species has been examined in the herbarium of The New York Botanical Garden but unfortunately nothing remains but the shells of the apothecia and loose ascospores. The species seems to be distinguished by its small spores.

DOUBTFUL SPECIES

Lachnellula cyphelloides (Ellis & Ev.) Sacc. Syll. Fung. **8**: 391. 1889; Peziza cyphelloides Ellis & Ev. Jour. Myc. **1**: 151. 1885. Although the writer has examined the type material, he is unable to find anything which would appear to belong to this genus.

Lachnellula theioidea (Cooke & Ellis) Sacc. Syll. Fung. **8**: 391. 1889; Peziza theioidea Cooke & Ellis, Grevillea **7**: 7. 1878. Reported from Michigan on material collected by A. H. Smith, determined by Geo. B. Cummins.

52. ERIOPEZIA (Sacc.) Rehm in Rab. Krypt.-Fl. **1**³: 695. 1888.

Tapesia subg. Eriopezia Sacc. Syll. Fung. **8**: 381. 1889.

Apothecia sessile, seated on a spiderweb-like subiculum, at first globose and closed, opening and finally becoming subdiscoid, externally clothed with hairs; asci subcylindric, 8-spored; spores ellipsoid, or elongated, simple; paraphyses filiform.

Type species, Peziza caesia Pers.

Distinguished from Tapesia by the pilose exterior of the apothecia.

1. Eriopezia prolifica (Ellis) Sacc. Syll. Fung. **18**: 73. 1906.

Peziza prolifica Ellis, Bull. Torrey Club **8**: 73. 1881.
Tapesia prolifica Sacc. Syll. Fung. **8**: 382. 1889.

Apothecia gregarious, sessile, minute, densely clothed with septate, coarse, spreading, cinereous hairs, cup-shaped; hymenium dark; asci clavate, reaching a length of 25 μ, 8-spored; spores 2-seriate, cylindric, curved, subhyaline, 2.5 $\times$ 7–8 μ.

On the end of a stick of white oak.

TYPE LOCALITY: Newfield, New Jersey.

DISTRIBUTION: Known only from the type locality.

53. ARACHNOPEZIZA Fuckel, Symb. Myc. 303. 1869.

Apothecia gregarious, seated on a thin, spiderweb-like, white, or yellowish mycelial subiculum, at first closed and rounded, opening and becoming patellate, or scutellate, externally clothed

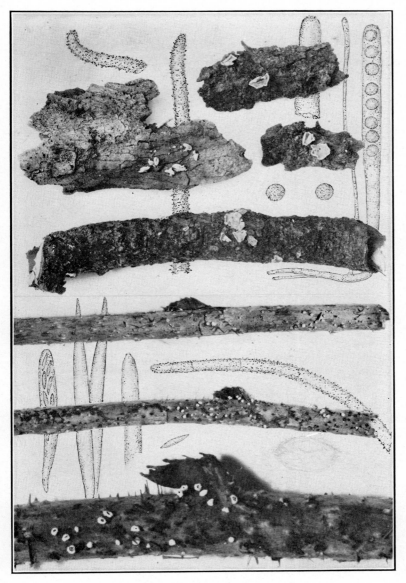

LACHNELLULA CHRYSOPHTHALMA
LACHNELLA BICOLOR

with fine bristly hairs; asci clavate, 8-spored; spores ellipsoid to fusoid, clavate, or filiform, becoming several-septate and often with an apiculus at each end, hyaline; paraphyses filiform, usually enlarged above.

Type species, *Peziza aurelia* Pers.

Spores, fusoid, apiculate 4–5 × 15–20 μ.	1. *A. aurelia.*
Spores clavate or filiform.	
Spores clavate, 3.5 × 40 μ.	2. *A. delicatula.*
Spores filiform, 2.5–3 × 65–75 μ.	
On wood and bark.	3. *A. aurata.*
On stems of *Arctostaphlos.*	4. *A. Arctostaphyli.*

1. **Arachnopeziza aurelia** (Pers.) Fuckel, Symb. Myc. 303. 1869. (PLATE 129.)

Peziza aurelia Pers. Myc. Eu. **1**: 270. 1822.
Peziza Wauchii Grev. Scot. Crypt. Fl. *pl. 139.* 1825.
Peziza candidofulva Schw. Trans. Am. Phil. Soc. II. **4**: 174. 1832.
Belonidium aurelia DeNot. Comm. Critt. Ital. **1**: 381. 1863.
Patellaria bicolor Curr. Trans. Linn. Soc. **24**: 494. 1864.
Polynema aurelium Fuckel, Symb. Myc. Nachtr. **1**: 49. 1871.
Lachnella aurelia Quél. Ench. Fung. 315. 1886.
Tapesia fulgens Hazsl. Zool.-Bot. Verh. 163. 1887.
Tapesia candidofulva Sacc. Syll. Fung. **8**: 385. 1889.
Belonidium fulgens Sacc. Syll. Fung. **8**: 500. 1889.

Apothecia gregarious, seated on a spreading, white, or yellowish mycelial web, sessile, at first rounded, then becoming scutellate, externally golden-yellow to pale-orange, clothed with fine hairs, reaching a diameter of 2–3 mm.; hymenium yellowish, a little paler than the outside of the apothecium; hairs slender, septate, reaching a length of 100 μ and a diameter of 2 μ, tapering to a slender point, collected into conical tufts which stand up about the margin like teeth; asci clavate, attenuated above, reaching a length of 70–90 μ and a diameter of 8–10 μ, 8-spored; spores fusoid, hyaline, becoming 3-septate, 4–5 × 15–20 μ, often with an apiculus at either end; paraphyses filiform, slightly enlarged above.

On leaves, soil, twigs and acorn-cups.

TYPE LOCALITY: Europe.

DISTRIBUTION: New York to Pennsylvania, Iowa and Manitoba; also in Europe.

ILLUSTRATIONS: Fuckel, Symb. Myc. Nachtr. **1**: *f. 35;* Grev. Scot. Crypt. Fl. *pl. 139;* Trans. Linn. Soc. **24**: *pl. 51, f. 15–16;* Boud. Ic. Myc. *pl. 520;* Mycologia **30**: 660, *f. 1.*

EXSICCATI: Ellis, N. Am. Fungi *59;* Rav. Fungi Car. **5**: *41.*

2. **Arachnopeziza delicatula** Fuckel, Symb. Myc. 304. 1869.

Belonidium delicatulum Sacc. Syll. Fung. **8**: 499. 1889.

Apothecia gregarious, or scattered, seated on a delicate, white, arachnoid subiculum, at first globose and closed, finally expanding, reaching a diameter of 1–2 mm.; hymenium concave, reddish-brown; asci clavate-cylindric, reaching a length of 80–100 μ and a diameter of 8–10 μ, 8-spored; spores elongated, clavate, slightly curved, simple, or becoming sparingly septate, reaching a length of 40 μ and a diameter of 3.5–4 μ; paraphyses filiform, slightly enlarged above, reaching a diameter of 3 μ.

On wood and bark.

TYPE LOCALITY: Europe.

DISTRIBUTION: Quebec and Michigan; also in Europe.

ILLUSTRATIONS: Papers Mich. Acad. Sci. **20**: *pl. 15, f. 1.* (as *Gorgoniceps delicatula* (Fuckel) Höhn.).

3. **Arachnopeziza aurata** Fuckel, Symb. Myc. 304. 1869.

Belonidium auratum Sacc. Michelia **1**: 66. 1879.
Tapesia aurata Massee, Brit. Fungus-Fl. **4**: 299. 1895.

Apothecia gregarious, sessile, at first closed, then expanding, externally yellowish, clothed with hairs, reaching a diameter of .5 mm., on a thin subiculum; hymenium a little darker than the outside of the apothecium; hairs long, cylindric, or tapering gradually toward the ends, septate, reaching a length of 60–85 μ and a diameter of 4 μ; asci clavate, the apex somewhat pointed, reaching a length of 96 μ and a diameter of 7 μ, 8-spored; spores filiform, or slightly clavate, becoming multiseptate, slightly bent, 2.5–3 × 65–75 μ; paraphyses very slender, hyaline, occasionally branched.

On wood or the inside of bark.

TYPE LOCALITY: Europe.

DISTRIBUTION: Ohio; also in Europe.

4. **Arachnopeziza Arctostaphyli** Cash, Mycologia **28**: 247. 1936.

Apothecia sessile, on sparse subicle of hyaline mycelium, at first subglobose, becoming patellate, cream-colored to light-buff when dry, .4–.7 mm. in diameter, attached at the margin by delicate, hyaline hairs; asci clavate, often arculate, narrowed above, attenuated toward the base, reaching a length of 100–115 μ and a diameter of 10–12 μ, 8-spored; spores irregularly fasciculate, many-guttulate, straight, or curved, 7-septate, 2.5–3 × 65–80 μ;

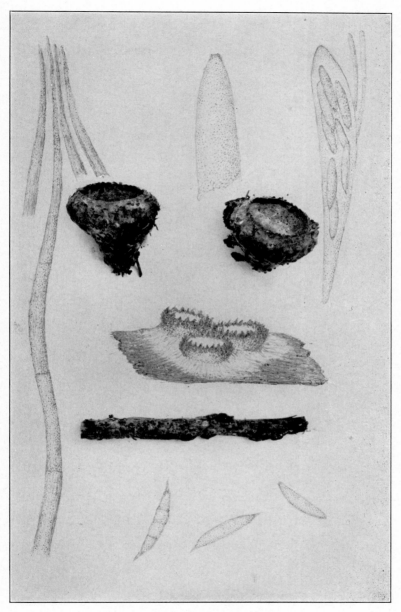

ARACHNOPEZIZA AURELIA

paraphyses filiform, branched about half way from the base, granular, hyaline.

On decorticated stems of *Arctostaphlos Tracyi.*

TYPE LOCALITY: Trinidad, California.

DISTRIBUTION: Known only from the type locality.

ILLUSTRATIONS: Mycologia **28**: 250, *f. 1.*

54. **HELOTIELLA** Sacc. Bot. Cent. **18**: 218. 1884.

Apothecia minute, sessile, or subsessile, usually bright-colored as in *Helotium,* or occasionally dark, clothed with hairs; asci cylindric, or subcylindric, usually 8-spored; spores ellipsoid, or fusoid, becoming 1-septate; paraphyses filiform.

Type species, *Helotium Citri* Penz.

On monocotyledonous plant tissues.
    On stems of *Andropogon.*   1. *H. aureococcinea.*
    On old cornstocks, *Zea.*   2. *H. pygmaea.*
Not on monocotyledonous tissues.
    On twigs of *Lonicera.*   3. *H. Lonicerae.*
    On rotten wood.
        Apothecia orange-colored.   4. *H. trabinelloides.*
        Apothecia white or whitish.
            Spores appendiculate.   5. *H. cornuta.*
            Spores not appendiculate.   6. *H. major.*
    On old paper.   7. *H. papyricola.*

1. **Helotiella aureococcinea** (Berk. & Curt.) Massee, Jour. Linn. Soc. **35**: 108. 1901.

Solenopezia aureococcinea Rehm, Ann. Myc. **2**: 352. 1904.
Patellaria aureococcinea Berk. & Curt. in Ellis, N. Am. Fungi 63. 1878. nom. nud.

Apothecia scattered, sessile, at first closed and subglobose, finally expanded, externally brownish-black, densely clothed about the margin with yellow hairs; hairs hyaline with transmitted light, reaching a diameter 5–6 μ and a length of 125 μ, blunt and slightly roughened; asci clavate, with the apex rounded, reaching a length of 60–70 μ and a diameter of 7–8 μ; spores fusoid, straight, becoming 1-septate, 2–2.5 × 15 μ; paraphyses filiform with the apex obtuse, hyaline reaching a thickness of 2–3 μ at their apices.

On decaying stems of *Andropogon.*

TYPE LOCALITY: Newfield, New Jersey.

DISTRIBUTION: Known only from the type locality.

ILLUSTRATIONS: Jour. Linn. Soc. **35**: *pl. 4, f. 5–7*.

EXSICCATI: Ellis, N. Am. Fungi *63* (as *Patellaria aureococcinea* Berk. & Curt. 1878).

2. **Helotiella pygmaea** Ellis & Ev. Proc. Acad. Sci. Phila. **1894**: 350. 1894.

Apothecia gregarious, minute, pallid, yellowish outside, sparingly clothed and the margin fringed with pale, slender hairs, convex-discoid when moist, .25–.35 mm. in diameter, when dry cup-shaped and yellowish, attached to the substratum by light-colored hairs; asci cylindric, reaching a length of 30 μ and a diameter of 6 μ, 8-spored; spores 2-seriate, fusoid, becoming 1-septate, and often slightly constricted, 3–3.5 × 10–14 μ; paraphyses not observed.

On old cornstalks, *Zea Mays*.

TYPE LOCALITY: Ohio.

DISTRIBUTION: Known only from the type locality.

3. **Helotiella Lonicerae** (Cash) Seaver, comb. nov.

Tapesia Lonicerae Cash, Mycologia **28**: 299.   1936.

Apothecia sessile, superficial, soft-fleshy, becoming patelliform, irregularly contorted when dry, the margin crenulate, inrolled, 1–2.5 mm. in diameter, exterior black, or dull livid-purple, densely setose at the margin; hairs brown, smooth, septate, 2.5–3 × 140–170 μ; hymenium pale-gray; asci cylindric, reaching a length of 65–75 μ and a diameter of 7–9 μ; spores obliquely 1-seriate, fusoid-clavate, 1-septate, 2–2.5 × 12–13.5 μ; paraphyses filiform, slightly inflated above.

On twigs of *Lonicera involucrata*.

TYPE LOCALITY: Mesa Lakes, Colorado.

DISTRIBUTION: Known only from the type locality.

ILLUSTRATIONS: Mycologia **28**: 302, *f. 1*.

Mr. Richard Korf who is working on the genus *Arachnopeziza* regards this as a true *Tapesia* and believes that it should not be included with the present genus.

4. **Helotiella trabinelloides** Rehm, Ann. Myc. **2**: 36.   1904.

Helotium trabinelloides Rehm, Hedwigia **26**: 82.   1887.
Solenopezia trabinelloides Sacc. Syll. Fung. **8**: 477.   1889.
Helotiella Nuttallii Ellis & Ev. Proc. Acad. Sci. Phila. **1894**: 351.   1894.
Dasyscypha trabinelloides Massee, Jour. Bot. **34**: 145.   1896.

Apothecia gregarious, sessile, flat-hemispheric, .5–1.5 mm. in diameter, cup-shaped when fresh, contracted and subspherical when dry, with only a small, round apical opening, orange-colored throughout, the margin fringed with hairs; hymenium watery-orange; hairs pale-yellow, straight, roughish; asci clavate-cylindric, sessile, reaching a length of 50–60 μ and a diameter of 5–6 μ, 8-spored; spores 2-seriate, ellipsoid, becoming 1-septate, 2.5–4 × 10–17 μ; paraphyses filiform, scarcely thickened above, with orange-colored granules.

On rotten wood.

TYPE LOCALITY: Europe.

DISTRIBUTION: West Virginia; also in Europe.

ILLUSTRATIONS: Jour. Bot. **34**: *pl. 357, f. 4.*

EXSICCATI: Ellis & Ev. N. Am. Fungi *3233.*

According to Massee (Jour. Bot. **34**: 146) *Helotiella Nuttallii* Ellis & Ev. is identical with *Helotium trabinelloides* Rehm distributed in his Ascomycetes *853.*

5. **Helotiella cornuta** (Ellis) Sacc. Syll. Fung. **8**: 474.   1889.

Peziza cornuta Ellis, Bull. Torrey Club **9**: 73.   1882.

Apothecia gregarious, minute, dull-white, flesh-colored when dry, sessile, thin, membranaceous, the margin sparingly fringed with straight, 4–5-septate, white, spreading hairs; asci clavate-cylindric, reaching a length of 50–60 μ and a diameter of 8–10 μ; spores 2-seriate, fusiform, becoming 1-septate, with a short, bristle-like appendage at either end, 3–4 × 12 μ; paraphyses filiform.

On decaying chestnut wood.

TYPE LOCALITY: Westchester, Pennsylvania.

DISTRIBUTION: Known only from the type locality.

Ellis writes: "I have not seen *P. arachnoidea* Schw. but that is said to be 'white villous,' while this in only ciliate."

6. **Helotiella major** Ellis & Ev. Proc. Acad. Sci. Phila. **1894**: 351.   1894.

Apothecia gregarious, often confluent, sessile, 1–2 mm. in diameter, externally dull-white, and densely clothed, especially about the margin, with short, pale, glandular hairs, expanded when fresh, the margin incurved when dry, attached by spreading, white filaments; asci cylindric or subcylindric, reaching a length of 65–70 μ and a diameter of 7–8 μ, 8-spored; spores mostly 2-

seriate, elongate-ellipsoid, slightly curved, becoming 1-septate, or occasionally 3-septate, 2.5–3 × 12–16 μ; paraphyses not observed.

On rotten wood.

TYPE LOCALITY: Ann Arbor, Michigan.

DISTRIBUTION: Known only from the type locality.

7. **Helotiella papyricola** Ellis & Ev. Proc. Acad. Sci. Phila. **1894**: 351. 1894.

Apothecia gregarious, sessile, or contracted below into a very short stipe, pale-orange, of a soft-waxy consistency, discoid when fresh, .2–.3 mm. in diameter, subspherical and nearly closed when dry, outside sparingly furfuraceous, margin fringed with short, spreading hairs; asci cylindric, very short-stipitate, reaching a length of 50–60 μ and a diameter of 12 μ, 8-spored; spores 2-seriate, fusoid, slightly curved, with two to four oil-drops, finally 1-septate, 3 × 12–15 μ; paraphyses filiform.

On old paper on the ground.

TYPE LOCALITY: Ann Arbor, Michigan.

DISTRIBUTION: Michigan and New Jersey.

EXCLUDED SPECIES

Helotiella Russellii (Berk. & Curt.) Sacc. Syll. Fung. **8**: 476. 1889; *Peziza Russellii* Berk. & Curt.; Berk. Grevillea **3**: 158. 1875. This is undoubtedly a *Nectria*. No material has been seen.

55. **ERINELLINA** Seaver, nom. nov.

Erinella Sacc. Syll. Fung. **8**: 507. 1889. Not *Erinella* Quél. 1886.
?Dasyscyphella Tranz. Hedwigia **38**: Beibl. (11). 1899.

Apothecia cup-shaped, or subdiscoid, externally clothed with hyaline, or subhyaline hairs; hymenium plane, or concave, variously colored; hairs rigid, or more often flexuous, smooth, or rough; asci cylindric to clavate, usually 8-spored; spores much elongated, fusiform to filiform, hyaline, or subhyaline, 3–many-septate; paraphyses filiform, clavate, or lanceolate.

Type species, *Peziza simillima* Berk. & Br.

The genus *Erinella* was established by Quélet and is a straight synonym of *Lachnella* as used here. Saccardo took up the name but used it in an entirely different sense thus creating a homonym. The new name above is proposed for the genus as treated by Saccardo.

Spores 100–120 μ long.
  Apothecia sessile, on *Pinus*.   1. *E. rhaphidospora.*
  Apothecia stipitate, on undetermined bark.   2. *E. calospora.*
Spores 40–65 μ long.
  Hymenium orange to scarlet.   3. *E. miniopsis.*
  Hymenium pallid to brown.
    Apothecia sessile.   4. *E. cervina.*
    Apothecia short-stipitate.   5. *E. appressa.*
Spores 16–40 μ long.
  On living leaves of *Persea*.   6. *E. maculosa.*
  On dead herbaceous stems or wood.
    Hairs yellowish, on herbaceous stems.   7. *E. Nylanderi.*
    Hairs white, on bark.   8. *E. longispora.*

1. **Erinellina rhaphidospora** (Ellis) Seaver, comb. nov.

Peziza rhaphidospora Ellis, Bull. Torrey Club **6**: 107. 1876.
Erinella rhaphidospora Sacc. Syll. Fung. **8**: 509. 1889.

Apothecia gregarious, sessile, minute, subconfluent, white-tomentose; hymenium plane, or slightly convex; asci cylindric-clavate, reaching a length of 100–140 μ and a diameter of 12–20 μ, 8-spored; spores nearly as long as the ascus (multiseptate?).
  On trunks of *Pinus*.
TYPE LOCALITY: Newfield, New Jersey.
DISTRIBUTION: New Jersey to Pennsylvania, and Bermuda.
EXSICCATI: Ellis, N. Am. Fungi *842*.

2. **Erinellina calospora** (Pat.) Seaver, comb. nov.

Erinella calospora Pat. Bull. Soc. Myc. Fr. **4**: 101. 1888.

Apothecia scattered, turbinate, short-stipitate, 2–3 mm. in diameter, externally white-tomentose; hymenium orange-yellow; hairs long, cylindric, flexuous; asci cylindric, substipitate, reaching a length of 120–150 μ and a diameter of 12–15 μ; spores nearly as long as the ascus, 2–3 × 90–120 μ, 6–14-septate, sub-hyaline; paraphyses simple, or branched, thickened above, septate, hyaline.
  On bark of some tree.
TYPE LOCALITY: Venezuela.
DISTRIBUTION: South America and the West Indies.
ILLUSTRATIONS: Bull. Soc. Myc. Fr. **4**: *pl. 18, f. 7.*

3. **Erinellina miniopsis** (Ellis) Seaver, comb. nov.

Peziza miniopsis Ellis, Bull. Torrey Club **8**: 66. 1881.
Erinella miniopsis Sacc. Syll. Fung. **8**: 510. 1889.
Dasyscyphella miniopsis Kanouse, Papers Mich. Acad. Sci. **23**: 151. 1938.

Apothecia gregarious, sessile, minute, less than 1 mm. in diameter, clothed with a white-tomentose coat of crisped hairs; hymenium concave, orange to scarlet; asci clavate-cylindric, reaching a length of 95 μ and a diameter of 12 μ, 8-spored; spores crowded, linear-lanceolate, 50–65 μ long, becoming multiseptate; paraphyses filiform, not thickened above.

On the outer bark of *Acer rubrum*.

TYPE LOCALITY: Newfield, New Jersey.

DISTRIBUTION: New Jersey to New Hampshire.

EXSICCATI: Ellis, N. Am. Fungi *563;* Barth. Fungi Columb. *2428;* Reliq. Farlow. *115.*

4. Erinellina cervina (Ellis & Ev.) Seaver, comb. nov.

Erinella cervina Ellis & Ev. Bull. Torrey Club **24**: 468.   1897.

Apothecia gregarious, sessile, at first subglobose, becoming urceolate, the margin incurved leaving only a small rounded opening, stag-colored; short-tomentose; hymenium concave, pallid; asci cylindric, or subcylindric, with a short, stem-like base, reaching a length of 75–85 μ and a diameter of 7–9 μ, 8-spored; spores fasciculate, acicular, attenuated toward each end, 2–2.5 $\times$ 50–60 μ; paraphyses stout, rather longer than the asci, scarcely enlarged above, 2–2.5 μ thick.

On decaying birch limbs.

TYPE LOCALITY: Dillon, Colorado.

DISTRIBUTION: Known only from the type locality.

The type specimen is in the herbarium of The New York Botanical Garden.

5. Erinellina appressa (Cash) Seaver, comb. nov.

Dasyscyphella appressa Cash, Univ. Iowa Stud. Nat. Hist. **17**: 217.   1937.

Apothecia obconic to infundibuliform, short-stipitate, .2–1 mm. in diameter .5 mm. high, waxy-membranous, ochraceous-tawny, pilose; hairs pale-brown, verrucose; hymenium cinnamon-brown; asci cylindric, narrowed above, reaching a length of 90–110 μ and a diameter of 7–9 μ, short-stipitate, 8-spored; spores fasciculate, filiform, 2–2.5 $\times$ 40–55 μ, subhyaline 1–5-septate; paraphyses filiform, 1.5 μ thick.

On woody stem.

TYPE LOCALITY: Panama.

DISTRIBUTION: Panama and Colombia.

ILLUSTRATIONS: Univ. Iowa Stud. Nat. Hist. **17**: *pl. 14, f. 2.*

This scarcely differs from *Erinella cervina* Ellis & Ev. so far as we can judge from the descriptions.

6. Erinellina maculosa (Ellis & Martin) Seaver, comb. nov.

Helotium maculosum Ellis & Martin, Am. Nat. **17**: 1284. 1883.

Apothecia sessile, .16 mm. in diameter, externally clothed with a few brown, bristle-like, faintly septate hairs arising from near the base; hymenium dull flesh-colored; asci clavate, reaching a length of 55 μ and a diameter of 12 μ, 8-spored; spores 2-seriate, broad-fusoid, becoming 3-septate, 4–5 μ 16–20 μ; paraphyses rather stout.

On pale-brown spots on living leaves of *Persea palustris*.

TYPE LOCALITY: Green Cove, Florida.

DISTRIBUTION: Known only from the type locality.

EXSICCATI: Ellis. N. Am. Fungi *1276*.

7. Erinellina Nylanderi (Rehm) Seaver, comb. nov.

Erinella Nylanderi Rehm in Rab. Krypt.-Fl. **1**[3]: 910. 1893.

Apothecia gregarious, sessile, globose, at first closed, expanding and becoming cup-shaped, externally yellow, or yellowish-brown, densely hairy, .3–2 mm. in diameter; hymenium faintly bluish, or reddish-gray; hairs pointed, septate, rough, greenish-yellow, reaching a length of 150 μ and a diameter of 4–5 μ; asci cylindric-clavate, slightly attenuated above, reaching a length of 80–90 μ and a diameter of 5–6 μ, 8-spored; spores elongated, fusiform, straight, or somewhat curved, becoming 3–5-septate, hyaline, 2 $\times$ 25–33 μ; paraphyses sparse, not strongly pointed, 3–4 μ in diameter.

On dead stems of herbaceous plants *Urtica* and *Impatiens*.

TYPE LOCALITY: Europe and Michigan.

DISTRIBUTION: New York; also in Europe.

Resembling and often confused with *Lachnella sulphurea*.

8. Erinellina longispora (Karst.) Seaver, comb. nov.

Lachnum longisporum Karst. Hedwigia **28**: 191. 1889.

?Erinella subcorticalis Pat. in Duss, Enum. Meth. Champ. Guadeloupe and Martin. 67. 1903.

?Dasyscyphella subcorticalis Cash, Mycologia **35**: 601. 1943.

Apothecia gregarious, or subsparse, sessile, or short-stipitate, white-tomentose, at first subglobose, then cup-shaped, .5 mm. in diameter; hymenium yellowish; asci cylindric-clavate, reaching a length of 60 μ and a diameter of 6–7 μ, 8-spored; spores fasciculate, fusoid-filiform, usually 3-septate, subhyaline, 2 × 18–40 μ; paraphyses 2 μ in diameter, acute at their apices.

On bark of some tree.

TYPE LOCALITY: Brazil.

DISTRIBUTION: Central to South America and the West Indies.

DOUBTFUL SPECIES

Dasyscyphella acutipila Cash, Univ. Iowa Stud. Nat. Hist. **17**: 216. 1937. On woody stems in Panama. Spores 1 × 33–45 μ. No specimens have been seen.

Erinella borealis Povah, Mycologia **24**: 241. 1932. Described from material collected on dead stems of *Chamaedaphne calyculata* on Isle Royale, Michigan. Later reported by Dr. B. Kanouse (Papers Mich. Acad. Sci. **20**: 74. 1935) as synonymous with *Dasyscyphella Cassandrae* Tranz.; *Erinella Cassandrae* Sacc. & Syd. in Sacc. Syll. Fung. **16**: 757. 1902. No material has been seen.

Erinella raphidophora (Berk. & Curt.) Sacc. Syll. Fung. **8**: 509. 1889; *Peziza raphidophora* Berk. & Curt.; Berk. Jour Linn. Soc. **10**: 368. 1868. First globose, then expanded, clothed with short, white hairs, the margin inflexed; asci elongate; spores linear, variable in length, but about 25 μ long. On rotten wood, Cuba and Venezuela.

Erinella similis Bres. Hedwigia **35**: 296. 1896. This species has been doubtfully reported from Porto Rico by the writer. Spores 2–3 × 35–4 μ.

56. **ECHINELLA** Massee, Brit. Fungus-Fl. **4**: 304. 1895.

Apothecia sessile, at first closed, then expanding and becoming cup-shaped; for some distance down the sides clothed with black, or brown hyphae; asci clavate, 8-spored; spores irregularly 2-seriate, hyaline, narrowly fusiform, becoming 3–many-septate; paraphyses filiform or slightly clavate.

Type species, *Peziza Vectis* Berk. & Br.

Distinguished from *Pirottaea* Sacc. by the elongated spores, and from *Erinellina* by the dark-colored hairs.

Echinella rhabdocarpa (Ellis) Seaver, comb. nov.

Peziza rhabdocarpa Ellis, Bull. Torrey Club **9**: 19. 1882.
Erinella rhabdocarpa Sacc. Syll. Fung. **8**: 510. 1889.

Apothecia scattered, sessile, .5 mm. in diameter, the margin clothed with short, black, fasciculate, obtusely clavate hairs; asci clavate-cylindric, reaching a length of 65–75 μ and a diameter of 10 μ; spores linear, multi-septate ?, nearly as long as the ascus.

On branches of *Comptonia asplenifolia*.
TYPE LOCALITY: Newfield, New Jersey.
DISTRIBUTION: New Jersey and Ontario, Canada.
EXSICCATI: Ellis, N. Am. Fungi *844*.

Family 5. CENANGIACEAE

Conidial stage present, variable in form, or unknown.

Apothecia usually occurring on woody plants, at first immersed and bursting through the outer bark, singly, or more often in cespitose clusters, cup-shaped, often opening with an irregular aperture, or discoid to patellate, pale to dark brownish-black, fleshy, waxy, or leathery, becoming horny when dry; asci usually broad-clavate, 4–8-spored; spores simple, ellipsoid, allantoid, rarely subglobose, or fusiform to filiform, then usually several-septate, hyaline, or colored; paraphyses filiform, often strongly enlarged above, the ends free or glued together forming an epithecium, hyaline, or colored.

Asci containing 4–8 ascospores.
  Ascospores simple.
    Apothecia tomentose.   1. VELUTARIA.
    Apothecia not tomentose.
      Ascospores remaining hyaline.
        Spores not over 14–20 μ long.   2. CENANGIUM.
        Spores 30 μ or more long.
          Apothecia hysteriform when young.   3. DERMATEOPSIS.
          Apothecia not hysteriform.   4. GODRONIOPSIS.
      Ascospores becoming brown.   5. SPHAERANGIUM.
  Ascospores becoming septate, with cross walls only.
    Asci with ascospores only.
      Spores 1-septate.
        Spores hyaline.   6. CENANGELLA.
        Spores becoming brown.   7. DERMATELLA.
      Spores more than 1-septate.
        Spores long fusoid to filiform.   8. GODRONIA.
        Spores ellipsoid, becoming tardily septate.
          Apothecia light-colored; conidia ellipsoid.   9. PEZICULA.
          Apothecia dark-colored; conidia fusiform. 10. DERMEA.
    Asci with both ascospores, and spermatoid bodies. 11. DURANDIELLA.
  Ascospores with cross and longitudinal walls (muriform). 12. MURANGIUM.
Asci containing only spermatoid bodies. 13. TYMPANIS.

1. **VELUTARIA** Fuckel, Symb. Myc. 300.   1869.

Schweinitzia Massee, Brit. Fungus-Fl. **4**: 134.   1895.

Apothecia erumpent, becoming apparently superficial, for the most part single, or in dense, cespitose clusters, at first closed, opening but remaining cupulate, externally clothed with a dense woolly growth, tan-colored, or light-brown; hymenium concave, or nearly plane; asci subcylindric, 8-spored; spores ellipsoid. simple, hyaline, or slightly colored; paraphyses filiform to clavate.

Type species, *Peziza rufoolivacea* Alb. & Schw.

The genus *Schweinitzia* was proposed by Massee based on *Cenangium phaeosporum* Cooke.   No authentic material of this species is available but the original drawings and notes of Massee are in the herbarium of The New York Botanical Garden.   In these notes he states "belongs to same genus as *P. rufoolivacea.*" Except for the brown spores the two species would seem to be identical.   The globose cells at the tips of the hairs in *Cenangium phaeosporum* have been noted in *P. rufoolivacea*, or material so determined.

Nannfeldt (Nova Acta Soc. Sci. Upsal. IV. **8**: 302) treats *Schweinitzia* as a doubtful synonym of *Velutaria*.   The present author would agree but would remove the doubt.   In fact, if as admitted by Massee, the two species are congeneric there was no reason for the founding of the genus *Schweinitzia* in the first place.

1. **Velutaria rufoolivacea** (Alb. & Schw.) Fuckel, Symb. Myc. 300.   1869.

Peziza rufoolivacea Alb. & Schw. Consp. Fung. 320.   1805.
Lachnea rufoolivacea Gill. Champ. Fr. Discom. 85.   1882.
Humaria rufoolivacea Quél. Ench. Fung. 291.   1886.
Lachnella rufoolivacea Phill. Brit. Discom. 275.   1887.
Schweinitzia rufoolivacea Massee, Brit. Fungus-Fl. **4**: 135.   1895.
Cenangium Rubi Bäumler, Ann. Nat. Hofmus. Wien **13**: 440.   1898.
Phaeangium Rubi Sacc. & Syd.; Sacc. Syll. Fung. **16**: 764.   1902.

Apothecia erumpent, either occurring singly, or several together in dense clusters and soon becoming apparently superficial, externally clothed with a dense coat of poorly developed and more or less disjuncted hairs often each with a large, globose apex which gives a dense mealy appearance, reaching a diameter of 1–3 mm.; hymenium concave, or nearly plane, dark-colored, slightly olivaceous to black; asci subcylindric, reaching a length of 110–120 μ and a diameter of 12–14 μ, 8-spored; spores 1-seriate, ellipsoid, 6–8 × 12–15 μ; hyaline, or faintly colored; paraphyses filiform, rather strongly enlarged above.

On dead branches of various kinds: *Quercus, Betula, Andromeda, Acer, Sassafras* and *Rubus*.

TYPE LOCALITY: Europe.

DISTRIBUTION: New Jersey to Oregon; also in Europe.

ILLUSTRATIONS: Alb. & Schw. Consp. Fung. *pl. 11, f. 4;* Phill. Brit. Discom. *pl. 8, f. 49.*

EXSICCATI: Ellis, N. Am. Fungi *69* (as *Dermatea lobata* Ell.)

DOUBTFUL AND EXCLUDED SPECIES

Velutaria cinereofusca (Schw.) Bres. has been regarded as a synonym of this species but is now regarded as a *Cyplella.* For details see page 275.

Velutaria griseovitellina Fuckel, Symb. Myc. 300. 1869. This species has been reported from Washington under the name of *Tapesina griseovitellina* Höhn. The 3-septate spores would place it in the genus *Belonium* as treated here.

Schweinitzia phaeospora (Cooke) Massee, Brit. Fungus-Fl. **4**: 135. 1895; *Cenangium phaeosporum* Cooke, Grevillea **12**: 44. 1883; *Phaeangium phaeosporum* Sacc. Syll. Fung. **16**: 765. 1902. No material available but apparently scarcely distinct from the preceding.

2. **CENANGIUM** Fries, Summa Veg. Scand. 364. 1849.

Apothecia occurring singly, or more often in cespitose clusters of variable number, erumpent, usually light-colored not black, at first closed, finally opening, deep cup-shaped, or becoming more shallow when expanded, the margin often incurved; asci usually 8-spored, of variable form; spores ellipsoid, or elongated, often allantoid, hyaline, simple; paraphyses filiform, simple, or branched, scarcely enlarged above, hyaline or subhyaline.

Type species, *Cenangium ferruginosum* Fries.

Occurring on coniferous trees.
   On branches of various conifers.

Apothecia 1–1.5 mm. in diameter, reddish-brown.	1. *C. Abietis.*
Apothecia 2–5 mm. in diameter, purplish-brown.	2. *C. atropurpureum.*
On needles of species of *Pinus.*	3. *C. acuum.*
Not on coniferous trees.	
Apothecia medium large; 5 mm. to 1 cm. in diameter.	
On branches of *Alnus.*	4. *C. furfuraceum.*

On branches of *Populus*, more rarely on *Fraxi-nus*.
Apothecia scattered, densely white-pruinose.   5. *C. pruinosum.*
Apothecia cespitose, not pruinose.   6. *C. populneum.*
Apothecia small, less than 5 mm. in diameter.
Spores medium large, 7–10 μ long.
On branches of *Ceanothus*.   7. *C. aureum.*
On leaves of *Yucca*.   8. *C. Yuccae.*
Spores small 4–6 μ long.
Apothecia accompanied by flask-shaped
conidial bodies.   9. *C. dimorphum.*
Apothecia not as above, forming tubercular
masses.   10. *C. tuberculiforme.*

1. **Cenangium Abietis** (Pers.) Rehm in Rab. Krypt.-Fl. **1**³: 227. 1889.

Peziza Abietis Pers. Syn. Fung. 671. 1801.
Cenangium ferruginosum Fries, Vet. Akad. Handl. **1818**. 361. 1818.
Dermatea Pini Phill. & Hark. Grevillea **13**: 22. 1884.

Apothecia gregarious, usually occurring in cespitose clusters, subsessile, at first nearly globose, later opening irregularly, expanding and becoming subturbinate, often irregular from mutual pressure, the margin inflexed when dry, coated externally with a reddish-brown powder, 1–1.5 mm. broad; hymenium concave, yellowish to olive-brown; asci cylindric-clavate, reaching a length 100–120 μ and a diameter 12–18 μ, 8-spored; spores ellipsoid, 5–7 × 10–14 μ, 1-seriate, or partially 2-seriate; paraphyses stout, simple, about 2 μ in diameter, slightly enlarged above, pale-yellowish.

On branches of various species of *Pinus*, *Abies*, *Picea* and probably other conifers.

TYPE LOCALITY: Europe.

DISTRIBUTION: New England to Oregon, California and Alabama; also in Europe.

ILLUSTRATIONS: E. & P. Nat. Pfl. **1**¹: 233, *f. E–G;* Grev. Scot. Crypt. Fl. **4**: *pl. 197;* Cooke, Handb. Brit. Fungi **2**: *f. 337.*

EXSICCATI: Ellis & Ev. N. Am. Fungi *2050, 2050* (B), Fungi Columb. *1416, 1514;* Clements, Crypt. Form. Colo. *517.*

2. **Cenangium atropurpureum** Cash & Davidson, Mycologia **32**: 734. 1940.

Apothecia erumpent, occurring singly, or two or three together from a stroma beneath the bark, subglobose, then cupulate, reaching a diameter of 2–5 mm. and of equal height, fleshy-leathery, substipitate, exterior furfuraceous, dull purplish-black to brownish-black when moist, darker when dry, margin lacerate, becoming inrolled when dry and triangular or hysteroid; hymenium light ochraceous-buff; asci cylindric, reaching a length of 70–85 μ and a diameter of 9–12 μ, 8-spored; spores broad-ellipsoid, irregularly 1-seriate, hyaline, simple, containing several granules, 5–8 $\times$ 9.5–11 μ; paraphyses hyaline, unbranched, enlarged at their apices, 3–5 μ in diameter.

On dead twigs *Pinus nigra*, *P. Mugho*, *P. pungens*, *P. rigida*, *P. virginiana*, *P. Taeda* and *P. sylvestris*.

TYPE LOCALITY: Sugar Loaf Mountain, Maryland, on *Pinus nigra*.

DISTRIBUTION: Maryland to Pennsylvania and Georgia.

ILLUSTRATIONS: Mycologia **32**: 729, *f. 1, A, B*.

3. **Cenangium acuum** Cooke & Peck; Cooke & Ellis, Grevillea **7**: 40. 1878.

Peziza Pinastri Cooke & Peck; Cooke, Bull. Buffalo Soc. Nat. Sci. **1**: 297. 1875. Not *Cenangium Pinastri* Hazsl. 1887.
Mollisia Pinastri Sacc. Syll. Fung. **8**: 327. 1889.

Apothecia erumpent through the epidermis of the host, at first closed, soon opening with an irregular aperture leaving the margin irregularly toothed, externally reddish-brown, reaching 1 mm. in diameter; hymenium concave or nearly plane, lighter than the outside of the apothecium; asci clavate, reaching a length of 80–100 μ and a diameter of 8–10 μ, tapering into an abrupt, stem-like base, 8-spored; spores fusoid, 3–4 $\times$ 12–14 μ; paraphyses filiform, enlarged above, reaching a diameter of 2–3 μ.

On needles of *Pinus Strobus* and other species of *Pinus*.

TYPE LOCALITY: Albany County, New York.

DISTRIBUTION: New York and New Jersey to British Columbia; also in Europe.

EXSICCATI: Ellis, N. Am. Fungi *367;* Rehm, Ascom. *822* (from Newfield, N. J.), *822 b* (from Lake Huron, Ontario); Sydow, Fungi Exotici Exsicc. *275* (from Lake Huron, Ontario); Rab.-Winter, Fungi Eu. *3365* (from Newfield, N. J.); Ellis & Ev. Fungi Columb. *642;* Barth. Fungi Columb. *4705.*

Pezizella ontariensis seems to differ in the color and size of the spores. However, it is suggestive of this species.

4. **Cenangium furfuraceum** (Roth) Sacc. Syll. Fung. **8**: 565. 1889.

Peziza furfuracea Roth, Catal. Bot. **3**: 257.   1806.
Phibalis furfuracea Wallr. Fl. Crypt. Germ. **2**: 447.   1833.
Dermatea furfuracea Fries, Summa Veg. Scand. 362.   1849.
Encoelia furfuracea Karst. Myc. Fenn. **1**: 218.   1871.

Apothecia erumpent in cespitose clusters of five or six each, or more rarely occurring singly, at first entirely closed, finally opening rather irregularly and expanded but with the margin usually incurved, externally whitish-furfuraceous, the individual apothecia reaching a diameter of 1 cm. or more; asci narrowly clavate, reaching a length of 80 μ and a diameter of 6–7 μ, 8-spored; spores narrow-ellipsoid, or allantoid, often with two small oil-drops, 2–4 $\times$ 8–10 μ; paraphyses filiform slightly enlarged above.

On branches of species of *Alnus*.

TYPE LOCALITY: Europe.

DISTRIBUTION: Newfoundland to Idaho, California, south to Iowa, and Pennsylvania; also in Europe.

ILLUSTRATIONS: Roth, Catal. Bot. *pl. 9, f. 2* (as "*Thelephora fimbriata*").

EXSICCATI: Shear, N. Y. Fungi *327;* Brenckle, Fungi Dak. *426.* Wilson & Seaver, Ascom. *52.*

5. **Cenangium pruinosum** (Ellis & Ev.) Seaver, comb. nov. (PLATE 130.)

Peziza pruinosa Ellis & Ev. Jour. Myc. **4**: 100.   1888.
Dermatea pruinosa Sacc. Syll. Fung. **8**: 555.   1889.

Apothecia gregarious, occurring in large numbers underneath the outer bark of the host, which rolls back exposing large blackened areas on which the apothecia are seated, the individual apothecia deep cup-shaped with the mouth at first constricted, expanding and reaching a diameter of 3–4 mm., externally covered with a white, granular coat, brown to brownish-black inside and out, the margin slightly crenate, when dry becoming boat-shaped or angular; asci narrow-clavate, reaching a length of 55 μ and a diameter of 4–5 μ, 8-spored; spores minute, allantoid, 2 $\times$ 8–10 μ; paraphyses filiform, very slender, enlarged above, with dark-brown tips.

On trunks of *Populus*.

CENANGIUM PRUINOSUM

TYPE LOCALITY: Colorado.

DISTRIBUTION: Colorado and Toronto, Canada.

The illustration accompanying this description was made from material collected by the writer and Mr. Bethel in 1910. A note in the herbarium of The New York Botanical Garden on a Bethel label 1898 states "this is abundant everywhere in the Mts. where aspen (*Populus tremuloides*) is found." It was apparently confused by Mr. Ellis with *Cenangium populneum*, although Ellis, himself, described it ten years before under the name indicated above.

6. **Cenangium populneum** (Pers.) Rehm in Rab. Krypt.-Fl. 1³: 220. 1889. (PLATE 131.)

Peziza populnea Pers. Tent. Disp. Fung. 35. 1797.
Peziza fascicularis Alb. & Schw. Consp. Fung. 315. 1805.
Peziza crispa Sow. Engl. Fungi *pl. 425.* 1814.
?Cenangium populinum Schw. Trans. Am. Phil. Soc. II. **4**: 239. 1832.
Dermatea fascicularis Fries, Summa Veg. Scand. 362. 1849.
Encoelia fascicularis Karst. Myc. Fenn. **1**: 217. 1871.
Cenangium fasciculare Karst. Acta Soc. Fauna Fl. Fenn. **2**⁶: 145. 1885.
Encoelia populnea Kirsch. Ann. Myc. **33**: 222. 1935.

Apothecia erumpent through the outer bark in cespitose clusters of six to twelve each, the clusters often 5 mm. or more in diameter, or more rarely solitary, the individual apothecia at first closed, finally opening and usually deep cup-shaped, often much contorted and twisted from mutual pressure, grayish-brown, reaching a diameter of 3–5 mm.; asci narrowly clavate, reaching a length of 75–90 μ and a diameter of 8–9 μ, 8-spored; spores narrowly ellipsoid, or allantoid, 3–4 × 12–14 μ; paraphyses slender, slightly enlarged above, reaching a diameter of 3 μ.

On dead branches of *Populus tremuloides* and *Populus grandidentata;* also reported on *Fraxinus*.

TYPE LOCALITY: Europe.

DISTRIBUTION: Massachusetts to Colorado, Iowa and Kansas; also Europe.

ILLUSTRATIONS: Alb. & Schw. Consp. Fung. *pl. 12, f. 2;* Rab. Krypt.-Fl. 1³: 215, *f. 2;* Bull. Lab. Nat. Hist. State Univ. Iowa **5**: *pl. 25, f. 1;* Sow. Engl. Fungi *pl. 425, f. 1.*

EXSICCATI: Ellis, N. Am. Fungi *1314;* Barth. Fungi Columb. *2209;* Brenckle, Fungi Dak. *209* (as var. *prunicolum* Rehm); Reliq. Farlow. *104.*

While this species usually occurs on *Populus tremuloides* one specimen from Canada was reported on *Populus grandidentata*. Albertini and Schweinitz (Consp. Fung. 315) report this species on *Salix alba* and on *Fraxinus.* A specimen in the Ellis collection is accompanied by the following unsigned note: "*Dermatea fascicularis* A. & S. Have found this again on ash; (you questioned its being an ash) when *120* was sent. Can this be a different species?" Apparently Ellis questioned the host. However we have in the herbarium of The New York Botanical Garden two other specimens reported on ash both by reliable collectors, one by S. H. Burnham at Hudson Falls, New York the fungus determined by the writer and one by R. F. Cain from Ontario, Canada, the fungus determinated by J. W. Groves. The spores in those specimens on ash agree with those from poplar so apparently this fungus is not restricted to *Populus*.

7. **Cenangium aureum** Ellis & Ev. Bull. Torrey Club **24**: 468. 1897.

Apothecia erumpent through transverse crevices in the bark, occurring singly, or two or three together, short-stipitate, cup-shaped, golden-yellow, externally flocose-furfuraceous, becoming nearly smooth, margin slightly incurved, 2–3 μ in diameter; stem short, stout, about 1 mm. long; asci clavate-cylindric, reaching a length of 90–110 μ and a diameter of 7–8 μ, 8-spored, gradually narrowed below; spores 1-seriate, ovoid to pyriform, 3.5 × 7–10 μ; paraphyses filiform, about as long as the ascus, scarcely thickened above.

On dead branches of *Ceanothus velutinus*.

Type locality: Bear Valley, Colorado (alt. 7000 ft.).

Distribution: Known only from the type locality.

Type material in The New York Botanical Garden has been examined, and is in excellent condition.

8. **Cenangium Yuccae** Clements, Crypt. Form. Colo. *518.* 1907.

Apothecia erumpent, scattered, at first closed, opening and becoming cup-shaped, externally densely furfuraceous, brown, reaching a diameter of 1 mm.; hymenium lighter, creamy; asci cylindric, reaching a length of 100–120 μ and a diameter of 16–12 μ, 8-spored; spores 1-seriate, ellipsoid, not granular within, about 8 × 10 μ paraphyses 2 μ in diameter, scarcely enlarged above, granular within, slightly yellowish.

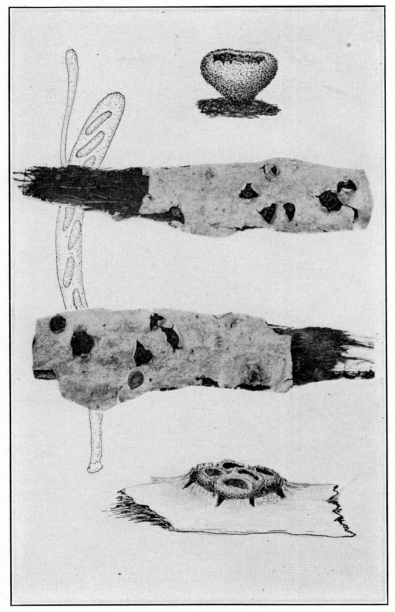

CENANGIUM POPULNEUM

Apotheciis sparsis, erumpentibus, primo clausis dein scutellatis, extus furfuraceis, brunneis, 1 mm. diam.; hymenio cremeo; ascis cylindraceis, 8-sporis; sporis ellipsoideis, 8 × 10 μ; paraphysibus filiformibus, 2 μ diam.

On leaves of *Yucca*.

TYPE LOCALITY: Mesa Verda, Colorado.

DISTRIBUTION: Colorado and California.

EXSICCATI: Clements, Crypt. Form. Colo. *518*.

A second collection of this species was obtained by the writer and Ellsworth Bethel in 1910. A third collection from Santa Anita Canyon, California was sent to the writer in 1939. This was at first thought to be an undescribed species but it was later found that Clements had distributed it in 1907 under the above name as a new species, without diagnosis.

9. **Cenangium dimorphum** Seaver, comb. nov.

Dermatea dimorpha Seaver, Mycologia **16**: 8. 1924.

Conidial stage described as consisting of club-shaped or flask-shaped bodies 1–2 mm. high, either occurring separately, or associated with and often fused to the base of the apothecia; pycnidia minute, more or less angular, about 2 μ in diameter.

Apothecia usually cespitose, sessile, or substipitate, apparently erumpent through the outer bark and becoming subsuperficial, shallow cup-shaped, with a purplish tint, furfuraceous, the hymenium somewhat darker, reddish, reaching a diameter of 1–2 mm.; asci clavate, reaching a length of 40 μ and a diameter of 4 μ, 8-spored; spores usually 1-seriate, ellipsoid, often with two small oil-drops; 2 × 4 μ; paraphyses very slender.

On the bark of some undetermined shrub or vine.

TYPE LOCALITY: Louisenhoj Estate on the island of St. Thomas.

DISTRIBUTION: St. Thomas of the Virgin Islands and Porto Rico; also in Venezuela and Brazil.

ILLUSTRATIONS: Mycologia **16**: *pl. 2, f. 1–4*.

In addition to the type, one specimen collected in Porto Rico by John Stevenson in February 1914 has been referred to this species, with which it agrees perfectly. Also one specimen collected by Dr. Carlos F. Chardon in Venezuela has been examined. The latter is said to have been collected on some dead *Citrus*. It was collected on a coffee plantation near San Cristobal. One specimen from the Farlow Herbarium collected by Rick in Brazil labeled *"Dermatea episphaeria"* Schw. is identical with our species but no authentic material of *Cenangium epi-sphaeria* Schw. has been seen.

10. **Cenangium tuberculiforme** Ellis & Ev. Proc. Acad. Sci. Phila. **1893**: 452. 1893.

Apothecia erumpent in dense clumps forming tubercular masses up to 1 cm. in diameter, individual apothecia light-yellow, furfuraceous-squamulose outside, irregularly cup-shaped with the margin incurved, reaching a diameter of 2–3 mm.; hymenium pale brick-red, contracted below into a short, stem-like base; asci cylindric-clavate, reaching a length of 40–50 μ and a diameter of 4–5 μ, 8-spored; spores obliquely 1-seriate, ellipsoid, hyaline, with two small oil-drops, 2–2.5 × 5–6 μ; paraphyses filiform, branched, scarcely enlarged above.

On dead twigs of *Ilex glabra*.

TYPE LOCALITY: Newfield, New Jersey.

DISTRIBUTION: New Jersey and Massachusetts.

EXSICCATI: Reliq. Farlow. *105*.

Excellent type material is in the herbarium at The New York Botanical Garden. The only other specimen seen is the one distributed by the Farlow Herbarium collected at East Billerica, Massachusetts by J. B. Rorer. This is a very conspicuous and interesting fungus.

DOUBTFUL AND EXCLUDED SPECIES

Many species of *Cenangium* have been reported from North America with meager descriptions and in numerous cases based on sterile material. Some of these are not ascomycetes at all. The names are recorded with the hosts, hoping that future collectors may be able to tie them up with something concrete.

Cenangium aeruginosum Schw. Trans. Am. Phil. Soc. II. **4**: 239. 1832. On decorticated wood. Asci and spores unknown.

Cenangium alboatrum Ellis & Ev. Bull. Torrey Club **24**: 136. 1897. On decorticated, decaying chestnut log. This does not appear to be a *Cenangium;* material very poor.

Cenangium alpinum Ellis & Ev. Bull. Torrey Club **24**: 468. 1897. On decorticated limbs of *Abies*. Spores described as allantoid, 2.5 × 10–14 μ.

Cenangium Andromedae (Schw.) Fries, Syst. Myc. **2**: 182. 1822; *Peziza Andromedae* Schw. Schr. Nat. Ges. Leipzig **1**: 123. 1822. Asci and spores unknown.

Cenangium apertum Schw. Trans. Am. Phil. Soc. II. **4**: 239. 1832. On dead branches of *Hydrangea*. Nothing is known of this species.

Cenangium asterinosporum Ellis & Ev. Bull. Torrey Club **10**: 76. 1883. Apothecia erumpent in cespitose cluster 2–3 mm. in diameter, individuals .5–1 mm. in diameter; spores ellipsoid or subpyriform, subhyaline, constricted in the center, becoming 3-septate and submuriform, 6–8 × 15–20 μ. New Jersey. A *Pezicula* ?

Cenangium Cassandrae Peck, Ann. Rep. N. Y. State Mus. **31**: 48. 1879. On *Cassandra calyculata*. Spores described as linear, 28–30 μ long. Compare *Godronia Cassandrae.*

Cenangium Castaneae Schw. Trans. Am. Phil. Soc. II. **4**: 239. 1832. On young branches of *Castanea*. Asci and spores unknown.

Cenangium clavatum Schw. Trans. Am. Phil. Soc. II. **4**: 238. 1832. No spores found. See *Dermea Cerasi.*

Cenangium compressum Schw. Trans. Am. Phil. Soc. II. **4**: 238. 1832. On *Betula*. Asci and spores unknown.

Cenangium conglobatum Ellis & Ev. Proc. Acad. Sci. Phila. **1895**: 428. 1895. On dead branches of some deciduous tree or shrub. Spores described as allantoid, 2–2.5 × 5–6.5 μ.

Cenangium contortum Berk. & Curt.; Cooke, Grevillea **21**: 75. 1893. On *Cornus*. Description inadequate. No material seen.

Cenangium confusum Schw. Trans. Am. Phil. Soc. II. **4**: 238. 1832. No specimen of this species could be found in the Schweinitz collection. Two specimens in the herbarium of The New York Botanical Garden under this name have 3-septate ascospores 3–4 × 35–45 μ and curved. Whether these are correctly named we do not know.

Cenangium Crataegi Schw. Trans. Am. Phil. Soc. II. **4**: 239. 1832. No material found in the Schweinitz collection at Philadelphia.

Cenangium Ellisii Sacc. Syll. Fung. **8**: 566. 1889; *Dermatea purpurea* Ellis, Bull. Torrey Club **6**: 108. 1875. Not *Dermatea purpurea* (Hedw.) Fries. 1849. Said to resemble *Dermatea furfuracea* but smaller. Ellis states "unfortunately the insects have nearly destroyed my spec." No specimens have been seen.

Cenangium enteroxanthum Schw. Trans. Am. Phil. Soc. II. **4**: 240. 1832. Asci and spores unknown. Description indicates that it may be one of the Hysteriales.

Cenangium episphaeria Schw. Trans. Am. Phil. Soc. II. **4**: 238. 1832. On *Diatrype Stigma* and other sphaeriaceous fungi. Asci and spores unknown, very doubtful. See *Cenangium dimorphum.*

Cenangium fatiscens Schw. Trans. Am. Phil. Soc. II. **4**: 239. 1832. On bark of *Syringa* and *Morus alba*. No asci or spores indicated. It may not be a Discomycete at all.

Cenangium fibrisedum Schw. Trans. Am. Phil. Soc. II. **4**: 239. 1832. On *Castanea*. Asci and spores unknown.

Cenangium Juglandis Berk. & Curt.; Berk. Grevillea **4**: 5. 1875. On *Juglans*. Description inadequate.

Cenangium leptospermum Berk. & Curt.; Berk. Grevillea **4**: 5. 1875. On *Abies*. Description inadequate.

Cenangium nigrofuscum Schw. Trans. Am. Phil. Soc. II. **4**: 239. 1832. On old wood. Asci and spores unknown.

Cenangium palmatum Schw. Trans. Am. Phil. Soc. II. **4**: 238. 1832. On bark of *Rhododendron maximum*. Asci and spores unknown. Very doubtful.

Cenangium patellatum Cooke, Grevillea **21**: 75. 1892. On *Acer*. Spores described as ellipsoid, tawny-brown, or amber, 5–6 × 18–20 μ. No material available.

Cenangium pezizoides Peck, Ann. Rep. N. Y. State Mus. **31**: 48. 1879. The spores suggest a *Dermea* but no septa observed.

Cenangium pyrinum Schw. Trans. Am. Phil. Soc. II. **4**: 239. 1832. Asci and spores unknown. Description indicates one of the Hysteriales. On *Pyrus*.

Cenangium punctoideum (Cooke) Sacc. Syll. Fung. **8**: 570. 1889; *Tympanis punctoidea* Cooke, Bull. Buffalo Soc. Nat. Sci. **3**: 30. 1875; *Phaeangium punctoideum* Sacc. & Syd. in Sacc. Syll. Fung. **16**: 765. 1902. No asci could be found in the type specimen.

Cenangium quercinum Schw. Trans. Am. Phil. Soc. II. **4**: 239. 1832; *Hysterium quercinum* Schw. Schr. Nat. Ges. Leipzig **1**: 49. 1822. Not a *Cenangium*.

Cenangium Rhois Schw. Trans. Am. Phil. Soc. II. **4**: 238. 1832. On branches of *Rhus glabra*. Asci and spores unknown.

Cenangium rigidum Schw. Trans. Am. Phil. Soc. II. **4**: 238. 1832. No apothecia found in the Schweinitz specimen. See *Dermea Prunastri*.

Cenangium Rosae Schw. Trans. Am. Phil. Soc. II. **4**: 239. 1832. On bark of *Rosa*. Asci and spores unknown.

Cenangium rubiginellum Sacc. Syll. Fung. **8**: 560. 1889. *Cenangium rubiginosum* Cooke; Rav. Fungi Am. *635*. Not *Peziza rubiginosa* Fries.

Cenangium sphaeriaemorphum Schw. Trans. Am. Phil. Soc. II. **4**: 238. 1832. On *Quercus alba*. Asci and spores unknown. May be a lichen.

Cenangium Spiraeae (Schw.) Sacc. Syll. Fung. **8**: 574. 1889. *Dermea Spiraeae* Schw. Trans. Am. Phil. Soc. II. **4**: 237. 1832. Schweinitz specimens examined May 21, 1931 and found worthless.

Cenangium Staphyleae Schw. Trans. Am. Phil. Soc. II. **4**: 238. 1832. On *Staphylea*. Asci and spores not known. Species very doubtful.

Cenangium sticticum (Berk. & Curt.) Sacc. Syll. Fung. **8**: 559. 1889; *Tympanis stictica* Berk. & Curt.; Peck, Grevillea **4**: 3. 1875. On *Salix babylonica*. Asci described as clavate; spores fusiform, curved, quadrinucleate. The species was also reported on *Prinus verticillata* and on *Quercus tinctoria*. The description is inadequate and material not available.

Cenangium tryblidioides Ellis & Ev. Bull. Torrey Club **4**: 136. 1897. On decorticated willow, *Salix*. Material weathered and unsatisfactory.

Cenangium triangulare Schw. Trans. Am. Phil. Soc. II. **4**: 238. 1832; *Coccomyces triangularis* Sacc. Syll. Fung. **8**: 750. 1889.

Cenangium tumorum Schw. Trans. Am. Phil. Soc. II. **4**: 239. 1832. On old wood. Asci and spores unknown.

Cenangium ustale (Berk. & Curt.) Sacc. Syll. Fung. **8**: 568. 1889; *Peziza ustalis* Berk. & Curt.; Berk. Grevillea **3**: 152. 1875. Description inadequate. No material available.

Cenangium Viburni (Schw.) Fries, Syst. Myc. **2**: 185. 1822; *Peziza Viburni* Schw. Schr. Nat. Ges. Leipzig **1**: 123. 1822. This may be a *Tympanis*. On *Viburnum*.

3. **DERMATEOPSIS** Nannf. Nova Acta Soc. Sci. Upsal. IV. **8**: 89. 1932.

Apothecia erumpent, thickly gregarious, or closely crowded, often cespitose, when young bilaterally compressed, hysteriform, opening with an irregular aperture leaving the margin often notched, or occasionally toothed, externally yellowish; hymenium darker, reaching a diameter of less than 1 mm.; asci 8-spored; spores long-fusoid; paraphyses filiform.

Type species, *Dermatea tabacina* Cooke.

1. **Dermateopsis tabacina** (Cooke) Nannf. Nova Acta Soc. Sci. Upsal. IV. **8**: 89. 1932.

Dermatea tabacina Cooke, Bull. Buffalo Soc. Nat. Sci. **3**: 24. 1877.
Godronia tabacina Seaver, Mycologia **37**: 350. 1945.

Apothecia as above; asci broad-clavate, reaching a length of 100 μ and a diameter of 30 μ, 8-spored; spores irregularly 2-seriate, very long, fusoid, straight, or curved, granular within, 8–10 × 60–75 μ; paraphyses filiform, enlarged above.

On branches of *Quercus alba* and *Quercus coccinea*.

TYPE LOCALITY: Newfield, New Jersey.

DISTRIBUTION: Known only from the type locality.

EXSICCATI: Ellis, N. Am. Fungi *146;* Rehm, Ascom. *359* (from Newfield, New Jersey); Roum. Fungi Selecti Exsicc. *1712* (from Newfield, New Jersey); Thüm. Myc. Univ. *1560* (from Newfield, New Jersey).

4. **GODRONIOPSIS** Diehl & Cash, Mycologia **21**: 243. 1929.

Apothecia sessile, or subsessile, seated on a black, blister-like subiculum which seems to be erumpent through the outer bark, at first closed, later cup-shaped, or subdiscoid, corky-leathery; asci clavate, 8-spored; spores simple, hyaline, fusiform; paraphyses filiform, slightly enlarged above, forming an epithecium.

Type species, *Peziza quernea* Schw.

1. **Godroniopsis quernea** (Schw.) Diehl & Cash, Mycologia **21**: 244. 1929.

Peziza quernea Schw. Schr. Nat. Ges. Leipzig **1**: 124. 1822.
Cenangium turgidum Fries, Syst. Myc. **2**: 186. 1822. Not *Cenangium turgidum* Duby. 1830.
Patellaria cenangiicola Ellis & Ev. Jour. Myc. **4**: 56. 1888.
Patellea cenangicola Sacc. Syll. Fung. **8**: 784. 1889.
Cenangium querneum Seymour; Thaxter, Mycologia **14**: 101. 1922.

Apothecia as described above, reaching a diameter of 1 mm., externally ridged the ends of the ridges giving rise to a toothed margin; hymenium brown; asci clavate to cylindric-clavate, reaching a length of 150 μ and a diameter of 13–16 μ, 8-spored; spores irregularly 2-seriate, simple, hyaline, subfusiform to fusiform, at first symmetrical and larger at one end, 6–10 × 30–47 μ; paraphyses filiform, hyaline at the base, enlarged above and light chestnut brown, 3–5.5 μ in diameter, the ends forming an epithecium.

On various species of *Quercus*.

TYPE LOCALITY: North Carolina.

DISTRIBUTION: New Hampshire to Missouri, Florida and Alabama.

ILLUSTRATIONS: Mycologia **21**: 245. *f. 1, 2.*

EXSICCATI: Rav. Fungi Car. **4**: *24;* Ellis & Ev. N. Am. Fungi *2148.*

5. **SPHAERANGIUM** Seaver, nom. nov.

Phaeangium Sacc. Syll. Fung. **16**: 764. 1902 (in part). Not *Phaeangium* Pat. 1894.

Apothecia as in *Cenangium;* asci clavate, 4–8-spored; spores large subglobose, or ellipsoid, at first hyaline, becoming brown; paraphyses forming an epithecium.

Type species, *Dermatea tetraspora* Ellis.

A note in the herbarium of The New York Botanical Garden by H. Rehm states that this species might be the type of a new genus, although no name was suggested. While the genus *Phaeangium* Sacc. was not founded on this species, it was included.

Asci 4-spored.   1. *S. tetrasporum.*
Asci 8-spored.
  Spores 19–22 × 40–50 μ.   2. *S. magnisporum.*
  Spores 20 × 26 μ.   3. *S. Tiliae.*

1. **Sphaerangium tetrasporum** (Ellis) Seaver, comb. nov.
  (PLATE 132.)

Dermatea tetraspora Ellis, Bull. Torrey Club **6**: 108. 1876.
Cenangium tetrasporum Sacc. Syll. Fung. **8**: 570. 1889.
Phaeangium tetrasporum Sacc. Syll. Fung. **16**: 765. 1902.

Apothecia erumpent but scarcely rising above the surface of the bark, occurring singly, or in cespitose clusters, externally

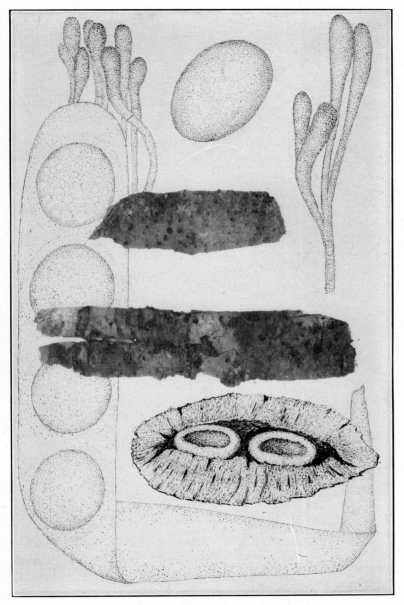

SPHAERANGIUM TETRASPORUM

light-brown, reaching a diameter of 2 mm.; hymenium concave, darker than the outside of the apothecium; asci clavate, reaching a length of 200–225 μ and a diameter of 25 μ, 4-spored; spores 1-seriate, very large and broad-ellipsoid to subglobose, 15–20 × 20–25 μ, becoming brown and very granular, often with several oil-drops; paraphyses slender, hyaline below, brown above, the ends adhering in an epithecium.

On dead limbs of *Quercus coccinea*.

TYPE LOCALITY: Newfield, New Jersey.

DISTRIBUTION: Known only from the type locality.

EXSICCATI: Ellis, N. Am. Fungi *70*.

2. **Sphaerangium magnisporum** (Cash & Stevenson) Seaver, comb. nov.

Phaeangium magnisporum Cash & Stevenson, Jour. Wash. Acad. Sci. **30**: 304. 1940.

Apothecia erumpent, occurring singly, or in groups of two or three each, closely surrounded by fragments of the ruptured bark, reaching a diameter of 1–1.5 mm., fleshy, subturbinate, or obconic, black, often contorted from mutual pressure, often falling out leaving cavities; hymenium brownish-black, rough; asci broad-clavate, abruptly narrowed at the base, reaching a length of 175–200 μ, and a diameter of 25–30 μ, 4-spored; spores 1-seriate, broad-ellipsoid, with one large oil-drop when young, surrounded by a thick, hyaline, gelatinous envelope, 19–22 × 40–50 μ (exclusive of the envelope); paraphyses numerous, branched and interwoven at their tips.

On branches of *Betula nigra*.

TYPE LOCALITY: Agricultural College Campus, Athens, Georgia.

DISTRIBUTION: Known only from the type locality.

ILLUSTRATIONS: Jour. Wash. Acad. Sci. **30**: 303. *f. 4*.

3. **Sphaerangium Tiliae** Seaver, sp. nov.   (PLATE 133.)

Apothecia erumpent through the outer bark which breaks irregularly, at first closed and clothed externally with white, furfuraceous granules, finally opening irregularly, often stellately, expanding and exposing the black hymenium; hymenium plane or nearly so, with a rough appearance apparently due to the exuded spores, reaching a diameter of 1–2 mm.; asci subcylindric, reaching a length of 240 μ, and a diameter of about 25 μ, with no extended base, 8-spored; spores 1-seriate, broad-ellipsoid, very granular within, at first hyaline, becoming dark-brown, 20 × 26 µ; paraphyses very slender, hyaline, about 2 µ thick.

Apotheciis erumpentibus, extus albido-furfuraceis, primo clausis dein apertis; hymenio nigro; ascis subcylindraceis, 8-sporis, 20–26 × 240 µ; sporis ellipsoideis, brunneis, 20 × 26 µ; paraphysibus filiformibus, 2 µ diam.

On branches of *Tilia heterophylla*.

TYPE LOCALITY: Agricultural College Campus, Athens, Georgia.

DISTRIBUTION: Known only from the type locality.

6. **CENANGELLA** Sacc. Bot. Cent. **18**: 248.   1884; Syll. Fung.
    8: 587.   1889.

Apothecia erumpent, urceolate to patellate, dark-colored, brown, or black, coriaceous; asci clavate, 4–8-spored; spores ellipsoid to fusoid, definitely 1-septate, hyaline; paraphyses filiform to clavate.

Type species, *Tympanis Ravenelii* Berk.

This genus was established by Saccardo in 1884 including two species *Cenangium Fraxini* Tul., based on *Tympanis Fraxini* Fries, which was based on *Peziza Fraxini* Schw. and *Cenangium dolosum* Sacc. The first, which would be the type, was apparently included through a misconception of the Schweinitz species since the spores do not agree at all with the above description (see Rehm, Ann. Myc. **11**: 166). The second is poorly known and we, therefore, suggest as the type of the genus *Tympanis Ravenelii* Berk. which was treated by Saccardo as *Cenangella Ravenelii* in 1889.

Spores very large, 14 × 40 µ.   1. *C. Ravenelii.*
Spores relatively small 3–4 × 12–15 µ.   2. *C. oricostata.*

1. **Cenangella Ravenelii** (Berk.) Sacc. Syll. Fung. **8**: 589.   1889.

Tympanis Ravenelii Berk. Grevillea **4**: 3.   1875.

Apothecia erumpent, single, or several together, dark-brown to black, scarcely exceeding 1 mm. in diameter; hymenium at first concave, becoming nearly plane; asci very broad-clavate, reaching a length of 130 µ and a diameter of 30–35 µ, tapering abruptly below, 8-spored; spores irregularly 2-seriate, ellipsoid, or fusoid, 1-septate and strongly constricted at the septum, densely filled with coarse granules, 14 × 40 µ, hyaline; paraphyses stout, reaching a diameter of 5–6 µ at their apices.

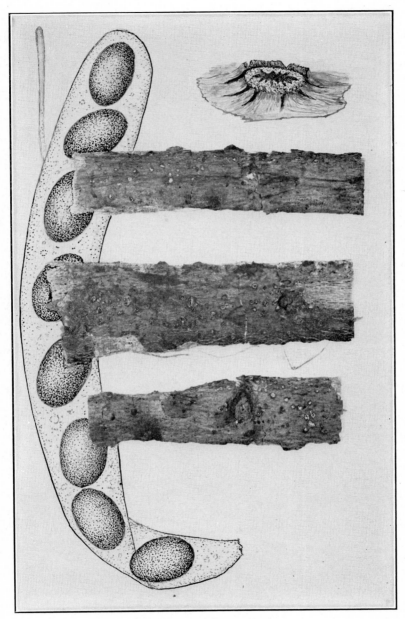

SPHAERANGIUM TILIAE

On branches of *Carpinus* and *Ilex*.

TYPE LOCALITY: South Carolina.

DISTRIBUTION: Pennsylvania to Missouri and Alabama.

EXSICCATI: Rav. Fungi Car. **4**: *66;* Rab.-Winter, Fungi Eu. *3464* (from Perryville, Missouri).

2. **Cenangella oricostata** Cash, Mycologia **28**: 298. 1936.

Apothecia coriaceous, subglobose, sessile, depressed, and radiately ridged at the apex, collapsing when old, emerging sometimes singly, more often in groups from cracks in the bark, on a thin, sclerotic layer, dark grayish-olive, olivaceous-black when dry, 1–2 mm. in diameter, not quite so high, slightly furfuraceous; hymenium pallid-grayish; asci cylindric, or subcylindric, short-stipitate, reaching a length of 130 μ and a diameter of 6–8 μ; spores fusoid-clavate, 1-seriate, or 2-seriate, hyaline, 1-septate, not constricted, upper cell broader, lower cell acute, 3–4 $\times$ 12–15 μ; paraphyses filiform, slightly swollen above and often curved at their apices.

On twigs of *Ribes Wolfii*.

TYPE LOCALITY: Mesa Lakes, Colorado.

DISTRIBUTION: Known only from the type locality.

ILLUSTRATIONS: Mycologia **28**: 302, *f. 4.*

DOUBTFUL AND EXCLUDED SPECIES

Cenangella abietina Ellis & Ev. Proc. Acad. Sci. Phila. **1895**: 429. 1895; *Phaeangella abietina* Sacc. & D. Sacc.; Sacc. Syll. Fung. **18**: 128. 1906. Based on material collected by Waghorn in Newfoundland. Identity uncertain.

Cenangella flavocinerea (Phill.) Sacc. Syll. Fung. **8**: 591. 1889; *Dermatea flavocinerea* Phill. Grevillea **7**: 23. 1878. No material has been seen.

Cenangella Hartzii Rostr. Medd. Grønl. **3**: 611. 1891. Apothecia densely gregarious, discoid, margin brown, reaching a diameter of .25–.5 mm.; hymenium black; asci clavate, reaching a length of 35–40 μ and a diameter of 7–9 μ, 8-spored; spores 2-seriate, ellipsoid, olivaceous, 1-septate, 4–5 $\times$ 7–9 μ. On decorticated branches of *Betula*. Known only from the type locality in Greenland.

Cenangella pruinosa Rostr. Medd. Grønl. **18**: 59. 1894. No material has been seen. On *Vaccinium uliginosum*.

Cenangella Rhododendri (Cesati) Rehm in Rab. Krypt.-Fl.1³: 230. 1889; *Peziza Rhododendri* Cesati; Rab. Bot. Zeit. **12**: 186. 1854; *Velutaria Rhododendri* Rehm, Ber. Naturh. Ver. Angsburg **26**: 63. 1881. Reported from Oregon. No material has been seen.

Cenangella thujina Ellis & Barth. in Ellis & Ev. N. Am. Fungi *3132*. 1894. This belongs with the Hysteriaceae and may be a *Glonium*.

Cenangella violacea Ellis & Ev. Proc. Acad. Sci. Phila. **1893**: 149.   1893. This belongs with the Hysteriaceae and may be a *Glonium*.   *Cenangella thujina* appears to be identical.

7. DERMATELLA Karst. Myc. Fenn. 1: 209.   1871.

Phaeangella Sacc. Syll. Fung. **18**: 128.   1906.

Apothecia erumpent, cupulate to discoid as in *Dermea;* asci cylindric to clavate, 4–8-spored; spores ellipsoid to fusoid, at first hyaline, becoming yellow, green, or brown, and 1–5-septate; paraphyses filiform to clavate.

Type species, *Peziza Frangulae* Pers.

Asci 4-spored.	1. *D. Frangulae.*
Asci 8-spored.	
Occurring on conifers, *Juniperus.*	2. *D. deformata.*
Occurring on deciduous trees.	
Spores 10–12 × 24–35 μ.	
On *Fraxinus.*	3. *D. Fraxini.*
On *Magnolia.*	4. *D. Magnoliae.*
Spores 6–8 × 12–24 μ.	
On branches of *Carya.*	5. *D. caryigena.*
On unidentified branches.	6. *D. montanensis.*

1. Dermatella Frangulae (Pers.) Karst. Myc. Fenn. 1: 209. 1871.

Peziza Frangulae Pers. Myc. Eu. **1**: 324.   1822.
Tympanis Frangulae Fries, Syst. Myc. **2**: 174.   1822.
Cenangium Frangulae Tul. Ann. Sci. Nat. III. **20**: 136.   1853.
Dermatea Frangulae Tul. Fung. Carp. **3**: 161.   1865.
Pezicula Frangulae Fuckel, Symb. Myc. 279.   1869.

Apothecia erumpent, usually occurring singly, or often several together, turbinate and prominent, at first closed, opening with an irregular margin, dark brownish-black, less than 1 mm. in diameter; hymenium lighter than the outside of the apothecium, becoming dark with age; asci clavate, reaching a length of 80–85 μ and a diameter of 12–14 μ, 4-spored; spores usually 1-seriate, ellipsoid, becoming 3-septate, at first hyaline, becoming brown, 8–10 × 20–22 μ; paraphyses rather stout, enlarged above to about 4 μ in diameter.

On branches of *Rhamnus purshiana.*

TYPE LOCALITY: Europe (on *Rhamnus frangula*).

DISTRIBUTION: Trinidad, California; also in Europe.

ILLUSTRATIONS: Fuckel, Symb. Myc. *pl. 4, f. 46;* Ann. Sci. Nat. III. **20**: *pl. 16, f. 1–8.*

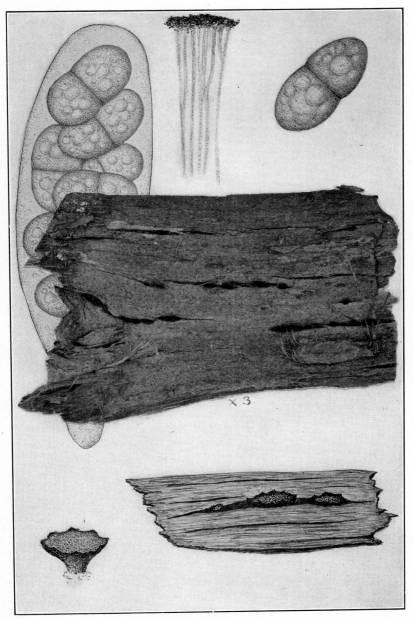

DERMATELLA DEFORMATA

Only one North American specimen of this species has been seen, collected by Harold E. Parks in California. American specimens agree well with European but the spores are a little larger than indicated in most descriptions. The spores are only pale-brown in our specimens. Fuckel describes and illustrates the spores as muriform but this has not been noted by other authors.

2. **Dermatella deformata** (Peck) Seaver, comb. nov. (PLATE 134.)

Cenangium deformatum Peck, Ann. Rep. N. Y. State Mus. **28**: 68. 1876.
Cenangella deformata Sacc. Syll. Fung. **8**: 593. 1889.
Phaeangella deformata Sacc. & D. Sacc.; Sacc. Syll. Fung. **18**: 128. 1906.

Apothecia erumpent, occurring singly, in small cespitose clusters, or with a serial arrangement, the individual apothecia at first rounded, gradually opening leaving the margin irregularly split and ragged, black and coarsely granular, reaching a diameter of 1.5 mm.; hymenium concave, lighter than the outside of the apothecium; asci broad-clavate, about 120 μ long and 35–45 μ broad, normally 8-spored, but often with part of the spores undeveloped; spores irregularly 2-seriate, broad-ellipsoid, 1-septate and deeply constricted at the septum, becoming dark-brown, 14–16 $\times$ 30–35 μ; paraphyses filiform, profusely branched, dark-brown with a greenish matrix.

On *Juniperus virginiana, Juniperus communis* and *Juniperus scopulorum.*

TYPE LOCALITY: Greenbush, New York.

DISTRIBUTION: New York, Montana and Wyoming and Bermuda.

ILLUSTRATIONS: Mycologia **34**: 520, *f. 2.*

EXSICCATI: Ellis & Ev. Fungi Columb. *1225.*

Excellent material of this species was collected by the writer and J. M. Waterston in Bermuda on Bermuda cedar, *Juniperus bermudiana.*

3. **Dermatella Fraxini** Ellis & Ev. Proc. Acad. Sci. Phila. **1893**: 149. 1893.

Apothecia gregarious, erumpent-superficial, occurring singly or with two or three grouped together, flesh-colored, reaching a diameter of .75–1 mm., contracted below and slightly turbinate; hymenium plane, or convex, similar in color to the outside of the apothecium; asci clavate, cylindric, reaching a length of 125–

150 μ and a diameter of 12–15 μ; spores partially 2-seriate, long-ellipsoid, becoming 1-septate, later 3–5-septate, 10–12 × 25–35 μ, at first hyaline, becoming brown; paraphyses filiform.

On bark of *Fraxinus*.

TYPE LOCALITY: London, Canada.

DISTRIBUTION: Known only from the type locality.

EXSICCATI: Ellis & Ev. N. Am. Fungi *2633*.

Type material has been examined and agrees with the above except that no asci were found in good condition, although spores were present in abundance. This may be due to the age of the material. Measurements were based on the original description. This looks like a *Pezicula* with a colored spore.

4. **Dermatella Magnoliae** (Berk. & Curt.) Seaver, comb. nov.

Cenangium Magnoliae Berk. & Curt.; Berk. Grevillea **4**: 5. 1875.

Apothecia erumpent in cespitose clusters 2–3 mm. in diameter, the individual apothecia turbinate, black, reaching a diameter of 1 mm.; hymenium concave, or nearly plane, black, or brownish-black; asci broad-clavate, reaching a length of 120 μ and a diameter of 26 μ, 8-spored; spores irregularly 2-seriate, broad-ellipsoid, often curved, or sausage-shaped, densely granular within, becoming 1-septate and brown, 9–12 × 24–30 μ; paraphyses filiform, slightly enlarged above, reaching a diameter of 3 μ.

On branches of *Magnolia glauca*, *Persea* and *Sassafras*.

TYPE LOCALITY: South Carolina.

DISTRIBUTION: South Carolina to Florida and Texas.

EXSICCATI: Rav. Fungi Am. *70*.

This was placed in the doubtful list by Groves. The brown 1-septate spores would place in the genus *Dermatella* as here treated.

5. **Dermatella caryigena** (Ellis & Ev.) Seaver, comb. nov.

Dermatea caryigena Ellis & Ev. Proc. Acad. Sci. Phila. **1893**: 148. 1893.

Apothecia erumpent in cespitose clumps of four to eight each, nearly black, reaching a diameter of .5–.75 μ, patelliform when fresh, collapsing when dry; hymenium plane, or slightly convex when fresh, similar in color to the outside of the apothecium; asci narrow-clavate, reaching a length of 75–90 μ and a diameter of 10 μ, 8-spored; spores 1-seriate, or irregularly crowded, ellipsoid, distinctly 3-septate, greenish-yellow to pale-brown, 6 × 12–15 μ; paraphyses filiform, profusely branched, olivaceous-brown.

On dead limbs of *Carya* (*Hickoria*).

TYPE LOCALITY: West Chester, Pennsylvania.

DISTRIBUTION: Known only from the type locality.

The only specimen seen is the type collection in the herbarium of The New York Botanical Garden.

6. **Dermatella montanensis** Ellis & Ev. Proc. Acad. Sci. Phila. **1893**: 148. 1893.

Apothecia erumpent-superficial, discoid, or subdiscoid, reaching a diameter of 1 mm., livid, subgelatinous when fresh; hymenium concave, black, with a thin, upturned margin when dry; asci broad-clavate, reaching a length of 72–80 μ and a diameter of 18–22 μ, 8-spored; spores irregularly 2-seriate, ellipsoid to clavate-ellipsoid, becoming 3-septate, slightly constricted at the septa, at first hyaline, becoming yellowish-brown, 7–8 $\times$ 24 μ; paraphyses, branched, enlarged at their apices, forming a dark-brown epithecium.

On small, dead limbs among driftwood.

TYPE LOCALITY: Sheridan, Montana.

DISTRIBUTION: Known only from the type locality.

DOUBTFUL AND EXCLUDED SPECIES

Dermatella populina Petrak, Ann. Myc. **20**: 196. 1922. On bark of *Populus tremuloides* from Idaho, Weir No. *18002*. Specimen of cotype material shows no apothecia.

Dermatella viticola Ellis & Ev. Proc. Acad. Sci. Phila. **1894**: 352. 1894. Apothecia convex-discoid and pale when moist, concave and flesh-colored when dry, then scarcely projecting above the surface of the bark, .5–.75 mm. in diameter; asci clavate-cylindric, reaching a length of 70 μ and a diameter of 12 μ, 8-spored; spores 2-seriate or partially so, ellipsoid, becoming 1–3-septate, 6–6.5 $\times$ 15–18 μ; paraphyses filiform. On dead shoots of *Vitis*, Nuttallburg, West Virginia. No material has been seen. It is probably not a *Dermatella* as used here.

8. **GODRONIA** Moug. Consid. Gen. Veg. Vosges 355. 1845.

Crumenula DeNot. Comm. Critt. Ital. **1**: 365. 1864.
Scleroderris (Fries) DeNot. Comm. Critt. Ital. **1**: 383. 1864.
Ascocalyx Naumov, Bolesni Rast. **14**: 138. 1925.
Atropellis Zeller & Goodding, Phytopathology **20**: 561. 1930.

Mougeot describes the genus as having apothecia which are coriaceous, or gelatinous like *Bulgaria*, and with the spores of a *Stictis*. The genus *Godronia* and the genus *Stictis* are strikingly similar and might easily be confused, notwithstanding the fact that they are placed in different orders. The genus is characterized by the erumpent apothecia and the very much elongated, or filiform spores which usually become 1–several-septate, although they may be for a long time without septa, as in other species of Cenangiaceae.

Type species, *Godronia Muhlenbeckii* Moug. & Lév.

On deciduous, woody plants.
  Spores relatively short, less than half the length of
    the ascus.
    On stems of *Ribes.*
      In congested masses; spores clavate.
        Spores becoming 3-septate.

Spores 20–38 μ long.	1. *G. Ribis.*
Spores 18–20 μ long.	2. *G. lobata.*
Spores simple, in specimens studied.	3. *G. tumoricola.*
Occurring singly; spores fusoid.	4. *G. Davidsoni.*

    Not on *Ribes.*
      Spores fusoid.

Spores 11–17 μ long; on *Lantana.*	5. *G. Lantanae.*
Spores 20–24 μ long.	
Apothecia turbinate; hymenium concave.	6. *G. turbinata.*
Apothecia patellate; hymenium plane or nearly so.	7. *G. fusispora.*
Spores 25–40 μ long.	
Apothecia with a laciniate border, on *Spiraea;* spores 3–4 × 25–25 μ.	8. *G. Spiraeae.*
Apothecia not laciniate on *Vitis;* spores 3–4 × 35–40 μ.	9. *G. viticola.*
Spores slender, rod-like; 1.5–2 × 10–12 μ.	10. *G. Lonicerae.*

  Spores relatively long, more than half as long as the
    ascus.
    On stems of woody plants.
      Spores long-fusoid.

Apothecia deeply concave; on *Salix.*	11. *G. fuliginosa.*
Apothecia slightly concave; on *Betula.*	12. *G. seriata.*

      Spores filiform or vermiform.

Spores short, 25–30 μ long; on *Kalmia.*	13. *G. Kalmiae.*
Spores 40–75 μ long.	
Spores slender, 1.5–2 μ thick.	
Apothecia cespitose, discoid.	14. *G. Nemopanthis.*
Apothecia scattered, urceolate.	15. *G. Urceolus.*
Spores 3–4 μ in diameter, nearly as long as the ascus.	
Apothecia scattered; on sage brush.	16. *G. montanensis.*

Apothecia in cespitose clusters.
Spores 3 × 65–70 μ; on *Cepha-
lanthus.*   17. *G. Cephalanthi.*
Spores 3–4 × 85–100 μ; on *Vi-
burnum.*   18. *G. viburnicola.*
Spores very long, nearly as long as the ascus.
On leaves of *Tetrazygia;* spores 250 μ long.   19. *G. parasitica.*
On branches of *Castanopsis.*   20. *G. Castanopsidis.*
On coniferous branches.
Paraphyses blue with transmitted light.
Spores 1–3-septate.   21. *G. pinicola.*
Spores simple or doubtfully septate.
Spores 1.5–3.5 × 32–63 μ.   22. *G. Zelleri.*
Spores 5–6 × 15–22 μ.   23. *G. sororia.*
Paraphyses not blue.
Hymenium light-colored.
Hymenium pink to brown; on *Picea in*
Alaska.   24. *G. Treleasei.*
Hymenium pale-yellow or whitish; on
Juniperus in Jamaica.   25. *G. jamaicensis.*
Hymenium grayish or black; on *Abies.*
Spores 8–12 μ long.   26. *G. Abietis.*
Spores 40–60 μ long.
On cankers on living branches.   27. *G. abieticola.*
On bleached bark.   28. *G. abietina.*

1. **Godronia Ribis** (Fries) Seaver, Mycologia **37**: 339. 1945.

Peziza ribesia Pers. Tent. Disp. Fung. 35. 1797.
Cenangium Ribis Fries, Syst. Myc. **2**: 179. 1822.
Tympanis Ribis Wallr. Fl. Crypt. Germ. **2**: 430. 1833.
Crumenula Ribis Karst. Fungi Fenn. *929.* 1870.
Scleroderris ribesia Karst. Myc. Fenn. **1**: 215. 1871.

Apothecia erumpent in cespitose clusters, 1–3 mm. in diameter from a stromatic base, each cluster consisting of four to
twelve apothecia, individual apothecia at first globose, or subglobose, short-stipitate, gradually expanding and becoming
shallow cup-shaped, with a notched, or fimbriate margin, reaching a diameter of 1–2 mm., blackish-brown; hymenium concave
pale-cinereous; asci clavate-cylindric, reaching a length of 90–
100 μ and a diameter of 7–8 μ, 8-spored; spores much elongated,
clavate, 3–4 × 20–38 μ, becoming 3-septate; paraphyses filiform,
slightly enlarged above.

Pycnidia, *Fuckelia Ribis*, present; pycnospores elongate-
clavate, 3–4 × 7–11 μ, usually containing two oil-drops.

On branches of species of *Ribes*.

TYPE LOCALITY: Europe.
DISTRIBUTION: Toronto, Canada; also in Europe.
ILLUSTRATIONS: Boud. Ic. Myc. *pl. 563;* Ann. Sci. Nat. III.
20: *pl. 16, f. 9–11;* Tul. Fung. Carp. **3**: *pl. 19, f. 1–9;* Rab. Krypt.-
Fl. **1**[3]: 200, *f. 1–2;* 201, *f. 6, 7;* E. & P. Nat. Pfl. **1**[1]: 255, *f. 187, G.*

2. **Godronia lobata** (Cash) Seaver, Mycologia **37**: 340.   1945.

Scleroderris lobata Cash, Mycologia **28**: 248.   1936.

Apothecia breaking through the bark, usually in clusters of
two to four, or rarely single, subglobose to cupulate, opening by
splitting into 4–6 lobes which fold over one another on drying,
reaching 1 mm. in diameter and height, externally blackish-
brown, smooth; hymenium concave, light olive-gray; asci
cylindric, short-stipitate, rounded and slightly narrowed above,
reaching a length of 90–115 μ and a diameter of 7–9; spores
1-seriate below, 2-seriate above, clavate, usually 3-septate,
3–4 $\times$ 18–20 μ; paraphyses filiform, simple, or branched near
the tip, slightly enlarged above.

On twigs of *Ribes Menziesii.*
TYPE LOCALITY: Spruce Cove, Trinidad, California.
DISTRIBUTION: Known only from the type locality.
ILLUSTRATIONS: Mycologia **28**: 250, *f. 3.*

3. **Godronia tumoricola** (Cash) Seaver, Mycologia **37**: 340.
1945.

Scleroderris tumoricola Cash, Mycologia **26**: 270.   1934.

Apothecia sessile, usually cespitose, or rarely single, cup-
shaped to nearly plane, triangular, or irregularly contorted when
dry, coriaceous, blackish-brown to black, .5–2 mm. in diameter;
hymenium concave to plane, drab, drying nearly black; asci
cylindric, narrowed above, reaching a length of 90–110 μ and a
diameter of 5–8 μ, 8-spored; spores 2-seriate, or irregularly 1-
seriate, clavate with the acute end below, first simple, later
septate?, 1.5–2 $\times$ 10–15 μ; paraphyses filiform, hyaline, simple,
or branched near the middle, enlarged above to 2 μ in diameter.

On swollen canker-like areas on twigs of *Ribes montigenum.*
TYPE LOCALITY: Mesa Lakes, Colorado.
DISTRIBUTION: Known only from the type locality.
ILLUSTRATIONS: Mycologia **26**: *pl. 32, f. 4.*

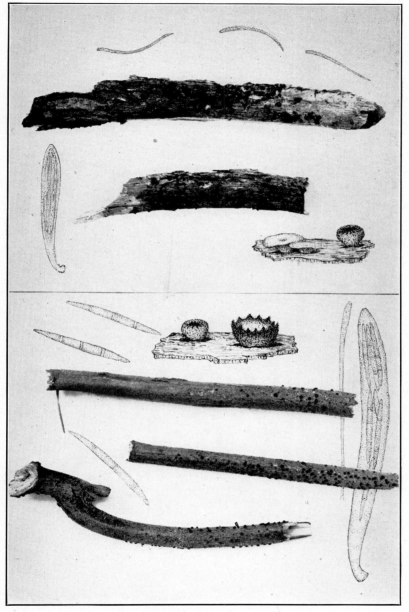

1. GODRONIA KALMIAE
2. GODRONIA SPIRAEAE

4. **Godronia Davidsoni** Cash, Mycologia **26**: 269.   1934.

Apothecia sessile, or substipitate, single, depressed-globose to urceolate, not exceeding 1 mm. in diameter and height, dark-colored, greenish-olive, opening with a circular opening, with a fimbriate margin; hymenium concave, smoky-gray; asci cylindric, gradually narrowed toward the base and above, 8-spored, reaching a length of 90–120 μ and a diameter of 5–7 μ; spores acicular-fusoid, parallel, or slightly twisted in the ascus, becoming 3-septate, 2.5–3 $\times$ 33–45 μ; paraphyses filiform, simple, hyaline, gradually enlarged above, reaching a diameter of 2–2.5 μ, often curved at the tips.

On stems of *Ribes Wolfii, R. bracteosum* $\times$ *R. nigrum.*
TYPE LOCALITY: Near Mesa Lakes, Colorado.
DISTRIBUTION: Colorado and Alaska.
ILLUSTRATIONS: Mycologia **26**: 269, *f. A; pl. 32, f. 3.*

5. **Godronia Lantanae** (Cash) Seaver, Mycologia **37**: 341.   1945.

Scleroderris Lantanae Cash, Mycologia **30**: 100.   1938.

Apothecia cespitose, sessile, cupulate to discoid, contorted by mutual pressure, reaching a diameter of 1–1.5 mm., furfuraceous, brown, the margin inrolled when dry, sometimes becoming hysteroid; hymenium concave to plane, brownish-black; asci clavate, rounded above, gradually narrowed toward the base, 8-spored, reaching a length of 50–55 μ and a diameter of 5 μ; spores 2-seriate, fusoid, straight, or slightly curved, becoming 1-septate, hyaline to pale-brownish, 1.5–2 $\times$ 11–17 μ; paraphyses simple, 2.5 μ in diameter at their apices.

On fallen branches of *Lantana Camara.*
TYPE LOCALITY: Kaluaaha Valley, Molokai, Hawaii.
DISTRIBUTION: Known only from the type locality.
ILLUSTRATION: Mycologia **30**: 99, *f. 4.*

6. **Godronia turbinata** (Schw.) Farlow; Thaxter, Mycologia **14**: 101.   1922.

Tympanis turbinata Schw. Trans. Am. Phil. Soc. II. **4**: 237.   1832.

Apothecia erumpent, turbinate, with the mouth strongly constricted, brownish-black, furfuraceous, reaching a diameter of 1 mm.; hymenium deeply concave, pallid; asci clavate, reaching a length of 110 μ and a diameter of 6–7 μ, apparently 8-

spored; spores often indistinct, fusiform, about $2 \times 20 \mu$, 3-septate.

On twigs of *Diervilla*.

TYPE LOCALITY: Bethlehem, Pennsylvania.

DISTRIBUTION: Pennsylvania and Maine.

EXSICCATI: Reliq. Farlow. *122*.

This description is based on material collected by Dr. R. Thaxter at Kittery Point, Maine. Material in the Schweinitz collection in Philadelphia is scant and uncertain.

7. **Godronia fusispora** (Ellis & Ev.) Seaver, Mycologia **37**: 342. 1945.

Dermatea fusispora Ellis & Ev. Proc. Acad. Sci. Phila. **1893**: 148. 1893.

Apothecia scattered, occurring singly, or occasionally two or three crowded together, externally subolivaceous, reaching a diameter of 1 mm.; hymenium plane or nearly so, with a slightly elevated margin, reddish; asci clavate, 8-spored, reaching a length of 70–75 μ and a diameter of 6–7 μ; spores narrow-fusoid, 2–3 $\times$ 20–24 μ; paraphyses filiform, slightly enlarged above, reaching a diameter of 2–3 μ.

On dead branches of *Betula* sp.

TYPE LOCALITY: Orono, Maine.

DISTRIBUTION: Known only from the type locality.

8. **Godronia Spiraeae** (Rehm) Seaver, Mycologia **37**: 342.   1945. (PLATE 135, FIG. 2.)

Scleroderris Spiraeae Rehm in Rab. Krypt.-Fl. **1**³: 1220.   1896.
?Belonidium Macounii Dearness, Mycologia **8**: 98.   1916.

Apothecia subsessile, thickly gregarious, occurring singly or rarely with two or three crowded together, subglobose, with the mouth constricted (in dried specimens) and laciniate, dark brownish-black, often with a greenish tint, reaching a diameter of 1 mm.; hymenium concave, obscured when dry by the incurved margin, freely exposed when moist, light-brown; asci cylindric-clavate, reaching a length of 80–85 μ and a diameter of 6–7 μ, 8-spored; spores fusoid, overlapping and subfasciculate, 3–4 $\times$ 25–35 μ, becoming 3-septate; paraphyses slender, slightly enlarged above.

On branches of *Spiraea salicifolia*.

TYPE LOCALITY: Europe.

DISTRIBUTION: New York and Toronto, Canada; also in Europe.

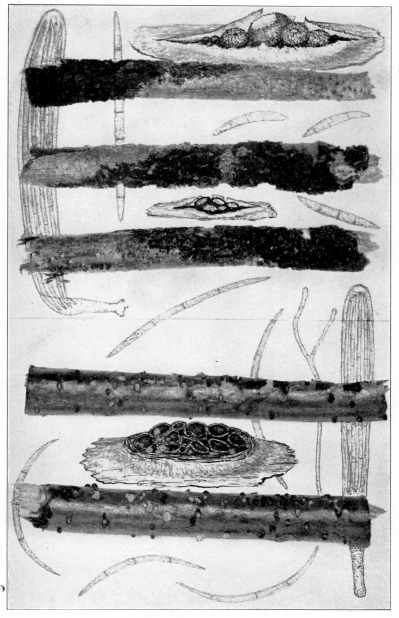

1. GODRONIA FULIGINOSA
2. GODRONIA SERIATA

ILLUSTRATIONS: Mycologia **37**: 337 (lower figure).

Described from the material in the herbarium of The New York Botanical Garden, collected by C. H. Peck at Hewitt's Pond in July (the year not indicated), at first thought to be new but later found to be as above.

Belonidium Macounii was described from material collected on dead stems of *Spiraea Menziesii* in Vancouver Island by John Macoun. Through the kindness of Dr. Dearness a portion of the type has been examined. It is a *Godronia*. The description of the spores fits this species but from the specimen examined they resemble more closely those of *Godronia Urceolus*.

9. **Godronia viticola** (Schw.) Seaver, Mycologia **37**: 343. 1945.

Peziza viticola Schw. Schr. Nat. Ges. Leipzig **1**: 123. 1822.
Tympanis viticola Fries, Syst. Myc. **2**: 176. 1822.
Peziza Schweinitziana Spreng. in L. Syst. Veg. ed. 16. **4**¹: 515. 1827.
Cenangium viticolum Sacc. Syll. Fung. **8**: 572. 1889.

Apothecia scattered, or gregarious, scutellate to subdiscoid, black, or blackish, soft and subgelatinous when moist, reaching a diameter of .15 mm.; asci clavate, tapering below into a slender, stem-like base, reaching a length of 85 μ and a diameter of 12 μ, 8-spored; spores fusoid, 1–3-septate, 4×35–40 μ; paraphyses filiform.

On bark of living grape vine, *Vitis* sp.

TYPE LOCALITY: North Carolina.

DISTRIBUTION: North Carolina to New Jersey.

EXSICCATI: Ellis, N. Am. Fungi *1317*.

10. **Godronia Lonicerae** Seaver, Mycologia **37**: 343. 1945.

Apothecia sessile, or subsessile, attenuated below, at first closed, externally dark-colored, striated near the margin and clothed with poorly developed, adpressed hairs, gradually expanding but with the margin constricted, .3 mm. diameter; hymenium dull but lighter than the outside of the apothecium; asci clavate, reaching a length of 40–45 μ and a diameter of 6–7 μ, 8-spored; spores slender, rod-like, 1.5–2×10–12 μ, containing several oil-drops; paraphyses filiform, 1.5–2 μ in diameter, slightly enlarged above, hyaline.

On branches of *Lonicera canadensis*.

TYPE LOCALITY: Lake Temagami, Toronto, Canada.

DISTRIBUTION: Known only from the type locality.

11. **Godronia fuliginosa** (Pers.) Seaver, Mycologia **37**: 344. 1945. (PLATE 136, FIG. 1.)

Sphaeria fuliginosa Pers. Obs. Myc. **2**: 68. 1799.
Cenangium difforme Fries; Moug. & Nest. Stirp. Crypt. *889.* 1826.
Cenangium fuliginosum Fries, Elench. Fung. **2**: 23. 1828.
Tryblidium difforme Pers.; Tul. Fung. Carp. **3**: 166. 1865.
Scleroderris fuliginosa Karst. Myc. Fenn. **1**: 216. 1871.
Lasiosphaeria striata Ellis & Ev. Proc. Acad. Sci. Phila. **1893**: 443. 1893.
Godronia Betheli Seaver, Mycologia **3**: 64. 1911.
Godronia striata Seaver, Mycologia **4**: 123. 1912.

Apothecia erumpent through the outer bark of the host, single, or occurring in clusters, often so numerous as to form congested masses many cm. in extent and often almost entirely surrounding the branches on which they grow, the individual apothecia at first nearly globose, opening at the top so as to leave an irregular margin, at maturity about 1 mm. broad and the same in height, brownish, furfuraceous externally and longitudinally striated; hymenium whitish, or bluish-white; asci clavate, reaching a length of 130 μ and a diameter of 7–8 μ, 8-spored; spores fasciculate in the ascus, subfiliform, tapering toward either end, sharp-pointed, 3–4 × 65–85 μ, becoming 7-septate at maturity and often slightly constricted at the septa, hyaline; paraphyses abundant, filiform.

Pycnidia often accompanying the apothecia, at first closed, becoming shallow cup-shaped, usually black; pycnospores fusiform, straight, or slightly curved, 3-septate, 3–4 × 28–30 μ.

On branches of *Salix.*

TYPE LOCALITY: Europe.

DISTRIBUTION: Ontario and Colorado; also in Europe.

ILLUSTRATIONS: Tul. Fung. Carp. **3**: *pl. 20, f. 1–4;* E. & P. Nat. Pfl. **1**[1]: 255, *f. 187, H–J;* Rab. Krypt.-Fl. **1**[3]: 201, *f. 3–7;* Mycologia **25**: *pl. 15* (upper figure).

This species was collected in abundance on willow at Tolland, Colorado by the author in company with Ellsworth Bethel in 1910 and described as a new species, *Godronia Betheli*. Later study revealed the fact that *Lasiosphaeria striata* Ellis & Ev. is identical but owing to the fact that it had been placed in the Sphaeriales it had been overlooked. The young apothecia are strongly constricted which doubtless accounts for the fact that Ellis placed it in that group. Later study has revealed the fact that both Ellis and Everhart's species and the species of the writer are identical with *Cenangium fuliginosum* of Fries.

1. GODRONIA JAMAICENSIS
2. GODRONIA CEPHALANTHI

Specimens erroneously reported on *Alnus* by Kauffman (Papers Michigan Acad. Sci. **1**: 109. 1923) have been examined and proved to be *Cyphella fasciculata* (Schw.) Berk. & Curt.

12. **Godronia seriata** (Fries) Seaver, Mycologia **37**: 345. 1945. (PLATE 136, FIG. 2.)

Cenangium seriatum Fries, Syst. Myc. **2**: 185. 1822.
Phacidium seriatum Fries, Elench. Fung. **2**: 131. 1828.
Tryblidium seriatum Fries, Sclerom. Suec. *161.*
Dermatea seriata Tul. Fung. Carp. **3**: 160. 1865.
Gelatinosporium betulinum Peck, Ann. Rep. N. Y. State Mus. **25**: 84. 1873.

Apothecia occurring in elongated clumps 3–4 mm. long and about 2 mm. broad, closely compressed together and often slightly irregular from mutual pressure, entirely black, reaching a diameter of .5 mm.; hymenium slightly concave, bordered by a slightly upturned margin, black; asci clavate, reaching a length of 95–110 μ and a diameter of 10–12 μ, 8-spored; spores long, fusiform and usually slightly curved when free, reaching a length of 45–60 μ and a diameter of 2–3 μ, 3-septate; paraphyses slender, branched, hyaline, or subhyaline.

The conidial stage accompanies the ascigerous and consists of a blackish stroma in which the pycnospores are produced; pycnospores fusiform, strongly curved and 3-septate, reaching a length of 40–45 μ from tip to tip and a diameter of 2–3 μ.

On *Betula lutea* and *Betula fontinalis*.

TYPE LOCALITY: Europe.

DISTRIBUTION: New York to Pennsylvania and Colorado; also in Europe.

ILLUSTRATIONS: Mycologia **25**: *pl. 15* (lower figure).

EXSICCATI: Ellis, N. Am. Fungi *537–537b;* Ellis & Ev. Fungi Columb. *853;* Shear, N. Y. Fungi *200.*

A fine collection of this species was obtained in Coal Creek Cañon, Colorado, by the writer and Paul F. Shope, in the summer of 1929 (No. *495*), and determined by Mr. W. W. Diehl as *Scleroderris seriata* (Fries) Rehm. This is the only perfect specimen of this in the herbarium of The New York Botanical Garden from America.

Material which seems to be a part of the type collection of *Gelatinosporium betulinum* Peck is found in our collection. A note apparently in the handwriting of C. H. Peck reads as follows: "Perhaps the same as *Sphaeronema seriatum* B. & C. possibly a condition of *Cenangium seriatum* Fr." This conclusion is undoubtedly correct since our material collected in Colorado shows both stages on the same stroma. The pycnospores agree very closely with those from Peck's type. No definite pycnidia could be detected.

13. Godronia Kalmiae (Rehm) Seaver, Mycologia **37**: 346. 1945. (PLATE 135, FIG. 1.)

Gorgoniceps Kalmiae Rehm, Ann. Myc. **2**: 353. 1904.

Apothecia gregarious, erumpent, at first cyathoid, short-stipitate, finally becoming irregularly discoid, externally reddish-brown, reaching a diameter of 1 mm.; hymenium at first concave, becoming plane, or convex, sordid-yellow; stem very short and stout, gradually expanding into the apothecium; asci cylindric-clavate, reaching a length of 40–50 μ and a diameter of 5–8 μ, 8-spored; spores filiform, straight, or curved when freed from the ascus, no septa apparent, 1 × 25–30 μ; paraphyses filiform, slightly enlarged above, 1–1.5 μ in diameter.

On decaying branches of *Vaccinium corymbosum*.

TYPE LOCALITY: North America (exact locality not given).

DISTRIBUTION: Known only from the type locality.

ILLUSTRATIONS: Mycologia **37**: 337 ((upper figure).

EXSICCATI: Ellis, N. Am. Fungi *147* (as *Dermatea Kalmiae* Peck). This material which was incorrectly named by Ellis was made the type of a new species by Rehm.

14. Godronia Nemopanthis (Peck) Sacc. Syll. Fung. **8**: 603. 1889.

Tympanis Nemopanthis Peck, Ann. Rep. N. Y. State Mus. **35**: 142. 1884.
Durandiella Nemopanthis Groves, Mycologia **29**: 75. 1937.

Apothecia occurring in cespitose clusters, or occasionally single, sessile, slightly narrowed below, reaching a diameter of 1 mm. and a height of 1.5 mm., circular in form, or becoming irregular from mutual pressure, dull-black, the consistency leathery to cartilaginous, or horny when dry; hymenium at first concave, becoming plane, or convex, black, or olivaceous when moist; asci cylindric to clavate-cylindric, tapering below into a stem-like base, reaching a length of 80–125 μ and a diameter of 7–9 μ; spores hyaline, filiform, septate, attenuated at the ends, straight, or variously curved, intertwined in the ascus, 1.5–2 × 50–85 μ; paraphyses hyaline, filiform, septate, branched, 1.5–2 μ

in diameter, scarcely enlarged above, forming a yellowish hymenium.

On dead stems of *Nemopanthes canadensis* (*Nemopanthes mucronata*).

TYPE LOCALITY: Grafton, Rensselaer County, New York.

DISTRIBUTION: New York to Maine and Ontario.

ILLUSTRATIONS: Mycologia **29**: 76, *f. 3;* 77, *f. 4–9.*

EXSICCATI: Ellis & Ev. N. Am. Fungi *2330;* Fungi Columb. *332;* Barth. Fungi Columb. *4539;* Reliq. Farlow. *121.*

15. **Godronia Urceolus** (Alb. & Schw.) Karst. Acta Soc. Fauna Fl. Fenn. **2**[6]: 144. 1885.

Peziza Urceolus Alb. & Schw. Consp. Fung. 332. 1805.
Peziza globularis Pers. Myc. Eu. **1**: 326. 1822.
Cenangium Urceolus Fries, Syst. Myc. **2**: 182. 1822.
Sphaeria uberiformis Fries, Syst. Myc. **2**: 491. 1823.
Tympanis Urceolus Wallr. Fl. Crypt. Germ. **2**: 425. 1833.
Mastomyces Friesii Mont. Ann. Sci. Nat. III. **10**: 135. 1848.
Cenangium globulare Fries, Summa Veg. Scand. 364. 1849.
Crumenula Urceolus DeNot. Comm. Critt. Ital. **1**: 363. 1863.
Cenangium urceolatum Ellis; Cooke & Ellis, Grevillea **6**: 9. 1877.
Godronia Cassandrae Peck, Ann. Rep. N. Y. State Mus. **39**: 50. 1886.
Cenangella urceolata Sacc. Syll. Fung. **8**: 591. 1889.

Apothecia scattered, or rarely five to seven closely crowded together, erumpent through the bark, usually with a short, thick stem, urceolate, reaching a diameter of .5–1.5 mm., externally brownish, or brownish-black; hymenium grayish, or pallid; stem very short and inconspicuous; asci clavate-cylindric, reaching a length of 90–125 μ and a diameter of 6–7 μ, 8-spored; spores filiform, many-septate, 1.5 $\times$ 50–75 μ; paraphyses filiform, 2 μ in diameter.

Reported on twigs of various kinds, *Alnus*, *Betula*, *Cassandra* and *Clethra*. American specimen on *Ribes* seems to agree.

TYPE LOCALITY: Europe.

DISTRIBUTION: Newfoundland to Alaska and Michigan; also in Europe.

ILLUSTRATIONS: Alb. & Schw. Consp. Fung. *pl. 3, f. 4;* E. & P. Nat. Pfl. **1**[1]: 234, *f. 178, G–J;* Rab. Krypt.-Fl. **1**[3]: 217, *f. 1–5;* Peck, Ann. Rep. N. Y. State Mus. **39**: *pl. 1, f. 16–20.*

EXSICCATI: Ellis, N. Am. Fungi *990* (as *Cenangium urceolatum*); Ellis & Ev. Fungi Columb. *742.*

Miss Daisy S. Hone in her Minnesota work (Minn. Bot. Studies **4**: 111. 1909) describes a variety *Godronia urceolata conferta* which was said to differ in the habitat on *Prunus pumila* as well as in the clustered habit of the apothecia and the slightly shorter spores.

Godronia Cassandrae described by Peck from material collected on *Cassandra calyculata* appears to be identical with the above.

Cenangium urceolatum Ellis is identical. The spores are described by Ellis as fusoid, 1-septate and $4 \times 15 \mu$. These were apparently conidia or pycnospores. Such spores have been found in connection with other specimens. The ascospores in this species are typical.

16. **Godronia montanensis** Seaver, Mycologia **37**: 348.   1945.

Apothecia scattered, erumpent, superficial, at first urceolate, later expanding and becoming subdiscoid, blackish and minutely verrucose, reaching a diameter of 1 mm.; hymenium concave, or nearly plane, pale-yellowish; asci cylindric, reaching a length of 175–185 μ and a diameter of 10–11 μ, probably 8-spored; spores filiform, nearly as long as the ascus, about 3 μ in diameter, many-septate and apparently breaking up into segments in the ascus, the segments about as long as broad; paraphyses filiform, about 1.5 in diameter.

On sage brush.

TYPE LOCALITY: Sheridan, Montana.

DISTRIBUTION: Known only from the type locality.

This interesting species was found sparingly associated with *Dermatella montanensis* Ellis & Ev. and was probably overlooked by Ellis. The species differs from others examined in the spores becoming disjuncted in the ascus as well as in the size of the spores and asci.

17. **Godronia Cephalanthi** (Schw.) Seaver, Mycologia **37**: 349. 1945.   (PLATE 137, FIG. 2.)

Peziza Cephalanthi Schw. Schr. Nat. Gez. Leipzig **1**: 123.   1822.
Cenangium Cephalanthi Fries, Syst. Myc. **2**: 188.   1822.
Scleroderris Cephalanthi Farlow; Thaxter, Mycologia **14**: 102.   1922.

Apothecia erumpent, usually in congested clusters several mm. long, the individual apothecia scutellate with the margin, dark brownish-black, reaching a diameter of 2 mm.; hymenium concave, pale-yellowish; asci cylindric, reaching a length of 65–70 μ and a diameter of 10–12 μ, 8-spored; spores filiform, about 3 × 65–70 μ, hyaline, becoming 5–7-septate; paraphyses filiform slender.

On *Cephalanthus occidentalis*.

TYPE LOCALITY: South Carolina.

DISTRIBUTION: New York and New Hampshire to South Carolina.

EXSICCATI: Reliq. Farlow. *103, 144.*

The description is drawn from material identified by C. H. Peck and W. G. Farlow. Specimens in the Schweinitz collection are immature and the spore characters therefore uncertain.

18. **Godronia viburnicola** Seaver, Mycologia **37**: 349. 1945.

Apothecia erumpent in cespitose clusters of two to ten each, the individual apothecia black, reaching a diameter of 1 mm., tapering below into a stem-like base; hymenium slightly concave, or nearly plane, similar in color to the outside of the apothecium; asci cylindric-clavate, reaching a length of 110 μ and a diameter of 10–12 μ, 8-spored; spores filiform, 3–4 × 85–100 μ; paraphyses slender, branched, slightly enlarged above, and brownish.

On *Viburnum cassinoides* and *Viburnum dentatum*.

TYPE LOCALITY: New Hampshire.

DISTRIBUTION: Known only from the type locality.

EXSICCATI: Reliq. Farlow. *154a, 154b.*

19. **Godronia parasitica** Seaver, Mycologia **24**: 354. 1932. (PLATE 138.)

Apothecia scattered on the under side of the living leaf, especially along the midrib, erumpent, at first globose, becoming expanded but with the mouth constricted, black, reaching a diameter of .3–.5 mm.; hymenium dingy, more or less concealed; asci clavate, reaching a length of 250–300 μ and a diameter of 27 μ; spores filiform, nearly as long as the ascus and about 4 μ thick, many-septate, the number of the septa difficult to determine but more than 50 have been counted, reaching a length of 250 μ; paraphyses slender and rather freely branched.

On leaves of *Tetrazygia longicollis*.

TYPE LOCALITY: Marmelade, Republic of Hayti.

DISTRIBUTION: Known only from the type locality.

ILLUSTRATIONS: Mycologia **24**: *pl. 9.*

This is described from material collected by Mr. George V. Nash, August 25, 1903. The species is distinguished by its huge asci and spores.

20. **Godronia Castanopsidis** Seaver, Mycologia **37**: 351. 1945.

Apothecia thickly gregarious, erumpent through the bark, sessile, usually occurring singly, irregularly rounded, externally blackish, reaching a diameter of 2–3 mm.; hymenium plane with a dark, elevated margin, pale-yellowish, or nearly white; asci subcylindric, tapering below into a stem-like base, attenuated above, reaching a length of 300 μ and a diameter of 16–18 μ, 8-spored; spores filiform, nearly as long as the ascus, hyaline, septate, 2 μ in diameter; paraphyses filiform, branched, pale-yellowish, scarcely enlarged above, 2–3 μ in diameter.

On dead branches of *Castanopsis chrysophylla*.

TYPE LOCALITY: Mount Shasta, California.

DISTRIBUTION: Known only from the type locality.

ILLUSTRATIONS: Mycologia **37**: 353 (lower figure).

21. **Godronia pinicola** (Reb.) Karst. Acta Soc. Fauna Fl. Fenn. **2**[6]: 144. 1885. (PLATE 139, FIG. 1.)

Peziza pinicola Reb. Fl. Neom. 385. 1804.
Peziza pinicola solitaria Fries, Syst. Myc. **2**: 113. 1822.
Heterosphaeria pinicola Fries, Summa Veg. Scand. 365. 1849.
Crumenula pinicola Karst. Myc. Fenn. **1**: 210. 1871.

Apothecia erumpent-superficial, at first rounded, expanding and becoming shallow cup-shaped to scutellate, black with a purplish tinge (purple by transmitted light) furfuraceous, or slightly hairy, reaching a diameter of 2–3 mm., sessile, or short-stipitate; hymenium concave, or nearly plane, similar in color to the outside of the apothecium; asci clavate, reaching a length of 120 μ and a diameter of 12 μ, 8-spored; spores in a fascicle near the end of the ascus or irregularly disposed, fusiform, hyaline, 3–4 $\times$ 24–36 μ, becoming 1–3-septate; paraphyses slender, branched and forming a purplish epithecium.

On living branches of *Pinus rigida*, *P. pungens* and *P. resinosa*.

TYPE LOCALITY: Europe.

DISTRIBUTION: New Hampshire to Pennsylvania; also in Europe.

ILLUSTRATIONS: Not. Soc. Fauna Fl. Fenn. **10**: *pl. 2, f. e;* Mycologia **18**: 182, *f. 1, B–C;* **37**: 357 (upper figure); Rab. Krypt.-Fl. **1**[3]: 217, *f. 1–5;* E. & P. Nat. Pfl. **1**[1]: 234, *f. 178, A–C.*

PLATE 138

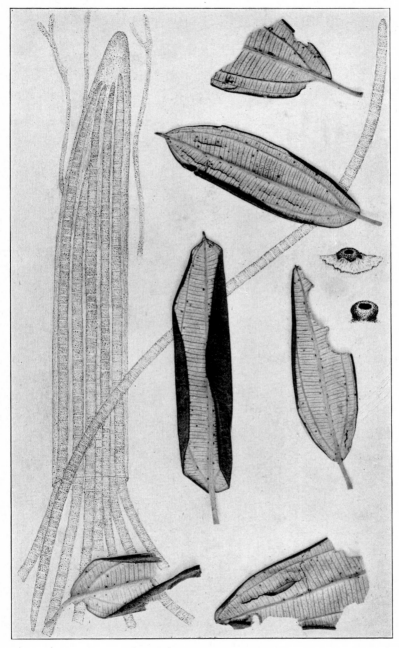

GODRONIA PARASITICA

This was listed by L. O. Overholts as a *Crumenula* and possibly *Crumenula pinicola* (Reb.) Karst. in Mycologia **18**: 181. It agrees reasonably well with the descriptions of that species except for the purplish color which was not mentioned by Karsten. It is, however, thought best to refer American specimens to that species.

22. **Godronia Zelleri** Seaver, Mycologia **37**: 354. 1945.

Atropellis pinicola Zeller & Goodding, Phytopathology **20**: 563. 1930. Not
 Godronia pinicola (Reb.) Karst. 1882.

Apothecia solitary, or gregarious, erumpent from outer cortical layers of bark, sessile, or on very short, central stalk, 2–4 mm. in diameter, at first closed, opening by stellate or irregular clefts, leaving rather fimbriate margins, expanding discoid, usually rolling up from two sides when drying, externally pruinose, black to fuscous-black; hymenium pruinose, black; asci clavate, hyaline, staining brown with iodine, 8-spored, 8–13 × 74–178 μ; spores filiform to acicular-clavate, hyaline continuous, guttulate, 1.5–3.5 × 36–63 μ; paraphyses hyaline, hair-like, flexuous, exceeding the length of the asci by 32–38 μ, tips slender, agglutinated, forming a dense epithecium with rosy and purplish tints in section.

Imperfect stage *Neofuckelia pinicola* Zeller & Goodding (Mycologia **27**: 464), usually associated with *A. pinicola*. Stromata erumpent, sometimes scattered, mostly gregarious, black-pulvinulate, sessile to short-stipitate, .8–1.2 mm. in diameter, containing 16–35 locules (pycnidia); conidiophores from entire inner surface of pycnidia, hair-like, simple, or branched; conidia hyaline, continuous, narrowly ellipsoid to bacillar, 1.7–3 × 8–11 μ.

On living branches and trunks of *Pinus lambertiana*, *Pinus monticola*, *Pinus Strobus*, and *P. contorta*.

TYPE LOCALITY: Oregon.

DISTRIBUTION: Oregon to California, Montana and British Columbia.

ILLUSTRATIONS: Phytopathology **20**: *pl. 1, f. G–M.*

23. **Godronia sororia** Karst. Acta Soc. Fauna Fl. Fenn. **2**[6]: 145. 1885. (PLATE 139, FIG. 2.)

Crumenula sororia Karst. Myc. Fenn. **1**: 211. 1871.
?Cenangium piniphilum Weir, Phytopathology **11**: 295. 1921.
Atropellis pinicola Lohman & Cash, Jour. Wash. Acad. Sci. **30**: 357. 1940.

Apothecia erumpent in congested masses 1 cm. or more in diameter, the individual apothecia black, or with a purplish tinge (decidedly purple when teased out), irregularly cup-shaped, closing when dry and often irregularly hysteriform, externally furfuraceous; reaching a diameter of 2–4 mm.; hymenium concave, similar in color to the outside of the apothecium; asci clavate, tapering very gradually into a long, stem-like base, reaching a length of 135 μ and a diameter of 12 μ, 8-spored; spores irregularly 2-seriate above, fusoid, granular within, often slightly constricted near the center and appearing pseudoseptate, 5–6 × 15–22 μ; paraphyses very slender, branched above and forming a purplish epithecium.

Forming cankers on trunks of *Pinus ponderosa*.

TYPE LOCALITY: Europe.

DISTRIBUTION: Idaho; also in Europe.

ILLUSTRATIONS: Phytopathology **11**: 294, *f. 1;* 295, *f. 2;* Mycologia **37**: 357, *f. 3* (lower figure).

The only American material seen was collected in Idaho by J. R. Weir who writes: "forming cankers on 16-year old pine, causing a black deposit to form." As pointed out by Karsten that species is similar to *G. pinicola* so far as apothecial characters are concerned. The specific name selected by Karsten doubtless indicates that it is a sister species to that one which was listed in the same paper. Both have the blue character with transmitted light which, however, was not mentioned by Karsten. Examination of authentic material of both species shows it to be present. The spores in this species are shorter and broader than in that one. It also has a black subiculum not noted in *G. pinicola*.

24. **Godronia Treleasei** (Sacc.) Seaver, Mycologia **37**: 355. 1945.

Scleroderris Treleasei Sacc. in Harriman Alaska Exped. **5**: 24. 1904.
Atropellis Treleasei Zeller & Goodding, Phytopathology **20**: 562. 1930.

Apothecia solitary, or gregarious, at first erumpent then entirely superficial, mostly sessile, at first pitcher-shaped, closed then scutellate, laciniately-dehiscent, 2.5–4 mm. in diameter, expanding to 3–5 mm. when moistened, outside and margins torn, dusky purplish-gray, carbonaceous, rugose; hymenium flatly-concave to convexly expanded, waxy, pinkish-cinnamon; asci clavate with obtusely acute apices, long-stipitate, 8–14 ×

1. GODRONIA PINICOLA
2. GODRONIA SORORIA

178 μ, spores fasciculate in the upper part of the ascus, mostly filiform, often somewhat clavate, hyaline, continuous, 2–2.5 × 42–60 μ; paraphyses filiform, with simple, or incurved, furcate tips which very slightly exceed tips of asci, hyaline.

On bark of *Picea sitchensis*.

TYPE LOCALITY: Alaska.

DISTRIBUTION: Known only from the type locality.

ILLUSTRATIONS: Harriman Alaska Exped. **5**: *pl. 3, f. 7 a–g;* Phytopathology **20**: *pl. 1, f. A–E.*

25. **Godronia jamaicensis** Seaver, Mycologia **37**: 356.   1945. (PLATE 137, FIG. 1.)

Apothecia scattered, erumpent through the outer bark, finally appearing quite superficial, sessile, becoming expanded and scutellate with a wavy margin, externally brownish-black and verrucose, or wrinkled, reaching a diameter of 2 mm.; hymenium plane, or slightly concave, surrounded by the up-turned, blackish margin, pale-yellowish, or whitish; asci clavate, reaching a length of 175 μ and a diameter of 10–12 μ, 8-spored; spores filiform, nearly as long as the ascus, 1.5–2 μ in diameter, no septa apparent; paraphyses filiform, freely branched.

On bark of *Juniperus*.

TYPE LOCALITY: Cinchona, Jamaica, altitude 4500–5400 feet.

DISTRIBUTION: Known only from the type locality.

ILLUSTRATIONS: Mycologia **37**: 353 (upper figure).

A liberal quantity of this material was collected by W. A. Murrill in 1908–1909.   The host species was not named but since only one species of *Juniperus* occurs in Jamaica it must have been that, *Juniperus lucayana*.

26. **Godronia Abietis** (Naumov) Seaver, Mycologia **37**: 356. 1945.

Fusisporium Berenice Berk. & Curt.; Berk. Grevillea **3**: 147.   1875.
Cenangium pithyum Berk. & Curt.; Berk. Grevillea **4**: 4.   1875.
Scleroderris pitya Sacc. Syll. Fung. **8**: 596.   1889.
Botryodiscus pinicola Shear, Bull. Torrey Club **34**: 313.   1907.
Pycnocalyx Abietis Naumov, Bull. Soc. Oural. Sci. Nat. Trud. Bur. Mykol. **35**: 34.   1915.   (Citation from Groves.)
Ascocalyx Abietis Naumov, Bolesni Rast. **14**: 138.   1925.   (Citation from Groves.)

Apothecia erumpent, usually in clusters of three to six arising from a rounded, black, basal stroma, circular, or slightly wavy in outline, narrowed below, .3–1 mm. in diameter, dull-black externally leathery to horny, becoming softer when moist; hymenium concave, becoming plane, smooth, gray to blackish, the margin infolded when dry; asci cylindric-clavate, short-stemmed, 8-spored, 9–11 × 65–100 μ; spores hyaline, ellipsoid to subclavate, irregularly 2-seriate, simple, becoming 1–3-septate, 4–5 × 14–22 μ; paraphyses hyaline, filiform, branched, 1.5–2 μ in diameter, scarcely enlarged above.

Pycnidia often arising from the stroma, at first almost globose, opening and becoming cup-shaped, reaching 2 mm. in diameter; pycnidial cavities immersed in the disc, ovoid, about 25–75 × 75–100 μ; conidiophores hyaline, tapering above, 1.5–2.5 × 8–12 μ; conidia hyaline, elongated to subfiliform, straight, or curved, simple, becoming 1–5-septate, 3–5 × 16–24 μ.

On species of *Abies*.

TYPE LOCALITY: Europe.

DISTRIBUTION: New Hampshire to Michigan and North Carolina; also in Europe.

ILLUSTRATIONS: Mycologia **28**: 452, *f. 1;* 454, *f. 2; 457, f. 3–6.*

27. Godronia abieticola (Zeller & Goodding) Seaver, comb. nov.

Scleroderris abieticola Zeller & Goodding, Phytopathology **20**: 565. 1930.

Apothecia single, or gregarious, from an erumpent stroma, short-stipitate, or subsessile, .5–1.2 mm. in diameter, at first closed, spherical, or ellipsoid, opening by a stellate, or irregular cleft, then cup-shaped to expanded, externally smooth to flaky, grayish-black to shiny-black; hymenium shiny-slate to shiny-black; asci clavate, reaching a length of 118–135 μ and a diameter of 9–14 μ, 8-spored, or rarely 4-spored; spores filiform with many oil-drops, finally distinctly 5–8-septate, 3.5–4.5 × 40–67 μ; paraphyses filiform, slightly enlarged above, somewhat agglutinated but not forming an epithecium.

On cankers on living branches and twigs of *Abies grandis* and *Abies amabilis*.

TYPE LOCALITY: Oregon.

DISTRIBUTION: Oregon.

ILLUSTRATIONS: Phytopathology **20**: *pl. 1, N–R.*

28. Godronia abietina (Ellis & Ev.) Seaver, comb. nov.

Scleroderris abietina Ellis & Ev. Am. Nat. **31**: 427. 1897.

Apothecia erumpent-superficial, black, subconical, about .5 mm. in diameter; hymenium plane, or convex, black; asci clavate-cylindric, reaching a length of 70 μ and a diameter of 15 μ, 8-spored; spores fasciculate, clavate-cylindric, containing numerous oil-drops, finally becoming multiseptate, 3–4 × 50–65 μ; paraphyses filiform.

On bleached bark of fir trees, *Abies* sp.

TYPE LOCALITY: Newfoundland.

DISTRIBUTION: Known only from the type locality.

Two fine specimens of this species are in The New York Botanical Garden both collected by Rev. A. C. Waghorne in Newfoundland and both apparently a part of the type specimen.

DOUBTFUL AND EXCLUDED SPECIES

Godronia rugosa Ellis & Ev. Jour. Myc. **8**: 70. 1902. This species was described by Ellis from material collected at Tuskegee, Alabama, August, 1900 (G. W. Carver *479*). The spores are described as fusoid, arcuate, 45–55 × 3–3.5 μ. Examination of the Ellis material showed such spores but no trace of asci. It is probably not an ascomycete.

Godronia Juniperi Rostr. Medd. Grønl. **3**: 611. 1891. Apothecia scattered, sessile, hard, black, reaching a diameter 1–2 mm.; asci reaching a length of 75–85 μ and a diameter of 7–8 μ; spores filiform, 2 × 35–40 μ; paraphyses filiform. On wood of *Juniperus*. Known only from the type locality in Greenland.

Godronia rhabdospora (Berk. & Curt.) Sacc. Syll. Fung. **8**: 602. 1889; *Tympanis rhabdospora* Berk. & Curt.; Berk. Grevillea **4**: 3. 1875. Spores said to be filiform. Reported on *Acer* from New England. No material seen.

Atropellis apiculata Lohman, Cash & Davidson, Jour. Wash. Acad. Sci. **32**: 297. 1942. Said to differ from other species on *Pinus* in that the apothecia are chocolate brown rather than blue as in other specimens on *Pinus*. No specimens have been seen.

Atropellis arizonica Lohman & Cash, Jour. Wash. Acad. Sci. **30**: 261. 1940. On stems of *Pinus ponderosa*. Said to differ from other species or *Pinus* mainly in the size of the spores.

Atropellis tingens Lohman & Cash, Jour. Wash. Acad. Sci. **30**: 257. 1940. On various species of *Pinus*. Said to be suggestive of *Atropellis pinicola* but differing mainly in spore size. No material has been seen.

9. **PEZICULA** Tul. Fung. Carp. **3**: 182. 1865.

Neofabraea Jackson, Oregon Agr. Exp. Station, Bienn. Crop. Pest and Hort. Rep. **1911–1912**: 187. 1913.

Conidial stage consisting of a basal fleshy stroma of the *Myxosporium* type on the surface of which the conidia are produced, or with definite conical, or flask-shaped pycnidia of the *Sphaeronema* form in which the conidia are produced; conidiophores variable in form and length, often very long; conidia broad-ellipsoid, usually granular and hyaline; microconidia often present.

Apothecia usually occurring in cespitose clusters on a stromatic base, sessile, or with a short, thick, stem-like base, usually light-colored, whitish, or yellowish, rarely exceeding 1 mm. in diameter, tubercular, or discoid, usually soft and fleshy; asci broad-clavate, usually 8-spored; spores ellipsoid, simple, or becoming tardily several-septate; paraphyses hyaline, or sub-hyaline and usually free, not agglutinated and not usually forming an epithecium.

Type species, *Peziza carpinea* Pers.

On non-coniferous hosts.
    On Betulaceae.
        On *Carpinus caroliniana*.   1. *P. carpinea*.
        On *Corylus rostrata*.   2. *P. corylina*.
        On *Alnus* or *Betula*.
            Spores small, fusoid.   3. *P. alnicola*.
            Spores large ellipsoid or ovoid.
                Spores 8–12 × 18–32 μ.   4. *P. aurantiaca*.
                Spores 7–10 × 15–20 μ.   5. *P. Alni*.
    On Rosaceae.
        On *Crataegus*.
            Spores 15–17 × 35–48 μ.   6. *P. crataegicola*.
            Spores 10–12 × 20–25 μ.   7. *P. olivascens*.
        Not on *Crataegus*.
            Apothecia at first club-shaped, on *Amelanchier*.   8. *P. pruinosa*.
            Apothecia at first subglobose.
                On *Rubus* spp.   9. *P. Rubi*.
                On *Rosa* sp.   10. *P. Brenckleana*.
    On Aceraceae.
        Spores ellipsoid 20–27 μ long.
            Conidial stage spike-like.   11. *P. spiculata*.
            Conidial stage not spike-like.   12. *P. acericola*.
        Spores subglobose 6–8 × 7–10 μ.   13. *P. spicata*.
    On Cornaceae.
        Spores 7–8 × 20 μ.   14. *P. cornicola*.
        Spores 10–13 × 28–34 μ.   15. *P. Corni*.
    On Salicaceae.
        Spores 10–12 × 25–30 μ.   16. *P. ocellata*.
        Spores 5–6 × 18–20 μ.   17. *P. Populi*.
    On various other hosts.
        Apothecia reddish-purple, on *Castanea*.   18. *P. purpurascens*.
        Apothecia ochraceous.
            On *Viburnum*.   19. *P. minuta*.
            On *Hamamelis*.   20. *P. Hamamelidis*.
            On *Rhamnus*.   21. *P. Morthieri*.
On coniferous hosts.
    On branches of *Pinus* and *Abies*.   22. *P. livida*.
    On leaves of *Abies*.   23. *P. phyllophila*.

1. **Pezicula carpinea** (Pers.) Tul. Fung. Carp. **3**: 183. 1865.
(PLATE 140.)

? Tubercularia fasciculata Tode, Fungi Meckl. **1**: 20.   1790.
Peziza carpinea Pers. Syn. Fung. 673.   1801.
Octospora paradoxa Hedwig, Obs. Bot. **1**: 13.   1802.
Peziza Betuli Alb. & Schw. Consp. Fung. 309.   1805.
Ditiola paradoxa Fries, Syst. Myc. **2**: 171.   1822.
Stictis Betuli Fries, Syst. Myc. **2**: 193.   1822.
Cycledum Carpini Wallr. Fl. Crypt. Germ. **2**: 512.   1833.
Dermatea carpinea Fries, Summa Veg. Scand. 362.   1849.
Discella discoidea Cooke & Peck; Peck, Ann. Rep. N. Y. State Mus. **28**: 58.
   1876.
Discula Peckiana Sacc. Syll. Fung. **3**: 675.   1884.
Ombrophila paradoxa Sacc. Syll. Fung. **8**: 620.   1889.
Ocellaria Betuli Rehm in Rab. Krypt.-Fl. **1³**: 136.   1888.
?Dermatella scotinus Morgan, Jour. Myc. **10**: 98.   1904.
Pezicula fasciculata Dearn. & House, Bull. N. Y. State Mus. **243**: 95.   1923.
Cryptosporiopsis fasciculata Petrak, Ann. Myc. **21**: 187.   1923.

Conidial stage consisting of a soft, fleshy, erumpent stroma
on the surface of which the conidia are produced; conidiophores
slender, often several times the length of the spore, strongly
swollen just below the point of attachment; conidia broad-
ellipsoid, simple, 10–12 × 20–24 μ.

Apothecia thickly gregarious, springing in cespitose clusters
from an immersed, fleshy stromatic base, the individual apothecia
tuberculate, or expanded and subdiscoid, with a short, stem-like
base, often distorted by mutual pressure, reaching a diameter of
1–3 mm., pale-yellowish; hymenium plane, or convex, similar in
color to the outside of the apothecium; asci clavate, reaching a
length of 150–200 μ and a diameter of 15–20 μ, 8-spored; spores
ellipsoid, straight, or slightly curved, granular within, for a long
time simple but often becoming 1–3-septate, 10–12 × 18–30 μ;
paraphyses slender, branched, enlarged above, reaching a diameter of 5 μ, not forming an epithecium.

On trunks and branches of *Carpinus caroliniana.*

TYPE LOCALITY: Europe.

DISTRIBUTION: Massachusetts to Missouri and Pennsylvania;
also in Europe.

ILLUSTRATIONS: Tode, Fungi Meckl. *pl. 4, f. 32;* Hedwig,
Obs. Bot. **1**: *pl. 8;* Rab. Krypt.-Fl. **1³**: 243, *f. 1–6;* Ann. Sci. Nat.
III. **20**: *pl. 16, f. 17, 18;* Mycologia **25**: *pl. 22.*

EXSICCATI: Ellis, N. Am. Fungi *67, 67* (*b*) (as *Dermatea
cornea*); Ellis & Ev. N. Am. Fungi *2623, 2741, 3333;* Shear, N. Y.

Fungi *93;* Rav. Fungi Car. **5**: *38;* Rab.-Winter, Fungi Eu. *3463* (From Missouri); Wilson & Seaver, Ascom. *91.*

2. **Pezicula corylina** Groves, Mycologia **30**: 47.   1938.

Excipula turgida Fries, Syst. Myc. **2**: 189.   1822.
Cenangium turgidum Duby, Bot. Gall. **2**: 736.   1830.
Catinula turgida Desm. Ann. Sci. Nat. III. **18**: 374.   1852.
Sphaeronema Coryli Peck, Ann. Rep. N. Y. State Mus. **24**: 85.   1872.

Conidial fruiting bodies erumpent, thickly scattered, or more or less in rows, mostly separate, sometimes two or three together, cylindric to cylindric-conic, or compressed when dry and subhysteriform, opening out widely when moist, .2–.5 mm. in diameter, .3–.5 mm. in height, black, or dark-olivaceous, hard, brittle, fleshy-leathery when moist; conidia ellipsoid, 8–10 × 17–28 µ.

Apothecia erumpent, scattered, or more or less in rows, separate, or cespitose, circular, sometimes crowded, sessile, narrowed below, pale-yellow, slightly pruinose when dry, much brighter when moist, close to sulphur-yellow, minute, .2–.5 mm. in diameter, .2–.4 mm. in height, soft-waxy in consistency, more fleshy when moist; hymenium at first concave, then plane to convex, slightly pruinose, pale-yellow to slightly reddish, margin at first forming a delicate lighter border, later disappearing; asci cylindric-clavate, short-stalked, usually 8-spored, occasionally 4-spored, 15–20 × 85–125 µ; spores oblong-ellipsoid, hyaline, straight, or slightly curved, simple, or 1-septate, irregularly biseriate, 6–10 × 15–27 µ; paraphyses hyaline, filiform, simple, or branched, septate, 2–2.5 µ in diameter, the tips swollen to 3.5 µ, forming a slight epithecium.

On *Corylus rostrata.*

TYPE LOCALITY: Bear Mountain, Temagami Forest Reserve, Ontario, Canada.

DISTRIBUTION: Pennsylvania to Washington and Canada; also in Europe.

ILLUSTRATIONS: Mycologia **30**: 48, *f. 1;* 49, *f. 2.*

EXSICCATI: Reliq. Farlow. *106* (as *Cenangium turgidum*); Ellis, N. Am. Fungi *949;* Ellis & Ev. N. Am. Fungi *1622.*

3. **Pezicula alnicola** Groves, Mycologia **32**: 120.   1940.

?Cenangium tennesseense Kanouse, Mycologia **33**: 462.   1941.

Conidial stage inconspicuous, developing beneath the outer bark splitting it but scarcely breaking through, the conidia

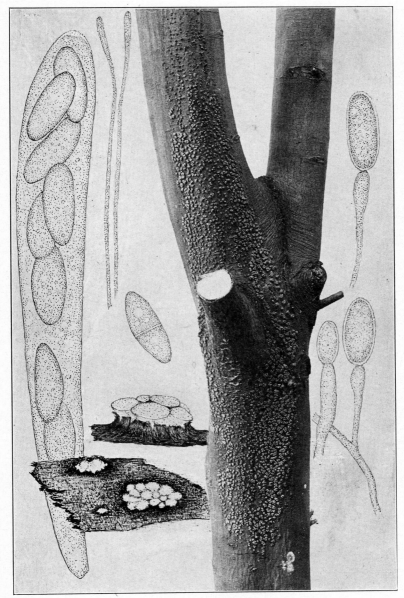

PEZICULA CARPINEA

emerging in whitish masses, the stroma usually circular, reaching a diameter of .6 mm., slightly conical t ᵼ cushion-shaped; conidiophores exposed on the upper surface, oᵼ ᵼn widely opening cavities, hyaline, cylindric, pointed, simple, or septate, not branched, 1.5–2.5 × 10–30 μ; conidia elongate-fusiform to subfiliform, hyaline, 5–7-septate, almost straight, sickle-shaped, or sigmoid, 4–5 × 35–53 μ.

Apothecia erumpent, scattered, or more often in clusters of five to ten or occasionally more, circular, or irregular from crowding, sessile, narrowed below, reaching a diameter of 1 mm., and half as high; hymenium concave to plane, or slightly convex, pinkish-cinnamon, slightly pruinose; asci cylindric to cylindric-clavate, short-stalked, 8-spored, reaching a length of 70–110 μ and a diameter of 8–12 μ; spores ellipsoid-fusoid, hyaline, simple, becoming 1–3-septate, straight, or slightly curved, 1-seriate, or becoming irregularly 2-seriate, 3.5–5 × 15–20 μ, rarely larger; paraphyses filiform, simple, or branched, 1.5–2 μ in diameter, swollen above to 5 μ and forming a slight epithecium.

On twigs and branches of *Alnus incana* and *Betula lutea*.

Type locality: Temagami Forest Reserve, Ontario, Canada.

Distribution: Ontario to Quebec; also in Europe (as *Pezicula Alni*).

Illustrations: Mycologia **32**: 114, *f. 3;* 116, *f. 8, 9,*

4. **Pezicula aurantiaca** Rehm, Ber. Bayer. Bot. Ges. **13**: 198. 1912.

Habrostictis aurantiaca Rehm, Ber. Naturh. Ver. Augsburg **26**: 67.   1881.
Ocellaria aurantiaca Rehm in Rab. Krypt.-Fl. 1³: 135.   1888.

Conidial stage forming beneath the outer bark raising and splitting it but scarcely breaking through but the conidia emerging in whitish masses, the stromata circular, or slightly elongated, reaching a diameter of 1 mm., obtusely conical, or cushion-shaped to discoid; conidiophores borne more or less exposed over the upper surface, or in widely open cavities, hyaline, cylindric to subclavate, simple, or occasionally branched, reaching a length of 15–25 μ (rarely more) and a diameter of 3–5 μ, sometimes swollen below the point of attachment to 7 μ; conidia ellipsoid to ovoid, hyaline, simple, or occasionally 1–3-septate, straight, or slightly curved, one end with a truncate apiculus, 12–16 × 25–40 μ; microconidia (in culture) filiform, straight, or curved, 1.5–2 × 6–14 μ.

Apothecia erumpent or subimmersed, occurring singly, or occasionally two or three in a cluster, circular, or somewhat irregular, sessile, scarcely narrowed below, 1–2 mm. in diameter and scarcely half as high; hymenium concave to plane, or slightly convex, brownish, or olive-brown, slightly pruinose; asci cylindric-clavate, reaching a length of 100–140 μ and a diameter of 15–20 μ, short-stalked; spores ellipsoid to ovoid, hyaline, straight, or slightly curved, simple, or occasionally 1–3-septate, irregularly 2-seriate, 8–12 × 18–32 μ; paraphyses filiform, hyaline, simple, or branched 1.5–2 μ in diameter, enlarged above to 5 μ.

On *Alnus crispa* var. *mollis*.

TYPE LOCALITY: Europe.

DISTRIBUTION: Toronto, Canada; also in Europe.

ILLUSTRATIONS: Mycologia **32**: 114, *f. 1–3;* 116, *f. 5.*

5. **Pezicula Alni** Rehm, Ber. Bayer. Bot. Ges. **13**: 199.   1912.

Dermatella quercina var. *Alni* Sacc. Syll. Fung. **8**: 490.   1889.
Dermatea Alni Rehm in Rab. Krypt.-Fl. **1**³: 252.   1889.

Conidial stage forming beneath the outer bark and splitting it but scarcely breaking through, slightly conical, or cushion-shaped, less than 1 mm. in diameter, blackish; conidiophores cylindric to clavate, simple, or branched, 2.5–4 × 12–30 μ, produced on the outer surface of the stroma; conidia ovoid to ellipsoid, or subfusiform, hyaline, simple, one end with a small, truncate apiculus, 5–7 × 14–18 μ.

Apothecia erumpent, mostly scattered, occasionally with two or three in a cluster, circular, or slightly elongated, sessile but narrowed below, reaching a diameter of 1 mm. and half as high; hymenium plane to convex, slightly pruinose, brownish-ochraceous; asci broad-clavate, 4–8-spored, mostly 8-spored, reaching a length of 100–130 μ and a diameter of 18–20 μ; spores ellipsoid, irregularly 2-seriate, at first simple, becoming 1–3-septate, 7–10 × 15–25 μ; paraphyses filiform 2–2.5 μ in diameter.

On species of *Alnus*, *Alnus crispa* var. *mollis*, *A. incana*.

TYPE LOCALITY: Europe.

DISTRIBUTION: Nova Scotia to Michigan and South Carolina; also in Europe.

ILLUSTRATIONS: Mycologia **32**: 114, *f. 2;* 116, *f. 6, 7.*

EXSICCATI: Rav. Fungi Car. **5**: *46* (as *Patellaria rhabarbarina*).

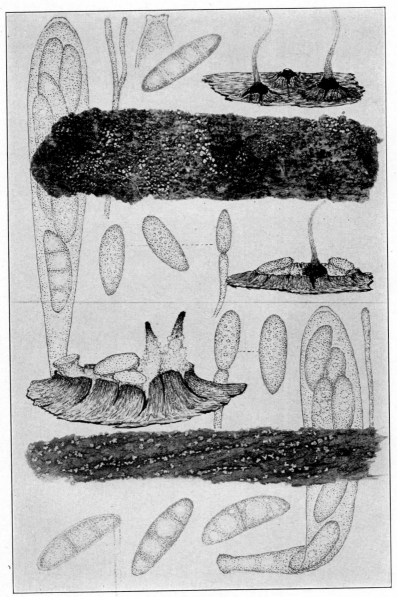

1. PEZICULA ACERICOLA
2. PEZICULA SPICULATA

6. **Pezicula crataegicola** (Durand) Groves, Mycologia **38**: 414. 1946.

Dermatea crataegicola Durand, Jour. Myc. **10**: 100. 1904.

Apothecia erumpent, occurring singly, or two to four together arising from a common stroma, sessile, or narrowed into a stem-like base .5–1 mm. in diameter, externally mealy-pruinose, rhubarb-color; hymenium plane or nearly so, or even slightly convex, yellowish-olivaceous; asci broad-clavate, attenuated into a short stem, at first filled with a homogeneous, granular protoplasm, reaching a length of 150–165 μ and a diameter of 25–30 μ, 8-spored; spores irregularly 2-seriate, or obliquely 1-seriate, hyaline, smooth, simple, oblong-ellipsoid, with a central oil-drop, 15–17 × 35–48 μ; paraphyses filiform, branched, enlarged above, olivaceous-yellow, 6–8 μ in diameter above, the ends free, not glued together to form an epithecium.

On branches of *Crataegus*.

TYPE LOCALITY: London, Ontario, Canada.

DISTRIBUTION: Known only from the type locality.

EXSICCATI: Ellis & Ev. Fungi Columb. *1917*.

This differs from *Pezicula olivascens* mainly in the size of the spores. This may be due to the immaturity of the material in *P. olivascens* which was collected very early in the season and the spores seemed poorly developed.

7. **Pezicula olivascens** (Rehm) Seaver, comb. nov.

Dermatea olivascens Rehm, Ann. Myc. **5**: 80. 1907.

Apothecia scattered, occurring singly, or two or three together, erumpent, subglobose becoming expanded, sessile, or with a very short, stem-like base, yellowish with an olivaceous tinge, .5–1.5 mm. in diameter; asci broad-clavate, at first densely granular within, reaching a length of 150 μ and a diameter of 20–25 μ, 8-spored; spores broad-ellipsoid, simple (in the specimens examined), irregularly 2-seriate, 10–12 × 20–25 μ, surrounded with much granular material, hyaline; paraphyses filiform, 2 μ in diameter, branched, slightly enlarged above, the ends free, not glued together into a distinct epithecium.

On dead branches of *Crataegus* still on the tree.

TYPE LOCALITY: Mt. Pleasant, Iowa.

DISTRIBUTION: Known only from the type locality.

ILLUSTRATIONS: Bull. Lab. Nat. Hist. State Univ. Iowa **6**: *pl. 38, f. 1.*

EXSICCATI: Rehm, Ascom. *1686;* Wilson & Seaver, Ascom. *2.*
This species was described from material collected in Mt.
Pleasant, Iowa in February. The ascospores appear to be more
or less immature. Whether they later became septate is uncertain. The species is very close to *Pezicula crataegicola*
(Durand) Groves but seems to differ in its smaller ascospores.
This apparent difference may be due to the immaturity of the
spores. Other characters seem to conform very closely with
that species.

8. **Pezicula pruinosa** (Peck) Farlow; Thaxter, Mycologia **14**:
    102. 1922.

Sphaeronema pruinosum Peck, Ann. Rep. N. Y. State Mus. **24**: 85.   1875.
Dermatea pruinosa Petrak, Ann. Myc. **20**: 196.   1922.   Not *D. pruinosa*
    (Ellis & Ev.) Sacc.   1889.
Lagynodella pruinosa Petrak, Ann. Myc. **20**: 207.   1922.

Conidial stage (*Sphaeronema*) consisting of conical, or spinu-
lose, black, pruinose pycnidia which spring from a stromatic
base, reaching a length of 1 mm.; conidia broad-ellipsoid, 8–9 ×
20–24 μ.

Apothecia scattered, erumpent, short-stipitate, occurring
singly or in cespitose clusters of three to five apothecia each
springing from a stromatic base, at first club-shaped, expanding
and becoming discoid, or subdiscoid above, pale cinnamon-
brown, reaching a diameter of 1 mm.; hymenium plane, or concave, similar in color to the outside of the apothecium; stem
short, almost as thick as the diameter of the apothecium, densely
albopruinose, reaching a length of 2 mm.; asci clavate, reaching
a length of 95–110 μ and a diameter of 18–21 μ, 8-spored; spores
1–2-seriate, subellipsoid, usually slightly curved, simple, or
becoming 1–3-septate, 8–9 × 22–25 μ; paraphyses filiform,
simple, or branched, reaching a diameter of 1 μ.

On *Amelanchier canadensis, A. alnifolia* and unnamed species
of *Amelanchier.*

TYPE LOCALITY: Sharon, Massachusetts.

DISTRIBUTION: New York, Massachusetts, North Dakota,
Idaho, and Montana.

EXSICCATI: Reliq. Farlow. *135 a. b.;* Brenckle, Fungi Dak. *450.*

9. **Pezicula Rubi** (Libert) Rab. in Fungi Eu. *2122.*   1876.

Patellaria Rubi Lib. Pl. Crypt. Ard. *231.*   1834.
Peziza ardennensis Mont. Ann. Sci. Nat. II. **5**: 287.   1836.

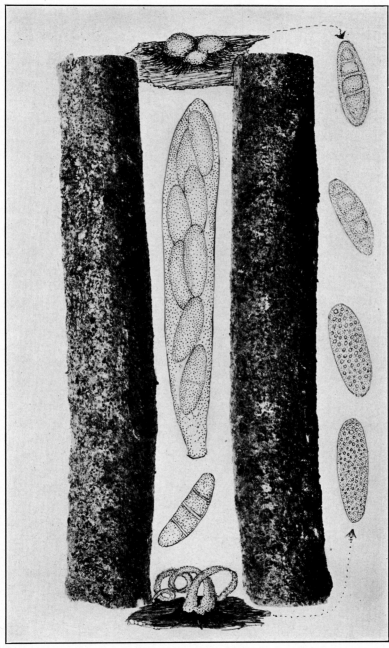

PEZICULA CORNICOLA

Peziza rhabarbarina Berk. in Hooker, Engl. Fl. **5**²: 197.   1836.
Lachnella rhabarbarina Fries, Summa Veg. Scand. 365.   1849.
Helotium Rubi Spree in Rab. Fungi Eu. *717*.   1865.
Pezicula rhabarbarina Tul.; Fuckel, Symb. Myc. 278.   1869.
Dermatea rhabarbarina Phill. Brit. Discom. 343.   1887.
Dermatea Rubi Rehm in Rab. Krypt.-Fl. **1**³: 258.   1889.

Apothecia usually occurring singly, at first subglobose, then expanding and becoming shallow cup-shaped, finally discoid, reaching a diameter of 1 mm.; hymenium plane, or convex, with a yellowish-brown, mealy coating; asci clavate, reaching a length of 90–110 μ and a diameter of 16–20 μ, 8-spored; spores irregularly 2-seriate, long-ellipsoid, becoming 1–2-septate, 6–7 $\times$ 18–24 μ; paraphyses slender, branched, the enlarged ends forming an epithecium and agglutinated together with a yellow matrix.

On *Rubus idaeus* and other species of *Rubus*.

TYPE LOCALITY: Europe.

DISTRIBUTION: Maine to West Virginia and Ohio; also in Europe.

ILLUSTRATIONS: Ann. Sci. Nat. II. **5**: *pl. 13, f. 5;* Boud. Ic. Myc. *pl. 560.*

EXSICCATI: Shear, N. Y. Fungi *95;* Reliq. Farlow. *114, 858.*

10. **Pezicula Brenckleana** Seaver, nom. nov.

Cenangium Rosae Sacc. Atti Soc. Veneto Sci. Nat. Padova **2**²: 160.   1873.
    Not *Cenangium Rosae* Schw.   1832.
Pezicula Rosae Sacc, Michelia **2**: 332.   1881.
Dermatea Rosae Rehm in Rab. Krypt.-Fl. **1**³: 259.   1889.

Apothecia occurring singly, or several in cespitose groups, at first rounded, opening and becoming shallow cup-shaped, then discoid, at first yellowish, becoming brown, reaching a diameter of .5–1 mm.; hymenium concave, plane, or slightly convex, yellowish-brown; asci clavate, reaching a length of 90–120 μ and a diameter of 16–18 μ, 8-spored; spores irregularly 2-seriate, ellipsoid, 7–9 $\times$ 20–24 μ, finally becoming 1–3-septate; paraphyses slender, enlarged above, the tips surrounded with a yellowish matrix.

On species of *Rosa*.

TYPE LOCALITY: Europe.

DISTRIBUTION: North Dakota; also in Europe.

ILLUSTRATIONS: Sacc. Atti Soc. Veneto Sci. Nat. Padova **2**²: *pl. 16, f. 19–22.*

EXSICCATI: Brenckle, Fungi Dak. *392.*

Only one specimen of this species has been seen from North America. While the spores and asci are slightly smaller than indicated in European descriptions, the specimens appear to be otherwise identical. While placed in the genus *Pezicula* by Saccardo, the conidial stage has not been observed. Since the name given this species is untenable we dedicate the species to Dr. Brenckle who collected the only specimen seen from America.

11. **Pezicula spiculata** Seaver, Mycologia **25**: 146. 1933. (PLATE 141, FIG. 2.)

The conidial stage (*Sphaeronema*) accompanies the apothecial both springing from a floccose stroma; pycnidia large and spike-like, black but covered with white flakes, reaching a length of 1.5 mm., swollen below; macroconidia 10 × 20–24 μ and borne on sporophores equalling or exceeding the length of the spore.

Apothecia cespitose in rounded or elongated clusters, or rarely occurring singly, reaching a diameter of 1 mm., pale-yellowish, sessile, or tapering into a short, stem-like base; hymenium slightly concave, or plane, not darker than the outside of the apothecium; asci broad-clavate, reaching a length of 120 μ and a diameter of 16–18 μ, 8-spored; spores irregularly 2-seriate, ellipsoid and often slightly curved, 8 × 24–27 μ, becoming 1–3-septate; paraphyses slender, slightly enlarged above, the ends free, branched, hyaline, about 2 μ in diameter.

On *Acer (spicatum?)*.

TYPE LOCALITY: Near Ithaca, New York.

DISTRIBUTION: Known only from the type locality.

ILLUSTRATIONS: Mycologia **25**: *pl. 23* (lower figure).

So far as the ascigerous stage of this species is concerned it can scarcely be distinguished from *Pezicula acericola*. However, the two are apparently entirely distinct in their conidial stages.

12. **Pezicula acericola** (Peck) Sacc. Atti Ist. Veneto VI. **3**: 725. 1885. (PLATE 141, FIG. 1.)

?Peziza cinnamomea D. C. in Pers. Myc. Eu. **1**: 268. 1822.
Nodularia acericola Peck, Ann. Rep. N. Y. Mus. **25**: 98. 1873.
Dermatea carnea Cooke, & Ellis, Grevillea **5**: 32. 1876.
?Dermatea cinnamomea Phill. Brit. Discom. 342. 1887. Not Cooke & Peck 1875.
Dermatea simillima Ellis & Ev. Proc. Acad. Sci. Phila. **1893**: 451. 1893.
Dermatea Alni f. *Aceris* Rehm in Rab. Krypt-Fl. **1**³: 252. 1889.

Apothecia erumpent in cespitose clusters of three to eight each, the individuals seldom exceeding 1 mm. in diameter, sessile or subsessile, pale-yellow at least when young; hymenium plane, or slightly convex, the margin rather indistinct, similar in color to the outside of the apothecium, becoming concave with age; asci clavate, reaching a length of 90–130 μ and a diameter of 15–20 μ, 4–8-spored, gradually tapering below into a slender, stem-like base; spores irregularly 2-seriate above, ellipsoid, straight, or curved, at first simple, often becoming 3–4-septate, 8 × 20–26 μ, hyaline, or subhyaline; paraphyses slender, enlarged above and often flexuose, hyaline, or slightly colored with age.

On dead branches of *Acer spicatum* and *Acer rubrum*.

TYPE LOCALITY: North Elba, New York.

DISTRIBUTION: New York to Newfoundland and Ontario, Canada.

ILLUSTRATIONS: Grevillea **5**: *pl. 75, f. 9* (as *Dermatea carnea*); Mycologia **25**: *pl. 23* (upper figure in part).

EXSICCATI: Ellis, Nova-Caesar, *56;* N. Am. Fungi *67;* Ellis & Ev. N. Am. Fungi *246;* Barth. Fungi Columb. *3420;* Reliq. Farlow. *112 a, b;* Thüm. Myc. Univ. *978* (from New York); Rehm, Ascom. *1901* (from Ontario, Canada).

In a previous paper (Mycologia **25**: 145. 1933) the writer concluded from herbarium observations that *Nodularia acerciola* Peck and *Tympanis acerina* Peck were identical, the latter being aged, discolored specimens of the former. The two were often found growing in the same clump; the ascospores were identical and both were accompanied by the *Sphaeronema* stage.

J. W. Groves, however, disagrees and regards *Tympanis acerina* Peck and *Nodularia acericola* Peck as two distinct species and treats *Sphaeronema acerinum* as the conidial stage of the former. He states that the conidial stage of *Pezicula acericola* is a very inconspicuous fleshy stroma developing beneath the outer bark.

Groves further states that the oblong-ellipsoid conidia of *Tympanis acerina* are similar in form to conidia of species of the related genus *Pezicula* and this has led to confusion regarding the specific identity and conidial relations of this species and of species of *Pezicula* occurring on *Acer*. This also suggests that *Dermea acerina* may be a discolored *Pezicula*.

13. **Pezicula spicata** Ellis & Ev. Bull. Torrey Club **25**: 506. 1898.

Apothecia scattered, usually occurring singly, erumpent, through the ruptured epidermis but scarcely protruding above its surface, .35–.5 mm. in diameter, dull watery-white, yellowish when dry; hymenium slightly concave when dry, the margin a little darker undulated; asci broad-clavate, reaching a length of 60–70 μ and a diameter of 12–15 μ, with a short, abrupt, stem-like base, 8-spored; spores irregularly 2-seriate, subglobose, 6–8 × 7–10 μ; paraphyses filiform, enlarged above, reaching a diameter of 3 μ.

On dead twigs of *Acer spicatum.*

TYPE LOCALITY: Near Ottawa, Canada.

DISTRIBUTION: Known only from the type locality.

14. **Pezicula cornicola** Seaver, Mycologia **29**: 337. 1937. (PLATE 142.)

Apothecia usually in cespitose clusters, individual apothecia sessile, reaching 1 mm. in diameter, pale-yellow; hymenium plane, or slightly convex; asci clavate, reaching a length of 100–120 μ and a diameter of 12–15 μ; spores partially 2-seriate, ellipsoid, straight, or slightly curved, about 7–8 × 20 μ, becoming tardily 1–3-septate; paraphyses filiform, slightly enlarged at their apices.

On bark of *Cornus* sp.

TYPE LOCALITY: East Hampton, New York.

DISTRIBUTION: Known only from the type locality.

ILLUSTRATIONS: Mycologia **29**: 336, *f. 2.*

This is associated with a *Myxosporium* which appears to be its conidial stage. The spores ooze out from the pycnidia in sausage-like streams, whitish in color. The conidia are ellipsoid, or slightly narrowed at one end, quite variable in size but often reaching a length of 40 μ and a diameter of 15 μ, densely filled with minute granules.

15. **Pezicula Corni** Petrak, Ann. Myc. **20**: 197. 1922. (PLATE 143.)

?Dermatea Corni Phill. & Hark. Grevillea **13**: 22. 1884.
Pezicula rhabarbarina f. *Corni* Ellis in Herb.

Apothecia solitary, or cespitose, at first rounded, becoming expanded and subdiscoid, with a mealy, brown covering, reaching a diameter of about .5–1 mm.; hymenium plane or nearly so,

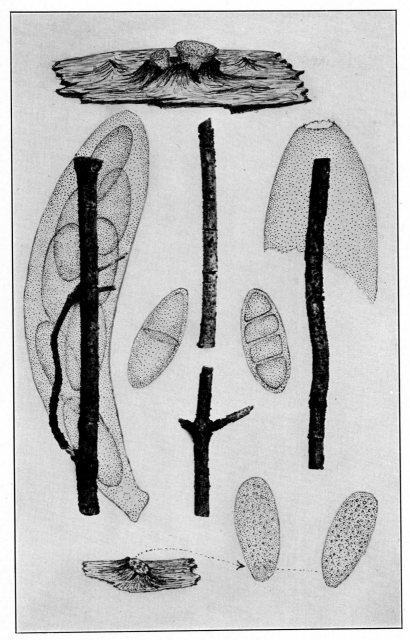

PEZICULA CORNI

yellowish to dark brownish-black; asci clavate, reaching a length of 120 μ and a diameter of 27 μ, 8-spored; spores ellipsoid, irregularly crowded in the ascus, 10–13 × 28–34 μ; paraphyses filiform, slightly enlarged above.

On *Cornus alternifolia*, *C. stolonifera*, and unnamed species of *Cornus*.

TYPE LOCALITY: Idaho.

DISTRIBUTION: California to Oregon and Ontario.

ILLUSTRATIONS: Mycologia 29: 335, *f. 1*.

EXSICCATI: Ellis & Ev. N. Am. Fungi *2809* (as *Pezicula rhabarbarina*); Reliq. Farlow. *857.*

This is accompanied by a *Myxosporium* which appears to be its conidial stage. The pycnospores are ellipsoid and densely filled with granules, 13–15 μ × 33-36 μ.

16. **Pezicula ocellata** (Pers.) Seaver, comb. nov. (PLATE 144, FIG. 1.)

Peziza ocellata Pers. Syn. Fung. 667. 1801.
Stictis ocellata Fries, Syst. Myc. **2**: 193. 1822.
Peziza Lecanora Schm. & Kunze, Deuts Schw. *174.* 1817.
Stictis Lecanora Fries, Syst. Myc. **2**: 193. 1822.
Phacidium Populi Lasch in Rab. Herb. Myc. ed. 2. *519.* 1857.
Ocellaria aurea Tul. Fung. Carp. **3**: 129. 1865.
Habrostictis ocellata Fuckel, Symb. Myc. Nacht. **1**: 38. 1871.
Dermatea cinnamomea Cooke & Peck; Cooke, Bull. Buffalo Soc. Nat. Sci. **3**: 24. 1875.
Dermatea inclusa Peck, Ann. Rep. N. Y. State Mus. **30**: 62. 1878.
Propolis ocellata Gill. Champ. Fr. Discom. 214. 1887.
Dermatea macrospora Clements, Bull. Torrey Club **30**: 87. 1903.
Ocellaria ocellata Seaver, Mycologia **3**: 65. 1911.
Pezicula eximia Rehm, Ann. Myc. **11**: 154. 1913.

Conidia borne on the surface of a fleshy, submerged stroma, large, ellipsoid, densely filled with granular matter, reaching a length of 30–35 μ and a diameter of 10–12 μ borne on a conidiophore about the length of the spore itself and strongly swollen just below the point of attachment.

Apothecia occurring singly, breaking through the bark, the broken edges of which form a ragged margin often extending above the margin of the cup, reaching a diameter of 1–2 mm., yellowish, with an irregularly notched margin which is often whitish; hymenium concave, or nearly plane, surrounded by the upturned margin of the apothecium, yellowish, or dull-orange; asci broad-clavate, reaching a length of 120 μ and a diameter of

24 μ, terminated abruptly at the base; spores ellipsoid, granular within, 10–12 × 25–30 μ, irregularly 2-seriate, hyaline, simple, or 1–3-septate; paraphyses enlarged above, reaching a diameter of 5–6 μ.

On branches of *Salix petiolaris, Salix discolor* and other species of *Salix* and *Populus.*

TYPE LOCALITY: Europe.

DISTRIBUTION: New York to Winnipeg, Washington and Colorado; also in Europe.

ILLUSTRATIONS: Tul. Fung. Carp. **3**: *pl. 18, f. 1–11;* Gill. Champ. Fr. Discom. *pl. 98;* E, & P. Nat. Pfl. **1**¹: 247, *f. 182. A–C;* Massee, Brit. Fungus-Fl. **4**: 22, *f. 34–35.*

EXSICCATI: Schm. & Kunze, Deuts. Schw. *174* (as *Peziza Lecanora*); Barth. Fungi Columb. *1918;* Brenckle, Fungi Dak. *535.*

This species has been placed in various genera and made the type of the genus *Ocellaria* by Tulasne which genus is usually treated with the Phacidiales. The writer can see no reason for treating it with that group. It has been collected in abundance by the writer in Colorado on some species of willow around Tolland at an altitude of 8,000 to 10,000 feet.

Through the kindness of John Dearness, the writer has been permitted to examine a part of the type collection of *Pezicula eximia* Rehm and finds it identical with the above.

17. **Pezicula Populi** (Thompson) Seaver, comb. nov.

Neofabraea Populi Thompson, Mycologia **31**: 458.   1939.

Apothecia thickly gregarious, usually occurring singly, or rarely two or three crowded together, entirely sessile, erumpent but scarcely rising above the surface of the substratum, surrounded with the upturned edges of the ruptured outer bark, scarcely exceeding .5 mm. in diameter, dark brownish-black; hymenium concave, or nearly plane, reddish-brown, with a slightly elevated, darker center which gives the hymenium an umbilicate appearance; asci clavate, apparently normally 8-spored but often with only 5 or 6 of the spores developed, reaching a length of 80 μ and a diameter of 12 μ; spores 2-seriate above, ellipsoid, usually with two oil-drops, hyaline, 5–6 × 18–20 μ, becoming indistinctly 3-septate; paraphyses enlarged above, hyaline.

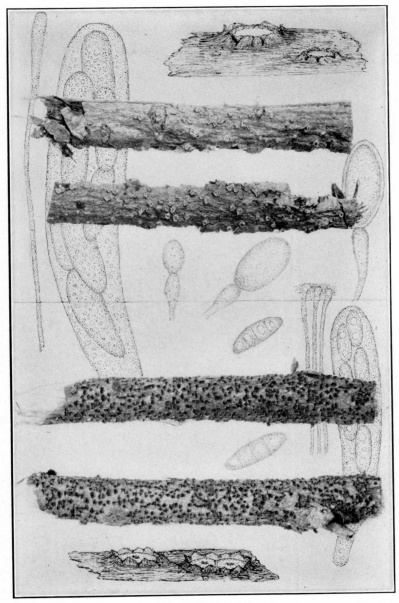

1. PEZICULA OCELLATA
2. PEZICULA POPULI

On *Populus grandidentata* and *Populus tremuloides*.

TYPE LOCALITY: Ontario, Canada.

DISTRIBUTION: Known only from the type locality.

ILLUSTRATIONS: Mycologia **31**: 457, *f. 1;* 460, *f. 2;* 462, *f. 3.*

This species is known from two collections by G. E. Thompson and communicated by H. S. Jackson, University of Toronto, Cryptogamic Herbarium *2054* and *2046*. Said to cause canker on living trees.

The conidial stage is reported by Thompson as a *Myxosporium* which forms definite cankers on young *Populus* branches.

18. **Pezicula purpurascens** (Ellis & Ev.) Seaver, Mycologia **34**: 414. 1942. (PLATE 145.)

Dermatea purpurascens Ellis & Ev. Jour. Myc. **4**: 100. 1888.
Ascoconidium Castaneae Seaver, Mycologia **34**: 414. 1942.

Conidia found associated with this species and possibly representing its perfect stage. Ascoconidiophores club-shaped reaching a length of 90 µ and a diameter of 12 µ, pale-brown, each containing one ascoconidium; ascoconidia broad-ellipsoid, reaching a length of 30–40 µ and a diameter of 9–10 µ, borne on slender stalk within the conidiophore, becoming disconnected, and finally discharged through the ruptured conidiophore, 3-septate, appearing brownish within the conidiophore but hyaline or subhyaline when discharged.

Apothecia scattered, erumpent, occurring singly, or two or three crowded together, sessile, or subsessile, externally reddish-purple, reaching a diameter of .75–1 mm.; hymenium plane, or slightly concave, dirty-white, becoming reddish-purple but lighter than the outside of the apothecium; asci cylindric-clavate, reaching a length of 120–140 µ and a diameter of 25–30 µ, 8-spored but some often undeveloped, 8–11 × 30–36 µ, hyaline or nearly so, ellipsoid with the ends strongly narrowed, becoming distinctly 3-septate, 9–11 × 30–36 µ; paraphyses slender, slightly enlarged above, reaching a diameter of 2–3 µ, often slightly colored.

The exterior of the apothecium is clothed with a palisade of appressed, poorly developed hairs which are dilutely purplish. It is this character which has suggested the specific name.

On dead limbs of chestnut, *Castanea dentata*.

TYPE LOCALITY: West Chester, Pennsylvania.

DISTRIBUTION: Pennsylvania and Massachusetts.

ILLUSTRATIONS: Mycologia **34**: 413, *f. 1.*
EXSICCATI: Ellis & Ev. N. Am. Fungi *2147.*

19. **Pezicula minuta** Peck, Bull. N. Y. State Mus. **1**²: 21. 1887.

Dermatea minuta Peck, Ann. Rep. N. Y. State Mus. **32**: 48. 1879.

Apothecia thickly scattered, occurring singly, or two or three together, rounded, becoming more or less turbinate, subochraceous, reaching a diameter of .25–.5 mm.; hymenium at first concave, becoming plane, similar in color to the outside of the apothecium; asci broad-clavate, 8-spored, reaching a length of 90 µ and a diameter of 20–22 µ; spores broad-ellipsoid, granular within, 10–12 × 20–25 µ; paraphyses filiform, slender.

On *Viburnum lantanoides.*

TYPE LOCALITY: Catskill Mountains, New York.

DISTRIBUTION: New York and Ontario, Canada.

Specimens in the herbarium of The New York Botanical Garden, a part of the type collection, have been examined.   In this material it is very difficult to find mature spores.   Whether they become septate could not be determined.

One specimen collected by Roy Cain, in Canada, No. *7162,* has been compared with the type and found identical.

20. **Pezicula Hamamelidis** Groves & Seaver, Canadian Jour. Res. **17**: 140. 1939. (PLATE 146.)

Conidial fruiting bodies minute, about .3–.4 mm. in diameter, erumpent, cushion-shaped, and containing one or more cavities lined with conidiophores 2.5–3 × 15–30 µ, swollen just below the point of attachment to 4 µ; conidia oblong-ellipsoid, hyaline, simple, straight, or slightly curved, 12–14 × 30–45 µ; microconidia borne in the same fruiting body, hyaline, filiform, straight, or curved, simple, 1.5–2 × 10–18 µ.

Apothecia erumpent, scattered, or more or less in rows, usually single, sometimes cespitose with two to six in a cluster, sessile, narrowed below, circular .3–1 mm. in diameter .2–.4 mm. high, pale ochraceous-yellowish, darker when dry, waxy; hymenium plane to convex, slightly pruinose; asci oblong-clavate, broad, short-stalked, 8-spored, reaching a length of 90–120 µ and diameter of 20–30 µ; ascospores oblong-ellipsoid to ovoid, hyaline, straight, or curved, irregularly 2-seriate, becoming 1–3-septate, 9–13 × 25–37 µ; paraphyses hyaline, filiform, simple, or

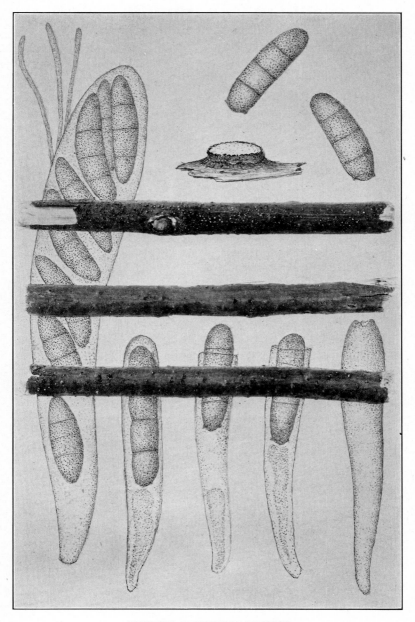

PEZICULA PURPURASCENS

occasionally branched, 2–3 μ in diameter, the tips swollen to 8 μ in diameter, forming a yellowish epithecium.

On branches of *Hamamelis virginiana*.

TYPE LOCALITY: Don Valley, Toronto, Canada.

DISTRIBUTION: Canada.

ILLUSTRATIONS: Canadian Jour. Res. **17**: 129, *f. 3, 1–2, f. 4, 5a, b*.

EXSICCATI: Sydow, Fungi Exotici Exsicc. *423* (as *Dermatea Rubi*).

The type specimen is intimately associated with a *Coryneum* which the writer took to be its conidial stage, but Groves decided that the conidial stage was a *Cryptosporiopsis* and not the *Coryneum* which often occurred in the same pustule with the *Pezicula*.

21. **Pezicula Morthieri** (Fuckel) Groves, Mycologia **39**: 329. 1947.

Sphaeria micula Fries, Elench. Fung. **2**: 101. 1828.
Micula Mougeotii Klotzs. Herb. Viv. Myc. 636. 1844.
Cenangium Morthieri Fuckel, Symb. Myc. 272. 1869.
Atractium Therryanum Sacc. Michelia **1**: 535. 1879.
Cenangella Morthieri Sacc. Syll. Fung. **8**: 592. 1889.
Dermatea micula Rehm in Rab. Krypt.-Fl. **1³**: 261. 1889.
Phaeangella Morthieri Sacc & D. Sacc; Sacc Syll. Fung. **18**: 128. 1906.
Sphaerographium niveum Dearn. & House, Bull. N. Y. State Mus. **266**: 89. 1925.

Conidial stage erumpent from the same stroma which produces the apothecia, cylindric-conic, reaching a height of 1 mm. and less than half as broad at the base, usually white, sometimes darkened; conidiophores septate, branched, 2–2.5 × 25–40 μ; conidia hyaline, filiform, straight, or curved, pointed at the ends, 3–4 × 45–65 μ, 1–3-septate.

Apothecia erumpent, occurring singly or two or three together, sessile, narrowed below, less than 1 mm. in diameter and height, yellowish-brown to reddish-brown; hymenium at first concave, becoming plane, or slightly convex, darker than the outside of the apothecium; asci cylindric-clavate, 8-spored, reaching a length of 65–100 μ and a diameter of 8–14 μ; spores ellipsoid-fusoid, hyaline, or pale-yellowish, straight, or slightly curved, 4–6 × 15–20 μ; paraphyses filiform, very slightly enlarged above, 1.5–3.5 μ in diameter.

On branches of *Rhamnus alnifolia*.

TYPE LOCALITY: Europe.

DISTRIBUTION: Quebec; also in Europe.

ILLUSTRATIONS: Mycologia **39**: 330, *f. 1, 2.*

Grove states that the position of this species is uncertain but places it in *Pezicula.*

22. **Pezicula livida** (Berk. & Br.) Rehm, Ascom. *462.* 1878.

Patellaria livida Berk. & Br. Ann. Mag. Nat. Hist. II. **13**: 466. 1854.
Dermatea abietina Auersw. in Rab. Fungi Eu. *1027*; Hedwigia **6**: 46. 1866.
Peziza eucrita Karst.; Nylander, Not. Soc. Fauna Fl. Fenn. **10**: 47. 1869.
Dermatea livida Phill. Brit. Discom. 340. 1887.
Dermatella livida Sacc. Syll. Fung. **8**: 490. 1889.
Dermatella eucrita Sacc. Syll. Fung. **8**: 491. 1889.
Durella livida Sacc. Syll. Fung. **8**: 795. 1889.
Myxosporium abietinum Rostr. Tids. Skow. **13**: 89. 1901.

Conidial stage consisting of a stroma on which the conidia are produced; conidia ellipsoid, $5 \times 12\text{--}16 \mu$; microconidia present, $1.5\text{--}2.5 \times 10\text{--}17 \mu$.

Apothecia occurring singly, or in small, cespitose clusters of usually three to six each, the clusters thickly gregarious, individual perithecia reaching a diameter of .5–1 mm., pale-yellowish; hymenium plane, or convex, similar in color to the outside of the apothecium, minutely roughened with the ends of the paraphyses; asci clavate, reaching a length of 100 μ and a diameter of 20 μ, 8-spored; spores partially 2-seriate, ellipsoid, straight, or more commonly curved, at first simple, often becoming 3–5-septate, $6\text{--}7 \times 20\text{--}31 \mu$; paraphyses slender below, strongly enlarged above, reaching a diameter of 5–7 μ.

On bark of *Pinus Strobus;* also reported on other species of *Pinus* and *Abies.*

TYPE LOCALITY: Europe.

DISTRIBUTION: New York and New England; also in Europe.

ILLUSTRATIONS: Not. Soc. Fauna Fl. Fenn. **10**: *pl. 1, f. 12:* Ann. Bot. **45**: *f. 1–5.* (pp. 77–80); Mycologia **24**: 424, *f. 1.*

According to Mary J. F. Gregor (Ann. Bot. **45**: 73. 1931) the above indicated species are synonyms.

23. **Pezicula phyllophila** (Peck) Seaver, comb. nov.

Dermatea phyllophila Peck, Ann. Rep. N. Y. State Mus. **31**: 47. 1879.

Apothecia minute, erumpent but often partially concealed by the epidermis which has a blistered appearance, yellowish-brown; hymenium pallid, or dingy-white, nearly plane; asci clavate,

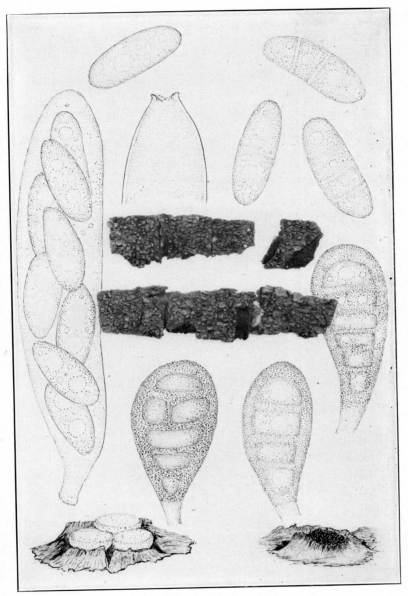

PEZICULA HAMAMELIDIS

8-spored, reaching a length of 80–90 μ and a diameter of 10–12 μ; spores broad-ellipsoid, 5–6 × 7–8 μ, usually containing one large oil-drop; paraphyses filiform, thickened above.

On lower surface of balsam leaves, *Abies balsamea*, while still on the tree; also on leaves of *Pinus rigida*.

TYPE LOCALITY: Summit, New York.

DISTRIBUTION: New York and (New Jersey?).

EXSICCATI: Ellis, N. Am. Fungi. *385*.

Type material of this species has been examined. It has all the appearance of a *Pezicula* except that the spores are small with no indication of septa. The type material is abundant.

DOUBTFUL AND EXCLUDED SPECIES

Pezicula atroviolacea Bres. Ic. Myc. **25**: *pl. 1237*. 1933; *Peziza atroviolacea* Bres. Fungi. Trid. **1**: 24. 1881; *Humaria atroviolacea* Sacc. Syll. Fung. **8**: 150. 1889; *Pachyella atroviolacea* Boud. Hist. Class. Discom. Eu. 51. 1907. This is certainly *Peziza clypeata* Schw. and not a *Pezicula*.

Pezicula Kalmiae (Peck) Sacc. Syll. Fung. **8**: 314. 1889; *Peziza Kalmiae* Peck, Ann. Rep. N. Y. State Mus. **25**: 99. 1873. No material seen. See *Trichopeziza Kalmiae*.

Beloniella marcyensis Kanouse, Papers Mich. Acad. Sci. **24**: 25. 1939. The erumpent habits, general description, and spores becoming tardily several-septate would indicate a *Pezicula*. No material has been seen.

Pezicula pallidula (Cooke) Rehm, Ber. Bayer. Bot. Ges. **13**: 199. 1912; *Dermatea pallidula* Cooke, Grevillea **16**: 70. 1888; *Scleroderris pallidula* Sacc. Syll. Fung. **8**: 598. 1889. Reported on branches of *Rhus venenata* from New Jersey. No material seen.

Pezicula Philadelphi (Schw.) Sacc. Syll. Fung. **8**: 315. 1889; *Peziza Philadelphi* Schw. Trans. Am. Phil. Soc. II. **4**: 177. 1832. Reported on *Philadelphus inodorus*. No material seen.

Pezicula viridiatra (Berk. & Curt.) Sacc. Syll. Fung. **8**: 315. 1889; *Peziza viridiatra* Berk. & Curt.; Berk. Jour. Linn. Soc. **10**: 369. 1868. Sessile, greenish-black, underneath granular, irregularly undulate; spores ellipsoid, finally subfuscous; spores ellipsoid, 5 μ long. This has been made a synonym of *catinella nigroolivacea* by Durand. On dead wood with *Polyporus*, Cuba.

10. DERMEA Fries, Syst. Orbis Veg. 114. 1825.

Dermatea Fries, Summa Veg. Scand. 362. 1849.

Apothecia occurring singly, or more often in cespitose clusters, often on a stromatic base, tubercular in form or discoid, more rarely scutellate, usually dark-colored, comparatively small, rarely exceeding 2 mm. and usually 1 mm. or less in diameter, coriaceous to subcarbonaceous; asci usually broad-clavate and

8-spored; spores usually comparatively large, occasionally minute, simple, or becoming tardily several-septate, the septation óften erratic even in the same species; paraphyses colored and their tips agglutinated into a dark-brown or blackish epithecium.

Type species, *Peziza Cerasi* Pers.

This genus grades rather closely into *Cenangium* but usually has smaller discoid apothecia and large often septate spores. The macroconidia in the various species which are produced in irregular pycnidial cavities in a fleshy stroma are fusiform, usually curved and septate. The conidial stage so far as observed belongs to the form genus *Micropera* (*Gelatinosporium*).

These studies are based partly on the work of Dr. J. W. Groves (Mycologia **38**: 351–431. 1946).

Apothecia 1.5 mm. or more in diameter.
  Not on coniferous hosts.
    On *Prunus*, apothecia distinct.   1. *D. Cerasi.*
    On *Betula*, apothecia often coalesced.   2. *D. molliuscula.*
  On coniferous hosts, *Abies* and *Tsuga*.   3. *D. balsamea.*
Apothecia mostly less than 1.5 mm. in diameter.
  Ascospores less than 5 μ in diameter.
    Conidia in beaked pycnidia, on *Viburnum*.   4. *D. Viburni.*
    Conidia not in beaked pycnidia.
      Asci more than 10 μ broad; on *Nemopanthes*.   5. *D. Peckiana.*
      Asci less than 10 μ broad.
        On *Amelanchier*, apothecia black.   6. *D. bicolor.*
        On *Sorbus*, apothecia brownish.   7. *D. Ariae.*
  Ascospores more than 5 μ in diameter.
    Conidia in beaked pycnidia.
      Conidia elongate-ellipsoid; on *Acer*.   8. *D. acerina.*
      Conidia elongate-fusiform; on *Prunus*.
        Pycnidia usually single.   9. *D. Padi.*
        Pycnidia usually cespitose.   10. *D. Prunastri.*
    Conidia not in beaked pycnidia.
      On coniferous hosts.
        Ascospores less than 14 μ long; on *Picea*. 11. *D. piceina.*
        Ascospores usually more than 14 μ long.
          On *Pinus*.   12. *D. pinicola.*
          On *Libocedrus*.   13. *D. Libocedri.*
      Not on coniferous hosts.
        Asci usually less than 15 μ in diameter.
          On *Hamamelis*.   14. *D. Hamamelidis.*
        Asci usually more than 15 μ in diameter.
          On *Fraxinus*.   15. *D. Tulasnei.*
          On *Chionanthus*.   16. *D. Chionanthi.*
          On *Morus*.   17. *D. Mori.*

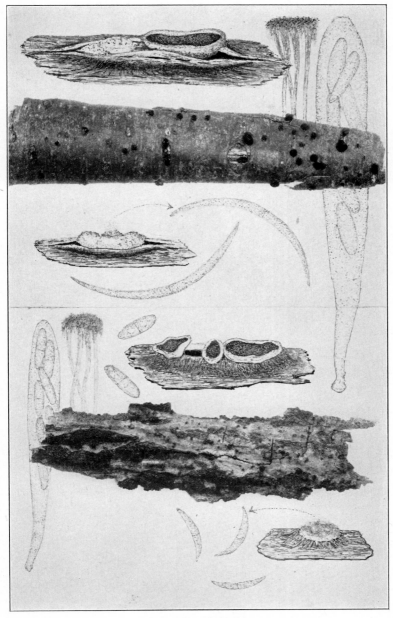

DERMEA CERASI
DERMEA BICOLOR

1. **Dermea Cerasi** (Pers.) Fries, Syst. Orb. Veg. 115. 1825.
(PLATE 147, FIG. 1.)

Peziza Cerasi Pers. Tent. Disp. Fung. 35. 1797.
Sphaeria dubia Pers. Ic. Pict. Fung. **4**: 48. 1806.
Cenangium Cerasi Fries, Syst. Myc. **2**: 179. 1822.
?Cenangium cerasorum Schw. Schr. Nat. Ges. Leipzig **1**: 118. 1822.
?Cenangium clavatum Fries, Syst. Myc. **2**: 179. 1822.
Cycledum Cerasi Wallr. Fl. Crypt. Germ. **2**: 512. 1833.
Micropera roseola Lév. Ann. Sci. Nat. III. **5**: 283. 1846.
Micropera drupacearum Lév. Ann. Sci. Nat. III. **5**: 283. 1846.
Micropera Cerasi Sacc. Atti Soc. Veneto Sci. Nat. Padova **2**²: 160. 1873.
Tympanis Cerasi Quél. Ench. Fung. 330. 1886.

Conidial stage erumpent, irregular in form, circular, or elongated to conical, usually with several flask-shaped bodies which open irregularly, reaching a height of 1 mm., whitish to yellowish, soft-waxy; conidiophores simple, or branched, tapering to a slender tip, $2-2.5 \times 10-25 \mu$; conidia sickle-shaped, or almost straight, fusiform, hyaline to faintly yellowish, simple, or 1-septate, $2.5-4.5 \times 35-65 \mu$; microconidia hyaline, filiform, straight, or curved, simple, $1-1.5 \times 12-23 \mu$.

Apothecia accompanying the conidial stage, single, or more often cespitose, narrowed below, 1–3 mm. in diameter and 1.5 mm. high, at first brownish, finally black, leathery to horny; hymenium at first concave, becoming plane, or convex, often slightly umbilicate, black, or slightly olivaceous; asci cylindric-clavate, tapering below into a stem-like base, reaching a length of $100-150 \mu$ and a diameter of $10-15 \mu$, 8-spored; spores ellipsoid-fusoid, hyaline to yellowish, straight, or slightly curved, simple, becoming 3-septate, $5-7 \times 15-25 \mu$; paraphyses simple, or branched, the tips swollen and forming a yellowish epithecium.

On branches of various species of *Prunus*.

TYPE LOCALITY: Europe.

DISTRIBUTION: Maine to Washington and South Carolina; probably widely distributed throughout North America; also in Europe.

ILLUSTRATIONS: Pers. Ic. Pict. Fung. *pl. 20, f. 1–2;* Tul. Fung. Carp. **3**: *pl. 19, f. 13–17;* E. & P. Nat. Pfl. **1**¹: 237, *f. 179 A–D;* Rab. Krypt.-Fl. **1**³: 242. *f. 1–6;* Mycologia **25**: *pl. 20,* (upper figure); **38**: 366, *f. 1, 2;* 383, *f. 27;* 395, *f. 42.*

EXSICCATI: Ellis, N. Am. Fungi *40, 989, 2812;* Ellis & Ev. N. Am. Fungi *2555;* Shear, N. Y. Fungi *94;* Barth. Fungi Columb. *4942;* Reliq. Farlow. *113;* Rav. Fungi Am. *246;* Fung. Car. **4**: *71.*

2. **Dermea molliuscula** (Schw.) Cash, Mycologia **29**: 304. 1937. (Plate 148.)

Cenangium molliusculum Schw. Trans. Am. Phil. Soc. II. **4**: 239. 1832.
Gelatinosporium fulvum Peck, Ann. Rep. N. Y. State Mus. **38**: 97. 1885.

Conidial stage, erumpent, circular, or transversely elongated, 1–4 mm. in diameter, .5–1 mm. high, blackish, waxy, usually containing several lobed, or flask-shaped cavities; conidiophores hyaline, simple, or branched, 1.5–2 × 15–30 μ; conidia hyaline, or pale-yellowish, subfiliform, ends pointed, sickle-shaped, or almost straight, simple, 1–1.5 × 7–12 μ.

Apothecia accompanying the conidial stage, erumpent, single, or cespitose in clusters of two to six, often crowded and sometimes fused, sessile, or substipitate, 1–3 mm. in diameter and 1–2 mm. high, tawny to almost black, leathery to horny; hymenium at first concave, becoming strongly convex, olivaceous-brown to black; asci cylindric-clavate, attenuated below, 8-spored, reaching a length of 100–150 μ and a diameter of 12–15 μ; spores narrow-ellipsoid to subfusiform, hyaline, becoming yellowish, straight, or slightly curved, 1-seriate, or irregularly 2-seriate, often becoming 1–3-septate, 4–7 × 15–22 μ; paraphyses filiform, simple, or branched, 1.5–2 μ in diameter, the tips swollen and embedded in a yellowish matrix forming a dark epithecium.

On branches of species of *Betula*.

Type locality: Mauch Chunk, Pennsylvania.

Distribution: New Hampshire to Pennsylvania and Canada.

Illustrations: Mycologia **25**: *pl. 21* (as *Dermea Betulae*); **32**: 742, *f. 2, 12, 13;* **38**: 379, *f. 21;* 386, *f. 39;* 395, *f. 47.*

3. **Dermea balsamea** (Peck) Seaver; Dodge, Mycologia **24**: 427. 1932. (Plate 149.)

Cenangium balsameum Peck, Ann. Rep. N. Y. State Mus. **38**: 101. 1885.
Cenangium balsameum var. *abietinum* Peck, Ann. Rep. N. Y. State Mus. **43**: 40. 1889.
Gelatinosporium abietinum Peck, Ann. Rep. N. Y. State Mus. **25**: 84. 1873.
Micropera erumpens Ellis & Ev. Proc. Acad. Sci. Phila. **1894**: 386. 1894.
Micropera abietina Höhn. Mitt. Bot. Inst. Wien **3**: 32. 1926.

Conidial stage erumpent to subimmersed, rounded to cylindric, or subconic, .5–1 mm. in diameter, .2–.5 mm. high, yellowish, or olivaceous to black, waxy, opening with one or more irregular cavities; conidiophores hyaline, simple, or branched, tapering

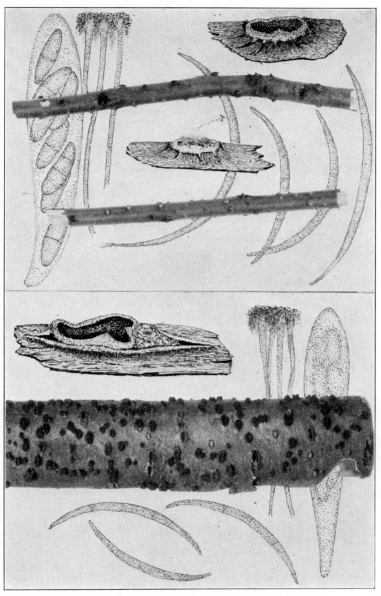

DERMEA MOLLIUSCULA

above, 2–2.5 × 15–25 μ; conidia elongate-filiform, pointed at the ends, hyaline to pale greenish-yellow, becoming 1–3-septate, sickle-shaped, or nearly straight, 4–5 × 50–90 μ; microconidia filiform, straight, or curved, 1–1.5 × 11–22 μ.

Apothecia erumpent, mostly single, sometimes cespitose, two to four in a cluster, sessile, or substipitate, circular, or undulated, 1–2.5 mm. in diameter .4–.8 mm. high, at first yellowish, or brownish, finally black, leathery to horny when dry; hymenium at first concave, then plane, or slightly convex, often umbilicate, light-brown to olivaceous-brown, or black; asci cylindric-clavate, tapering below, 8-spored, reaching a length of 90–150 μ and a diameter of 14–16 μ; spores ellipsoid-fusiform, hyaline or slightly yellowish, becoming 1–3-septate, straight, or curved, irregularly 2-seriate, 6–10 × 20–35 μ; paraphyses filiform, usually much branched, 1.5–2 μ in diameter, the tips very slightly swollen and glued together forming a yellowish epithecium.

On branches of *Abies balsamea* and *Tsuga canadensis*.

TYPE LOCALITY: Caroga Lake, New York.

DISTRIBUTION: Maine to Virginia and Canada.

ILLUSTRATIONS: Mycologia 24: 424, *f. 1, pl. 10, pl. 11, e. f.;* 38: 366, *f. 8, 9;* 383, *f. 30;* 400, *f. 54.*

EXSICCATI: Shear, N. Y. Fungi *328;* Reliq. Farlow. *102.*

4. Dermea Viburni Groves, Mycologia 32: 745.   1940.

Sphaeronema hystricinum Ellis, Bull. Torrey Club 6: 106.   1876.
Sphaerographium hystricinum Sacc. Syll. Fung. 3: 597.   1884.
Sphaerographium hystricinum var. *Viburni* Dearn. & House; House, Bull. N. Y. State. Mus. 197: 35.   1917.

Conidial fruiting bodies erumpent, scattered, or more or less in rows, single, or two or three arising from the same basal stroma, cylindric-subulate, dark-brown to black, often with a reddish tinge, .3–.5 mm. in diameter at the base, and the beaks about 1 mm. long, leathery to horny, basal stroma containing a single more or less elongated cavity about 150–200 μ in diameter; conidiophores cylindric, occasionally branched, tapering above to a slender tip, 2–2.5 × 15–30 μ; conidia elongate-fusiform, to subfiliform, hyaline, sickle-shaped, or almost straight, one end more attenuated than the other, 1–3-septate, 2.5–4 × 30–45 μ; microconidia not observed.

Apothecia erumpent, separate, or in small clusters of two to six each, sessile, slightly narrowed below, circular, or slightly undulated, .3–1 mm. in diameter, .2–.5 mm. high, dark-brown to black, leathery to horny, softer when moist; hymenium black, at first concave, becoming plane to convex; asci cylindric-clavate, short-stalked, 8-spored, reaching a length of 50–75 μ, and a diameter of 8–12 μ; spores ellipsoid fusoid, hyaline, becoming slightly yellowish, straight, or slightly curved; simple, or 1-septate, 3.5–5.5 × 14–18 μ; paraphyses, hyaline, filiform, much branched, 1.5–2 μ in diameter, the tips swollen to 3 μ and glued together, forming a yellowish epithecium.

On various species of *Viburnum*.

TYPE LOCALITY: Hatchley, Ontario.

DISTRIBUTION: Vermont and Ontario, Canada.

ILLUSTRATIONS: Mycologia **32**: 737. *f. G, H;* **38**: 379, *f. 24;* 386, *f. 38;* 398, *f. 49.*

EXSICCATI: Ellis, N. Am. Fungi *337* (as *Sphaeronema hystricinum*); Reliq. Farlow. *198a, 198b* (as *Sphaerographium hystricinum*).

5. **Dermea Peckiana** (Rehm) Groves, Mycologia **29**: 67. 1937.

Cenangium Peckianum Rehm, Ann. Myc. **13**: 3.   1915.
Sphaeronema stellatum Ellis, Bull. Torrey Club **6**: 107.   1876.
Sphaerographium stellatum Sacc. Syll. Fung. **3**: 598.   1884.
Micropera Nemopanthis Peck, Ann. Rep. N. Y. State Mus. **46**: 31.   1893.
Micropera stellata Jacz. Nouv. Mem. Soc. Imp. Nat. Moscow **15**: 366.   1898.

Conidial stage erumpent, rounded, often somewhat capitate, circular to more or less transversely elongated, .5–2 mm. in diameter, 1 mm. high, upper surface uneven and wrinkled around the openings of the cavities, pale-yellowish to black, often with a greenish cast, subfleshy, softer than the apothecia; conidiophores hyaline, cylindric, sometimes branched, tapering above, 2–2.5 × 20–40 μ; conidia hyaline, elongate-fusiform to subfiliform, sickle-shaped, ends pointed, becoming 1–3-septate, 2.5–4.5 × 40–60 μ; microconidia hyaline, filiform, straight, or curved, simple, 1.5 × 8–13 μ.

Apothecia arising from the old conidial stroma, cespitose in clusters up to fifteen each, .3–.8 mm. in diameter, circular, or undulated, narrowed below, dark-brown to black; asci cylindric-clavate, tapering below, 8-spored, reaching a length of 75–110 μ and a diameter of 9–12 μ; spores ellipsoid-fusiform, hyaline to pale-yellow, straight, or slightly curved, becoming 1–3-septate, 3–4 × 12–18 μ; paraphyses hyaline, filiform, simple, or branched 1.5–2 μ in diameter, the tips swollen to 3–5 μ and glued together forming a yellowish epithecium.

DERMEA BALSAMEA

On branches of *Nemopanthes mucronata* and *Ilex verticillata*.
Type locality: Munith, Michigan.
Distribution: Michigan and Canada.
Illustrations: Mycologia **29**: 68, *f. 1*, 77, *f. 4–9;* **38**: 379, *f. 18*, 25; 386, *f. 35;* 400, *f. 56*.
Exsiccati: Ellis & Ev. N. Am. Fungi *3042;* Ellis & Ev. Fungi Columb. *332*.

6. **Dermea bicolor** (Ellis) Groves, Mycologia **35**: 460. 1943.

Tympanis bicolor Ellis, Am. Nat. **17**: 193. 1883.
Cenangium bicolor Sacc. Syll. Fung. **8**: 572. 1889.
Cenangium dichroum Sacc. Syll. Fung. **8**: 1143. 1889.
Patinella Brenckleana Sacc. Mycologia **12**: 203. 1920.
Dermea Brenckleana Seaver, Mycologia **25**: 142. 1933.

Conidial fruiting bodies more or less immersed, splitting the bark and slightly erumpent, circular to elongated, or angular, .2–.8 mm. in diameter, .2–.4 mm. high, yellowish, soft-waxy, containing one to several lobed cavities which open irregularly exposing the greenish to yellowish spore masses; conidiophores lining the cavity, hyaline, cylindric, unbranched, tapering above, 1.5–2.5 × 15–30 μ; conidia hyaline fusiform, sickle-shaped, or almost straight, pointed at the ends, simple, or 1-septate, 2.5–4 × 15–25 μ; microconidia not observed.

Apothecia erumpent, usually single, sometimes in more or less elongated clusters, circular, or undulated, sessile, narrowed below, .5–1.5 mm. in diameter, .5–1 mm. high, yellowish or greenish when moist, finally dark-brown to black, leathery to horny, softer when moist; hymenium concave to plane, or slightly convex, greenish when young, becoming dark-brown to black on drying; asci cylindric-clavate, tapering below to a short stalk, 8-spored, reaching a length of 60–80 μ and a diameter of 8–10 μ; spores ellipsoid-fusiform, straight, or slightly curved, simple, or 1-septate, hyaline, becoming yellowish brown, irregularly 2-seriate, 3–4.5 × 12–16 μ; paraphyses filiform, simple, or branched, 1.5–2 μ in diameter, the tips scarcely swollen but more or less glued together forming an epithecium.

On branches of *Amelanchier*.
Type locality: Decorah, Iowa.
Distribution: Iowa to North Dakota.
Illustrations: Mycologia **25**: *pl. 20* (lower figure); **35**: 460, *f. 1, 2;* 462, *f. 3;* **38**: 379, *f. 19, 20;* 383, *f. 33;* 395, *f. 45*.

7. **Dermea Ariae** (Pers.) Tul.; Karst. Myc. Fenn. **1**: 224.   1871.

Peziza Ariae Pers. Myc. Eu. **1**: 325.   1822.
Tympanis Ariae Fries, Syst. Myc. **2**: 175.   1822.
Sphaeria Cotoneastri Fries in Kunze & Schm. Myk. Hefte **2**: 46.   1823.
Sphaeria Cotoneastri Sorbi Fries, Syst. Myc. **2**: 494.   1823.
Tympanis inconstans Fries, Summa Veg. Scand. 400.   1849.
Cenangium Ariae Tul. Ann. Sci. Nat. III. **20**: 136.   1853.
Cenangium inconstans Fuckel, Symb. Myc. 268.   1869.
Sphaeronema pallidum Peck, Ann. Rep. N. Y. State Mus. **25**: 85.   1873.
Cenangium subnitidum Cooke & Phill.; Cooke, Grevillea **3**: 186.   1875.
Micropera Sorbi Sacc. Michelia **2**: 628.   1882.
Micropera Cotoneastri Sacc. Syll. Fung. **3**: 605.   1884.
Phaeangella subnitida Massee, Brit. Fungus.-Fl. **4**: 137.   1895.

Conidial fruiting bodies erumpent, usually single, occasionally two or three together, bluntly conical, about .25–.35 mm. in diameter and .25–.5 mm. high, reddish-brown to olivaceous, or black, horny, becoming softer when moist, containing a single ovoid cavity; conidiophores hyaline, cylindric, simple, or branched, attenuated at the tips, 1.5–2 × 20–40 μ; conidia hyaline to pale yellowish-green, fusiform, sickle-shaped, or occasionally almost straight, ends pointed, simple, or 1-septate, 2–4 × 15–25 μ; microconidia not observed.

Apothecia erumpent, single, or in clusters of two to four, circular to undulated, sessile, narrowed below, .4–1 mm. in diameter, .2–.4 mm. high, dark reddish-brown to black, leathery to horny, softer when moist; hymenium concave to plane, black; asci cylindric-clavate, narrowed into a short stalk, 8-spored, reaching a length of 70–100 μ and a diameter of 8–10 μ; spores ellipsoid-fusoid, hyaline to pale-yellowish, simple, or becoming 1–3-septate, straight, or slightly curved, irregularly 2-seriate, 3–5 × 12–18 μ; paraphyses hyaline, filiform, simple, or branched, 1.5–2.5 μ in diameter, the tips swollen to 5 μ and glued together forming a yellowish epithecium.

On various species of *Sorbus*.

TYPE LOCALITY: Europe.

DISTRIBUTION: Quebec to Ontario, Canada; also in Europe.

ILLUSTRATIONS: Mycologia **38**: 379, *f. 22, 23;* 383, *f. 34;* 398, *f. 53.*

EXSICCATI: Ellis & Ev. Fungi Columb. *571* (conidia only).

8. **Dermea acerina** (Peck) Rehm, Ber. Bayer. Bot. Ges. **13**: 197. 1912.

Sphaeronema acerinum Peck, Ann. Rep. N. Y. State Mus. **24**: 86.   1872.
Sphaeronema nigripes Ellis, Bull. Torrey Club **6**: 107.   1876.
Tympanis acerina Peck, Ann. Rep. N. Y. State Mus. **31**: 48.   1879.
Scleroderris acerina Sacc. Syll. Fung. **8**: 599.   1889.
?Lecanidion acericolum Atk.; Peck, Ann. Rep. N. Y. State Mus. **49**: 24.   1896.
?Patellaria acericola Atk.; Butler, Mycologia **32**: 810.   1940.

Conidial bodies erumpent, single, or cespitose in small clusters, or more often in long rows, subulate, basal stroma subglobose to ovoid .2–.5 mm. in diameter, dark-brown to black, leathery to horny, softer when moist, the beak slender, tapering, straight, or curved, reaching a length of 1.5 mm. and 100–150 μ in diameter at the base, paler and sometimes translucent toward the tip; conidiophores hyaline, cylindric, simple, 2×20–$40\ \mu$, swollen below the point of attachment; conidia ellipsoid, hyaline, straight, or sometimes slightly curved, one end with a truncate apiculus, 5–8×15–$25\ \mu$; microconidia hyaline, filiform, simple, straight, or curved, 1–2×6–$10\ \mu$.

Apothecia erumpent, single, or in rows, circular, or undulate, .4–1 mm. in diameter, .2-.5 mm. high, sessile, narrowed below, black, or dark-brownish, leathery to horny, softer when moist; hymenium at first concave, becoming plane, or slightly convex, slightly paler than the outside of the apothecium; asci cylindric-clavate, short-stalked, 8-spored, reaching a length of 85–125 μ and a diameter of 13–16 μ; spores ellipsoid-fusiform, hyaline, becoming yellowish, simple, or 1–3-septate, straight, or sometimes slightly curved, irregularly 2-seriate, 5–6×13–$20\ \mu$; paraphyses hyaline, filiform, simple, or branched, 1.5–2 μ in diameter, the tips slightly swollen and glued together forming a yellowish epithecium.

On various species of *Acer, A. Saccharum, A. saccharinum, A. rubrum.*

TYPE LOCALITY: Greenbush, New York.

DISTRIBUTION: New York to Michigan and Virginia.

ILLUSTRATIONS: Mycologia **38**: 379, *f. 16;* 386, *f. 36;* 398, *f. 50.*

EXSICCATI: Ellis, N. Am. Fungi *947;* Ellis & Ev. Fungi Columb. *2086, 3585;* Reliq. Farlow. *143a, b.*

9. **Dermea Padi** (Alb. & Schw.) Fries, Summa, Veg. Scand. 362. 1849.

Peziza Cerasi Padi Alb. & Schw. Consp. Fung. 345.   1805.
Cenangium Cerasi Padi Fries, Syst. Myc. **2**: 180.   1827.
Tympanis Padi Quél. Ench. Fung. 330.   1886.
Cenangium Padi Sacc. Michelia **2**: 84.   1880.
Sphaeria padina Sacc. Michelia **2**: 84.   1880.
Sphaeronema brunneoviride Auersw.; Sacc. Syll. Fung. **3**: 186.   1884.
Cryptosporium brunneoviride Jacq. Nouv. Mem. Soc. Imp. Moscow **15**: 369. 1898.

Conidial fruiting bodies erumpent, mostly single, occasionally two or three in a cluster, cylindric to conic, .5–1.3 mm. in height and .2–.5 mm. in diameter, opening at the tip, dark reddish-brown to black, leathery to horny, softer when moist, the basal stroma containing one cavity in which the conidiophores are produced; conidiophores cylindric, tapering to a slender point, 2–3 × 25–50 μ; conidia elongate, fusiform to subfiliform, pointed at the ends, sickle-shaped to almost straight, simple, or 1-septate, hyaline, 2.5–4 × 20–35 μ; microconidia hyaline, bacilliform, straight, or slightly curved, simple.

Apothecia erumpent, single, or cespitose in clusters of two to six, sessile, narrowed below, circular, or undulated, .5–1.5 mm. in diameter and about .3–1 mm. high, dark reddish-brown to black, leathery to horny, softer when moist, hymenium at first concave, then plane to convex, black, or dark-brown; asci cylindric-clavate, tapering to a short stalk, 8-spored, reaching a length of 85–100 μ and a diameter of 15–15 μ; spores ellipsoid-fusoid, hyaline, becoming yellowish, straight, or slightly curved, simple, or 1–3-septate, 5–7 × 15–20 μ; paraphyses hyaline, fili-form, simple, or branched, 1.5–2 μ in diameter, the tips swollen to 3 μ and glued together forming a yellowish epithecium.

On species of *Prunus*, *P. domestica*, *P. Padus*, *P. spinosa*, *P. virginiana*.

TYPE LOCALITY: Europe.

DISTRIBUTION: New York, also in Europe.

ILLUSTRATIONS: Mycologia **38**: 366, *f. 3, 4; 383, f. 29; 395, f. 44.*

10. **Dermea Prunastri** (Pers.) Fries, Summa Veg. Scand. 362. 1849.

Peziza Prunastri Pers. Tent. Disp. Fung. 35.   1797.
Ceratostoma spurium Fries, Obs. Myc. **2**: 338.   1818.

Cenangium Prunastri Fries, Syst. Myc. **2**: 180. 1822.
Cenangium Prunastri rigida Fries, Syst. Myc. **2**: 180. 1822.
?Cenangium rigidum Schw. Trans. Am. Phil. Soc. II. **4**: 238. 1832.
Tympanis Prunastri Wallr. Fl. Crypt. Germ. **2**: 427. 1833.
Sphaeronema spurium Sacc. Syll. Fung. **3**: 186. 1884.
Phaeangella Prunastri Massee, Brit. Fungus-Fl. **4**: 137. 1895.
Dermatella Prunastri Dowson, New Phytologist **12**: 207. 1913.

Conidial bodies erumpent, occasionally single, more often cespitose, cylindric, or cylindric-conic to subulate, 1–2 mm. in height, .2–.4 mm. in diameter at the base, arising from a more or less circular, or transversely elongated basal stroma, black to greenish, or olivaceous when moist, hard, horny, brittle, softer when moist, with a single, ovoid cavity which is frequently in the tip of the beak only but may extend into the basal stroma also; conidiophores hyaline, cylindric, tapering to a slender point, simple, 2.5–3 × 20–35 μ; conidia elongate-fusiform, hyaline, slightly greenish, simple, almost straight, or slightly sickle-shaped, the ends acute, 5–7 × 20–35 μ; microconidia hyaline to yellowish, filiform, straight, or slightly curved, simple, ends rounded, 1.5 × 7–10 μ.

Apothecia erumpent, single, or cespitose, sessile, narrowed below, circular, or undulated, .5–1 mm. in diameter .2–1 mm. high, dark-brown to black, leathery to horny, fleshy when moist; hymenium concave to plane, or slightly convex, black; asci cylindric-clavate, short-stalked, 8-spored, reaching a length of 90–125 μ and a diameter of 12–15 μ; spores ellipsoid-fusiform, hyaline, becoming yellowish, simple, or 1–3-septate, straight, or slightly curved, irregularly 2-seriate, 5–7.5 × 15–25 μ; paraphyses hyaline, filiform, simple, or branched, 1.5–2 μ in diameter, the tips swollen to 2.5–3 μ and glued together forming a yellowish epithecium.

On various species of *Prunus*.

TYPE LOCALITY: Europe.

DISTRIBUTION: New Hampshire to Washington and South Carolina.

ILLUSTRATIONS: Phill. Brit. Discom. *pl. 10, f. 66;* New Phytologist **12**: 209, *f. 1, 2;* 211, *f. 3;* Mycologia **38**: 366, *f. 5, 6;* 383, *f. 28*, 395; *f. 43.*

EXSICCATI: Ellis & Ev. Fungi Columb. *3118* (as *Dermatea Cerasi*).

11. **Dermea piceina** Groves, Mycologia **38**: 404.   1946.

Conidial bodies erumpent, mostly single, minute, black, or greenish-black, .1–.3 mm. in diameter, almost globose, opening above and the spores emerging in a whitish, or pale-greenish mass or cirrhus; conidiophores lining the cavity, hyaline, cylindric, tapering to a tip, simple, or 1–3-septate; microconidia hyaline, filiform, simple, strongly curved, ends rounded, 1–1.5 × 9–15 μ.

Apothecia erumpent, single, or cespitose in clusters of two to six, circular, or slightly undulated, sessile, narrowed below and substipitate, .5–1.5 mm. in diameter, .5–1 mm. high, dark reddish-brown to almost black; asci cylindric-clavate, short-stalked, 8-spored, reaching a length of 75–100 μ and a diameter of 14–16 μ; spores ellipsoid, or ovoid, hyaline, becoming brownish, simple, or 1–3-septate, irregularly 2-seriate, 6–8 × 12–18 μ; paraphyses hyaline, filiform, simple, or branched, 1.5–2 μ in diameter, the tips swollen to 3 μ, forming a slight epithecium.

On branches of *Picea glauca.*

TYPE LOCALITY: Petawawa Forest Experiment Station, Ontario, Canada.

DISTRIBUTION: Known only from the type locality.

ILLUSTRATIONS: Mycologia **38**: 375, *f. 11, 12;* 383, *f. 32;* 398, *f. 52.*

12. **Dermea pinicola** Groves, Mycologia **38**: 403.   1946.

Conidial stage a rounded, fleshy stroma about .5 mm. in diameter, containing a single cavity; conidiophores, hyaline, cylindric, simple, or branched, 3–4 × 20–45 μ; conidia elongate-fusiform, to subfiliform, 1–3-septate, hyaline, sometimes yellowish, mostly sickle-shaped, or nearly straight, one end more acute than the other, 4–6 × 30–50 μ.

Apothecia erumpent, single, or cespitose in clusters of two to six each, circular, or slightly undulate, sessile, narrowed below, .3–.8 mm. in diameter, .2–.4 mm. high, dark reddish-brown to black, leathery to horny, softer when moist; hymenium concave, becoming plane, or slightly convex, the margin slightly raised; asci cylindric-clavate, short-stalked, 8-spored, reaching a length of 70–100 μ and a diameter of 14–20 μ; spores ellipsoid-fusiform, hyaline, straight, or slightly curved, simple, or 1-septate, irregularly 2-seriate, 5–7.5 × 18–20 μ; paraphyses hyaline, filiform, simple, or branched, 1.5–2 μ in diameter, the tips swollen to 3 μ and glued together forming an epithecium.

On branches of *Pinus Strobus*.

TYPE LOCALITY: West mainland, Lake Temagami, Ontario, Canada.

DISTRIBUTION: Vermont to Ontario.

ILLUSTRATIONS: Mycologia **38**: 375, *f. 15;* 398, *f. 51 a–b*.

13. Dermea Libocedri Groves, Mycologia **38**: 382. 1946.

Conidial fruiting bodies erumpent, separate, black, minute, .2–.3 mm. in diameter and about the same in height, rounded to subcylindric, opening at the tip, containing a single ovoid, or slightly lobed cavity lined with the conidiophores; conidiophores hyaline, cylindric, pointed at the tip, simple, or branched near the base, occasionally septate, 2.5–3.5 × 20–30 μ; conidia elongate-filiform, hyaline, sickle-shaped, pointed at the ends, 1–3-septate, 4–6 × 42–65 μ; microconidia hyaline, filiform, simple, straight, or curved, ends not pointed, 1–1.5 × 10–18 μ.

Apothecia erumpent, single, or in clusters of two to four each, sessile, slightly narrowed below, circular, or slightly undulated, .3–.5 mm. in diameter, .2–.3 mm. high, dark brownish-black, fleshy-leathery when moist; hymenium concave to plane, black; asci cylindric-clavate, short-stalked, 8-spored, reaching a length of 75–100 μ and a diameter of 12–17 μ; spores ellipsoid-fusoid, hyaline, 1–3-septate, straight, or slightly curved, irregularly 2-seriate, 6–8 × 15–20 μ; paraphyses hyaline, filiform, simple, or branched, 1.5–2 μ in diameter, the tips swollen to 3–4 μ and forming an epithecium.

On *Libocedrus decurrens*.

TYPE LOCALITY: Darlingtonia, California.

DISTRIBUTION: Known only from the type locality.

ILLUSTRATIONS: Mycologia **38**: 366, *f. 7;* 383, *f. 31;* 400, *f. 57*.

14. Dermea Hamamelidis (Peck) Groves, Mycologia **32**: 743. 1940.

Patellaria Hamamelidis Peck, Ann. Rep. N. Y. State Mus. **33**: 32. 1880.
Lecanidion Hamamelidis Sacc. Syll. Fung. **8**: 800. 1889.
Dermatella Hamamelidis Ellis & Ev. Proc. Acad. Sci. Phila. **45**: 149. 1893.
  (as a new species).
Dermatella Hamamelidis Durand, Bull. Torrey Club **29**: 464. 1902.

Conidial fruiting bodies, minute, about .15–.20 mm. in diameter, erumpent through and splitting the outer bark, appearing as thickly scattered, blister-like elevations in the bark, the spores emerging through the cracks in the bark when moist; conidiophores simple, hyaline, tapering above to a slender tip, 2 × 10–25 μ; conidia elongate-fusiform to subfiliform, hyaline, simple, or 1-septate, straight, or curved, one end often narrower than the other, 4.5–6 × 18–25 μ; no microconidia observed.

Apothecia erumpent, more or less in rows, single, or in small clusters, circular, or somewhat undulated, sessile, narrowed below, .3–.8 mm. in diameter, .2–.4 mm. high, dark reddish-brown to black, leathery to horny, softer when moist; hymenium concave to plane, or finally convex; asci cylindric-clavate, short-stalked, 8-spored, reaching a length of 80–120 μ and a diameter of 12–15 μ; spores ellipsoid-fusiform, hyaline to yellowish, straight, or slightly curved, irregularly 2-seriate, simple, or 1–3-septate, 5–7.5 × 15–20 μ; paraphyses filiform, hyaline, simple, or branched, 1.5–2 μ in diameter, the tips slightly swollen and glued together, forming a yellowish epithecium.

On branches of *Hamamelis virginiana*.

TYPE LOCALITY: Clinton, Catskill Mountains.

DISTRIBUTION: New Hampshire to Pennsylvania and Canada.

ILLUSTRATIONS: Mycologia **32**: 737, *f. 1, E, F;* 742, *f. 6–8;* **38**: 379, *f. 17, 25;* 386, *f. 41;* 395, *f. 46.*

EXSICCATI: Ellis & Ev. N. Am. Fungi *2634;* Ellis & Ev. Fungi Columb. *2016.*

15. **Dermea Tulasnei** Groves, Mycologia **38**: 399.   1946.

Cenangium Fraxini Tul. Ann. Sci. Nat. III. **20**: 140.   1853.   Not *Peziza Fraxini* Schw.   1822.
Cenangella Fraxini Sacc. Syll. Fung. **8**: 590.   1889.
Dermea Fraxini Rehm, Ber. Bayer. Bot. Ges. **13**: 196.   1912.

Conidial bodies erumpent, rounded to short-conical, .15–.50 mm. in diameter and .2–.4 mm. high, in elongated clusters up to 1 mm. in length, dark reddish-brown to black, softer than the apothecia, containing one, or sometimes several, more or less ovoid, simple to slightly chambered cavities which open irregularly at the top; conidiophores cylindric, simple, or branched, attenuated above, 4–5 × 25–50 μ; conidia hyaline to yellowish-green, fusiform, or subfiliform, simple, or occasionally septate, strongly curved to almost straight, 6–8 × 40–50 μ.

Apothecia erumpent, single, or in small clusters, circular, or slightly undulate, sessile, slightly narrowed below, .5–1 mm. in diameter, .2–.6 mm. high, waxy-leathery to horny, softer when moist; hymenium concave to plane, or slightly convex, dark reddish-brown to almost black; asci cylindric-clavate, tapering into a short stalk, 8-spored, reaching a length of 85–115 μ and a diameter of 14–20 μ; spores hyaline to pale yellowish-green, ellipsoid-fusiform, straight, or slightly curved, simple, or 1–3-septate, irregularly 2-seriate, 6–10 × 15–22 μ; paraphyses hyaline, filiform, usually branched, the tips swollen and glued together forming a yellowish epithecium.

On species of *Fraxinus*, *Fraxinus nigra*.

TYPE LOCALITY: Europe.

DISTRIBUTION: Canada; also in Europe.

ILLUSTRATIONS: Mycologia **38**: 375, *f. 13, 14;* 386, *f. 37;* 400, *f. 55.*

16. **Dermea Chionanthi** Ellis & Ev. Proc. Acad. Sci. Phila. **45**: 148. 1893.

Conidial fruiting bodies erumpent, single, black, about .2–.4 mm. in diameter and .2 mm. high, irregularly rounded to slightly conical, opening widely at the top, fleshy-membranous, containing a single cavity on the base of which the conidiophores are borne; conidiophores hyaline, cylindric, simple, or occasionally branched, tapering above to a tip on which the conidia are borne, 3.4 × 12–20 μ; conidia elongate-fusiform to subfiliform hyaline, curved, sickle-shaped to almost straight, simple, or 1-septate, bluntly pointed at the ends, 5–7 × 25–35 μ; no microconidia observed.

Apothecia erumpent, gregarious, often more or less in rows, single, or cespitose, sessile, slightly narrowed below, circular, or undulated, .4–1 mm. in diameter, .2–.4 mm. high, dark-reddish, or olivaceous-brown to black, leathery, or horny, softer when moist; hymenium at first concave, becoming plane to slightly convex, reddish-brown to black; asci cylindric-clavate, narrowed below into a rather short stalk, 8-spored, reaching a length of 90–110 μ and a diameter of 15–20 μ; spores ellipsoid to ellipsoid-fusoid, hyaline, becoming yellowish, simple, often becoming 1–3-septate, straight, or slightly curved, irregularly 2-seriate, 6–7 × 18–25 μ; paraphyses hyaline, filiform, simple, or occasionally branched, 1.5–2.5 μ in diameter, the tips swollen to 3–4 μ in diameter and forming an epithecium.

On branches of *Chionanthus virginica*.

TYPE LOCALITY: Wilmington, Delaware.

DISTRIBUTION: Delaware and Maryland.

ILLUSTRATIONS: Mycologia **38**: 375, *f. 10;* 398, *f. 48.*

EXSICCATI: Ellis, N. Am. Fungi *2635;* Ellis & Ev. Fungi Columb. *2423.*

17. **Dermea Mori** Peck, Bull. N. Y. State Mus. **157**: 46.   1912.

Apothecia erumpent and usually occurring singly, patellate, reaching a diameter of 1–2 mm., black, or blackish; hymenium plane or nearly so, similar in color to the outside of the apothecium; asci broad-clavate, reaching a length of 80–90 μ and a diameter of 20–25 μ, 8-spored; spores irregularly 2-seriate, hyaline, apparently simple, ellipsoid, or slipper-shaped, 8–10 × 20–30 μ; paraphyses slender, hyaline below but ends forming a brown epithecium.

On dead twigs of Russian mulberry, *Morus alba tartarica.*

TYPE LOCALITY: Concordia, Kansas.

DISTRIBUTION: Known only from the type locality.

Description drawn from part of original collection sent from Albany by Dr. H. D. House.   From the material examined it seems to be properly placed in the genus.

DOUBTFUL AND EXCLUDED SPECIES

Dermatea crypta Cooke, Grevillea **16**: 70.   1888.   Ellis states: "This appears to be the same thing I have published in Torr. Bull. as *Dermatea olivacea* on *Ilex glabra* Newfield, N. J. 1875".

Dermatea cucurbitaria Cooke; Ellis, N. Am. Fungi *68; Tryblidium cucurbitaria* Rehm, Ber. Naturh. Ver. Augsburg **26**: 78.   1881.   This is regarded as a *Tryblidium.*

Dermatea Cydoniae Schw. Trans. Am. Phil. Soc. II. **4**: 237.   1832.   On *Cydonia.*   According to Groves, this is not a discomycete.

Dermatea Eucalypti Cooke & Hark.; Cooke, Grevillea **9**: 130.   1881. This according to Groves is apparently not a *Dermea* but it is impossible to place it.

Dermatea ferruginea (Cooke & Ellis) Rehm, Ann. Myc. **2**: 353.   1904; *Patellaria ferruginea* Cooke & Ellis, Grevillea **5**: 91.   1877.   Groves states that this is probably a *Pezicula* but its identity is uncertain.   Ellis, N. Am. Fungi *148* under this name is on stems of *Desmodium.*

Dermatea olivacea Ellis, Bull. Torrey Club **6**: 133.   1876.   According to Groves this is probably a *Pezicula.*   The species, however, needs more study from fresh material.   It was originally reported on *Ilex glabra.*   See N. Am. Fungi *851.*   The name as pointed out by Groves is invalid having been previously used.

Dermatea puberula Durand, Jour. Myc. **10**: 101.   1904.   This according to Groves is not a *Dermea* and its position is uncertain.

Dermatea Xanthoxyli Peck, Ann. Rep. N. Y. State Mus. **31**: 47.   1877. This is *Thyronectria pyrrhochlora* (Auersw.) Sacc.

11. **DURANDIELLA** Seaver, Mycologia **24**: 261.   1932.

Durandia Rehm, Ann. Myc. **11**: 166.   1913.   Not *Durandia* Bockel, 1896.

Apothecia erumpent, coriaceous, black as in *Tympanis;* asci clavate, containing many minute, spermatoid, spore-like bodies as in *Tympanis* and in addition normal ascospores; paraphyses filiform.

Type species, *Peziza Fraxini* Schw.

In a previous paper (Mycologia **37**: 336) this genus was regarded as a synonym of *Godronia*.   Groves (Mycologia **29**: 79) believes that this should be retained as a distinct genus.   It might be characterized as a *Tympanis* with ascospores.

We have a similar situation in *Scoleconectria* of the Hypocreales in which the normal spores are often obscured by the numerous spermatoid, spore-like bodies.

1. **Durandiella Fraxini** (Schw.) Seaver, comb. nov.

Peziza Fraxini Schw. Schr. Nat. Ges. Leipzig **1**: 123.   1822.
Tympanis Fraxini Fries, Syst. Myc. **2**: 174.   1822.
?Sphaeronema Fraxini Peck, Ann. Rep. N. Y. State Mus. **29**: 71.   1878.
?Sphaerographium Fraxini Sacc. Syll. Fung. **3**: 598.   1884.
Durandia Fraxini Groves, Mycologia **29**: 78.   1937.
Godronia Fraxini Seaver, Mycologia **37**: 350.   1945.

Apothecia erumpent in clusters of three to ten, or rarely occurring singly, black, or blackish, reaching a diameter of 1 mm., circular, or irregular from mutual pressure; hymenium plane or nearly so, similar in color to the outside of the apothecium; asci clavate, reaching a length of 120–150 μ and a diameter of 10–12 μ, 8-spored; spores filiform, attenuated at either end, septate, 2.5–3 × 50–90 μ, accompanied by minute sprematoid, spore-like bodies; paraphyses filiform.

The apothecia often accompanied with *Sphaerographium Fraxini* which appears to be its conidial stage.

On branches of *Fraxinus americana.*

TYPE LOCALITY: North Carolina.

DISTRIBUTION: North Carolina to Massachusetts, Ontario and Ohio.

EXSICCATI: Barth. Fungi Columb. *3885;* Rehm, Ascom. *2027* (from Canada); Reliq. Farlow. *155 a–b.*

12. **MURANGIUM** Seaver, gen. nov.

Apothecia as in *Cenangium*, gregarious, or congested; asci broad-clavate, 8-spored; spores very large, ellipsoid, hyaline, or brown, becoming muriform.

Apotheciis erumpentibus; ascis clavatis; sporis muriformibus.

Type species, *Cenangium Sequoiae* Plow.

1. **Murangium Sequoiae** (Plow.) Seaver, comb. nov.

Cenangium Sequoiae Plow.; Phill. Grevillea **7**: 23. 1878.
Scleroderris Sequoiae Sacc. Syll. Fung. **8**: 596. 1889.

Apothecia erumpent through the crevices of the bark, usually
in linear groups, often congested, turbinate, at first closed, opening with an irregular aperture; hymenium concave, or nearly
plane; asci very broad-clavate, reaching a length of 140 μ and a
diameter of 30–40 μ, 8-spored, tapering abruptly into a stem-like
base; spores irregularly disposed in the ascus, broad-ellipsoid,
the contents granular, becoming 1–3-septate, then muriform,
16 $\times$ 33–36 μ; paraphyses filiform.

On bark of *Sequoia gigantea* and *Juniperus virginiana*.

Type locality: California.

Distribution: South Carolina and California.

Exsiccati: Ellis, N. Am. Fungi *671;* Rav. Fungi Am. *634.*

13. **TYMPANIS** Tode, Fungi Meckl. **1**: 23.   1790; Fries, Syst. Myc. **2**: 173.   1822.

Apothecia erumpent as in *Dermea*, single, or cespitose,
usually hard and coriaceous, dark-brown, or black; hymenium
concave to plane at maturity; asci clavate, filled with numerous
spermatoid, spore-like bodies; paraphyses filiform.

Type species, *Tympanis saligna* Tode.

The genus *Tympanis* comprises a number of species in which
normal ascospores are absent or undeveloped. Ascospores sometimes develop in addition to the spermatoid bodies in which case
the species is transferred to some other genus depending upon the
nature of the ascospores (see *Durandiella Fraxini*); the species
listed here are those in which no definite ascospores have been
observed.

On coniferous hosts, *Pinus, Abies*.	1. *T. Pinastri.*
On nonconiferous hosts.	
On species of Alnus.	2. *T. alnea.*
On species of *Populus*.	3. *T. spermatiospora.*
On *Oxydendrum*.	4. *T. Oxydendri.*
On miscellaneous hosts.	
Spermatoid bodies .5 $\times$ 1–2 μ.	5. *T. conspersa.*
Spermatoid bodies 1.5 $\times$ 4–5 μ.	6. *T. fasciculata.*

1. **Tympanis Pinastri** (Pers.) Tul. Fung. Carp. **3**: 151.   1865.

Cenangium Pinastri Pers. Obs. Myc. **2**: 83.   1799.
Phacidium Pinastri Fries, Scler. Suec. *55*.   1820.
Cenangium Pinastri Fries, Syst. Myc. **2**: 184.   1822.
Tryblidium Pinastri Fries, Summa Veg. Scand. 369.   1849.
Cenangium laricinum Fuckel, Symb. Myc. 270.   1869.
Tryblidiopsis Pinastri Karst. Myc. Fenn. **1**: 262.   1871.
Micropera Pinastri Sacc. Michelia **2**: 104.   1880.
Cenangella Pinastri Sacc. Syll. Fung. **8**: 588.   1889.

Apothecia scattered, erumpent, occurring either singly, or
in cespitose clusters on a stroma, short-stipitate, at first closed,
expanding and becoming turbinate, black, or blackish, reaching
a diameter of 1 mm.; hymenium concave, becoming plane, or
convex, lighter than the outside of the apothecium, cinereous;
asci clavate, reaching a length of 120–130 μ and a diameter of
12 μ, filled with minute, spore-like bodies, 1 $\times$ 2–2.5 μ; paraphyses filiform, branched at their apices, reaching a diameter of 3 μ.

The pycnidial stage, *Micropera Pinastri*, is described with
pycnospores fusoid-falcate, 5–7 $\times$ 50–60 μ.

On *Abies balsamea, Pinus rigida* and *Pinus Strobus*.

TYPE LOCALITY: Europe.

DISTRIBUTION: Newfoundland to Washington and Colorado.

ILLUSTRATIONS: Tul. Fung. Carp. **3**: *pl. 19, f. 10–12;* Rab.
Krypt.-Fl. **1**³: 245, *f. 1*.

EXSICCATI: Reliq. Farlow. *156 a–b;* Rav. Fungi Car. **3**: *63*.

2. **Tympanis alnea** (Pers.) Fries, Syst. Myc. **2**: 174.   1822.

Peziza alnea Pers. Syn. Fung. 673.   1801.
Tympanis conspersa alnea Karst. Myc. Fenn. **1**: 227.   1871.

Apothecia erumpent in cespitose clusters of three to twenty
each, on a rounded, black stroma 3–4 mm. in diameter, the
individual apothecia short-stipitate, reaching a diameter of
.3–.5 mm., black but often covered with a whitish, powdery coat;
asci cylindrical-clavate, reaching a length of 150–160 μ and a
diameter of 15–20 μ, filled with numerous spore-like bodies;
paraphyses branched enlarged at their tips where they reach a
diameter of 6 μ, brown, forming a brown epithecium.

On branches of *Alnus, Alnus tenuifolia* and *Alnus incana*.

TYPE LOCALITY: Europe.

DISTRIBUTION: Newfoundland to New York and Montana.

This species is very similar to *Tympanis cespitosa* and was
regarded as a variety by Karsten.   The specimen in Schweinitz,

North American Fungi *1961* is a *Godronia*. Other specimens, however, do not show the *Godronia* type of spore so the species is allowed to remain in the present genus.

3. **Tympanis spermatiospora** (Nyl.) Nyl. Not. Soc. Fauna Fl. Fenn. **10**: 70. 1869.

?Dermea populnea Schw. Trans. Am. Phil. Soc. II. **4**: 237. 1832. Not *Peziza populnea* Pers. 1797.
Patellaria spermatiospora Nyl. Not. Soc. Fauna Fl. Fenn. **4**: 125. 1858.
Cenangium populinum Fuckel, Symb. Myc. 268. 1869.
Tympanis populina Sacc. Bot. Cent. **18**: 247. 1884.
Cenangium spermatiosporum Sacc. Syll. Fung. **8**: 560. 1889.
?Cenangium Schweinitzii Sacc. Syll. Fung. **8**: 576. 1889.

Apothecia erumpent, occurring either singly, or more often in dense, cespitose clusters, the individual apothecia short-stipitate, at first closed and knob-like, opening and gradually expanding and becoming shallow cup-shaped, or turbinate, black, reaching a diameter of 1 mm.; hymenium concave, or nearly plane, brownish-black; asci clavate, reaching a length of 75–80 μ and a diameter of 12 μ, densely filled with spore-like bodies 1 × 3–3.5 μ; paraphyses slender, enlarged above, the ends adhering and forming a brown epithecium.

On dead branches of *Populus*, especially *Populus tremuloides*.

TYPE LOCALITY: Europe.

DISTRIBUTION: Newfoundland to New York, Idaho, Colorado and Iowa; also in Europe.

4. **Tympanis Oxydendri** Ellis & Ev. Proc. Acad. Sci. Phila. **1894**: 352. 1894.

Apothecia erumpent in cespitose clusters of three to five each, the individual apothecia expanding and becoming subturbinate, reaching a diameter of .5–.75 mm.; hymenium becoming plane, or convex, lighter than the outside of the apothecium, greenish-black when dry; asci cylindrical-clavate, reaching a length of 110–130 μ and a diameter of 15–18 μ, filled with minute spore-like bodies 1 × 2.5–3 μ; paraphyses slender, abundant, freely branched, hyaline.

On dead limbs of *Oxydendrum arboreum*.

TYPE LOCALITY: Nuttalburg, West Virginia.

DISTRIBUTION: Known only from the type locality.

EXSICCATI: Ellis & Ev. N. Am. Fungi *3043;* Fungi Columb. *247.*

This closely resembles *Tympanis fasciculata* but is distinguished by the greenish hymenium.

5. **Tympanis conspersa** Fries Syst. Myc. **2**: 175. 1822.

Sphaeria cespitosa Tode, Fungi Meckl. **2**: 41. 1791.
Sphaeria Aucupariae Pers. Syn. Fung. 51. 1801.
Peziza Pyri Pers. Syn. Fung. 671. 1801.
Sphaeria conspersa Fries, Vet. Akad. Handl. **1817**: 112. 1817.
Cenangium Aucupariae Fries, Syst. Myc. **2**: 181. 1822.
Tympanis Aucupariae Wallr. Fl. Crypt. Germ. **2**: 427. 1833.
Peziza Aucupariae Pers. Myc. Eu. **1**: 327. 1822.

Apothecia erumpent in cespitose clusters of ten to twenty each, seated on a fleshy stroma, at first closed, gradually expanding and becoming turbinate, with a short, thick stem, reaching a diameter of .5–1 mm., black but often covered with a whitish mealy coat, semicartilaginous when fresh, horny when dry; hymenium plane, or slightly concave, blackish; asci cylindric-clavate, reaching a length of 175 μ and a diameter of 18–20 μ, filled with minute, spore-like bodies .5 $\times$ 1–2 μ; paraphyses slender, branched, brown above and forming a brown epithecium.

On branches of various kinds, *Malus, Tilia* and *Prunus*.

Type locality: Europe.

Distribution: New York and New Jersey to Massachusetts, also in Europe.

Illustrations: Tode, Fungi Meckl. **2**: *pl. 14, f. 113;* Grev. Scot. Crypt. Fl. *pl. 338;* Phill. Brit. Discom. *pl. 11, f. 67;* Rab. Crypt.-Fl. **1**³: 245, *f. 2.*

Exsiccati: Ellis, N. Am. Fungi *66;* Barth. Fungi Columb. *4588;* Reliq. Farlow. *153.*

6. **Tympanis fasciculata** Schw. Trans. Am. Phil. Soc. II. **4**: 237. 1832.

Apothecia erumpent in fascicles of two to six apothecia each, the individual apothecia becoming expanded and subturbinate, black, reaching a diameter of .5 mm.; hymenium concave, or nearly plane, similar in color to the outside of the apothecium; asci cylindric-clavate, reaching a length of 130 μ and a diameter of 15 μ, filled with minute, spore-like bodies 1.5 $\times$ 4–5 μ; paraphyses very slender, branched.

On branches of *Viburnum Lentago;* also on *Rhus* and *Cornus.*

Type locality: Bethlehem, Pennsylvania.

DISTRIBUTION: New Jersey and Pennsylvania to Ontario.

EXSICCATI: Ellis, N. Am. Fungi *65;* Rehm, Ascom. *423* (from Newfield, New Jersey).

For additional information on *Tympanis* see Groves, Mycologia **41**: 59–76.   1949.

DOUBTFUL AND EXCLUDED SPECIES

Tympanis cinerascens Schw. Trans. Am. Phil. Soc. II. **4**: 237.   1832.   On unidentified branches, Easton [Pennsylvania].

Tympanis gyrosa Berk. & Curt.; Berk. Grevillea **4**: 3.   1875.   The species was described from sterile specimens.   No authentic material has been seen.

Tympanis plicatocrenulata (Schw.) Sacc. Syll. Fung. **8**: 580.   1889; *Peziza plicato crenata* Schw. Schr. Nat. Ges. Leipzig **1**: 123.   1822.   Material in the Schweinitz collection was very scant and looked more like a *Tapesia*.

Tympanis saligna Tode, Fungi Meckl. **1**: 24, *pl. 4, f. 37*.   1790.   Reported on very scant material.

Tympanis pithya Karst. Myc. Fenn. **1**: 228.   1871.   This species has been reported on *Pinus albicaulis* at Horse Camp on Mt. Shasta.   See Groves.

POSITION DOUBTFUL

HAEMATOMYXA Sacc. Bot. Cent. **18**: 250.   1884.

Apothecia subglobose, finally subpatellate, irregularly subcerebriform, gelatinous; asci broadly clavate, 8-spored; spores becoming muriform, brown.

Type species, *Haematomyces vinosus* Cooke & Ellis.

Spores large 8–10 × 25–30 μ.   1. *H. vinosa.*
Spores small 8–10 × 18–20 μ.   2. *H. ascoboloides.*

1. Haematomyxa vinosa (Cooke & Ellis) Sacc. Bot. Cent. **18**: 250.   1884.   (PLATE 150, FIG. 1.)

Haematomyces vinosus Cooke & Ellis, Grevillea **4**: 179.   1876.
Dothiora rufa (Ellis & Ev. in herb.); Rehm, Ann. Myc. **10**: 397.   1912.
Haematomyxa rufa Rehm, Ann. Myc. **10**: 397.   1912.

Apothecia at first subglobose, becoming patellate and irregularly gyrose, wine-colored to blackish; less than 1 mm. in diameter; asci broad-clavate, 8-spored; spores broad-ellipsoid, slightly constricted in the middle, becoming muriform; paraphyses sparse.

On decorticated oak.

TYPE LOCALITY: Newfield, New Jersey.

DISTRIBUTION: Known only from the type locality.

The species shows some affinity with the Patellariaceae.

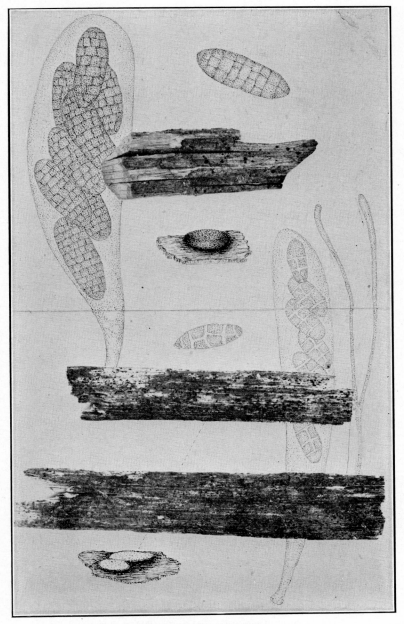

1. HAEMATOMYXA VINOSA
2. HAEMATOMYXA ASCOBOLOIDES

2. **Haematomyxa ascoboloides** Ellis & Ev. Bull. Torrey Club **27**: 60. 1900. (PLATE 150, FIG. 2.)

Apothecia orbicular, fleshy-tremelloid, convex-discoid and wine-colored when fresh, concave and nearly black when dry, .5–.75 mm. in diameter; asci clavate-cylindric, reaching a length of 80–100 μ and a diameter of 12–15 μ. 8-spored; spores 1-seriate, ellipsoid, becoming 3-septate and muriform, 8–10 $\times$ 18–20 μ; paraphyses filiform, branched.

On weathered oak.

TYPE LOCALITY: Kansas.

DISTRIBUTION: Known only from the type locality.

SARCOMYCES Massee, Jour. Myc. **6**: 178. 1891.

Apothecia subgelatinous, subsessile, occurring singly, or several in a cluster, reaching a diameter of 2.5 cm.; asci cylindric, 8-spored; spores becoming muriform and brown.

Type species, *Tremella vinosa* Berk. & Curt.

1. **Sarcomyces vinosa** (Berk. & Curt.) Massee, Grevillea **20**: 14. 1891.

Tremella vinosa Berk. & Curt. in herb.; Massee, Jour. Myc. **6**: 179. 1891.

Apothecia as above; spores 1-seriate, becoming 3-septate and muriform, 8–10 $\times$ 21–24 μ; paraphyses numerous, not thickened above, 2–5 μ thick.

TYPE LOCALITY: Venezuela.

DISTRIBUTION: Venezuela and South Carolina.

Massee states that this species has much the appearance and habitat of *Bulgaria inquinans* but of a dark-purple color. It is recorded from South Carolina but no material has been seen. It would doubtless belong with our Ascotremelleae.

EXPLANATION OF PLATES

(Continued from previous volume)

PLATE 75

(frontispiece)

Ciborinia bifrons. A branch of *Populus tremuloides* showing healthy and infected leaves. Below, photograph of three sclerotia somewhat reduced; also enlarged photographs of two sclerotia with apothecia. The enlarged photographs were made by Paul F. Shope from fresh material collected in Colorado.

PLATE 76 (P. 10)

Mitrula phalloides. Upper figure, habitat from leaves submerged in water. Center, *Ombrophila Clavus* on oak leaf. Lower figure, several ascophores, about natural size. Photographs furnished by L. R. Hesler (in part).

PLATE 77 (P. 28)

Trichoglossum velutipes. Photographs of a group of ascophores, about natural size with drawings of an ascus with spores, tips of paraphyses and tip of a ruptured ascus; also one of the setae from the ascophore. Photographed from material collected in New Jersey by Mr. Fred R. Lewis.

PLATE 78 (P. 30)

Trichoglossum hirsutum. This is regarded as a capitate form of the above. Photograph of a clump of sporophores, about natural size or slightly enlarged. Drawings of an ascus with spores, the end of a ruptured ascus, the tip of a paraphysis, one spore removed from the ascus and one of the setae. From material collected near Arden, New York, by Fred R. Lewis.

PLATE 79 (P. 32)

Trichoglossum Wrightii. Photograph of ascophores from material collected in Bermuda by H. H. Whetzel, F. J. Seaver, and L. Ogilvie, Jan. 1926, about natural size, with drawings of portion of ascus with spores and tip of seta. Setting improvised from Bermuda limestone and soil.

PLATE 80 (P. 34)

Trichoglossum Farlowii. Photograph of ascophores, about natural size, with drawings of ascus with spores and paraphyses; also one of the setae. Photographed from material collected during foray at Ithaca, August, 1931.

PLATE 81 (P. 36)

UPPER FIGURE (1). *Spathularia velutipes.* Photograph of ascophores, about natural size, with drawings of spores.

LOWER FIGURE (2). *Spathularia clavata.* Photograph of ascophores, about natural size, with drawings of ascus with spores and paraphysis. Photographs furnished by the Dominion National Museum, Canada.

PLATE 82 (P. 38)

UPPER FIGURE (1). *Leotia lubrica.* Photographed from material collected in The New York Botanical Garden, about natural size, with drawings of ascus, spores, and paraphysis.

374

LOWER FIGURE (2). *Cudonia circinans.* Photographed by Paul F. Shope from material collected in Colorado in 1929, about natural size, with drawings portion of ascus, spores, and paraphysis.

PLATE 83 (P. 50)

Monilinia fructicola. Above, mummified fruits with conidia and young apothecia; also drawings of conidia. Below, mummified fruits with mature apothecia, about natural size; also drawings of ascus with spores and paraphyses. Photographs furnished by Edwin E. Honey.

PLATE 84 (P. 54)

Monilinia Seaveri. Center, branches of wild cherry showing leaves infected with the conidial stage of the fungus *Monilia Seaveri.* Above, drawing of the conidia. Below, several seeds with the apothecial stage, about natural size. Left, ascus with spores and paraphysis.

PLATE 85 (P. 58)

UPPER FIGURE (1). *Seaverinia Geranii.* Photograph of the rootstocks of wild *Geranium* with growth of *Botrytis.* Center, drawing of a cluster of conidia with some immature at the extreme left.

LOWER FIGURE (2). *Seaverinia Geranii.* Photograph of a clump of apothecia. Left, drawing of ascus with spores. Below, three spores germinating.

PLATE 86 (P. 60)

Streptotinia Arisaemae. 21, concentrically arranged sclerotia from conidia on potato agar, natural size; 22, conidia with typically twisted branches; 23, upper surface of leaf of *Arisaema triphyllum,* natural size; 24, *Botrytis streptothrix* on *Orontium aquaticum,* structure of sclerotium. From H. H. Whetzel.

PLATE 87 (P. 68)

Martinia panamaensis. 1, conidia, developed on potato dextrose agar; 2, two plantings of sclerotia showing tendency to aggregate and coalesce; 3, stick with apothecia developed from sclerotia on bark in moist chamber, twice enlarged; 4, apothecia developed from sclerotia grown on potato dextrose agar. From H. H. Whetzel.

PLATE 88 (P. 70)

Ciborinia bifrons. Above, photograph of poplar leaves, showing shotholes, where the sclerotia had dropped out. Below, photograph of a number of sclerotia producing apothecia. Photographs made from dried specimens, about natural size.

PLATE 89 (P. 72)

Ciborinia Whetzelii. Above, sclerotia on leaf of *Populus tremuloides.* Below, apothecia arising from sclerotia, natural size. From H. H. Whetzel.

PLATE 90 (P. 74)

Sclerotinia tuberosa. Above, photograph of several apothecia from New York City, about natural size. Below, several apothecia removed from the ground, with diagram showing position of sclerotia in the soil. Left, drawing of ascus with spores and paraphysis.

PLATE 91   (P. 76)

Sclerotinia Caricis-ampullaceae. Upper figure, left, part of the type specimen in *Carex inflata* from Finland. Center, clump of apothecia attached to a
sclerotium, part of which is broken off, from specimen collected by Dr. C. L.
Porter in Medicine Bow Mt. Swamp, Wyoming. Right, apothecia on sclerotium attached to leaves of *Carex aquatilis*, from Sweden. Lower figure, apothecia in natural setting in Medicine Bow Mountains. Photographed by Dr.
C. L. Porter, one-half natural size. From H. H. Whetzel.

PLATE 92   (P. 78)

Sclerotinia Duriaeana. Upper left-hand corner, photographs of host with
several sclerotia. Upper right-hand corner, drawings of ascus with spores and
paraphysis. Below, photographs of apothecia, about natural size. Photographs furnished by H. H. Whetzel.

PLATE 93   (P. 84)

Ciboria amentacea. Photograph of alder catkins bearing apothecia, about
natural size. At the left, drawing of an ascus with spores and paraphysis.
Below, drawing of one apothecium, much enlarged. Above, portion of ascus
showing pore and three spores isolated. Photographed from material collected
in Oregon by J. R. Keinholz in 1932.

PLATE 94   (P. 86)

UPPER FIGURE (1). *Ciboria pseudotuberosa.* Photograph of apothecia on
old acorns, with drawing of an ascus with spores and paraphyses.
LOWER FIGURE (2). *Ciboria Caucus.* Several apothecia on catkins of
Salix, collected in Hood River region, Oregon by J. R. Keinholz.

PLATE 95   (P. 88)

Ciboria carunculoides. Upper figure, stromatized fruits of mulberry,
Morus, the sclerotial stage. Below, photographs of apothecia arising from
sclerotia, about natural size. From H. H. Whetzel and F. A. Wolf.

PLATE 96   (P. 90)

Sclerotinia Smilacinae. Photographs of two clumps of apothecia, about
natural size. Photographs furnished by H. H. Whetzel. Above, drawing of
an ascus with spores and paraphyses.

PLATE 97   (P. 92)

Midotis Westii. Center, photograph of several clumps of apothecia, about
twice natural size. Above, drawing of an apothecium, several times enlarged
with ascus and spores. Below, several apothecia in different stages of development.

PLATE 98   (P. 96)

Podophacidium xanthomelum. Center, photograph of apothecia showing
habitat, somewhat enlarged. Left, ascus with spores and tip of paraphysis.
Right, portion of empty ascus. Above, three spores. Below, drawing of two
apothecia much enlarged. Photographed from material collected at the
Mycological Foray in Quebec during the summer of 1938.

PLATE 99 (P. 98)

UPPER FIGURE (1). *Midotis versiformis.* Photographs of apothecia in various stages of development, about natural size; also drawings of ascus with spores and paraphysis. Photographs furnished by the Dominion National Museum, Canada.

LOWER FIGURE (2). *Chlorociboria strobilina.* Photographs of spruce cone with apothecia, about natural size; also one scale removed with apothecia. Upper right-hand corner, drawings of two apothecia, enlarged about two diameters; also drawing of ascus with spores and paraphysis. Photographed from material collected in Michigan by Dow V. Baxter.

PLATE 100 (P. 100)

Kriegeria Seaveri. Photograph of type material collected in Montana by J. R. Weir, somewhat enlarged. At the right, drawings of branches bearing apothecia, much enlarged; also several apothecia removed. At the left, drawing of ascus with spores and paraphyses. Below, two spores removed from the ascus.

PLATE 101 (P. 102)

Kriegeria enterochroma. Near center, photograph of branch, much enlarged, showing apothecia. Below, drawings of apothecia removed. At the left, drawing of an ascus with spores and paraphysis. At the right, drawings of three spores removed from the ascus. Below, drawings of apothecia, much enlarged. Photographed from material collected by H. S. Jackson in Ontario, Canada.

PLATE 102 (P. 104)

Kriegeria cedrina. Center, photograph of two branches of cedar with apothecia, from material collected by F. A. Wolf in North Carolina. Below, three apothecia, very much enlarged. Left, drawing of ascus with spores and tips of paraphyses. Right, tip of ruptured ascus showing ascostome and below two mature spores.

PLATE 103 (P. 108)

UPPER FIGURE (1). *Calycina bolaris.* Photographs of several apothecia, about natural size. At the left, drawing of ascus with spores and paraphysis. At the right, several spores. Photographs furnished by H. H. Whetzel.

LOWER FIGURE (2). *Calycina macrospora.* Photograph of several apothecia, about natural size, with drawings of ascus with spores and paraphysis. Photograph furnished by the Dominion National Museum, Canada.

PLATE 104 (P. 112)

UPPER FIGURE (1). *Helotium virgultorum.* Photographs of apothecia on alder roots, about natural size. Lower right-hand corner, one apothecium enlarged. Near center, drawing of ascus with spores and paraphysis. Above, three isolated spores.

LOWER FIGURE (2). *Helotium citrinum.* Photographs of dead wood with the often confluent, apothecia, about natural size. At the left, drawing of ascus with spores and paraphysis; also two isolated spores and one apothecium enlarged. Photographed from material collected in Colorado by F. J. Seaver and Paul F. Shope in 1929.

PLATE 105   (P. 120)

Helotium nyssicola. Above, photographs of a number of seeds of *Nyssa sylvatica* with apothecia, removed from the soil, about natural size. Below, photographs of apothecia as they appear above the soil. At the left, drawing of ascus with spores and paraphysis. Near center, drawings of three spores isolated. Photographed from type material collected near the Museum building of The New York Botanical Garden.

PLATE 106   (P. 124)

Helotium rhizicola. Photographs of roots of *Polygonum* and *Collinsonia* with apothecia, about natural size. At the right, drawing of ascus with spores and paraphysis; also three isolated spores. Photographed from type material collected in The New York Botanical Garden.

PLATE 107   (P. 128)

UPPER FIGURE (1). *Helotium Conocarpi.* Photograph of part of leaf with apothecia, about four times enlarged. Right, ascus with spores and paraphysis, portion of empty ascus and free spores. Below, drawings of two apothecia, much enlarged.

LOWER FIGURE (2). *Helotium atrosubiculatum.* Several apothecia on portion of leaf, about three times enlarged. Below, drawing of apothecium and spores. Left, ascus with spores and paraphysis.

PLATE 108   (P. 134)

Helotium cudonioides. Below, photograph of a group of apothecia on rotten wood, about two and one-half times enlarged. Above, drawings of three apothecia in different stages and from different angles. Left, ascus with spores and paraphysis. Right, empty ascus showing ascostome.

PLATE 109   (P. 178)

Belonioscypha lactea. Photograph of apothecia on *Liriodendron* much enlarged. Photograph furnished by L. O. Overholts from Pennsylvania. Above, drawings of several apothecia, enlarged, also an ascus with spores and paraphysis and three spores isolated.

PLATE 110   (P. 180)

UPPER FIGURE (1). *Gorgoniceps aridula.* Photographs of several scales from cones of *Pinus pungens* collected in Pennsylvania by L. O. Overholts. At the left, drawing of an ascus with spores. Above, drawing of one apothecium, enlarged; also one spore isolated.

LOWER FIGURE (2). *Gorgoniceps confluens.* Photographs of rotten wood bearing apothecia, with drawings of three apothecia below, much enlarged. At the right, an ascus with spores. Above, one spore, isolated. Photographed from type material collected in Bermuda by Stewardson Brown, N. L. Britton, and F. J. Seaver in the winter of 1912.

PLATE 111   (P. 184)

Apostemidium vibrisseoides. Photograph of sticks bearing apothecia, about natural size. Below, drawing of one apothecium with protruding spores; also clump of cells from outside of apothecium. At the left, drawing

of an ascus with spores and paraphysis. Above, portion of ascus with protruding spores. Photograph furnished by F. L. Drayton, from material collected in New York State.

Plate 112   (P. 192)

Pestalopezia Rhododendri. Center, photograph of leaf of *Rhododendron
maximum* infected with *Pestalotia* sp. and the apothecia of *Pestalopezia Rhododendri.* Lower right corner, photograph of several apothecia, much enlarged.
Left, an ascus and paraphysis. Upper right-hand corner, one spore of the
Pestalotia associate.

Plate 113   (P. 218)

Mollisiella ilicincola. Left, photograph of a stick bearing apothecia of
Mollisiella ilicincola on the hysterothecia of *Tryblidiella rufula*, much enlarged.
Lower right, two hysterothecia, much enlarged, bearing apothecia. Above,
drawing of an ascus with paraphysis and a clump of short hairs from the outside of the apothecium of the *Mollisiella.*

Plate 114   (P. 220)

Catinella nigroolivacea. Above, photograph of apothecia on rotten wood,
collected by W. A. Murrill in the island of Jamaica, about natural size. Below
photograph of rotten wood with spothecia from material collected in Nebraska
by Leva B. Walker, about natural size. Left, drawing of ascus with spores and
paraphysis. Right, diagram of a section of an apothecium. Center, drawing
of a portion of hair from substratum.

Plate 115   (P. 230)

Stamnaria americana. Photograph of stems of *Equisetum* with apothecia,
about natural size. At the left, drawing of ascus with spores and paraphysis.
At the right, drawing of a clump of apothecia; also one apothecium removed
and two spores isolated. Photographed from material collected in Ohio by
J. H. Schaffner.

Plate 116   (P. 232)

Stamnaria Thujae. Photograph of infected branch of *Thuja.* Upper
right, drawing of ascus with spores and paraphysis. Upper left, drawing of
infection on leaves. Below, two empty asci and isolated spores. Also drawing
of a clump of apothecia, much enlarged.

Plate 117   (P. 234)

Upper figure (1). *Holwaya gigantea.* Photograph of bark bearing apothecia, about natural size. Above, several apothecia removed, also drawing
of one apothecium, enlarged. At the left, an ascus with spores. Photographed
from material collected in New Jersey by B. O. Dodge in 1909.

Lower figure (2). *Phaeobulgaria inquinans.* Photograph of apothecia
on bark from dried specimen in herbarium collected by F. S. Earle at Reading,
Connecticut in 1902. At the left, drawing of an ascus with spores. Above,
two spores isolated and drawing of one apothecium, enlarged.

Plate 118   (P. 236)

Ascotremella faginea. Photograph of apothecial mass, about natural size.
Upper left-hand corner, an ascus with spores. At the right, several spores
isolated. Photograph furnished by the Dominion National Museum, Canada.

PLATE 119   (P. 238)

Ascotremella turbinata. Photographs of apothecia in various positions and stages, about natural size. Upper left-hand corner, an ascus with spores and paraphysis. Upper right-hand corner, portion of ascus with spores; also two spores, isolated. At the right, diagram of section of an apothecium. Near center, sketch of a cluster of spore-like bodies observed in the tissue of the apothecium. Photographs supplied by H. M. Fitzpatrick from material collected at Ithaca, New York.

PLATE 120   (P. 240)

Lachnella corticalis. Center, photograph of a piece of bark of *Populus tremuloides* with apothecia, twice enlarged. Above, drawing of apothecia, much enlarged. Left, ascus with spores and paraphysis. Right, cluster of hairs. Below, four spores, isolated.

PLATE 121   (P. 248)

Lachnella Agassizii. Photograph of apothecia on a blister-rust canker on *Pinus Strobus*, taken during moist weather at Dannemora, New York, about natural size. Photograph furnished by Walter H. Snell. At the right, drawing of an apothecium much enlarged; also hairs from the outside of the apothecium. At the left, drawing of an ascus with spores and paraphysis; also two spores isolated.

PLATE 122   (P. 252)

Lachnella crucifera. Center, photograph of apothecia, about natural size. Upper left, drawing of three apothecia, much enlarged. Left, asci and paraphyses. Right and below, hairs and crystals from the outside of the apothecium.

PLATE 123   (P. 254)

Lachnella Willkommii. Photograph of Japanese larch branch with apothecia, about natural size. At the right, ascus with spores and paraphysis and two spores, isolated. At the left, drawing of three apothecia, enlarged; also hairs from the outside of the apothecium. Photograph furnished by Walter H. Snell from Rhode Island.

PLATE 124   (P. 260)

Lachnella Solenia. Drawing of a cluster of apothecia, several times enlarged and two isolated apothecia, greatly magnified. Also drawings of ascus with spores and paraphysis and hairs from the outside of the apothecium. Cluster near center showing the two kinds, the hyaline tipped and the brown. Drawn from material in the herbarium of The New York Botanical Garden which is apparently part of the type.

PLATE 125   (P. 264)

UPPER FIGURE (1). *Lachnella sulphurea.* Photograph of stems showing apothecia, somewhat enlarged. Photograph furnished by H. H. Whetzel. At the right, ascus with spores and paraphysis. At the left, drawings of hairs from outside of apothecium.

LOWER FIGURE (2). *Lachnella pygmaea.* Photograph of roots bearing apothecia, about natural size, from material collected in New York City. Near center, drawings of ascus with spores and paraphysis; also hair from outside of the apothecium.

Plate 126 (P. 268)

Lachnella arida. Photograph of apothecia on coniferous trunk, about natural size. Photograph by Paul F. Shope from fresh material in the field at Middle Boulder Canon, Colorado, summer, 1929. Upper left-hand corner, drawing of an ascus with spores and paraphysis. Lower left-hand corner, drawing of portion of hairs from outside of apothecium.

Plate 127 (P. 282)

Diplocarpa Curreyana. Center, photograph of a group of apothecia, four times enlarged. Above, drawings of two apothecia enlarged, also asci with spores and paraphyses with their conidium-like apices. Below, two apothecia, much enlarged, and hairs from the outside of the apothecium.

Plate 128 (P. 284)

Upper figure (1). *Lachnellula chrysophthalma.* Photographs of apothecia from dried material collected in Wyoming during the summer meeting of Botanical Society of America, 1929, about natural size. At the right, drawing of ascus with spores and paraphysis; also portion of ascus. Near center, drawing of poritions of hairs from apothecium.

Lower figure (2). *Lachnella bicolor.* Photographs of *Rubus* bearing apothecia, both natural size and enlarged. At the left, drawings of ascus with spores and paraphyses; also ascus showing pore. At the right, drawing of hair from apothecium; also apothecium enlarged. Photographed from material collected on Pikes Peak, Colorado by F. J. Seaver, Paul F. Shope and F. E. Clements in 1929.

Plate 129 (P. 286)

Arachnopeziza aurelia. Photograph of two acorns and a twig with apothecia, about natural size and drawing of three apothecia, much enlarged. Upper right, an ascus with spores and paraphyses, and portion of empty ascus. Left, clump of hairs from apothecia. Below three spores.

Plate 130 (P. 300)

Cenangium pruinosum. Photograph of two clumps of apothecia, about natural size. Below and above, drawings of apothecia greatly enlarged. Left, ascus with spores and paraphysis and loose spores below.

Plate 131 (P. 302)

Cenangium populneum. Photograph of several clumps of apothecia, somewhat enlarged. Below, clump of apothecia bursting through the bark, much enlarged. Above, one apothecium, much enlarged. Left, drawing of ascus with spores and paraphysis.

Plate 132 (P. 308)

Sphaerangium tetrasporum. Center, photographs of branches with apothecia, somewhat enlarged. Below, two apothecia, much enlarged. Left, ascus with spores and paraphyses. Above, one spore, isolated, and branching paraphyses.

Plate 133 (P. 310)

Sphaerangium Tiliae. Center, photograph of portion of three branches, about natural size. Above, one apothecium, much enlarged. Left, drawing of ascus with spores and paraphysis.

382   THE NORTH AMERICAN CUP-FUNGI

PLATE 134   (P. 312)

Dermatella deformata.   Center, photograph of wood with apothecia, about three times natural size.   Below, drawing of apothecia on wood with one isolated, much enlarged.   Upper left, ascus with spores.   Above one spore, isolated; also tips of paraphyses.

PLATE 135   (P. 318)

UPPER FIGURE (1).   *Godronia Kalmiae.*   Photograph of branches showing apothecia.   Below, drawing of apothecia in various stages of development; also an ascus with spores.   Above, three spores removed.   Photographs and drawings from material distributed in North American Fungi *147.*

LOWER FIGURE (2).   *Godronia Spiraeae.*   Photographs of branches bearing apothecia.   Above, two apothecia, enlarged.   At the right, drawing of ascus with spores and paraphysis.   Above, drawing of two apothecia, enlarged. At the left, three ascospores, isolated.   Photographed and drawn from material collected by C. H. Peck (no date) and listed as *Godronia urceolata.*

PLATE 136   (P. 320)

UPPER FIGURE (1).   *Godronia fuliginosa.*   Photograph of willow branches bearing apothecia and conidia.   Above, drawing of clump of apothecia, enlarged.   At the left, drawing of one ascus with spores; also one ascospore, isolated.   Near the center, conidia producing bodies.   At the right, three conidia, isolated.   Photographed and drawn from material collected in Colorado by F. J. Seaver and Ellsworth Bethel in 1910.

LOWER FIGURE (2).   *Godronia seriata.*   Photograph of branches of *Betula fontinalis* bearing apothecia and conidia, about natural size.   Near center, drawing of clump of apothecia much enlarged.   At the right, drawing of an ascus with spores and paraphysis.   Above, two ascospores, isolated.   Photographed and drawn from material collected in Coal Creek Canon, Colorado by F. J. Seaver and Paul F. Shope in 1929.

PLATE 137   (P. 322)

UPPER FIGURE (1).   *Godronia jamaicensis.*   Photograph of two fragments of wood with apothecia, about natural size; also one apothecium much enlarged.   Left, an ascus with spores and paraphysis.   Also one loose spore.

LOWER FIGURE (2).   *Godronia Cephalanthi.*   Photograph of branch with apothecia, somewhat enlarged.   Right, one apothecium, much enlarged; also portion of ascus with spores and paraphysis.

PLATE 138   (P. 328)

Godronia parasitica.   Photographs of several leaves with apothecia, about natural size.   At the right, two apothecia enlarged.   At the left, portion of an ascus with spores and paraphyses.   Diagonally across the plate, one spore. Photographed from type collected in the island of Hayti, by George V. Nash, in 1903.

PLATE 139   (P. 330)

UPPER FIGURE (1).   *Godronia pinicola.*   Photographs of a piece of branch with apothecia, about natural size.   Above, drawing of apothecia, much enlarged.   Left, ascus with spores, paraphysis, tip of ascus and hair from apothecium.   Below three loose spores.

LOWER FIGURE (2). *Godronia sororia*. Photograph of piece of bark with apothecia, about natural size. Above, two apothecia, much enlarged. Right, ascus with spores. Below, two loose spores and portion of two hairs from apothecium.

PLATE 140   (P. 336)

Pezicula carpinea. Photograph of branch showing apothecia, considerably reduced. To the left, sketch of apothecia with an ascus with spores and paraphyses. To the left, pycnospores in various stages of development.

PLATE 141   (P. 338)

UPPER FIGURE (1). *Pezicula acericola*. Near the center photograph of bark showing apothecia, about natural size. To the left, an ascus with spores and paraphysis and the end of a ruptured ascus. Upper right-hand corner, sketch of *Sphaeronema acericola*. Lower right-hand corner, stroma showing both apothecial and conidial stages produced on the same stroma.

LOWER FIGURE (2). *Pezicula spiculata*. Near the center, photograph of bark showing apothecia and pycnidia, about natural size. Upper left-hand corner, sketch of stroma showing apothecia and pycnidia and pycnospores. Right-hand side, an ascus with spores and paraphysis. Below, drawing of mature ascospores.

PLATE 142   (P. 340)

Pezicula cornicola. Photograph of two branches showing apothecia. Above, sketch of apothecium, much enlarged. Center, an ascus with spores. Above right, two ascospores. Below, a stroma with exuding conidiospores. Below right, two conidia.

PLATE 143   (P. 344)

Pezicula Corni. Center, photographs of several twigs showing apothecia. Above, sketch of two apothecia, much enlarged. Background, drawings of an ascus with spores; also two spores removed and the end of a ruptured ascus. Below, drawing of a sorus and two conidia.

PLATE 144   (P. 346)

UPPER FIGURE (1). *Pezicula ocellata*. Photograph of willow branches showing apothecia, enlarged about one-third. Above, drawing of two apothecia, enlarged. At the left, an ascus with spores and paraphysis. Below, drawing of conidia in various stages. Photographed from material collected in Colorado by F. J. Seaver and Ellsworth Bethel in 1910.

LOWER FIGURE (2). *Pezicula Populi*. Photograph of branches of poplar bearing apothecia, enlarged about one-third. Below, drawing of several apothecia, enlarged. At the right, drawing of an ascus with spores and paraphyses; also two ascospores, isolated. Photographed and drawn from type material collected in Ontario, Canada, by G. E. Thompson in 1930.

PLATE 145   (P. 348)

Pezicula purpurascens. Photograph of chestnut branches bearing apothecia and conidia, about natural size. At the left, drawing of an ascus with spores and paraphyses. Above, drawing of one apothecium, enlarged; also two ascoconidia. Below, several ascoconidiophores showing stages in the development and discharge of the ascoconidium.   Photographed and drawn from type material collected at West Chester, Pennsylvania in 1888.

Plate 146   (P. 350)

Pezicula Hamamelidis.   Photograph of bark of *Hamamelis* bearing apothecia and conidia, natural size.   Below at the left, drawing of three apothecia, enlarged.   At the right, cluster of conidia; also three conidia isolated.   Above, tip of ascus showing rupture; also four ascospores at different stages.   Photographed from type collected in Ontario by H. S. Jackson.

Plate 147   (P. 352)

Upper figure (1).   *Dermea Cerasi.*   In the center, photograph of branch showing apothecia about natural size, with enlarged sketch of an apothecium; also stromata, and pycnospores.

Lower figure (2).   *Dermea bicolor.*   In the center, photograph of branch showing apothecia from type material, about natural size.   Above, large drawing of apothecia.   To the left, an ascus with spores and paraphyses. Below, sketch of stroma and pycnospores.

Plate 148   (P. 354)

Upper figure (1).   *Dermea molliuscula.*   Photograph of twigs of *Betula lutea* showing apothecia, about natural size.   Above, enlarged sketch of apothecia.   To the left, an ascus with spores and paraphyses.   In the center, sketch of a stroma and pycnospores.

Lower figure (2).   *Dermea molliuscula.*   In the center, photograph of branch of *Alnus* showing apothecia, about natural size.   Above, enlarged sketch of an apothecium and stromata.   To the right, an ascus with spores and paraphyses.   Below pycnospores.

Plate 149   (P. 356)

Dermea balsamea.   Photographs of branches of hemlock bearing apothecia and conidia.   Below, drawing of portion of branch, enlarged, showing cirrhi of exuding conidia.   Above, drawing of one apothecium, enlarged; also ascus with spores and paraphysis, three isolated ascospores, one conidiospore and three microconidia.   Photographs and drawing of branch furnished by B. O. Dodge from material collected at Scarsdale, New York.

Plate 150   (P. 372)

Upper figure (1).   *Haematomyxa vinosa.*   Photograph of wood bearing apothecia, enlarged about one-half.   At the left, drawing of an ascus with spores. Below, one apothecium, enlarged.   Above, one spore, isolated. Photographed from material in the Ellis herbarium, collected at Newfield, New Jersey.

Lower figure (2).   *Haematomyxa ascoboloides.*   Photograph of herbaceous stems bearing apothecia, enlarged about one-third.   Below, drawing of two apothecia, enlarged.   At the right, an ascus with spores and paraphysis. Above, one spore, enlarged.   Photographed from type material collected at Tacoma Park, Maryland in 1889.

Abies amabilis
  Godronia abieticola
Abies balsamea
  Dermea balsamea
  Lachnella Agassizii
  Lachnella resinaria
  Pezicula phyllophila
  Tapesia balsamicola
  Tympanis Pinastri
Abies grandis
  Godronia abieticola
Abies sp.
  Cenangium Abietis
  Cenangium alpinum
  Cenangium leptospermum
  Belonium inconspicuum
  Ciboria rufofusca
  Godronia abietina
  Godronia Abietis
  Helotium fumosellum
  Lachnella arida
  Lachnellula chrysophthalma
  Pezicula livida
Acanthonitschkea coloradensis
  Lachnella Acanthonitschkeae
Acer floridanum
  Pyrenopeziza leucodermis
Acer leucoderme
  Pyrenopeziza leucodermis
Acer rubrum
  Ciboria acerina
  Dermea acerina
  Erinellina miniopsis
  Pezicula acericola
  Pycnopeziza sympodialis
Acer saccharinum
  Ciboria acerina
  Dermea acerina
Acer Saccharum
  Dermea acerina
Acer spicatum
  Pezicula acericola
  Pezicula spicata
  Pezicula spiculata

Acer sp.
  Belonidium fuscopallidum
  Godronia rhabdospora
  Helotium albovirens
  Helotium albuminum
  Helotium fraternum
  Helotium luteovirescens
  Helotium naviculasporum
  Helotium translucens
  Holwaya gigantea
  Lachnella Rhytismatis
  Mollisia paullopuncta
  Pezizella floriformis
  Velutaria rufoolivacea
Adiantum sp.
  Lachnella aspidicola
Alnus crispa mollis
  Pezicula Alni
  Pezicula aurantiaca
Alnus incana
  Ciboria amentacea
  Pezicula Alni
  Tympanis alnea
Alnus oregona
  Helotium seminicola
Alnus tenuifolia
  Tympanis alnea
Alnus sp.
  Cenangium furfuraceum
  Ciboriella rufescens
  Godronia Urceolus
  Helotium alniellum
  Helotium fastidiosum
  Helotium umbrinum
  Hyaloscypha alniseda
  Lachnum alneum
  Lachnum hyalinellum
  Mollisia Teucrii
  Orbilia luteorubella
  Trichopeziza punctiformis
Amelanchier alnifolia
  Pezicula pruinosa

385

Amelanchier canadensis
Monilinia Amelancheris
Pezicula pruinosa
Amelanchier Cusickii
Monilinia gregaria
Amelanchier intermedia
Monilinia Amelanchieris
Amelanchier oblongifolia
Monilinia Amelanchieris
Amelanchier sp.
Dermea bicolor
Fabraea maculata
Monilinia Amelanchieris
Ammophila longifolia
Pyrenopeziza Ellisii
Andromeda sp.
Dasyscypha fuscidula
Trichopeziza marginata
Velutaria rufoolivacea
Andropogon furcatus
Trichopeziza distincta
Andropogon scoparius
Helotium planodiscum
Andropogon sp.
Belonium Andropogonis
Belonium culmicola
Helotiella aureococcinea
Lachnella albotestacea
Mollisia atriella
Anemone nemorosa
Sclerotinia tuberosa
Angelica atropurpurea
Mollisia Angelicae
Antennaria plantaginifolia
Mollisia lanaria
Apios tuberosa
Mollisia apiophila
Apium sp.
Sclerotinia minor
Apocynum androsaemifolium
Pyrenopeziza Dearnessii
Aquilegia coerulea
Dasyscypha tuberculiformis
Arabis furcata
Belonium arabicolum
Aralia racemosa
Helotium minimum
Aralia sp.
Trichopeziza setigera

Archontophoenix Alexandrae
Helotium atrosubiculatum
Arctium Lappa
Lachnella canadensis
Arctostaphylos pungens
Trichopeziza labrosa
Arctostaphylos Tracyi
Arachnopeziza Arctostaphyli
Calloria nitens
Arisaema triphyllum
Streptotinia Arisaemae
Aronia sp.
Monilinia Aroniae
Artemisia vulgaris
Pyrenopeziza Artemisiae
Artemisia sp.
Pyrenopeziza Absinthii
Arundinaria macrosperma
Dasyscypha Arundinariae
Dasyscypha caulicola
Asclepias sp.
Mollisia Asclepiadis
Aspidium sp.
Lachnella aspidicola
Aster puniceus
Orbilia assimilis
Aster sp.
Trichopeziza brevipila
Azalea sp.
Ovulinia Azaleae
Bambusa vulgaris
Gorgoniceps jamaicensis
Betula alba
Helotium Friesii
Betula fontinalis
Godronia seriata
Betula lutea
Godronia seriata
Verpatinia duchesnayensis
Betula nigra
Sphaerangium magnisporum
Betula sp.
Cenangella Hartzii
Cenangium compressum
Dermea molliuscula
Godronia fusispora
Godronia Urceolus
Helotium Linderi
Tapesia lividofusca

Tapesia secamenti
Veltutaria rufoolivacea
Bidens sp.
Peziza heterocarpa
Pyrenopeziza Absinthii
Bigelovia graveolens
Tapesia tumefaciens
Bignonia sp.
Pezizella leguminum
Brassica sp.
Helotium brassicaecolum
Pezizella brassicaecola
Bryum sp.
Helotium turbinatum
Calamagrostis canadensis
Lachnella agrostina
Calamagrostis sp.
Belonidium aurantiacum
Belonium intermedium
Calluna sp.
Cenangella Ericae
Caltha palustris
Verpatinia calthicola
Camellia japonica
Sclerotinia Camelliae
Carex ampullacea
Sclerotinia Caricis-ampullaceae
Carex aquatilis
Sclerotinia Caricis-ampullaceae
Carex crinita
Dasyscypha crinella
Sclerotinia Duriaeana
Sclerotinia longisclerotialis
Carex exsiccata
Sclerotinia paludosa
Carex flava
Sclerotinia Duriaeana
Carex hystericina
Sclerotinia Duriaeana
Carex inflata
Sclerotinia Caricis-ampullaceae
Carex interior
Sclerotinia longisclerotialis
Carex lacustris
Pyrenopeziza multipunctoidea
Carex limosa
Ombrophila limosa
Carex nebraskensis
Sclerotinia Duriaeana

Carex prairea
Sclerotinia Duriaeana
Sclerotonia longisclerotialis
Carex retrorsa
Sclerotinia Caricis-ampullaceae
Sclerotinia Duriaeana
Sclerotinia longisclerotialis
Carex riparia lacustris
Sclerotinia Duriaeana
Carex stricta
Sclerotinia Duriaeana
Carex utriculata
Calloria caricinella
Carex vesicaria
Sclerotinia longisclerotialis
Carex sp.
Belonioscypha miniata
Belonium caricincolum
Helotium citrinulum
Helotium turgidellum
Mollisia epitypha
Mollisa euparaphysata
Pyrenopeziza caricina
Pyrenopeziza Caricis
Pyrenopeziza cervinula
Pyrenopeziza doryphora
Pyrenopeziza yogoensis
Carpinus caroliniana
Pezicula carpinea
Carpinus sp.
Cenangella Ravenelii
Helotium ammoides
Tapesia lividofusca
Carya ovata
Lambertella Hicoriae
Carya sp.
Dermatella caryigena
Helotium fructigenum
Cassandra calyculata
Cenangium Cassandrae
Godronia Urceolus
Castanea dentata
Pezicula purpurascens
Castanea vesca
Ciboria americana
Castanea sp.
Belonium basitrichum
Dasyscypha translucida
Dasyscypha vixvisibilis

Castanopsis chrysophylla
  Belonium Parksi
  Coronellaria Castanopsidis
  Godronia Castanopsidis
Catalpa sp.
  Pezizella leguminum
Caulophyllum thalictroides
  Orbilia Caulophylli
Ceanothus velutinus
  Cenangium aureum
Cecropia sp.
  Helotium Cecropiae
Celtis sp.
  Lachnella succina
Cephalanthus occidentalis
  Godronia Cephalanthi
  Lambertella Cephalanthi
Cephalanthus sp.
  Pyrenopeziza Cephalanthi
Chamaedaphne calyculata
  See **Cassandra**
  Erinella borealis
  Helotium Cassandrae
Chionanthus virginica
  Dermea Chionanthi
Citharexylum spinosum
  Lambertella Jasmini
Clethra alnifolia
  Godronia Urceolus
  Lachnella albopileata subauata
Coccoloba sp.
  Lambertella tropicalis
Collinsonia canadensis
  Helotium rhizicola
Comptonia asplenifolia
  Echinella rhabdocarpa
Conocarpus erectus
  Helotium Conocarpi
Cornus alternifolia
  Pezicula Corni
Cornus Amomum
  Lachnella Corni
Cornus florida
  Mollisia miltophthalma
  Trichopeziza roseoalba
Cornus stolonifera
  Pezicula Corni
Cornus sp.
  Cenangium contortum
  Helotium propinquum

  Monilinia Corni
  Pezicula cornicola
  Tympanis fasciculata
Corydalis Brandegei
  Dasyscypha Bakeri
Corydalis sp.
  Calloria coccinea
Corylus rostrata
  Pezicula corylina
Corylus sp.
  Helotium epiphyllum
Crataegus punctata
  Monilinia Johnsoni
Crataegus sp.
  Cenangium confusum
  Pezicula crataegicola
  Pezicula olivascens
Crocus sp.
  Stromatinia Gladioli
Cucurbita sp.
  Orbilia Cucurbitae
Cydonia sp.
  Dermea Cydoniae
  Fabraea maculata
Daucus Carota
  Sclerotinia intermedia
Delphinium sp.
  Hyalopeziza ciliata
  Phialea pallida
Desmodium sp.
  Dermea ferruginea
Diatrype Stigma
  Cenangium episphaeria
Diatrype sp.
  Orbilia coccinella
Diatrypella sp.
  Helotium episphaericum
Dichaena sp.
  Helotium strumosum
Dicranopteris pectinata
  Lachnella Dicranopteridis
Dicranum flagellare
  Helotium destructor
Diervilla sp.
  Godronia turbinata
Elymus canadensis
  Cyathicula alpina
Equisetum hyemale
  Lachnella inquilina

Equisetum robustum
Stamnaria americana
Equisetum sp.
Stamnaria americana
Erigeron sp.
Helotium nigrescens
Mollisia erigeronata
Mollisia exigua
Pyrenopeziza Absinthii
Trichopeziza carneorubra
Eriophorum callitrix
Lachnum carneolum
Eryngium sp.
Dasyscypha eryngiicola
Erythronium americanum
Ciborinia Erythronii
Eschscholtzia californica
Helotium Eschscholtziae
Eucalyptus globulus
Mollisia subcornea
Eucalyptus sp.
Lachnella atropurpurea
Orbilia Eucalypti
Pezizella carneorosea
Eupatorium ageratoides
Lachnella Solenia
Eupatorium maculatum
Dasyscypha longipila
Eupatorium purpureum
Lachnella Eupatorii
Eupatorium sp.
Belonium bicolor
Helotium herbarum
Fagus americana
Ascotremella faginea
Fagus grandifolia
Helotium albopunctum
Fagus sp.
Calycina petiolorum
Helotium naviculasporum
Mollisia caesia
Festuca tenella
Pyrenopeziza Ellisii
Fomes fomentarius
Helotium mycetophilum
Fragaria sp.
Fabraea Earliana
Fraxinus americana
Durandiella Fraxini

Fraxinus nigra
Dermea Tulasnei
Fraxinus sp.
Cenangium populneum
Dasyscypha puberula
Dermatella Fraxini
Dermea Tulasnei
Helotium luteovirescens
Freesia sp.
Stromatinia Gladioli
Galium sp.
Pyrenopeziza nigritella
Pseudopeziza repanda
Garrya elliptica
Belonium Parksi
Harknessiella purpurea
Lachnella tautilla
Gaultheria Shallon
Cyathicula aquilina
Lachnella Gaultheriae
Mollisia Gaultheriae
Pestalopezia brunneopruinosa
Gaylussacia dumosa
Trichopeziza venturioides
Gentiana sp.
Pseudopeziza Holwayi
Geranium maculatum
Seaverinia Geranii
Gladiolus sp.
Botryotinia Draytoni
Stromatinia Gladioli
Gleditsia triacanthos
Helotium midlandense
Orbilia rubrococcinea
Gleichenia sp.
Lachnella Gleicheniae
Glyceria nervata
Belonium Glyceriae
Halesia carolina
Lachnella Halesiae
Hamamelis virginiana
Dermea Hamamelidis
Pezicula Hamamelidis
Helianthus annuus
Mollisia lilacina
Helianthus sp.
Pyrenopeziza Absinthii
Hydrangea sp.
Cenangium apertum

Hypnum sylvaticum
Peziza hypnicola
Hypoxylon sp.
Helotium episphaericum
Hysterium sp.
Mollisiella ilicincola
Ilex glabra
Cenangium tuberculiforme
Dermea crypta
Dermea olivacea
Ilex opaca
Lachnella pulveracea
Pezizella aquifoliae
Ilex verticillata
Cenangium sticticum
Dermea Peckiana
Ilex sp.
Cenangella Ravenelii
Impatiens sp.
Erinellina Nylanderi
Helotium herbarum
Iris sp.
Botryotinia convoluta
Helotium nigromaculatum
Mollisia Iridis
Iva xanthifolia
Lachnella Ivae
Pyrenopeziza Absinthii
Jasminum gracile
Lambertella Jasmini
Juglans regia
Mollisia abdita
Juglans sp.
Cenangium Juglandis
Juncus alpinus
Mollisia alpina
Juncus effusus californica
Sclerotinia juncigena
Juncus sp.
Lachnella diminuta
Mollisia stictoidea
Jungermania sp.
Helotium destructor
Juniperus bermudiana
Dermatella deformata
Juniperus communis
Dermatella deformata
Kriegeria juniperina
Juniperus lucayana
Godronia jamaicensis

Juniperus scopulorum
Dermatella deformata
Juniperus virginiana
Murangium Sequoiae
Juniperus sp.
Godronia Juniperi
Kalmia angustifolia
Trichopeziza Kalmiae
Kalmia sp.
Ovulinia Azaleae
Pezicula Kalmiae
Lactuca sp.
Sclerotinia minor
Lantana camara
Godronia Lantanae
Larix europaea
Lachnella Hahniana
Lachnella oblongospora
Lachnella occidentalis
Lachnella Willkommii
Larix laricina
Lachnella oblongospora
Lachnella occidentalis
Larix leptolepis
Lachnella occidentalis
Larix sp.
Helotium laricinum
Lachnella arida
Lachnella occidentalis
Ledum groenlandicum
Cenangella Ericae
Leonurus cardiaca
Helotium fumosum
Libocedrus decurrens
Dermea Libocedri
Kriegeria alutipes
Linum Lewisii
Pyrenopeziza californica
Liquidambar sp.
Mollisia vulgaris sanguinella
Pseudohelotium sacchariferum
Lithospermum canescens
Calloria Lithospermi
Lonicera canadensis
Godronia Lonicerae
Lonicera involucrata
Helotiella Lonicerae
Lonicera sp.
Mollisia complicatula

Lysichiton camtschatcense
Ombrophila Lysichitonis
Magnolia glauca
Ciborinia gracilipes
Dermatella Magnoliae
Pyrenopeziza protrusa
Magnolia grandiflora
Dasyscypha hystricula
Magnolia sp.
Belonium phlegmaceum
Dasyscypha albopileata
Holwaya gigantea
Mollisia fumigata
Mollisia glenospora
Orbilia diaphanula
Malus sp.
Ciboria aestivalis
Tympanis conspersa
Marchantia polymorpha
Cyathicula Marchantiae
Medicago sativa
Sclerotinia sativa
Medicago sp.
Pseudopeziza Jonesii
Pseudopeziza Medicagnis
Melilotus alba
Sclerotinia sativa
Melilotus officinalis
Sclerotinia sativa
Melilotus sp.
Pseudopeziza Medicaginis
Pseudopeziza Meliloti
Meliola sp.
Trichobelonium leucorrhodinum
Morus alba
Cenangium fatiscens
Ciboria carunculoides
Pezizella conchella
Morus alba tatarica
Dermea Mori
Myirangium sp.
Mollisiella ilicincola
Myrica cerifera
Dasyscypha callochaetes
Myrica Gale
Ciboria acerina
Dasyscypha sulphurella
Trichopeziza myricacea

Myrica sp.
Trichopeziza alboviridis
Trichopeziza punctiformis
Narcissus
Botryotinia narcissicola
Nardus sp.
Lachnum Nardi
Nemopanthes canadensis
Godronia Nemopanthis
Nemopanthes mucronata
Dermea Peckiana
Godronia Nemopanthis
Nyssa sylvatica
Helotium nyssicola
Sclerotinia nyssaegena
Oenothera sp.
Mollisia Oenotherae
Olea europaea
Lachnella fasciculata
Osmunda sp.
Mollisia tenella
Peziza frondicola
Trichopeziza Osmundae
Ostrya virginiana
Ciboria acerina
Ostrya sp.
Midotis Westii
Oxydendrum arboreum
Tympanis Oxydendri
Paludella squarrosa
Mitrula gracilis
Panax quinquefolium
Sclerotinia Panacis
Pastinaca sp.
Pezizella Pastinacae
Patellaria sp.
Mollisiella ilicincola
Pedicularis bracteosa
Sclerotinia coloradensis
Pedicularis groenlandica
Sclerotinia coloradensis
Pedicularis sp.
Cyathicula alpina
Peltigera canina
Calloria Mülleri
Persea palustris
Erinellina maculosa
Persea sp.
Dermatella Magnoliae

Philadelphus inodorus
Pezicula Philadelphi
Phragmites sp.
Lachnella albotestacea
Mollisia hydrophila
Physocarpus capitatus
Belonium Parksi
Phytolacca sp.
Orbilia pulviscula
Picea glauca
Dermea piceina
Picea Mariana
Lachnella Agassizii
Lachnum hyalinellum
Tapesia lividofusca fallax
Picea pungens
Lachnella oblongospora
Picea rubra
Lachnella Agassizii
Picea sitchensis
Godronia Treleasei
Picea sp.
Cenangium Abietis
Chlorociboria strobilina
Gorgoniceps Pumilionis
Helotium aurantium
Helotium sulphuratum
Lachnella arida
Lachnella Ellisiana
Lachnellula chrysophthalma
Pinus Banksiana
Lachnellula chrysophthalma
Pinus contorta
Godronia Zelleri
Pinus inops
Peziza conorum
Pinus Lambertiana
Godronia Zelleri
Pinus monticola
Godronia Zelleri
Lachnella Agassizii
Pinus Mugho
Cenangium atropurpureum
Pinus Murrayana
Lachnella fuscosanguinea
Pinus nigra
Cenangium atropurpureum
Pinus ponderosa
Atropellis arizonica
Godronia sororia

Pinus pungens
Cenangium atropurpureum
Godronia pinicola
Gorgoniceps aridula
Lachnella oblongospora
Pinus resinosa
Godronia pinicola
Gorgoniceps ontariensis
Pinus rigida
Cenangium atropurpureum
Godronia pinicola
Pezicula phyllophila
Tympanis Pinastri
Pinus Strobus
Cenangium acuum
Dermea pinicola
Godronia Zelleri
Lachnella Agassizii
Mollisia Scoleconectriae
Pezicula livida
Tympanis Pinastri
Pinus sylvestris
Cenangium atropurpureum
Pinus Taeda
Cenangium atropurpureum
Pinus virginiana
Lachnella oblongospora
Pinus sp.
Atropellis apiculata
Atropellis tingens
Cenangium Abietis
Cenangium acuum
Erinellina rhaphidospora
Lachnella Ellisiana
Lachnella pulverulenta
Pezicula livida
Platanus sp.
Helotium luteovirescens
Podophyllum peltatum
Septotinia podophyllina
Polycodium stamineum
Monilinia Polycodii
Polygonum sp.
Stromatinia Rapulum
Polygonum virginianum
Helotium rhizicola
Polygonum viviparum
Pseudopeziza Bistortae
Polygonum sp.
Mollisia Polygoni

Polyporus igniarius
Peziza mycogena
Polyporus Stevensii
Mollisia incrustata
Polyporus sp.
Mollisia incrustata
Mollisia vulgaris myceticola
Orbilia coccinella
Orbilia epispora
Populus alba
Pseudopeziza Populi-albae
Populus canadensis
Pycnopeziza sympodialis
Populus grandidentata
Cenangium populneum
Pezicula Populi
Populus Tremula
Helotium Friesii
Populus tremuloides
Belonium aggregatum
Cenangium populneum
Cenangium pruinosum
Ciboria Caucus
Ciborinia bifrons
Dermatella populina
Lachnella populicola
Pezicula Populi
Pycnopeziza sympodialis
Tympanis spermatiospora
Populus sp.
Helotium Friesii
Helotium gemmarum
Mollisia complicatula
Orbilia luteorubella
Pezicula ocellata
Tympanis spermatiospora
Potentilla sp.
Mollisia Dehnii
Pyrenopeziza coloradnsis
Sclerotinia fallax
Prinus verticillata
See **Ilex**
Cenangium sticticum
Prunus avium
Lambertella Pruni
Prunus demissa
Monilinia demissa
Prunus domestica
Dermea Padi

Prunus emarginata
Monilinia emarginata
Prunus Padus
Dermea Padi
Prunus serotina
Monilinia Seaveri
Prunus spinosa
Dermea Padi
Prunus virginiana
Dermea Padi
Monilinia Padi
Prunus spp.
Dermea Cerasi
Dermea Prunastri
Monilinia fructicola
Monilinia laxa
Tympanis conspersa
Pseudotsuga taxifolia
Lachnella Agassizii
Lachnella ciliata
Lachnella oblongospora
Lachnella Pseudotsugae
Pseudotsuga sp.
Lachnella Hahniana
Psoralea macrostachya
Orbilia myriospora
Pteridium aquilinum
Lachnaster miniatus
Pteridium aquilinum pubescens
Calloria cremea
Pteris sp.
Lachnella pteridicola
Lachnella Pteridis
Pyrus sp.
Cenangium pyrinum
Fabraea maculata
Quercus agrifolia
Helotium furfuraceum
Phillipsia nigrella
Quercus alba
Dermateopsis tabacina
Lachnella capitata
Quercus coccinea
Dermateopsis tabacina
Helotium strumosum
Sphaerangium tetrasporum
Quercus laurifolia
Helotium castaneum
Quercus Prinus
Pyrenopeziza prinicola

Quercus tinctoria
  Cenangium sticticum
Quercus sp.
  Belonium basitrichum
  Calycina petiolorum
  Chlorosplenium olivaceum
  Ciboria pseudotuberosa
  Dasyscypha epixantha
  Dasyscypha uncinata
  Godroniopsis quernea
  Helotium ferrugineum
  Helotium midlandense
  Helotium puberulum
  Helotium sordidatum
  Helotium translucens
  Holwaya gigantea
  Lachnella capitata
  Lachnella ciliaris
  Lachnella pollinaria
  Lachnella succina
  Phaeobulgaria inquinans
  Pyrenopeziza prinicola
  Tapesia culcitella
  Trichopeziza comata
  Trichopeziza obscura
  Trichopeziza punctiformis
  Velutaria rufoolivacea
Ranunculus sp.
  Fabraea Ranunculi
  Pseudopeziza singularis
Rhamnus alnifolia
  Pezicula Morthieri
Rhamnus purshiana
  Belonium Parksi
  Dermatella Frangulae
Rhododendron maximum
  Cenangium palmatum
  Helotium castaneum
Rhododendron roseum
  Monilinia Azaleae
Rhododendron sp.
  Cenangella Rhododendri
  Monilinia Azaleae
  Ovulinia Azaleae
  Pestalopezia Rhododendri
Rhus venenata
  Pezicula pallidula
Rhus sp.
  Tympanis fasciculata

Ribes bracteosum x nigrum
  Godronia Davidsoni
Ribes Menziesii
  Godronia lobata
Ribes montigenum
  Godronia tumoricola
  Tapesia ribicola
Ribes prostratum
  Lachnella albolabra
Ribes Wolfii
  Cenangella oricostata
  Godronia Davidsoni
Ribes sp.
  Godronia Ribis
  Godronia Urceolus
  Helotium discretum
  Pseudopeziza Ribis
Ricinus communis
  Botryotinia Ricini
Robinia Pseudo-acacia
  Helotium erraticum
Rosa sp.
  Cenangium Rosae
  Fabraea Rosae
  Pezicula Brenckleana
  Tapesia Rosae
Rubus idaeus
  Pezicula Rubi
Rubus nutkanus
  Dasyscypha scabrovillosa
Rubus odoratus
  Pyrenopeziza lacerata
Rubus villosus
  Mollisia alabamensis
Rubus sp.
  Lachnella subochracea
  Pezicula Rubi
  Pyrenopeziza Rubi
  Velutaria rufoolivacea
Sabal serrulata
  Mollisia Sabalidis
Sabal sp.
  Gorgoniceps confluens
Salix alba
  Cenangium populneum
Salix babylonica
  Cenangium sticticum
Salix discolor
  Ciboria amentaceae
  Ciboria amenti

Tilia heterophylla
 Pyrenopeziza minuta
 Sphaerangium Tiliae
Tilia sp.
 Ascotremella faginea
 Helotium luteovirescens
 Helotium naviculasporum
 Holwaya gigantea
 Trichopeziza Tiliae
 Tympanis conspersa
Tragopogon porrifolius
 Sclerotinia intermedia
Trifolium arvense
 Pseudopeziza Trifolii-arvensis
Trifolium sp.
 Pseudopeziza Trifolii
Trigonella sp.
 Pseudopeziza Medicaginis
Tryblidiella sp.
 Mollisiella ilicincola
Tsuga canadensis
 Dermea balsamea
 Lachnella Agassizii
 Orbilia Fairmani
Tsuga sp.
 Dasyscypha chamaeleontina
Tulipa sp.
 Sclerotinia sativa
Typha latifolia
 Mollisia epitypha
Typha sp.
 Helotium turgidellum
Ulmus americana
 Mollisia lilacina
 Pseudohelotium fibrisedum
Ulmus sp.
 Mollisia caespiticia
 Trichopeziza leonia
Umbellularia californica
 Midotis plicata
Urtica canadensis
 Trichopeziza urticina
Urtica sp.
 Calloria fusarioides
 Dasyscypha caulicola
 Erinellina Nylanderi

Vaccinium corymbosum
 Godronia Kalmiae
 Monilinia Vaccinii-corymbosi
Vaccinium myrtilloides
 Trichopeziza coarctata
Vaccinium Oxycoccos
 Monilinia Oxycocci
Vaccinium parvifolium
 Belonium Parksi
Vaccinium uliginosum
 Cenangella pruinosa
Vaccinium sp.
 Monilinia Urnula
Valsa sp.
 Helotium episphaericum
Veratrum californicum
 Sclerotinia Veratri
Viburnum cassinoides
 Godronia viburnicola
 Lambertella Viburni
Viburnum dentatum
 Godronia viburnicola
Viburnum lantanoides
 Pezicula minuta
Viburnum Lentago
 Dasyscypha Lentaginis
 Tympanis fasciculata
Viburnum sp.
 Cenangium Viburni
 Dermea Viburni
Vitis vulpina
 Chlorosplenium salviicolor
Vitis sp.
 Dermatella viticola
 Godronia viticola
 Helotium pullatum
 Lachnella ascoboloidea
 Trichopeziza penicillata
Webera nutans
 Mitrula muscicola
Xylaria sp.
 Helotium episphaericum
Yucca sp.
 Cenangium Yuccae
Zea Mays
 Helotiella pygmaea

INDEX TO ILLUSTRATIONS

The following species are illustrated in this volume. The colored frontispiece is in bold face type.

397

INDEX TO RECOGNIZED GENERA

INDEX TO GENERA AND SPECIES

(*Synonyms in italics*)